T0142879

Signals and Communication Technology

More information about this series at http://www.springer.com/series/4748

Igor Izmailov · Boris Poizner
Ilia Romanov · Sergey Smolskiy

Cryptology Transmitted Message Protection

From Deterministic Chaos up to Optical Vortices

Igor Izmailov
Tomsk State University
Tomsk
Russia

Ilia Romanov
Tomsk State University
Tomsk
Russia

Boris Poizner
Tomsk State University
Tomsk
Russia

Sergey Smolskiy
Moscow Power Engineering Institute
Moscow
Russia

ISSN 1860-4862 ISSN 1860-4870 (electronic)
Signals and Communication Technology
ISBN 978-3-319-80728-7 ISBN 978-3-319-30125-9 (eBook)
DOI 10.1007/978-3-319-30125-9

Printed on acid-free paper

This Springer imprint is published by Springer Nature
The registered company is Springer International Publishing AG Switzerland

Authors, with tender feeling, dedicate this book to their near relations who patiently share all severities of creative activities: Valentina D. Izmailova, Valeriy P. Izmailov, Elza A. Zakharova, Nadezhda F. Romanova, Vladimir G. Romanov, Natalia N. Smolskaya, Michael S. Smolskiy, Ekaterina S. Smolskaya

Acknowledgments

Authors consider necessary to recall their professional genealogy, to remember everything that owes *scientific schools of thoughts* which educated them: on oscillation theory and fundamental radio electronic engineering in Tomsk University and in Moscow Power Engineering Institute.

Authors heartily appreciate colleagues from different research and educational organizations for manifold—in contents, style, form—synergy and help at all stages of investigations generalized in this book. First of all, to: V.P. Aksenov, V.M. Anikin, R.G. Barantsev, V.M. Bogachev, F.Yu. Kanev, M.V. Kapranov, A.P. Kokhanenko, V.N. Kuleshov, A.P. Kuznetsov, S.P. Kuznetsov, A.V. Lyachin, A.L. Magazinnikov, V.V. Negrul, N.M. Ryskin, A.V. Savel'eva, A.L. Sanin, St.M. Shandarov, D.A. Shergin, E.E. Slyadnikov, E.A. Sosnin, F.I. Starikov, O.V. Tikhomirova, D.I. Trubetskov, G.M. Utkin, S.N. Vladimirov, V.F. Vziatyshev.

Authors are thankful to Springer Science+Business Media and personally to Dr. Habil. Claus E. Ascheron, Senior Editor Physics for support of this book and valuable advices.

Contents

About the Authors

Igor V. Izmailov was born on August 25, 1976 in the city Alma-Ata (Kazakhstan). In 1993, he graduated from secondary school. In 1999, he graduated from Faculty of Radio Physics (FRP) of State Tomsk University (STU) and was recommended for further education, Ph.D. course. In 2002, he graduated with a Ph.D. degree from STU and defended his Ph.D. thesis in Optics.

At present, Igor V. Izmailov works at the Department of Quantum Electronics and Photonics in FRP, STU as Associate Professor and in Siberian Physical-Engineering Institute of STU as Senior Researcher.

During this time, he participated in a series of new scientific and applied research directions.

Igor V. Izmailov introduced and developed the concept of spatial deterministic chaos. It turns out to be important for two-dimensional images protection. In the context of information protection means, he developed the route-operator formalism. It allows providing the construction design of radio electronics and optical devices for message deciphering if they were subjected by masking influence of the chaotic signal source with known construction. Having developed the system approach to oscillating-wave devices, he constructed the poly-discipline axiomatic procedure of investigations of the dynamic systems. In this context, he generalized concepts "order", "order parameters", suggested the concept of "the evolution equivalency property (isodynamism)", "likening", etc. This allows determination of unbiased gradation of similarity of function pairs, signals, transfer characteristics and development of principles and scenarios of variofication (i.e., increasing of diversity) of dynamic systems in engineering. As the one of key tools of variofication, he offered and developed the conception of the "self-changing" (self-controlled) nonlinear transfer characteristic. In this concept we can demonstrate that some classical optical and radio frequency devices of nineteenth–twentieth centuries are unidentified precedents of the "self-changing" nonlinearity.

Having extended a series of advanced ideas to the solution of singular and adaptive problems, Igor V. Izmailov offered an operation principle and construction of the topologic charge detector for optical vortices as well as a position sensor (detection-finder) of the optical vortex. He modernized the Fried algorithm in problems of phase front reconstruction of the laser beam experienced distortions in the turbulent atmosphere. Having suggested a program of new research and application direction ("quantum-synergy cyto-informatics"), he set a question about the functional analog of cyto-skeleton microtubule (in optical range), offered a variant of its constructions, and constructed the process model in this analog intended for data processing.

Igor V. Izmailov is a co-author or author of more than 290 publications in the above-mentioned directions: four books, 106 scientific papers, and two textbooks popularized these new approaches.

Being a student, he was the winner of the medal of Russian Ministry of Education and Science «For the best scientific students' work in natural, engineering and humanitarian sciences» (1999). He is a laureate of Tomsk region competition in education and science (2001), of Prize of State Duma of Tomsk region for young researchers (2003), and of STU Prize for the best scientific work of young researchers (2002 and 2004).

Boris N. Poizner was born on May 5, 1941 in Tomsk city (Siberia, Russia). He graduated in 1958 the secondary school and Radio Physics Faculty of the State Tomsk University (STU) in 1963 and began to work at Department of Radio Electronics in STU. In 1970 he graduated with a Ph.D. degree from STU and defended his Ph.D. thesis in Radio Physics.

At present, he works as Professor of Department of Quantum Electronics and Photonics of Radio Physics Faculty of STU.

During this time, he participated in a series of new scientific and applied direction of investigations.

He investigated the forced synchronization of multi-frequency UHF oscillators by the sinusoidal signal and experimentally demonstrated a possibility of the radiation spectrum control of the UHF oscillator. He revealed the similarity and differences in multi-frequency oscillator (a laser, a reflex klystron, a Gunn diode) reactions on the external sinusoidal influence. He described how the natural line width of the laser radiation, noises, mode competition degree in these devices affect the interference results in nonlinear dynamics from outside. He discovered conditions when a weak sinusoidal signal induces the multi-frequency of chaotic mode, and peculiarities of these phenomena at mutual synchronization. He suggested (1974) the multi-stable element on the base of multi-frequency oscillator controlled by the external sinusoidal signal and an approach of laser parameters determination with the help of synchronization mode. Together with colleagues—medical experts—he offered an experiment on variation of bioelectric activity of the cerebrum of the laboratory animal under the influence of low-intensive laser radiation and participated in this experiment (1975).

Among applied problems, which were solved by him, are the following: modulation of radiation intensity of two laser junctions coupled through the upper level; system of automatic control of the laser local oscillator and the transmitter; methods of materials research and control of two-dimensional and three-dimensional objects by means of holographic interferometry including diffusing face; correction of the wave front of the light beams (these results are defended by patent and authors certificates); adjustment of the gas laser without using a photo-receiver; and the Fourier-like Fresnel hologram.

Boris N. Poizner performed a series of shaping simulations in the transverse plane of the laser beam in the ring interferometer, which contains the Kerr nonlinearity and discovered a mode of "cyclic self-reorganization of optical systems. He developed the concept of spatial deterministic chaos and offered the operation principle and the construction of the topologic charge detector of the optical vortices. Having developed the system approach to oscillating-wave devices, he constructed the poly-discipline axiomatic procedure of investigations of the dynamic systems. For variofication of engineering systems, he offered and developed the concept of "self-changing" (self-controlled) nonlinear transfer characteristics showing that some unidentified classical optical and radio frequency devices of nineteenth–twentieth centuries are unidentified precedents of the "self-changing" nonlinearity.

Boris N. Poizner performed a series of researches on scientometrics revealing the dynamics of holography development, laser physics and engineering as well as the structure of inter-discipline links of quantum electronics in 1960s–1990s.

Having developed the "order from a chaos," Boris N. Poizner developed the ideas of self-organization subject and a replicator as the latent trans-discipline concept in the modern science. He participated in the creation of the laser model for the creative work (1997). Generalization of these results allows the extending of non-mathematical information theory to the creative work including scientific researches and the activity of scientific schools of thoughts. Under his participation, the conception of the six creativity levels (and relevant leaders) is constructed, particularly, accentuated personalities in institutions.

He is a co-author of more than 450 publications including 22 scientific books and textbooks, 10 authors certificates, and more than 120 papers in scientific journals.

Boris N. Poizner was awarded the following titles: «Honored worker of Higher Education of Russian Federation» (2005), «Honored worker of Higher Education of RF» (2001), he was decorated by the Sign of Ministry of Education "For Development of Research Activity of Students» (2005). He is a Honored Member of scientific section of Books and Graphics of a House of Scientists of Russian Scientific Academy (2012).

Ilia V. Romanov was born on June 5, 1981 in Ust-Ilimsk town (Siberia, Russia). He graduated the secondary school in 1998 and Radio Physics Faculty of State Tomsk University (STU). After that, he began to work as Engineer-researcher of STU. He graduated with a Ph.D. degree from STU and defended his Ph.D. thesis in Radio Physics.

At present, he works in the Company "Tomion" as the Chief-Engineer.

During this time, he participated in a series of new scientific and applied directions of investigations.

He researched a problem of chaotic oscillation application for the task of secretive transmission of analog and digital signals in the communication system on high frequencies. He experimentally demonstrated a possibility to control of chaotic oscillation spectrum and scenarios of transition to chaos of the UHF chaotic oscillator. He experimentally showed a possibility to transmit audio- and video signals in the communication system using chaotic oscillations oscillator with a nonlinearity of parabola composition type in the transmitter. He experimentally and theoretically proved an influence degree of destabilizing factors (noises and distortions in the communication channel, non-identity of transmitter and receiver parameters etc.) to data transmission quality in the communication system. He theoretical investigated a possibility of application of the quartz optical fiber with a touch of CdHgTe as the nonlinear element of the chaotic oscillator on the base of the nonlinear fiber ring interferometer capable to operate in the dynamic chaos mode. He performed experimental research of charge carriers mobility in the thin films of organic semiconductor. The method offered by him for the analysis of experimental data allows determination of mobility values and its dispersion of the main and non-main charge carriers basing on the analysis of transfer electric luminescence in OLED. He experimentally investigated with colleagues a possibility of early diagnostics of the caries with the help of terahertz radiation. At present, he occupies by the development of scientific equipment for investigation of the Earth ionosphere.

Ilia V. Romanov is a co-author of 21 publications in the scientific journals, has two authors certificates, prepared 25 report in Russian and foreign conferences.

Sergey M. Smolsky was born on January 2, 1946 in Moscow (Russia). In 1964 he graduated secondary school with a medal and entered Moscow Power Engineering Institute (Technical University) (MPEI). He graduated from Radio Engineering Faculty of MPEI on specialty Radio Physics and Electronics in 1970 and had got a recommendation to the Ph.D. course. During the Ph.D. course at MPEI he had got the scholarship named after Lenin and he defended the Ph.D. thesis in 1974 on the theory of high-frequency transistor oscillators. After finishing the Ph.D. course he worked at

Radio Transmitters Department of MPEI as the Senior Researcher, the Head of Research Lab, the Head of Research Division, and after he moved to position of MPEI as Deputy Vice-Rector on research activity and after as MPEI Vice-Rector on international relations. In 1993 he defended the Doctor of Sciences thesis on Radar and Radio Navigation Systems, in 1994 he became Full Professor, and in 1995 he was elected as Head of Radio Receiver Department. From 2006, he works as Professor of this department and Deputy Director of the MPEI Institute of Radio Engineering and Electronics.

During this time, he worked in a series of significant scientific and applied problems. In particular, he participated in the development of the theory of autonomous and synchronized transistor self-oscillating systems, in development of methods of mathematical modeling and effective algorithms for analysis of self-oscillating systems. He actively has developed theory and implementation of autodyne (self-oscillating mixers) systems for the short-range radar, researches in the field of medical electronics, radio monitoring systems, self-similar processes in telecommunications, investigations of chaotic oscillations and signals. Professor Sergey M. Smolskiy is the recognized expert in these (and other) areas.

Sergey M. Smolskiy has the academic degree of Doctor of Engineering Sciences. He was elected as an academician of several Russian and International Academies: International Academy of Electrical Sciences, International Academy of Informatization, International Academy of Higher Educational Institutions, a member of IEEE, The Honor Doctor of three foreign universities.

Being the famous scientist in the field of nonlinear oscillating systems and radar devices, information science, and medical instrumentation, Sergey M. Smolskiy during past years was actively and fruitfully working in the scientific school of thoughts created with his participation: development of the theory and practical implementation of short-radar systems and self-similar systems; improvement of radio monitoring and chaotic signal theory; researches in the field of medical instrumentation, etc.

He is the author of more than 300 scientific publications including 12 books, textbooks including those published by Artech House Publishers, John Wiley and Sons, Springer US, and Scientific Research Publishers.

Under his scientific supervision there have been two doing Doctor of Sciences theses and 11 doing Ph.D. theses. He takes part in conferences with reports and reviews and with lecture course in foreign universities. Some his pupils became heads of departments in different Russian universities and in different companies.

Professor Sergey M. Smolskiy made a serious contribution in holding the All-Russia and International conferences and symposiums on the actual problems of the theory and implementation oscillating and radar systems.

He is a member of Russian Popov Scientific-Engineering Association of Radio Engineering, Electronic and Communication and the Head of Section "Radio

Engineering Methods in Power Engineering", he is a scientific supervisor and a member of organizing committees of international scientific conferences.

He is Honor Radio Engineer of Russia; he was awarded the Gold Medal of Popov «For services in development of radio electronics and communication». He is Honor Worker of Russian Higher Education, was awarded by Polish order, Ryazansky Medal from Academy of Space Research.

Abbreviations

AA	Adjustable amplifier (linear)
AC	Alternating current
ADC	Analog–digital converter
AFC	Amplitude-frequency characteristic
BC	The bias circuit
CCD	Charge-coupled device
CCFS	Channel of "coordinating-forced" synchronization
CCMS	Channel of "coordinating-mutual" synchronization
CCS	Channel of "coordinating" synchronization
CFS	Channel of forced synchronization
CMS	Channel of mutual synchronization
CS	Channel of synchronization
CTP	Chain of the transposition points
DAC	Digital–analog converter
DC	Direct current
DL	Delay line
DM	Discrete map
DNRI	Double-circuit nonlinear ring interferometer
EDFA	Erbium-doped fiber amplifier
EDFRL	Erbium-doped fiber ring laser
FBL	Feedback loop
FC	Functional converter
HPF	High-pass filter of the first order
IC	Information channel
LF	Fried's radius
LPF	Low-pass filter of the first order
NBICS	Nano-bio-info-cogno-socio technologies
NE	Nonlinear element
NM	Nonlinear medium

NRI	Nonlinear ring interferometer
ODE	Ordinary differential equation
PhM	Phase modulator
RF	Radio frequency
S	Informational signal
SCI	Source of chaotic influence
SD	Screw dislocation
SIC	Synchro-information channel
SNR	Signal/noise ratio
SSCI	Source of synchronized and chaotic influence
SSI	Source of synchronization influence
TC	Transfer characteristic
UHF	Ultra high frequency

Introduction

Protection of transmitted messages is becoming wider and wider with an avalanche-like growing wave of scientific and engineering publications and (moreover) a rise of an appropriate business branch. When thinking about the origin of this problem of information security trying to find out when its multi-century history started, one should go as far back as the fourteenth century B.C. Scientist Gelb [1] informs us that at that time people from the Near East passing from hieroglyphs of the ancient Egyptians to linear writing that used abstract and simple conventional signs formed of lines instead of the images easily recognizable by the addressee. These signs were drawn on clay surfaces or carved in stone.

The following step in this long way was the so-called semitic writing (the end of the second millennium B.C.), containing 22 letters only. In [2] Friedrich points out the principal innovation: these 22 signs transmit not the meaning, as in previous types of writing, but only the sounding. According to the Russian orientalist Diakonov (1996), this literal writing had the special "type of merchants' cryptographic writing" as its predecessor. It was contrived by the merchants from Phoenicia as the country at the eastern shore of the Mediterranean Sea. In other words, the cryptography writing served as a "progenitress" of the alphabet writing and has had 34 centuries of its history.

The cryptology (from ancient Greek κρυπτος—secretive, vieled + λογοσ—science) has high antiquity and arose answering, first, the needs of commerce, and later conforming to the problems of state and military management. At present, it has acquired a total new status determined by the new challenges of the information society. Society development, as the modern laws of nonlinear system dynamics [3] specifically serves as the stimulus for the ongoing globalization in the different areas of economy, consumption, daily life, and entertainment,is in turn simultaneously "urged forward" by the globalization.

Many social philosophers, public figures, and politicians do commonly see the information society as a way to maximal openness, transparency, wideness, and mixed-level contacts. Undoubtedly, the deeper the penetration of the world community by the communication channels, forming the "central nervous system" of the mankind, the higher the level of openness. But the absolute openness of **everybody to everybody** is, unfortunately, a beautiful utopia.

As we know, the internal evolution driver is the competition of the purposeful systems. In the real conditions of competitive relationships, the absolute informational openness leads to the vulnerability of practically all social, political, economical, and public structures and makes many aspects of the private life unsafe. For instance, the modern innovational *modus vivendi* in the post-industrial society implies that even at the early stages of the project development the sharing to some extent of the key information expressing the essence of the innovation is inevitable. But, a leak of the progressive ideas and technical or/and technological details vital for the project is fatal indeed exactly at these first stages of the original product development. Therefore, it is very likely that nowadays, in comparison with the historical past, there is a much more acute need for the **culture of the secrets protection**. And the existence of the extremist organizations around the world only aggravates the situation. Naturally, the culture of secrets protection requires theoretical, engineering, and organizational support at all levels of communications in the society.

At present, development of methods and devices for information protection represents an interdisciplinary problem. Its solution is being carried out by the specialists in cryptography [4], information science [5], quantum mechanics [6], radio electronics [7–12], optics [13, 14], and nonlinear dynamics [15].

Each of these sciences suggests its own set of the unique methods for the information protection, and they, as a rule, supplement each other. Thus, radio electronics and optics, as the sciences studying oscillations and waves, use ideas and methods of nonlinear dynamics (the other names for it are Synergetics, Complexity Science, Deterministic Chaos Theory). The demand for its ideas and methods is caused by the fact that at the last third of the twentieth century the limited non-periodic motions in the nonlinear systems with a small number of degrees of freedom were discovered. Such motions are called stochastic or chaotic motions. They were discovered in mechanical, radio electronic, optical, chemical, biological, geophysical, cosmological, social, and economical cognitive systems [3, 15–26]. The book of Mainzer [3] is an excellent example of the brief description, presented from the unified synergetic positions, of the complicated phenomena in physical, biological, and social realities.

Along with this, the chaos theory happens to be productive for the description and the explanation of the phenomena forming the subject matter for specific sciences. For instance, both geophysics and astrophysics study the problem of the Earth dynamo. The specialists managed to explain the paleomagnetic data for the last 600 million years by the chaotic dynamics of the fields appearing during the rotation of the Earth by its ferric-nickel core [27, 28]. No less important are the nonlinear dynamics problems referring to the medical diagnostics of heart, cerebrum, breathing system, and gastrointestinal tract pathologies.

Another example. In the atmosphere with turbulence, which is a variety of chaotic dynamics, vortices often arise. Realization of this fact gave rise to the singular optics at the end of 1980s. It concentrates on the problems of generation, propagation, and detection of the laser beams with the optical vortices [29–32]. One of the fields of application for chaos and vortex is the information technology—information security, in particular.

Such wide, multidirectional, and interdisciplinary investigation of the deterministic (dynamic) chaos as a whole has confirmed the researchers that chaos is the promising ground for the development of the new systems for information processing and storage. It is sufficient to note that the human brain operates in the chaotic mode [3, 33, 34]. Investigations in this area are directed towards the creation of the mathematical models, in which the information processing using chaos is realized. As such models, there exist neuro-like structures such as one-dimensional and multidimensional maps, one-dimensional (chains) and two-dimensional (lattices) arrays of the chaotic cells.

Both in radio electronics and in optics, recently a number of important practical problems, for the solution of which we need sources of the signals with wide spectral line, high spectral density and occupying a wide frequency range has been formulated. Although the standard requirement for the oscillator of radio frequency range and for the lasers was, and still is, the highest coherency of the output signal, short-term and long-term time stability of this signal and a low level of amplitude and phase noises providing narrowness of its spectral line (are also now of great importance).

The increased demand for the sources of the wide-band noise-like oscillations is caused by the fact that they provide the potential for the development of the modern systems for the counter-radio measures, radio masking, the noisy radar technologies, confidential communications, the ultra-speed radio communications, cryptographic structures, devices for the non-traditional influence on the biological objects, different types of special purpose devices [7–12, 35–37]. Since fiber optical communication lines now form a significant and a fast-growing segment of the global communication systems, the need arises for the oscillators of the deterministic chaos in the optical range [38–40].

Thus, since the 1990s, in radio electronics and later in optics, not only the development of principles and devices for information protection, relying mainly on the chaos phenomena, has taken place, but also the study has been carried out with the optical vortex as information carrier. The authors believe that there are reasons to consider these lines of research as referring to *nonlinear-dynamic cryptology* [41].

In the book now presented, the authors describe their experience of the theoretical investigation of the properties of the deterministic chaos generators in radio and optical range for the protected communication systems based on them, as well as the models of generators and communication devices for the radio range.

This book has six chapters. The first one reviews the general situation in the chosen field of study, the other four contain original material.

Chapter 1 includes the examples of the dynamic systems with the complicated behavior in radio electronics and optics. There are discussed the general conception, the principles and the basic concepts of the classical and the nonlinear dynamic cryptology and the classification of the common communication systems exploiting the dynamic chaos. The general information is presented on the reasons why and how the optical vortices emerge and on their properties and characteristics.

Chapter 2 introduces the outcome of the theoretical research on the radio electronic oscillator of the deterministic chaos with the nonlinearity in the form of the parabolas combination. The experimental installation is described, along with the results of its different modes testing. Various dynamic modes are identified and the possible scenarios of transition to the chaos in the generator are also discussed. The variants of nonlinear ring interferometers (with Kerr nonlinearity) as the chaos sources are theoretically examined; their mathematical models are made; and the processes that go on in them with their features and properties are studied. The concept of the "spatial" deterministic chaos is introduced.

In Chapter 3, the structure and the mathematical model are described for the radio electronic system for the data transmission with the nonlinear mixing (adulteration) of the informational signal in the transmitter and with the synchronous chaotic response in the receiver. The numerical modeling is performed and the factors affecting the data transmission quality are revealed. The experimental installation is described providing the communication exploiting the chaos generator. The principles of the signal/noise ratio measurements and the estimation of the data transmission channel capacity are stated. The results of the real transmission sessions for the analogous and the digital signals are presented.

Chapter 4 introduces the conception of the nonlinear dynamic cryptology that suggests the means for the information protection in the nonlinear systems. Electromagnetic field transformations in the optical cipherer and the decipherer are described. The equation for the deciphering error is obtained and the simulations of both the secret image transmission and the system "cracking" are carried out. The conditions for the increased stability of the "cracking" are identified.

In Chapter 5, the principles are suggested for how to determine the topological charge of the optical vortex and its localization; the mathematical model is made of the vortex charge detector operation. Simulations of the vortex recognition are carried out in the presence of different opposing factors, and their influence is studied on the probability of the data transmission errors.

In Chap. 6 outlines the ways for the further investigations and their prospects. The introduced additions to the cryptosystem classification allow conducting the further synthesis of the potentially feasible cryptosystems. The conception of controlled/self-controlled nonlinearity is suggested, as well as the principles of its application in order to increase the stability of the confidential communication systems.

Finally, "Conclusions" formulate the problems to obtain for nonlinear dynamic cryptosystems for correct prototypes of known definitions of algorithms for polynomial and exponential complexity. So, the nontrivial task will be necessary for solution in the future.

Authors are aware that the reader cannot find the perfect cipher in this book. The following thesis of F. Zimmermann (creator of the famous *PGP* cipher) can serve as a consolation for us: "Everyone who thinks that he discovered "not-cracking" ciphering scheme, is either the unbelievingly rare genius, or is simply naive and unexperienced…" (cited according to [42]).

References

1. Gelb IJ. A study of writing. Chicago & London: The University of Chicago Press; 1963.
2. Friedrich I. Geschichte der Schrift (unter besonderer berücksichtigung ihrer geistigen entwicklung). Heidelberg. Carl Winter Universitätsverlag; 1966.
3. Mainzer Kl. Thinking in complexity: The computational dynamics of matter, mind, and mankind. Springer; 2004.
4. Koblitz N. A course in number theory and cryptography. NY: Springer; 1994.
5. Wagner NR. The laws of cryptography with java code. USA: Univ. of Texas San Antonio; 2003.
6. Bouwmeister D, Elkert A, Zeilinger A. The physics of quantum information. NY: Springer; 2000.
7. Hasler M, Vandewalle J. Special issue on communications, information processing and control using chaos. Int J Circuit Theory Appl. 1999;27(6):525–531.
8. Rulkov NF, Sushchik MM, Tsimring LS, et al. Digital communication using chaotic-pulse-position modulation, IEEE Trans Circuits Syst I Fundam Theory Appl. 2001;48(12):1436–1444.
9. Volkovsky AR, Rul'kov NF. Synchronous chaotic response of the nonlinear oscillation system as a principle of detection of the informational chaos component (in Russian). Lett J Tech Phys. 1993;19(3):72–77.
10. Yang T, Chua LO. Impulse control and synchronization of nonlinear dynamical systems and application to secure communication. Int J Bifurcations and Chaos. 1997;7(3); 645–664.
11. Yang T, Yang LB, Yang CM. Cryptoanalyzing chaotic secure communications using return maps. Phys Lett A. 1998;245(6):495–510.
12. Kuznetsov SP. Dynamical chaos and uniformly hyperbolic attractors: from mathematics to physics, Uspekhi Fizicheskikh Nauk. 2011;181(2):121–149.
13. Garcia-Ojalvo J, Roy R. Parallel communication with optical spatiotemporal chaos. IEEE Trans Circuits Syst I Fundam Theory Appl. 2001;48(12):1491–1497.
14. Mirasso CR, Mulet J, Masoller C. Chaos shift-keying encryption in chaotic external-cavity semiconductor lasers using a single-receiver scheme. IEEE Photonics Technol Lett. 2002;14 (4):456–458.
15. Anichchenko VS, Astahov VV, Neiman AV, Schimansky-Geier L. Nonlinear dynamics of chaotic and stochastic systems. Berlin: Springer-Verlag; 2007.
16. Blekhman II. Synchronization of dynamic systems. Moscow: Nauka Publ.; 1971. (in Russian).
17. Rabinovich MI, Trubetskow DI. Oscillations and waves in linear and nonlinear systems. Dordrecht: Kluwer Acad. Press; 1989.
18. Loskutov AYu, Mikhailov AC. Introduction to synergetics. Moscow: Nauka Publ.; 1990. (in Russian).
19. Landa PS. Nonlinear Oscillations and waves in dynamical systems. Dordrecht: Cluver Academic Publ.; 1996.
20. Chaotic Dynamics and Transport in Classical and Quantum Systems. NATO Science Series. Series II: Mathematics, Physics and Chemistry. vol. 182, Dordrecht. Kluwer Academic Press, 2005.
21. Galor O. Discrete dynamical systems. Berlin: Springer; 2007.
22. Anikin VM, Golubentsev AF. Analitycal models of deterministic chaos. Moscow: FizMatLit; 2007. (in Russian).
23. Haken G. Brain dynamics. Introduction to models and simulations. Springer-Verlag; 2007.
24. Kapranov MV, Tomashevsky AI. Regular and chaotic dynamics of systems with discrete time. Moscow: MPEI Publ.; 2010. (in Russian).
25. Kuznetsov SP. Dynamical chaos and uniformly hyperbolic attractors: from mathematics to physics. Uspekhi Fizicheskikh Nauk. 2011;181(2):121–149.
26. Dmitriev AS. Chaos generation. Moscow: Tekhnosfera; 2012. (in Russian).

27. Ershov SV, Malinetskii GG, Ruzmaikin AA. A generalized two disk dynamo model. Geophys astrophys fluid dynamics. 1989;47:251–277.
28. Potapov AB, Gizzatulina SM, Ruzmaikin AA, Rukavishnikov VD, Malineskii GG. Dimension of geomagnetic attractor from data on length of day variations, Phys of the Earth and Planetary Int. 1990;59:170–181.
29. Vasnetsov M, Staliunas K. Optical vortices (Horizons in world physics. vol. 228). Nova Science Publ. Inc.; 1999.
30. Masajada J. Optical vortices and their application to interferometry: Publ. Institute of Physics Wrocław University of Technology, 2004 (Prace Naukowe Instytutu Fizyki Nr 36, Monografie Nr 25).
31. Saleh BEA, Teich MC. Fundamentals of photonics. Wiley and Sons Ltd.; 2007.
32. Tyson RK. Adaptive Optics Progress. Publ.: InTech, 2012. doi:10.5772/46199; Progress in Optics. vol. 53, 2009, Elsevier. doi:10.1016/S0079-6638(08)00205-9.
33. Haken G. Brain dynamics. Introduction to models and simulations. Springer-Verlag; 2007.
34. Evin IA. Brain synergetics. Moscow-Izhevsk. Regular and Chaotic Dynamics Publ.; 2005. (in Russian).
35. Gulyaev YuV, Kislov VYa, Kislov VV. New signal class for information transmission—wideband chaotic signals. Rep Acad Sci. 1998;359(6):750–754. (in Russian).
36. Vladimirov SN, Perfiliev VI. Source of over broadband noise oscillations to diagnostic of electronic devices. Microwave and Telecommunication Technology (CriMiCo’2004): Proc. 14-th Int. Conf. (September 13–17, 2004, Sevastopol), p. 642–644.
37. Vladimirov SN, Perfiliev VI. Microwave chaotic waveform source. International Conference on actual problems of electronic instrument engineering (APEIE’2004): Proc. 7-th Int. Conf. (September 21–24, 2004, Novosibirsk). vol. 4, p.104–107.
38. Garcia-Ojalvo J, Roy R. Spatiotemporal communication with synchronized optical chaos. 2000. p. 4. http://www.lanl.gov/abs/nlin.CD/0011012. Accessed 6 Nov 2000.
39. Napartovich AP, Sukharev AG. Synchronizing a chaotic laser by injecting a chaotic signal with a frequency offset. J Exp Theor Phys. 1999;88(5):875–881. (in Russian)
40. Sivaprakasam S, Shore KA. Critical signal strength for effective decoding in diode laser chaotic optical communications. Phys Rev E. 2000;61(5):5997–5999.
41. Izmailov IV, Shulepov MA. Simulation of signal enciphering by means of nonlinear ring interferometer and decoding. in Optoelectronic information systems and processing (11–15 September 2000, Vladivostok, Russia) Yuri N. Kulchin, Oleg B. Vitrik, Editors, Proceedings of SPIE. vol. 4513, p. 46–51 (2001).
42. Yashchenko VV. Introduction to cryptography. Sankt-Peterburg: Piter Publ.; 2001. (in Russian).

Chapter 1
Deterministic Chaos Phenomenon from the Standpoint of Information Protection Tasks

To understand better the problems of the cryptic information transmission by means of the deterministic chaos, it is expedient at first to refresh the concept and the main categories of the classical cryptology. We consider not only the cryptographic methods for the messages protection, but the physical and the steganographic methods as well, that cannot be neglected in any way. To implement these methods we can use optical vortices as data transfer carriers in a communication system. We shall touch upon the reasons for their appearance and their features. Then it is useful to give some examples of dynamic systems in radiophysics and optics with the so-called *complex* behavior. After that, we must describe a number of principles of the nonlinear-dynamic cryptology. The well-known classification of the communication systems employing the dynamic chaos will help us to systematize a variety of its structural and hardware implementations.

1.1 Principles and Concepts of the Classical Cryptology as the Traditional Strategy of Information Protection

To discuss the information protection problems, it is expedient to remind some terms and concepts of the classical cryptology.

Information is words in the some finite *alphabet A*.

Message and *information* are considered as synonyms.

Word is any finite length sequence constituted out of the *letters of the A alphabet*.

Language L in the alphabet A is $L \subset A^*$, where A^* is a set of all finite words in the A alphabet (Λ is an empty word).

For instance, if the A alphabet is a set of *figures* ($A = \{0,..,9\}$), then some *numbers* may be used as words in the alphabet A, and either their number or their

I. Izmailov et al., *Cryptology Transmitted Message Protection*,
Signals and Communication Technology, DOI 10.1007/978-3-319-30125-9_1

set (sequence) forms information, while the set of all the three-figure numbers may serve as the L language.

The main methods of information protection are the following [1, p. 8].

(1) *Physical*. They make the communication channel inaccessible for the enemy.
(2) *Steganographic* (from Greek στεγανος – covering, protecting + γραφω – write). Their aim is to hide the very fact of the information transmission. Examples are: in antiquity, it was accepted to write on the slave's shaved head, and when the hair grow again, the steganogram is ready [2]; writing by milk; adding information bits into the files or the digital images; the method of micro-dots; placing information into noises; acrostich (from Greek ακρος – edge + στιχος – row).
(3) *Cryptographic* (from Greek κρυπτος – secretive + γραφω – write, i.e. a cryptogram). Main fields of cryptographic methods application are the transmission of confidential (from Latin *confidentia* – trust) information via communication channels; the authentication of messages transmitted; saving and storing information on the different carriers in the ciphered form. We shall discuss them later. Let us now remind some definitions.

Cryptology (from Greek κρυπτος – secretive + λογος – science) is a science consisting of the two branches: cryptography and cryptoanalysis [1, p. 15]. Their aims are really the opposite ones.

Cryptographic is a science about (mathematical) methods of the information transformation (*ciphering*) with the purpose of protection from the enemy threats (the non-authorized access).

Such methods and ways of information transformation are called *ciphers*. In other words, ciphers are cryptographic methods. They are based on mathematical approaches.

Ciphering (*coding*) is a process of cipher application to the information being protected, i.e. transformation of the protected information (*an open text*) into the ciphered message (*the cipher-text, the cryptogram*) by means of the rules contained in the cipher.

Deciphering is a process, which is inverse to ciphering, i.e. the transformation of the *cryptogram* into the *open text* by means of the rules, contained in the cipher.

A key is an exchangeable element or the varying cipher parameter used for the (de)-ciphering of the specific message. Or (in brief): a *key* is information necessary for the unhampered ciphering and deciphering of the texts. Sometimes, under the key we can understand the various values of the exchangeable element of the varying cipher parameter.

The cryptoanalysis is a science (and its application practice) about methods and ways of opening ("splitting", "breaking") the ciphers.

Let us give the main "postulate" of the cryptographic: the enemy knows about the *cryptosystem* everything besides the specific *secure key*, which was used for all the interceptive *cryptograms* (the channel is available to everyone; the information is protected but not the fact of its transmission). The diagram of the interaction of the authorized users of the communication channel (a sender, a recipient) and the enemy in cryptologic situation is shown in Fig. 1.1.

Let us designate the languages of messages, cryptograms and keys by symbols *X*, *Y*, *Z*, accordingly. Then algorithms of ciphering and deciphering according to the key $z \in Z$ can be denoted by: E_z, D_z and are defined as images E_z: $X \rightarrow Y$, D_z: $Y \rightarrow X$ (Fig. 1.1).

Algorithms E_z and D_z should have some features. For instance, the reversibility feature: $D_z(E_z(x)) = x$; $E_z(D_z(y)) = y$. It ensures recovering of the initial message from the cryptogram. Moreover, the function E_z: $X \rightarrow Y$, $(y(x))$ should be a *function with secrecy*.

To introduce functions with secrecy, it is necessary to remind the concept of the algorithm complexity.

The algorithm complexity is a time of algorithm execution depending on the data sizes (word length). We can distinguish the *exponential* and *polynomial* complexity.

The function with secrecy K (*the trap*) is the function F_K: $X \rightarrow Y$ depending upon the *K* parameter and possessing three properties:

(a) for any *K* and *x*, there is the polynomial algorithm for $F_K(x)$ calculation;
(b) when *K* is unknown, there is no polynomial algorithm for F_K inversion;
(c) when *K* parameter is known, there exists the polynomial algorithm for F_K inversion.

The problem of the functions with secrecy is open yet, since it is not easy to prove the (b). At present, we can assume that for the functions used in cryptography the presence of the property (b) remains a complicated and a long studied mathematical problem.

Now we can determine the *cryptosystem* as a quadruple: $\langle X, Y, Z\,(E_Z, D_Z) : z \in Z\rangle$. It is accepted to distinguish the symmetric and asymmetric cryptosystems. And it is assumed that the key *z* consists of a key for ciphering *e* and a key for deciphering *d*, i.e. $z = (e, d)$.

In *symmetric* cryptosystems, it is unnecessary to distinguish a key for the ciphering and a key for the deciphering, since the keys *e* and *d* are closed (secretive) and, as a rule, coincide. Here we presuppose that *z* is calculated according to *x*, *y* with exponential complexity.

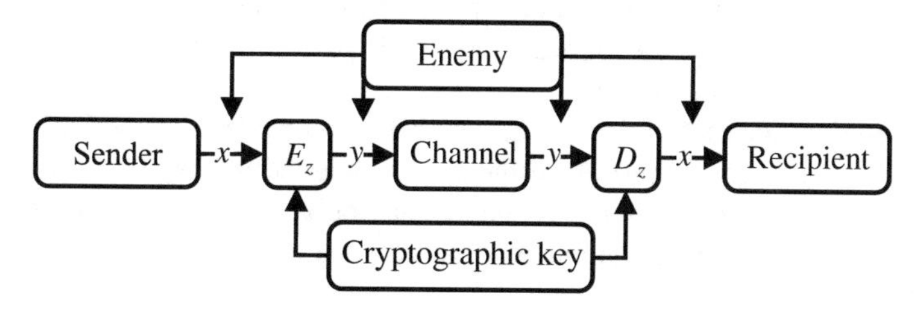

Fig. 1.1 Generalized diagram of authorized users interaction (a sender, a recipient) and the enemy in cryptologic situation. Here *X*, *Y*, *Z* are the message languages *x*, cryptograms *y* and keys *z*; E_z, D_z are algorithms of ciphering and deciphering according to the key $z \in Z$

In *asymmetric* cryptosystems the keys e and d do not coincide. Thus, the key d is closed, and e is open ($E_e{:}X \rightarrow Y$, $D_d{:}Y \rightarrow X$; $D_d(E_e(x)) = x$, $E_e(D_d(y)) = y$). Due to this, any person may create a cryptogram, but it can be read only by the owner of the secure key. We assume here that d can be calculated according to e or x, y, e with exponential complexity.

In accordance with [1, p. 20], the application of the functions with secrecy in the cryptographic allows:

(1) arranging the ciphered messages exchange employing the open communication channels only, i.e. it is possible not to use the secretive communication channels for the preliminary key exchange;
(2) including the complicate mathematical problem into the task of cipher opening and thus increasing a validity of cipher crypto-resistance;
(3) solving new cryptographic problems, which differ from ciphering (an electronic digital signature etc.).

We should note that the unified cipher, which is suitable for all the cases, does not exist. A choice of the ciphering method depends upon the peculiarities of information and its value, as well as upon engineering, technological and financial capabilities of its owners concerning information protection. First of all, we emphasize an existence of a huge variety of types of information to be protected: documents, telephone calls and messages, television programs, different kinds of computer data, etc. Each type of information has its own peculiarities strictly influencing the ciphering methods choice. Volumes of information and the required transmission rate, a possibility/necessity of storage of ciphered information, the required time for secrecy keeping (years, days, hours), and the assumed scientific and engineering equipment of the enemy, all these aspects are of great significance [1, c. 13].

The cipher crypto-resistance is an ability to withstand attacks on the cipher (i.e. cryptoanalysis). It depends upon a pair: a danger and an attack.

The concept of the cipher crypto-resistance is the central concept for the cryptography. It is rather easy to understand it but the obtaining of the strict provable estimations of the crypto-resistance for each specific cipher is not the solved problem yet. The reason for that is the absence of the mathematical apparatus necessary for the solution. Therefore, the specific cipher crypto-resistance can be estimated only by means of various attempts of its opening, and it depends as well upon the qualification of the crypto-analysts [1, p. 14]. Such a procedure is often called the *crypto-resistance test.*

We should notice that this problem is rightfully considered as not solved yet, at least, until the effective approach to the specific cipher opening is offered.

The threat to a cryptosystem (a cipher) is associated with the solution of some particular task achieved by the enemy. We can distinguish between the following types of the threats.

1. The threat of the cipher opening (disclosure, “breaking”): the enemy learns the secrete key.

2. The threat to the confidentiality (secrecy, privacy): having known the cryptogram y, but not knowing the secrete key, the enemy gets to know the message x.
3. The threat to integrity:

   3.1. The threat of cryptogram destruction;
   3.2. The threat to authenticity (identity);

      (a) The threat of cryptogram substitution;
      (b) The threat of cryptogram transmission imitation;

   3.3. The threat of substitution of the published keys.

The attack on the cryptosystem (cipher) is an attempt to realize the conceived threat.

Let us give an example of classification of attacks arranged by their aims and implementation conditions

(1) The attack when the cryptogram is known (in presence of the known cipher-text).
(2) The attack in presence of the known open text (message). Evidently, we assume that the appropriate cryptogram is known.
(3) The simple attack with a choice of the open text. The enemy has messages chosen in advance and receives the appropriate cryptograms (*midnight*, or *coffee break attack*).
(4) The adaptive attack with a choice of the open text. This attack is similar to the attack in p. 3, but in contrast to it, the enemy has a possibility to choose messages transmitted by him during or after transmission of his previous messages.
(5) The simple attack with a choice of the cipher-text.
(6) The adaptive attack with a choice of the cipher-text.
(7) The attack with a choice of the text.

One can distinguish between the two types of crypto-resistance, at least: (1) the theoretical one (informational); (2) the practical one (computational).

An estimation of the least length of the cryptogram, which corresponds to the only key, may serve as the *theoretical crypto-resistance* example.

Let us explain the above-mentioned case. The thing is that the same cryptogram y can be received from different messages x_1, x_2 applying different keys z_1, z_2. As length y decreases, this possibility may completely disappear. Hence, different cryptograms y_1, y_2 will always correspond to the different messages x_1, x_2 and to different keys z_1, z_2. In other words, y uniquely gives the pair (x, z). It means that a cipher can be definitely recovered, if the only one cryptogram with a small length is known. The presence of the last cryptogram is the sufficient condition for the cryptosystem breaking.

The distance for uniqueness is the least length of the cryptogram, according to which we can unambiguously recover the key (for the cryptogram).

The practical crypto-resistance expresses itself in the number/quality of resources (time, a computer, finances, personnel qualification etc.), which are necessary for the enemy for cipher cracking (determination of the key). The larger number of all the possible keys serves as a pre-condition for the high crypto-resistance.

The absolutely crypto-resistant (theoretically) cipher is such a cipher, by means of which the set of keys for any cryptogram that can be used from different messages coincides with a set of all the cryptosystem keys. In other words, for any cryptogram we can select the message for any key.

According to the known Shannon theorem [1, pp. 258–264], the simultaneous fulfillment of the three following conditions is a necessary and a sufficient requirement for the absolutely crypto-resistant cipher:

(1) complete key randomness (equiprobability of letters in the key);
(2) key length is no less than the message length;
(3) one-time application of the key (or of any of its parts).

With such a cipher, the enemy will not gain from the interception of any number of the opened messages and cryptograms. In other words, a priori probability of the key cracking equals to a posteriori one. We can choose, as an example, the algorithm of information ciphering presented in the binary form and consisting of the bitwise modulo 2 ($\oplus$) addition of the message x and the key z:

The message x	01011010101
the key z	1010111011001101001110…
the cryptogram $y = x \oplus z$	11110100011…

In Fig. 1.2, the example of classification of information protection systems, which is accepted in the cryptography, is given. Some of the ciphers presented in this figure, are described in the publications [1–3].

The physical and steganographical protection methods were defined earlier. They can be different but, in our opinion, for these aims we can use the optical vortices. As it will be clear from the following section, and, moreover, from the Chap. 5, the reason for it is that the information is coded by the phase front shape and not by the level of the light beam intensity. Therefore, for the enemy the correct reception will be difficult in the cases, when his photo-receiver is considerably shifted relative to the light beam axis, and, especially, if the enemy analyzes the scattered radiation of the information-bearing light beam. We can generally state that from the properties of the vortical beams there follow the advantages of the "vortical" connection: it possesses physical crypto-resistance to the threat of the enemy's interception of the lateral scattered radiation, and it provides the electromagnetic compatibility of a large number of correspond pairs in the one part of wavelength and in the one space area.

In the next section, we shall give the minimal necessary (to understand Chap. 5) information on vortices, i.e. explaining spiral dislocations of the beam wave front.

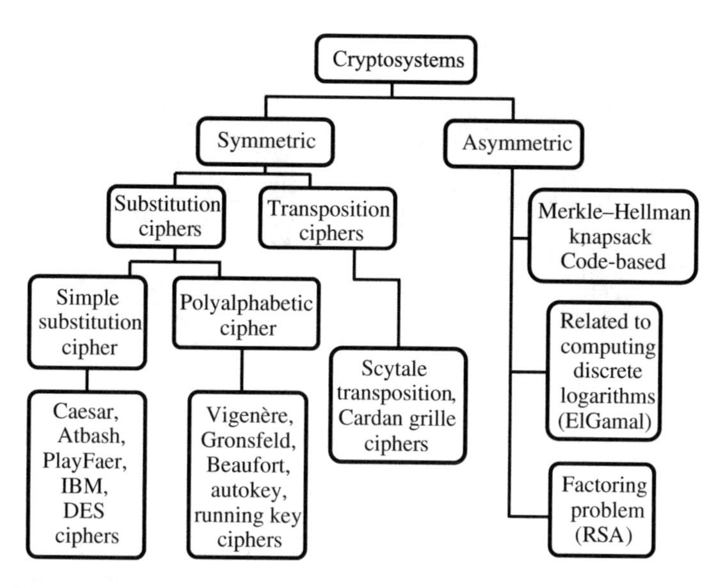

Fig. 1.2 Classification of ciphers

1.2 The Optical Vortex as a Product of the Beam Perturbation and the Data Carrier in the Communication System

We know that the light beams not only transfer energy but also possess linear and angular moments of the momentum. And the complete angular moment may contain a spin component associated with its polarization (caused by the photon spin presence), and the orbital component associated with the spatial (transverse) intensity profile and a phase [4]. Values of the field angular moment are rather large in the case of the field with a spiral dislocation of the wave front, i.e. in the presence of the optical vortex (Fig. 1.3). In the beam, which carries the optical vortex (the vortex beam, the singular beam), the field complex amplitude in the vortex vicinity is described by the equation [5]

$$A = C \cdot r^{V_d} \exp(jV_d \vartheta), \tag{1.1}$$

where C is a constant, V_{d} is the topological charge of the optical vortex, $\mathbf{r} \equiv (x,y)$ is a radius-vector of the point in the transverse section of the light beam, $r \equiv |\mathbf{r}|$ is the distance from the vortex center, ϑ is the azimuth angle, $j = (-1)^{1/2}$ is an imaginary unit. It is usually understood that at $r = 0$, the wave front *singularity* takes place: the intensity equals to zero in this point and the phase value is not defined. Many publications are devoted to the different aspects of the investigation of the light fields with vortices (for instance, [5–23]).

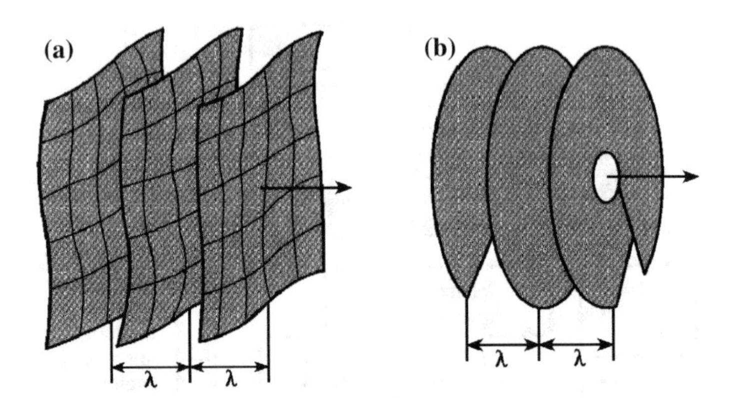

Fig. 1.3 A structure of wave fronts without spiral (screw) dislocations (**a**) and at its presence (**b**) (based on [5])

The orbital angular moment of such vortices equals to $\hbar V_d$ in terms of per photon, where $\hbar$ is Planck's constant, divided by 2π. We can unambiguously associate the value of V_d with the value of data bits (or bytes). For example, we can associate the value of $V_d = 0$ with the value of the "logic zero" ('0'), and to the value $V_d = 1$ there can be attributed the "logic unit" ('1'). Therefore, in publications [24, 25], the orbital angular moment for the information coding and processing was suggested to be used in the optical communication systems also. In the simplest case, to determine the orbital angular moment it is enough to know V_d [25].

In other words, indeed, we can theoretically use the optical vortex as a data carrier in the communication system (either singular optical communication or the vortical one). Then, of course, the several questions will arise: how to construct the appropriate communication system, what devices would constitute it, and how will phase and amplitude distortions influence its operation while propagation in the optically non-uniform medium takes place. Some answers to these questions will be found in Chap. 5.

Naturally, the reader must be interested to know the answer to the questions about the origin, the nature and the ways of the optical vortex generation, etc. Many publications appeared on this theme during the two last decades, and some of them we have already mentioned above. But it is only worth reminding that optical vortices are capable of appearing in the light beam with an initially smoothed phase front (for instance, the Gaussian beam or the beam with the spherical front) then passing through a medium with the non-uniform refraction index (i.e. through the medium with optical irregularity). And vice versa, it is the irregularity that may lead to the transformation of the vortical beam into the beam with smoothed front (disappearance of dislocation). We should note that there is a variety of different dislocations of the beam wave front (i.e. many dislocations in the one transverse section of the beam).

It is clear that not each and every irregularity fits as a reason for the appearance (or disappearance) of the optical vortex. Let, for example, imagine the Gaussian beam passing through the phase shield, ensuring the non-uniform phase incursion in the transverse beam section, the one that obeys to the law $V_d\vartheta$. Owing to the light wave diffraction on the shield, and at some distance from it, the "model" vortex of V_d-order is formed in the center of the beam (i.e. with a charge V_d). An expression (1.1) is its approximate model. From the above mentioned it is obvious that the irregularities of the transparent medium can essentially influence the operation of the vortical communication system leading to the generation of "false" vortices or, on the contrary, to the disappearance of the "information" vortices.

We know that the optical irregularity can have the complicated spatial structure and sometimes leads to the speckle-field appearance. In these cases, if it does not change in time, one often speaks about an optical turbulence. The last is related to the deterministic chaos in the spatially distributed dynamic systems. Therefore, we can now proceed to the sources both of the temporal and of the spatially-temporal chaos (chaotic electromagnetic oscillations and waves), and then—to the discussion of the approaches to the protection of the messages being transmitted.

1.3 Examples of Dynamic Systems in Radiophysics and Optics with Complicated Behavior

1.3.1 Examples of Radio Physical Systems with Complicated Behavior

The differently systematized surveys of chaotic phenomena in radiophysics and electronics, as well as the results of non-equilibrium systems researches (including the tests of the self-oscillating ones), are presented, for example, in the following studies [26–36].

According to the synergetic conceptions, the presence of no less than 1.5 degree of freedom (i.e. three dynamic variables) is necessary (but not sufficient) for the appearance of the deterministic chaos in the dynamic system. Apparently, due to this condition, a variety of the common radio physical chaotic generators has exactly these (minimally required) one and a half degrees of freedom. Moreover, the chaotic generator is often designed as a modification of the generator with one degree of freedom to which some nonlinear element is added. In this case the nonlinearity of the initial generator is of the secondary importance from the point of view of chaos generation (some nonlinear elements will be discussed in this section, some will be grouped in Sect. 1.3.2 of this chapter, and an attempt to discover the nonlinearity nature and develop the general approach to its synthesis can be found in Chap. 6).

Classical generators are Teodorchik–Kaptsov system, its modification in the Anishchenko–Astakhov system, Kiyashko–Pikovskiy–Rabinovich, Dmitriev–Kislov, Chua systems. Here we are not to go into the description of the chaotic oscillation of the Neimark pendulum [27], as well as the maser and laser (the Lorenz–Haken model (1975) [37–39]), and the chaotic oscillation based on the semiconductor lasers [40], and some other issues.

Although it is sufficient for the chaos generation that the system has one and a half degree of freedom, at the turn of the 21-st century, the diversification of dynamic chaos sources is developing, which goes beyond the matters discussed in our study. So, some comparisons of the properties of the dynamic systems with one and a half and two and a half degrees of freedom are performed in this study [41, pp. 114, 115]. It is noted that in the chaotic generator with 2.5 degree of freedom we can achieve the band-pass of the chaotic signals. This is important in the context of the signal distortion minimization in the transmission channel caused by he finite width of the pass-band of this path.

Recent publications of the results of the investigation carried out in Saratov State University and supervised by S.P. Kuznetsov have attracted our attention [42–46]. Here the information is presented about design and analysis of the systems, in which the chaotic dynamic is provided by the presence of the uniform hyperbolic attractor in the system phase space. Thus structural stability of the system is produced, i.e. its features in chaotic mode are not sensible to the system parameter variations. Such property of the attractor is called robustness, and the chaos produced is known as the robust chaos. This must appear to be a valuable advantage when there are involved the problems of the confident communication systems based on chaos. The point is (it will be clearer from the further consideration) that those systems should function in the chaotic mode, and the values of their parameters play the role of keys.

Now we are to consider in brief the basic generators capable of demonstrating chaotic types of oscillations.

Kiyashko–Pikovskiy–Rabinovich generator (1978) represents an electronic system in which the tunnel diode is included in the tuned circuit (Fig. 1.4a), the volt-ampere characteristics of which have an *N*-type view (Fig. 1.4c), and the equivalent circuit of the chaotic generator is presented in Fig. 1.4b. When the current I achieves the value I_m, the tunnel diode switches to the section II; the switching speed relates to the natural capacity value of the c_d diode. Since the diode resistance increases, the scattering (dissipation) of electric energy, equal to the product VI, increases as well in the LCR-circuit. Due to this, both the current I in the circuit and the voltage U decrease. As a result, at some moment the diode inverse switching occurs into the section I characterized by the small VI value. According to [32], the principal possibility of the deterministic chaos appearance in the system is

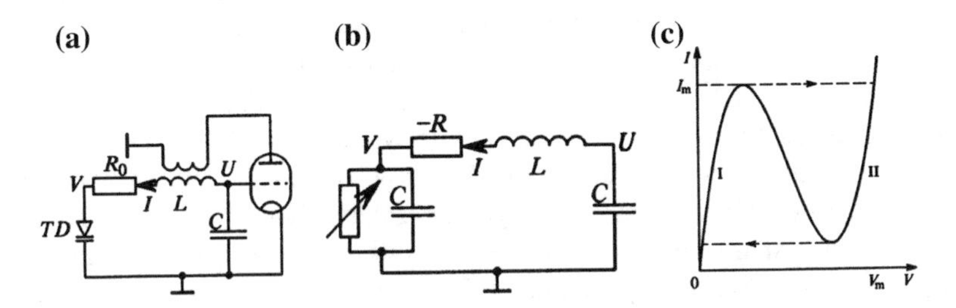

Fig. 1.4 The oscillator of chaotic oscillations Kyashko–Pikovskiy–Rabinovitch (**a**), its equivalent circuit (**b**) and the volt-ampere characteristics of the tunnel diode (**c**) [32]

grounded on the fact that in the area of small values of V, the dynamics of I (at some set of system parameters) is locally stable with respect to the perturbation of the initial conditions. Thus, the nonlinearity of electronic valve characteristic is of no fundamental significance.

The following dynamic equations correspond to the equivalent circuit of the device

$$L\frac{dI}{dt} = U - V + RI,$$
$$C\frac{dU}{dt} = -I,$$
$$C_d\frac{dV}{dt} + I_m f(V/V_m) = I.$$

In such a system there are both regular and chaotic oscillations (Fig. 1.5). The phase portraits of attractors in some typical points of the parameter plane are shown on the peripheral part of the figure.

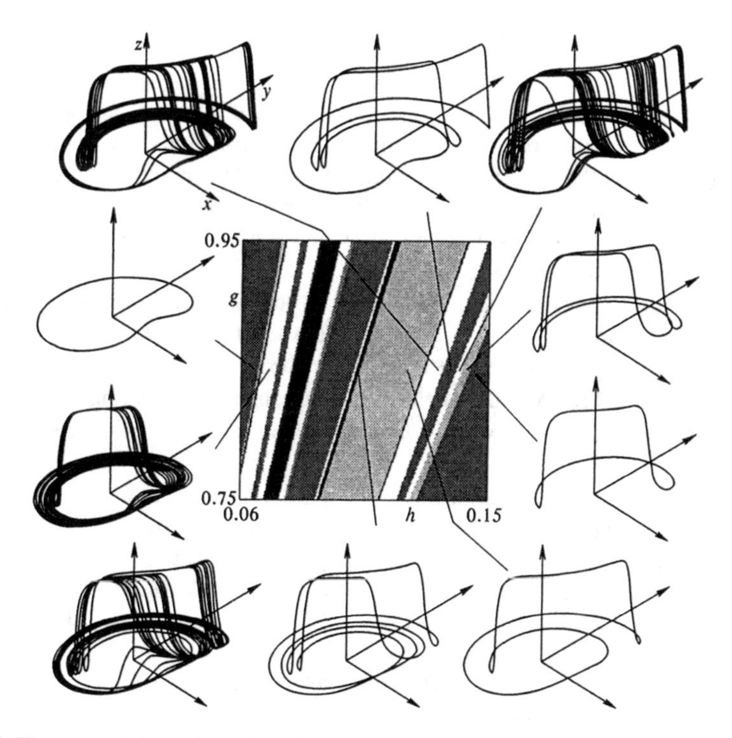

Fig. 1.5 The map of dynamic and models and phase portrait of attractors for the Kiyashko–Pikovskiy–Rabinovitch oscillator on the parameter plane ($h = 0.5(\frac{C}{L})^{1/2}$, $g = V_m(\frac{C}{L})^{0,5}\frac{1}{I_m}$) for $\varepsilon = g\frac{C_d}{C} = 0.2$ and for nonlinear characteristics of the tunnel diode $f(z) = 8.592z - 22z^2 + 14,408z^3$. The chaos areas are shown in white. On phase portraits $x \sim I$, $y \sim U$, $z \sim V$ [32]

The Dmitriev–Kislov ring generator. In 1960s–1970s at the Institute of Radio Engineering and Electronics of the Academy of Sciences of the USSR, under the supervision of V.Ya. Kislov, there were investigated the ways of the effective noise generation based on the traveling-wave tubes. The offered circuit consisted of two such tubes. The first tube functioned as an amplifier, the second was used as a nonlinear element ensuring the delayed feedback [32, 47].

In order to design the simplest model-prototype, the authors considered a circuit closed into the ring constituted of nonlinear amplifier, *RLC*-filter and the inertial element as depicted in Fig. 1.6 [48, 49]. In other words, the created generator (the Dmitriev–Kislov generator) is characterized by the element with the specific nonlinear transfer function playing the part of an amplifier. Therefore, this device is capable of demonstrating the complex dynamics (Fig. 1.7). It is describes by the system of differential equations:

$$\begin{aligned} T\frac{dx}{dt} + x &= M \cdot z \cdot \exp\left(-z^2\right), \\ \frac{dy}{dt} &= x - z, \\ \frac{dz}{dt} &= y - \frac{z}{Q}. \end{aligned} \tag{1.2}$$

Here x characterizes a signal at the inertial element output, z is a signal at input of an amplifier. Parameters T and Q define the relaxation time of the inertial element and the Q-factor of the *RLC*-filter, accordingly. The M parameter defines the gain of the amplifier. The equation of the second order $\ddot{z} + \dot{z} + /Q + z = x$ in Fig. 1.6 is transformed into a pair of the last equations of the first order in the model (1.2) by means of the variable y change.

The map of dynamic modes drawn on basis of numerical solutions of (1.2) at Q = 10 is shown in Fig. 1.7. The phase portraits of attractors in some typical point of the parameter's plane are depicted on the periphery of this figure (as in Fig. 1.5).

When the M bifurcation parameter passes through a bifurcation value M_0 = 1, the bifurcation occurs in the system, namely—there takes place the loss of stability

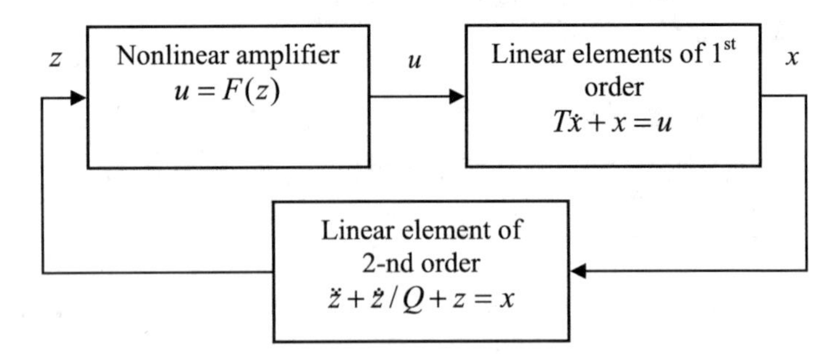

Fig. 1.6 The structural diagram of the Dmitriev–Kislov oscillator capable to be a source of the deterministic chaos [32]

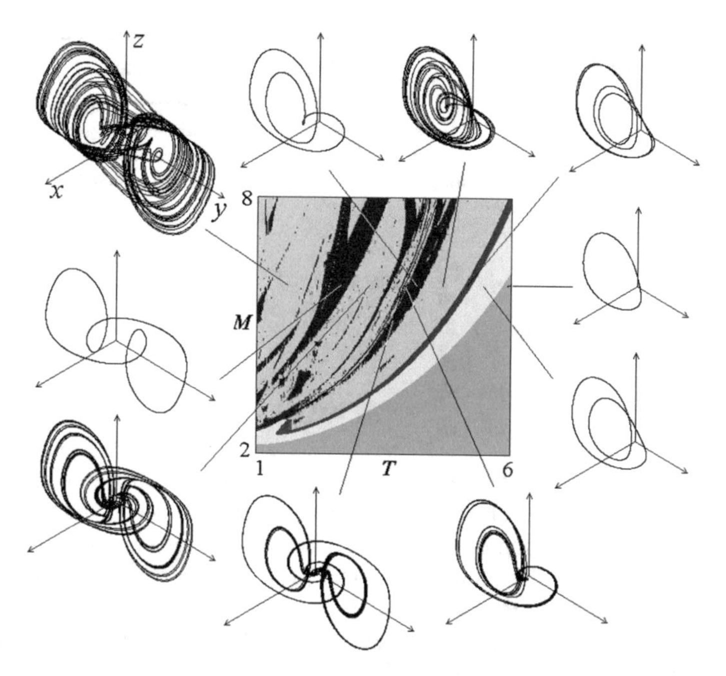

Fig. 1.7 The map of dynamic modes on the parameter plane (T, M) and phase portraits of attractors for the Dmitriev–Kislov oscillator at $Q = 10$ [32]

of the equilibrium condition at the origin. As a result, the pair of symmetrically located stable states $x = Qy = z = \pm\sqrt{\ln(M)}$, which move away from the origin at M increase. When $M_1 = \exp(\frac{(T+Q)(1+TQ)}{2TQ^2})$, both these positions lose stability and become the instable focuses. The limit cycle arises in the vicinities of each equilibrium position. These two cycles are equivalent partners for each other, so that we can follow any of them. For further growth of the M parameter there are the doubling-period bifurcation set and the transfer to the chaos. Then, at larger values of M, the attractor combination with symmetric partner results and the formation of the unified symmetric attractor occurs.

The generator with inertial nonlinearity as the "chaos-gene" element. Historically, its probable first version is the so-called Teodorchik–Kaptsov system (1945–1974 [27]), including thc thermistor as an element with inertial nonlinearity. If instead of the thermistor we use the full-wave quadratic detector with a RC-filter, its output voltage controlling the gain of the linear amplifier, then we get the Anishenko–Astakhov system, 1981 [30], shown in Fig. 1.8 [30, 32]) etc. The following three differential equations serve as the model of this system:

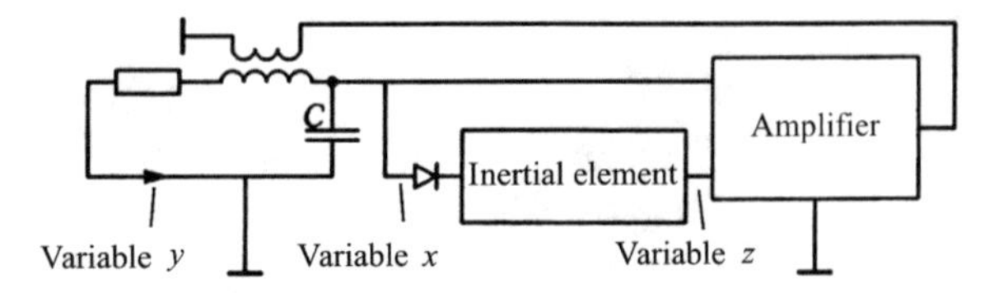

Fig. 1.8 The structural diagram of the Anishenko–Astakhov oscillator [32]

$$\frac{dx}{dt} = mx + y - xz,$$
$$\frac{dy}{dt} = -x,$$
$$\frac{dz}{dt} = -gz + I(x)x^2,$$

where the Heaviside function $I(x) = 0$ at $x \leq 0$, $I(x) = 1$ at $x > 0$; $g > 0$ is a parameter defined by the relaxation time of the inertial element; the m parameter defines the transfer function of the amplifier accounting for the oscillation energy, which is dissipated in the ohmic resistance [32]. Some operation modes of this generator are reflected in Fig. 1.9.

The Chua circuit. The chaotic generator of RF range is often constructed on basis of the Chua circuit (Fig. 1.10) [30].

The analysis of the chaotic generator circuit obtained as a result of the decomposition of the Chua circuit into two subsystems (Fig. 1.11) is described in the reviews [50, 51] and in the study [36]. Dynamics of the chaotic generator obeys to the system of three differential equations

$$\begin{aligned} C_1(\frac{dV_{C1}}{dt}) &= (V_{C2} - V_{C1})/R_1 - I_{N_R}(V_{C1}), \\ C_2(\frac{dV_{C2}}{dt}) &= (V_{C1} - V_{C2})/R_2 + I_L, \\ L(\frac{dL_L}{dt}) &= -V_{C2}. \end{aligned} \tag{1.3}$$

Here[1] $i \equiv I_{N_R}(V_{C1}) = G_a V_{C1} + 0.5(G_b - G_a)[|V_{C1} + E| - |V_{C1} - E|]$, G_b and G_a are tangent of slope angles of the piecewise volt-ampere characteristic of $I_{N_R}(V_{C1})$ for $|V_{C1}| < |E|$ and $|V_{C1}| > |E|$, accordingly, and the value of E gives values of V_{C1}, at which this slope angle changes (Fig. 1.11b). The meaning of the other variables is clear from Fig. 1.11b, c.

We can modify any parameters in the model (1.3) in order to control the oscillating modes, but it is more convenient to fulfill this control by means of resistors R_1, R_2 changing. We should point out that, when $R_1 = R_2$, these equation coincide with the standard model of the Chua circuit [50, 51].

[1]As we can predetermine, the magnitude difference (not the sum of this) should be in the formula (5.4) for $I_{N_R}(V_{C1})$ in [36].

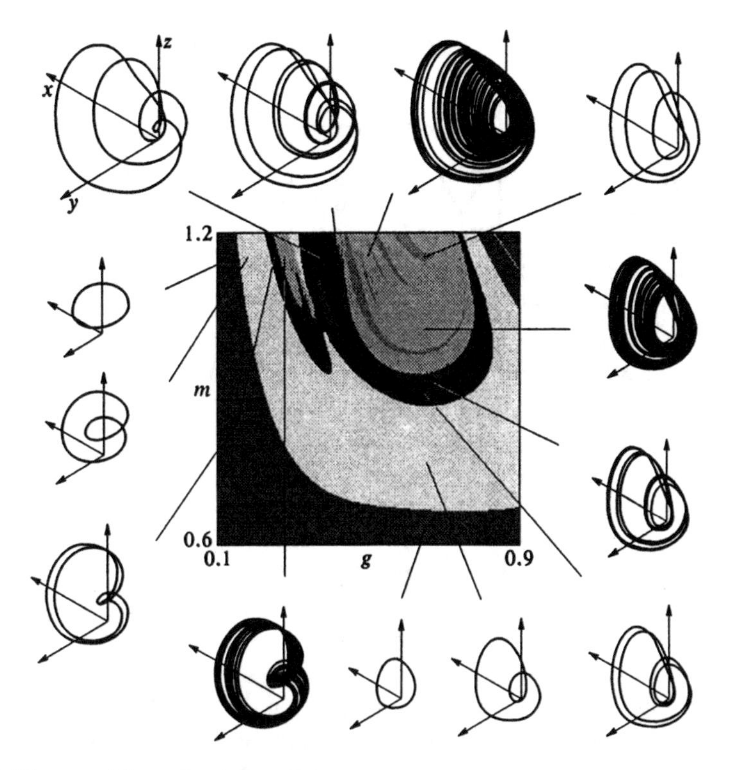

Fig. 1.9 The map of dynamic modes and phase portraits of attractors in some typical points of the Anishenko–Astakhov oscillator [32]

Fig. 1.10 The Chua oscillator circuit (**a**), the volt-ampere characteristics of the nonlinear element (**b**) [30]

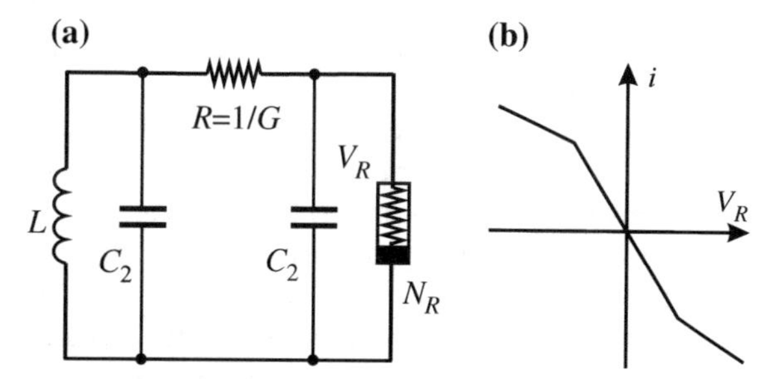

If to introduce the variable change $V_{C1} = Ex$, $V_{C2} = Ey$, $I_L = Ez/R_2$, $T = t/R_2C_2$ and the new designations $\alpha = R_2C_2/R_1C_1$, $\beta = C_2R_2^2/L$, $a = 1 + R_1G_a$, $b = 1 + R_1G_b$, $h(x) = ax + 0,5(b - a)[|x + 1| - |x - 1|]$, then it is easy to obtain the equation system in dimensionless quantities[2] from (1.3):

[2]We would like to note that in comments to the formula (5.29) in [30] the first term in $h(x)$, probably, should be ax (not bx).

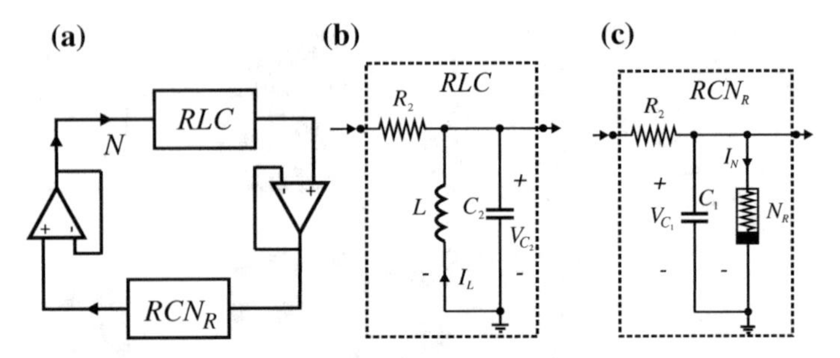

Fig. 1.11 The chaotic oscillator circuit (**a**), obtained by decomposition of the Chua circuit into subsystems: *RLC* pass-band filter (**b**) and RLN_R low-frequency filter (**c**) [50]

$$
\begin{aligned}
\frac{dx}{dT} &= \alpha[y - h(x)], \\
\frac{dy}{dT} &= x - y + z, \\
\frac{dz}{dT} &= -y\beta.
\end{aligned}
\tag{1.4}
$$

The chaotic attractor of the double scroll type is typical for the complex dynamics in the model (1.3). It was is presented in Fig. 1.12, together with the signal shape on the C_2 capacitance and its power spectrum.

As it is mentioned in the study [30, p. 132], the Chua circuit is involved into the solution of fundamental and applied problems of the deterministic chaos theory. It demonstrates not only the properties peculiar for the Rössler model and for the generator with inertial nonlinearity, but a series of specific features related to the attractor symmetry.

In publications devoted to the problem of confidential communication systems design, the functioning of two Chua circuits is widely discussed, including its conditions and the synchronous modes characteristics. Thus, authors of the study [52] theoretically investigate how the possibility of synchronization depends upon the depth and frequency of the capacitance modulation for the different values of the generator parameters related through the same capacitance. The dependence of the threshold value of the synchronizing signal is revealed from the capacitance coupling factors.

In [53], the synchronization of one Chua circuit is studied experimentally by the sequence of chaotic signal pulses (with frequency about ten kilohertz), obtained as "segments" of the continuous chaotic signal of the other similar circuit. It demonstrates that synchronization in the presence of noise persists until the Chua circuit parameters detuning reaches 2 %. The optimal points for the connection of the synchronization channel to the Chua circuits are found.

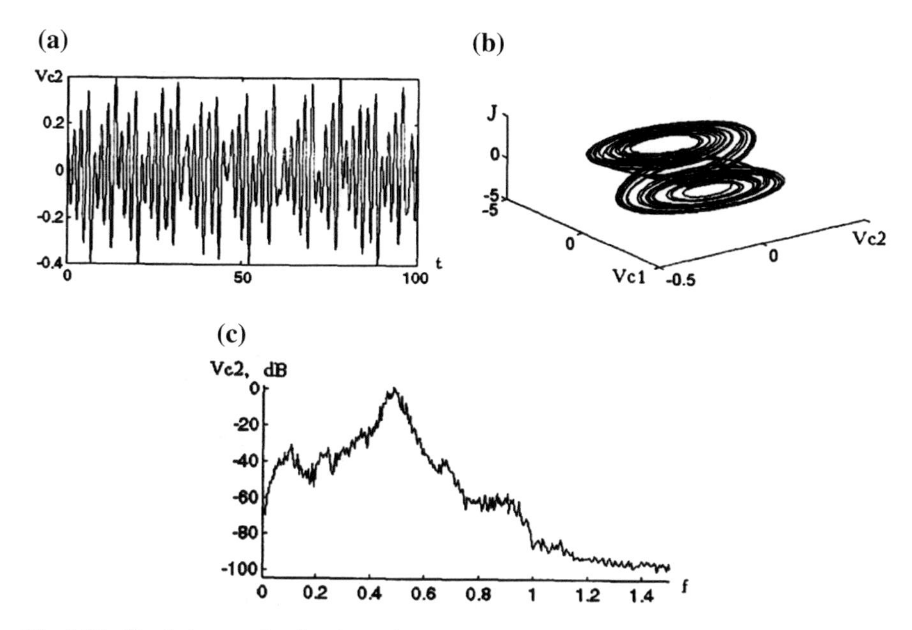

Fig. 1.12 Simulation results for dynamic equations of the Chua circuit, which subjected by decomposition (at $G_a = -1.143$ mS; $G_b = -0.714$ mS; $E = 1$ V; $R = R_2 = 1$ Ohm; $L = 0.0625$ H, $C_1 = 0,1024$ F, $C_2 = 1$ F): V_{C2}, **a** the phase portrait, **b** the power spectrum of the chaotic signal V_{C2} [50]

In addition to the above considered chaotic generators with the minimal necessary dimension of the phase space, we are now to review the dynamic systems with more than three dimensions.

The chaotic generators based on the solid-state structures. In publications [54–56] the compact chaotic self-oscillating system based on the solid-state structures is analyzed. The typical active elements here are the bipolar or field-effect transistors. Transition to the chaos in such systems occurs through the destruction of the two-dimensional torus. The structural diagram of these chaotic generators (Fig. 1.13) is similar to those that has been discussed above: Figs. 1.6 and 1.11a.

Dimension growth of the phase space in such devices is ensured by the increase in order and in number of filters in the ring structure. In the upshot, the phase space dimension of such a dynamic system with lumped parameters equals to the sum of orders of included filters. For instance, when we use the filter shown in Fig. 1.14, the dimension is equal to five. Nevertheless, such frequency-selected systems restrict the signal spectrum in the required frequency band. We would like to note that in publication [57] the stable generation is demonstrated of chaotic oscillations with the spectral density maximum in the frequency range 2.8–3.8 GHz.

The next logical stage in dimension expansion of the system phase space is a transition to distributed systems, for example, to systems with a delay in the feedback loop. In this case, it is accepted to speak that the phase space dimension is *infinite*, because the dynamic system state is given not only by a finite number of

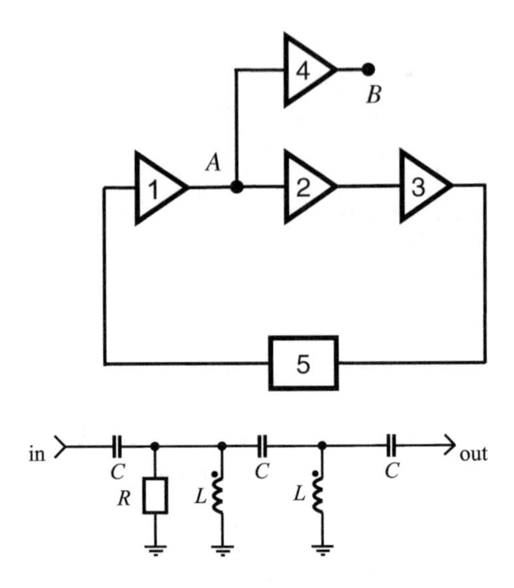

Fig. 1.13 Block-diagram of the oscillating system: *1*, *2*, *3*, *4* are amplifiers; *5* is a frequency-selective system; *A* is a point of the output signal from the feedback network *B* is an output [57]

Fig. 1.14 The frequency-selective system in the feedback network of the oscillating system [57]

parameters but by a function of the continuous argument (or even several functions). Then, evidently, these functions are giving *a continuum* of numerical values of physical quantities.

On equivalence of the influence of the delay line and the inertial element on the system dynamics. In publications [47, 49, 58] it is demonstrated, theoretically and experimentally, the existence of the chaotic modes in oscillating systems with delayed feedback. The structural diagram of such an generator includes the nonlinear amplifier, the delay line and a filter is shown in Fig. 1.15a.

The model of such a generator, as we already mentioned, represents the system with the infinite number of degrees of freedom from the point of view of mathematics. It is clear that the Z signal at output of the ideal delay line relates to the X signal at its input through the equation $Z(t) = X(t - \tau)$.

But, from the physical point of view, a number of degrees of freedom n can be considered as limited and equal to $n \approx \Delta f \tau$, where Δf is the filter pass-band, τ is a delay time. The thing is that the delay line can be approximately considered as a block consisting of a successive connection of n inertial (aperiodic) networks. This block (1.5) is called the inertial delay of n-order. Signals at input and output of the block are connected by the equation [59]:

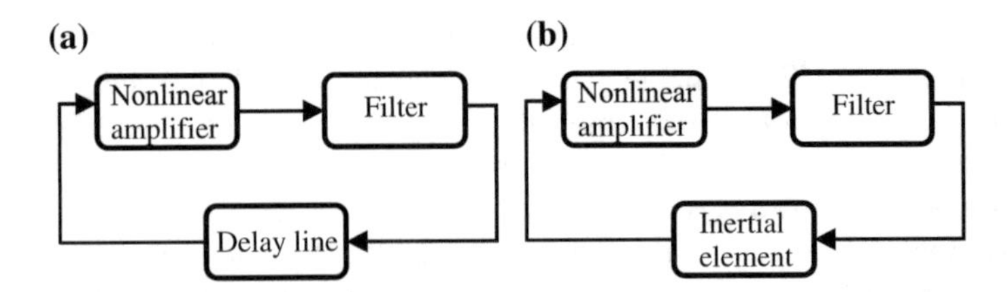

Fig. 1.15 The block-diagram of the oscillator with delayed feedback (**a**) and the oscillator with inertial network (**b**)

$$\left(\frac{\tau}{n}\frac{\mathrm{d}}{\mathrm{d}t}+1\right)^{n} Z = X. \tag{1.5}$$

The inertial network of the first order in electronic systems can be implemented in the form of the *RC*-filter. Naturally, while substituting the delay line for the block with inertial delay (Fig. 1.15b), not all properties of the system in Fig. 1.15a are saved. However, substituting the delay line for the inertial delay even of the not high order often slightly affects the main properties of the system.

Let us review in detail one of the typical representatives of chaotic generators with delayed feedback.

The chaotic generator based on the self-oscillating system with delay. The structural diagram of the chaotic generator with delayed feedback is shown in Fig. 1.16 [60, 61]. The feedback network is formed by the serial *RLC*-tuned circuit with normalized resonant frequency $\omega_0 = 1$ and the damping factor ϵ. The line of the delay за the signal by τ time is the second element.

The mathematical model of the generator under investigation with the delayed feedback is given by the system of differential equations

$$\begin{cases} \dot{x}_1(t) = x_2(t), \\ \dot{x}_2(t) = -x_1(t) + \varepsilon\left[\frac{\partial F(x_0(t-\tau))}{\partial x_0} x_2(t-\tau) - x_2(t)\right], \end{cases} \tag{1.6}$$

where x_1 corresponds to x in Fig. 1.16.

We choose the following function as the function $F(x)$, which describes the nonlinear transfer characteristic of the amplifier in Fig. 1.16 [61]:

$$\begin{aligned} F(x) &= kx\exp(-k^2x^2/g), \\ g &= \begin{cases} 2e, & x \geq 0, \\ 2e(1-\beta), & x<0, \end{cases} \end{aligned} \tag{1.7}$$

where the β parameter characterizes the asymmetry degree of the nonlinearity: $0 \leq \beta \leq 1$. For $\beta = 0$, the nonlinearity $F(x)$ is the odd symmetric function (Fig. 1.17), and, when $0 < \beta \leq 1$, $F(x)$ loses symmetry, and the asymmetry degree increases, if $\beta \to 1$. In contrast to the widely used cubic polynomial, the function $F(x)$ is suitable for modeling and experimental realization, as it is smooth and stays limited for any values of the argument $x(t)$: $|F(x)| \leq 1$.

The electrical circuit [62] and the transfer function (transfer (volt-volt) characteristic) of the nonlinear element capable of realization of the dependency close to (1.7), is used in the laboratory experiment in Fig. 1.18. The operation principle of the nonlinear element is the following. With small signal at the device input, both transistors are closed, and the gain is equal to unity. When the input signal increases, then, depending on its polarity, one transistor opens and the gain of the

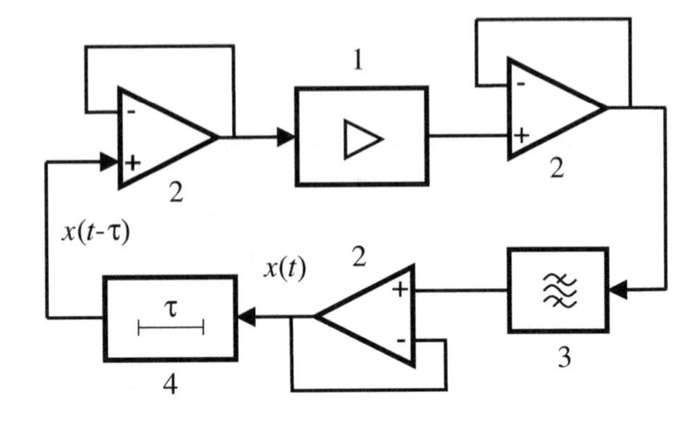

Fig. 1.16 The structural diagram of the oscillator with delayed feedback; *1* is a nonlinear amplifier, *2* is the buffer amplifier, *3* is the *RLC*-pass-band filter, *4* is a delay line (τ) [61]

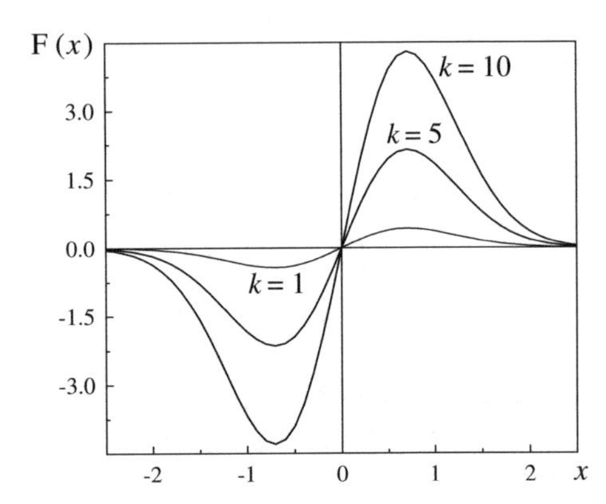

Fig. 1.17 The function used in the model (1.7) for different values of gain k of the linear amplifiers; $\beta = 0$ [61]

circuit can be reduced. We can control asymmetry degree of the nonlinear element by the variable resistor in the emitter circuit.

The performance capabilities of the chaotic generator (Fig. 1.16) and the sequence of modes leading to chaotic oscillations in it are demonstrated in Fig. 1.19. This figure compares the modeling results (figures) and the laboratory experimental ones (pictures from the oscilloscope screen). In the laboratory breadboard, the generator includes the sequential tuned circuit with resonant frequency $f_0 = 33$ kHz and the Q-factor $Q = 1/\varepsilon \approx 2$. The delay line (with the delay time $\tau = 10$ μs) in the feedback network ensures the phase shift $2\pi/3$ for the frequency f_0.

It is impossible not to mention another class of the chaos sources with the infinite-dimension phase space: microwave generators containing the electron flows. It is reasonable to name them as chaotic generators of the ring type based on the nonlinear interaction of two waves: the charge density in the electron (sometimes, in ionic) beam and the direct or reverse electromagnetic wave. Such devices usually operate in the large signal mode. In UHF range, these processes are frequently realized in the traveling wave tube or in the backward-wave tube [63–71]. The signal delay often contributes to the arising of the chaotic mode, which is

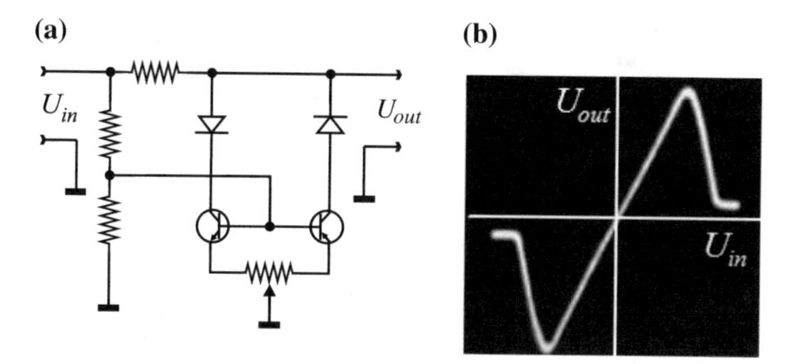

Fig. 1.18 The electrical circuit (**a**) and the transfer function (**b**) of the nonlinear element (laboratory experiment) [61]

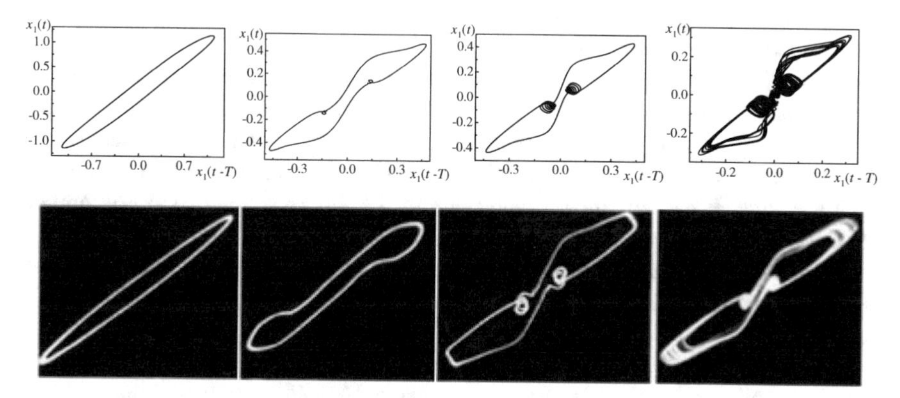

Fig. 1.19 Phase portraits of oscillating modes during the transition from the periodic motion to the chaos in the oscillating circuit with delay ($T \equiv \tau$) in feedback: modeling (upper row) and laboratory experiment (lower row) [61]

associated with a distributed character of interaction of waves and/or with arranged external feedback.

1.3.2 Designs of Nonlinear Elements

Historically, the first chaotic generators were obtained in radio electronic devices with the negative differential resistance. Nonlinear elements can be created on basis of the following semiconductor devices: tunnel diodes, single scan avalanche-injection transistors, the avalanche thermal and photo transistors, dinistors, thyristors etc. However, the operation area of such devices lies within the limits of pulse circuits and devices. Their effective application in the analog engineering is limited due to such features as impossibility of controlling of switch-on and switch-out of

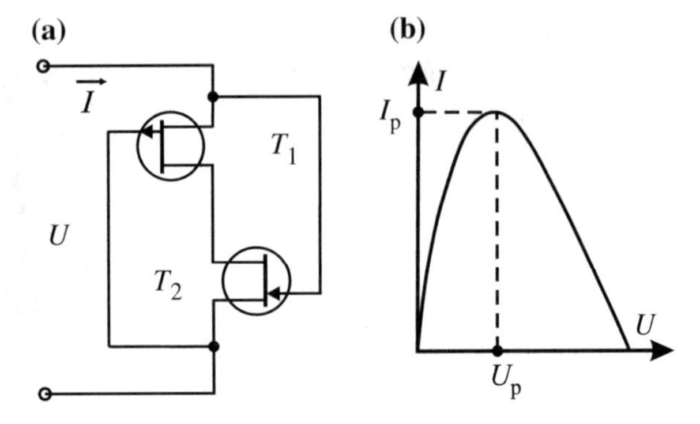

Fig. 1.20. The electric circuit of the Λ-diode (**a**) and its volt-ampere characteristic (**b**)

current, the very large negative dynamic resistance (modulo), the comparably large parameter range, as well as difficulties in obtaining of required parameters of the volt-ampere characteristics (Fig 1. 20b).

In practice, in designing of devices in the interests of development of the ring oscillating systems with chaotic behavior, the devices are used where the negative dynamic resistance is achieved with the help of circuit solutions based on the discrete devices. The Λ-diode (lambda-diode) [72, 73] can serve as an example of such a device. It has got such a name due to the Λ-like shape of the volt-ampere characteristics (Fig. 1.20).

Composite transistor nonlinear elements. One of the electric circuit versions is shown in Fig. 1.20 together with the typical volt-ampere characteristic of the Λ-diode.

The current of the device grows at first with increase of voltage of positive polarity applied to the diode. At some voltage $U = U_p$ it achieves the maximal value (U_p is equal approximately to the voltage of the channel overlapping of any of the field-effect transistors), and after that it decreases. At U equaled to a sum of channel overlapping voltages of both transistors, they will close, and a current through the device is defined by the inverse leakage currents. At further voltage increase the diode remains in the closed condition up to the descriptive voltage of the one of transistors [28].

The similar circuit of the Λ-like characteristic implementation has a series of disadvantages: small peak current $I_p(U_p)$, absence of electronic control of the volt-ampere characteristic parameters, limited regulation range of the negative differential resistance. Modifications of Λ-diode circuit allow elimination of some part of these disadvantages.

For example, one of the field-effect transistors can be changed to MOS-transistor with the induced channel (Fig. 1.21 [28]). The operation principle of the nonlinear element modified in such a manner consists in the following: if some negative voltage U_g will be applied to the gate of the transistor T_1, the area of p-channel will be formed in the transistor, through which the current flows when the negative voltage between the source and the drain is present. Current growth with U_{IN} increase continues until reaching the certain value of $I = I_p$. For small voltage

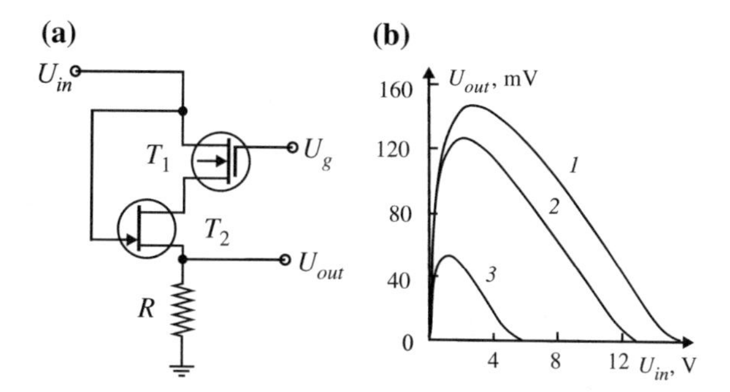

Fig. 1.21 The electric circuit of the nonlinear element (**a**) and the family of transfer (volt-volt) characteristics of the nonlinear element (**b**) [28]

values, the current flowing through transistors increases proportionally to the applied voltage growth, due to the feedback presence between the drain of T_1 and the gate of T_2, with growth of input voltage the transistor T_2 begins to close because of the conductance channel narrowing. Simultaneously, the total current through the device achieves saturation and begins to decrease: the volt-ampere characteristics section arises with the negative differential resistance.

The value of the maximal current through the lambda-diode, and, hence, the dynamic range of the nonlinear element is defined by opportunities of the induced channel of the T_1 transistor. The more voltage value U_g on the gate, the wider the region of conductance of the *p*-channel of T_1, and hence, the more the dynamic range of tunings. In the area of negative input voltages and also at complete closing of the T_2 transistor, the input signal is determined by the level of inverse currents of T_1 and T_2 transistors. Within the frequency range up to hundreds of kilohertz, this element is the non-inertial device, since typical terms of spatial charges dispersion near *p*-*n*-junctions are by several orders less than the typical term of oscillation period in the system.

Some circuit solutions of elements with the negative differential resistance are presented in [74–77], based on the application of different types of transistors (bipolar, field-effect with the isolated gate).

Nonlinear elements using piecewise characteristics. As it will be clear from the further description, one of the requirements to the chaotic generator in the context of application in the data transmission system is repeatability of the chaotic modes in different generator variants. For this, in particular, it is required to ensure coincidence of the transfer characteristics of nonlinear elements. In the case of usage of elements constructed on basis of discrete devices with smooth transfer characteristics, the selection of two similar elements represents the complicated problem. This reason is that the one of the main disadvantages of application of the discrete devices as nonlinear elements is the huge dispersion of their parameters and instability of characteristics due to the varying device operation conditions.

One can solve this problem by application of nonlinear elements with repeatable transfer characteristics. The piecewise functions are the examples of mentioned characteristics (functions), including several linear segments. Then, one proceeds from the possibility of a rather easy reproduction of the transfer characteristics of

Table 1.1 Nonlinear elements on the base of integrated circuits

	Circuit solution	Transfer characteristics
[78]	R_1 I + V R_1 R_2 –	I V –
[41]	22 kΩ V_{in} 1 22 kΩ R1 15 kΩ 15 kΩ 15 kΩ 15 kΩ 2 3 10 kΩ 10 kΩ 10 kΩ 100 kΩ 4 7,5 kΩ V_{out} 5 R2	V_{out} V_{in}
[79]	2R R R R 2R R R V_C G J D S I_{IN} I_D V_{IN} + – $I_{VCCS} = -\frac{2I_{DSS}}{V_P} V_{DS}$ I_{IN}	4.0mA I_{IN} 0A -4.0mA I(in) -1.0V -0.5V 0V 0.5V 1.0V V_{IN}
[80]	Input "X(t)" OP1 OP2 Output "α f(X)" Rn1 Rn2 Rn3 D2 D1 OP3 "α" Rn6 Rn4 Rn5	0.5 0.3 0.1 -0.1 -0.3 -0.5 f(x) -2.0 -1.0 0.0 1.0 2.0 X
[81]	I + V – S₃ S₄ S₅ S₂ S₁ AD633JN AD633JN AD633JN AD633JN AD633JN	I x10^-3 1,5 1,0 0,5 0 -0,5 -1,0 -1,5 -5 -4 -3 -2 -1 0 V 1 2 3 4 5

the nonlinear element at each of its parts being linear ones. Nevertheless, the very fact of piecewise characteristics application is an insufficient condition for the accurate reproduction of chaotic generators. For instance, in Table 1.1, nonlinear elements have piecewise signal characteristics, but in the one of them, the diodes are used as the nonlinear elements having the significant dispersion of volt-ampere characteristic shape. In other words, the development of nonlinear elements should be based on the precision element base or on circuit solutions permitting the fine adjustment, capable of compensating the lack of coincidence of parameter values of separate elements.

Obviously, during the final analysis, the final form of piecewise characteristic is important. It should ensure both the generation of the chaotic signals and the fulfillment of synchronous chaotic response stable against slight deviation of generator parameters and certain other factors.

Further, we would like to get acquaintance with the chaotic generators of optical wavelength range.

1.3.3 The Nonlinear Ring Interferometer as an Example of the Optical System with Complex Behavior

The last decades are marked by the growing interest attracted by the phenomena of structure formation and chaotization in the optical systems including lasers. The publications [82–90] demonstrate the diversity of possible problems.

In this context, the nonlinear ring interferometer (a resonator containing the medium where there is the optical Kerr effect) and mathematical models of its processes become convenient means in the study of the self-organization and chaotization structure phenomena. They appear as a result of nonlinear refraction, the feedback and the large-scale transformation of the optical field: see for instance, [91–106].

In this investigation area the publication of K. Ikeda et al. is considered as the pioneer paper [107], which gives birth to a series of theoretical and experimental researches. K. Ikeda et al. described a dynamic system with delay τ, i.e. with the infinite number of degrees of freedom: the four-mirror ring interferometer containing the saturated absorber (Fig. 1.22). As Landa notes [27], such a ring resonator is the specific case of some *heredity* systems, the dynamic of which can be described by the equation (in approximation of the spatial character of the light field):

$$\frac{dx(t)}{dt} = -bx(t) + f(x(t-\tau)),$$

where $f(x)$ is the nonlinear function of the light field intensity x.

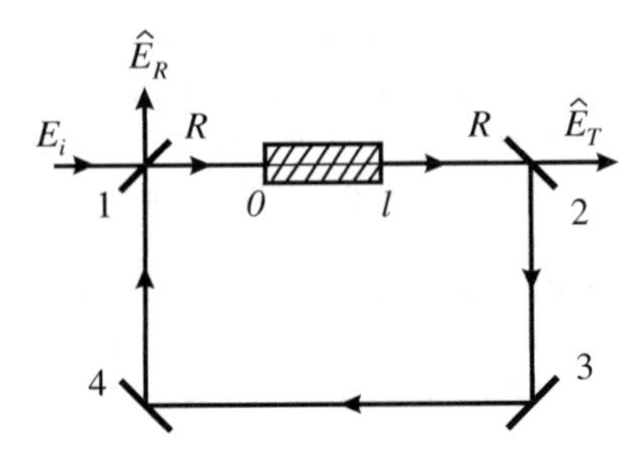

Fig. 1.22 The diagram of the ring resonator with the absorbing medium [27]

Let the crystal be a nonlinear medium, in which there is the optical Kerr effect, i.e. the crystal refraction index depends linearly on the intensity $|E_{n+1}|^2$ of the light field with complex amplitude E_{n+1} (for simplicity assumed dimensionless). Then, while neglecting the spatial character of the light field and assuming that the crystal response is instantaneous, the following two-dimension discrete map, named after Ikeda, is true

$$E_{n+1} = A + BE_n \exp(i|E_{n+1}|^2 + i\varphi),$$

where A—is a parameter defining the laser radiation field, which passes inside the resonator, B is the loss (transfer) factor of the radiation amplitude for the complete resonator bypass, φ is the field phase incursion passed through the ring interferometer (without the account of addition $|E_{n+1}|^2$ to φ caused by the variation of the refraction index due to the Kerr effect) [32]. This map demonstrates both the regular and the chaotic behavior (Fig. 1.23).

Further studies in this field stimulated appearance of a series of publications [96, 108–112], where issues of structure formation in the nonlinear ring interferometer (NRI) are discussed. The diagram of the nonlinear ring interferometer with the two-dimension feedback and with large-scale transformation of the optical field in the feedback loop is shown in Fig. 1.24a.

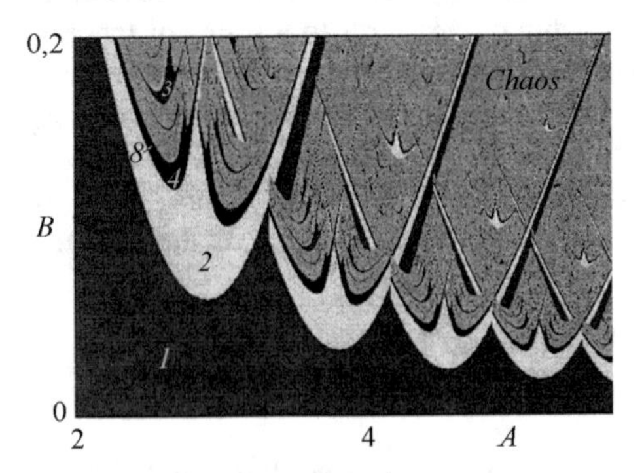

Fig. 1.23 The map of dynamic modes of the Ikeda map on the parameter plane (A, B) for $\varphi = 0$ [32]. Here numbers *1*, *2*, *3*, *4*, *8* mark parameters areas corresponding to oscillations with relevant period values

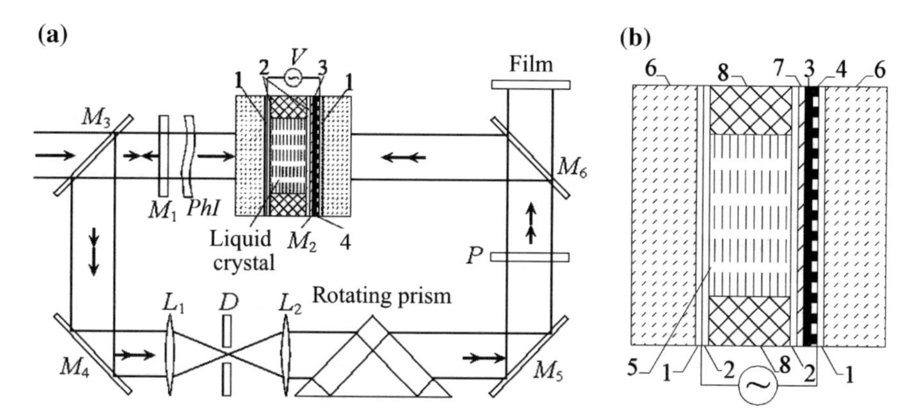

Fig. 1.24 The optical system of the nonlinear interferometer with the two-dimension feedback (**a**). The structure of the liquid crystal—photoconductor with the phase modulation of reflected light (**b**). There M_1, M_3, M_6 are light dividers; M_2 is the dielectric mirror of the transparent; M_4, M_5 are mirrors; L_1, L_2 are lens; D is a diaphragm; P is a polarizer; *PhI* is a phase irregularity; *1* are transparent electrodes; *2* are orientate coverings; *3* is a light-blocking layer; *4* is a photo-conducting layer; *5* is a liquid crystal; *6* are substrates; *7* is the dielectric mirror; *8* is a spacer [109]

According to [109], the nematic liquid crystals (from Greek νημα—thread) are the effective nonlinear medium allowing the nonlinear response in the field of the helium-neon laser operation. It is known that the molecules in this medium are put in order in such a manner, that, in spite of the absence of the order in the location of these gravity centers, there is a long-range order in the orientation of their longer axes. Therefore, directions of the molecule longer axes change regularly (i.e. in no chaotic manner) in the space forming a spatial spiral. This spatial structure may have a period corresponding to the light wavelength and ensure the strong interaction with the light field [113].

In order to obtain the nonlinear effects in the ring interferometer, we can use the liquid-crystal phase transparent. Having a strong nonlinear effect, liquid crystals allows modulation of the optical parameters of the light wave by relatively weak controlling fields, although it relates to noticeable increase of the relaxation time of molecule polarization. Having applied the external electric field, we may effect molecule orientation of liquid crystals. The elastic forces prevent the molecule deviation from the initial orientation set by the conditions at the layer boundaries. When choosing the appropriate technology for the substrate processing and applying the external field, we may obtain any molecule orientation of liquid crystals by thus controlling the permittivity anisotropy.

The liquid crystal, located in the ring interferometer, has the cubic nonlinearity (i.e. its polarization is the third-order polynomial with respect to the optical field strength E, or, in other words the nonlinearity of the Kerr type: the refraction index of the liquid crystal n depends upon the optical field strength according to the following law

$$n = n_0 + n_2|\mathbf{E}|^2, \tag{1.8}$$

where n_0 is not-perturbed refraction index, n_2 is the nonlinear refraction factor.

Along with liquid crystals suitable for the experimental demonstrations, it is expedient to use the solid-state low-inertial Kerr media, for instance, GaAs, InSb etc. presented in the diagram in Fig. 1.6 from [109], as well as fullerenes as the nonlinear medium.

The coherent light wave $\mathbf{E}(\mathbf{r}, t)$, having fallen into the interferometer (Fig. 1.24a), passes the thin layer of the nonlinear layer and obtains the nonlinear (in its origin) phase incursion proportional to the radiation intensity. Fields reflected from mirrors M_2 and M_1 interfer. The element G represents a device performing the linear transformation of the optical field: a rotation in the transverse plane, compression, stretching, a shift of the laser beam, the Fourier filtrating of spatial harmonics.

It is easily seen that an interferometer is the open (flowing) nonlinear complex system including the loop of two-dimensional feedback. Interconnection between optical fields in the different points of the transverse section of the laser beam is performed by means of diffraction, due to the beam limitation and/or molecule diffusion of the nonlinear liquid crystal, as well as due to the linear large-scale (non-local) transformations of the light field in the feedback loop. A presence of nonlinearity in the system in the combination with the feedback (of ring type) provides a possibility for the arising of the positive feedback and, hence, for the formation of spatial-temporal structures in the transverse section of the laser beam.

As the mathematical model of processes in the interferometer we can use the following equations:

$$\tau_{\mathrm{n}} \frac{\partial U(\mathbf{r}, \mathbf{t})}{\partial t} = -I(\mathbf{r},t) + D_{\mathrm{e}} \Delta U(\mathbf{r},t) + K\{1 + \gamma \cos[U(\mathbf{r}',t) + \Phi(\mathbf{r}') + \varphi_0]\}, \tag{1.9}$$

Here τ_{n} is relaxation time of the nonlinear part of the refraction index $n(r, t)$ of the medium with extension l; $U(\mathbf{r}, t)$ is the nonlinear phase incursion in the nonlinear medium (NM); D_{e} is the effective diffusion factor normalized to a square of the input laser beam radius and the NM relaxation time; $\gamma = 2\kappa R \exp(-\alpha l/2)$ is a loss characteristic; κ is the field attenuation factor in the feedback loop (FL); α is the Bouguer absorption index in NM; R is the mirror reflection factor in intensity; $\varphi_0 = lkn_0$, $\Phi(\mathbf{r}')$ are constant phase incursions; $K = n_2 lk(1 - R)I_{\mathrm{in}}$ is a nonlinearity factor; $\mathbf{r} = (x, y)$ are coordinates of transverse section points of the laser beam normalized to its input radius; k is a wave-number; I_{in} is an intensity of the laser beam in NRI input.

Connection between $\mathbf{r}$ and $\mathbf{r}'$ is determined by a character of the field spatial transformation in FL. The diffusion of NM particles is taken into account phenomenologically by the Laplace operator according to the transverse coordinates.

Examples of structures observed in laboratory experiments are shown in Figs. 1.25, 1.26 and 1.27.

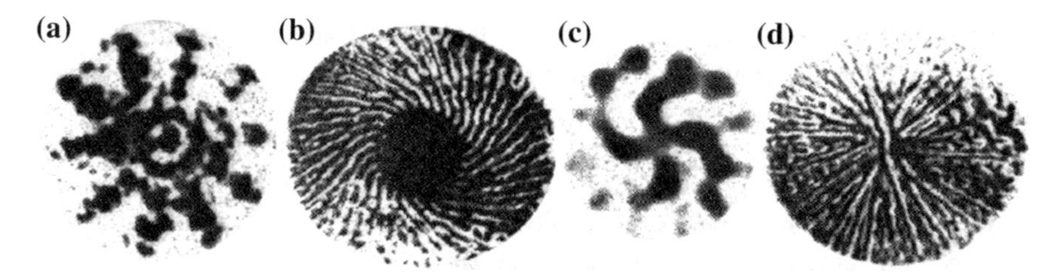

Fig. 1.25 Various types of multi-petailed structures [109]

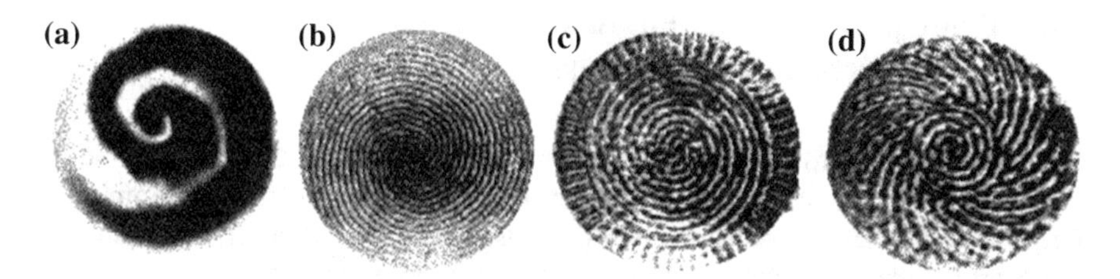

Fig. 1.26 Structures arising at field rotation and its scale variation [109]

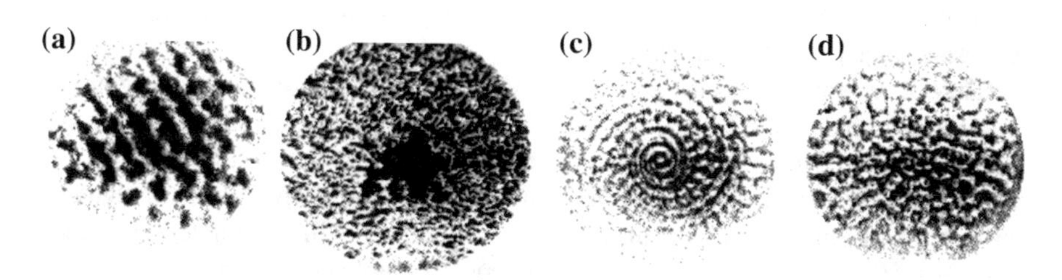

Fig. 1.27 Stochastization of the light field, which destroys the structure regularity [109]

The model (1.9) is true, if a series of limitation is met:

- an input field of the interferometer is monochromatic, plane-polarized, unmodulated neither in phase, nor in amplitude (neither in the space, nor in time);
- characteristics of the nonlinear medium (n_2, D_e, τ_n) are uniform and non-varying in time;
- the approximation of large loss is suitable (which makes necessary to assume γ as small), the one that authors in [96, 108, 109, 111] state as an approximation of one bypass (which, generally speaking, is not the same);
- the field energy re-distribution is absent in the transverse section at large-scale beam transformation in FL;
- the field propagation time in feedback loop (FBL) is negligibly small compared to the time of the noticeable variation of the nonlinear phase incursion.

Listed limitations narrow the variety of real optical systems described by the model (1.9). For the solution of various applied problems, related, for example, to information processing (chaotic communications), adaptive optics and registration of input radiation characteristics, it is necessary to construct more complicated mathematical models. This will be performed by us in Chap. 4.

It is expedient now to pass from the description of deterministic chaos generators to some communication systems employing such generators.

1.4 Principles of Information Protection by the Deterministic Chaos

Traditionally, the information transmission is performed with the help of modulation or shift-keying of the one of the periodic process parameter. The amplitude, frequency and phase modulation are widely used for the sine carrier; modulation of pulse amplitude, duration and time position are used for the pulse carrier and appropriate types of shift-keying. With the development of the theory and the methods of generation of the complicated (noise-like) signals, a possibility of application of the pseudo-random and noisy signals as the information process-carrier [114–116] arises.

Chaotic synchronization. An interest to chaotic oscillations as to the potential information carrier arose after the publication of the pioneer researches on synchronization of dynamic systems with the deterministic chaos [117–119]. It was shown that chaotic motions $x_1(t)$ and $x_2(t)$ of two (and more) *identical* dynamic systems at strong mutual or unidirectional coupling will completely coincide: $|x_1(t) - x_2(t)| \to 0,\ t \to \infty$. Then this tendency is stable concerning perturbations. The term "chaotic synchronization" (or "synchronization of the chaos") appears. It means the synchronization of dynamic systems, which are connected in the single or double-directional manner by means of chaotic signals (i.e., at least, the one of them should function in the deterministic chaos mode). It becomes clear that chaotic signals can be used as the carrier for information transmission. Thus, we do not need additional equipment to achieve and maintain the synchronization between transmitting and receiving sides.

We will not forget that studying of the synchronization phenomena has had a great history. It goes back to the beginning of the natural sciences in the 17-th century: "the watchs sympathy" was discovered and described in 1665 by Chr. Huygens. The traditional synchronization theory, i.e. based on the application of the regular signals, operate with such concepts and terms as frequency capture and frequency pulling [26].

Regarding the chaotic systems, the synchronization theory is today in the making. Many authors suggest the discussion of the synchronization, including chaotic, in the wider meaning. They start from the fact that at each possible combination of oscillation mode types in master and slave systems, the situation may occur, when the dynamics of the slave system will reproduce the definite characteristics of the master system dynamics. In this case we speak about the (forced) synchronization in some generalized sense. Evidently, together with such one-directional effect, it is possible to have the mutual interaction of two oscillating system equally influencing each other. Then we speak about mutual synchronization: the definite dynamic characteristics coincide [120]. Owing to some inaccuracy of the words "definite characteristics", the term "synchronization" and particularly "chaotic synchronization" is interpreted differently in various publications. As a rule, it requires additional explanations in each specific case.

Thus, it is acceptable to distinguish between the following types of the chaos synchronization: the full chaotic synchronization [119, 121]; the phase synchronization [122], which is experimentally implemented in [123]; the generalized synchronization [124], experimentally observed in [125]; the synchronization with delay, which offered and investigated in [126, 127]; the synchronization induced by a noise [128–133]. Each type of them is characterized by its own advantages and disadvantages. For example, some require the high chaotic generator identity. It is clear that the chaotic synchronization phenomenon is defined by two mechanisms. The first one is produced by the system instability and it is responsible for the process chaotization in the coupling systems. The second mechanism is caused by dissipative character of the motion, it forces phase trajectories of the dynamic system to compress and to approach each other along stable directions.

In the synchronization context, the situation is typical when both systems are generators, i.e. amplifiers with positive feedback of necessary intensity. But we can well speak about synchronous behavior of both pendulums—two passive (i.e. non-oscillating) systems, and about synchronous behavior of the generator and the passive system it is affecting. In the last case, tending to emphasize a specificity of physical processes, one can speak about the (chaotic) synchronous response [134], a well as about the passive (chaotic) synchronization [135], together with the expression "(chaotic) synchronization".

In view of practical importance of this case, we explain reasons of such a behavior through the prism of the decomposition of some self-oscillating system [41, 119].

The chaotic synchronous response. Let it be an initial self-oscillating system, which is described by the equation

$$\frac{\mathrm{d}\mathbf{U}}{\mathrm{d}t} = f(\mathbf{U}), \tag{1.10}$$

where $\mathbf{U}$ is a vector belonging to n-dimensional space: $\mathbf{U} \in \mathbf{R}^n$. If the system structure is such that it can be represented in the form of two subsystems V and W (U → (V,W)), then (1.10) can be rewritten as

$$\frac{\mathrm{d}\mathbf{V}}{\mathrm{d}t} = g(\mathbf{V}, \mathbf{W}), \quad \frac{\mathrm{d}\mathbf{W}}{\mathrm{d}t} = h(\mathbf{V}, \mathbf{W}), \tag{1.11}$$

Fig. 1.28 Decomposition of oscillating system [41]

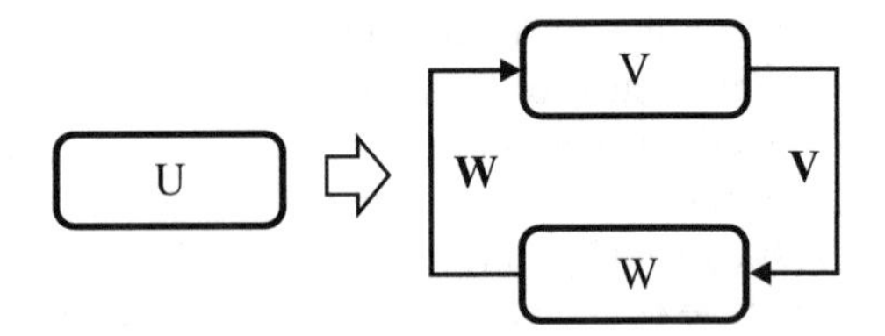

where $\mathbf{V} = (U_1, U_2, \ldots, U_m)$, $\mathbf{W} = (U_{m+1}, U_{m+2}, \ldots, U_n)$, $g = (f_1(\mathbf{U}), f_2(\mathbf{U}), \ldots, f_m(\mathbf{U}))$, $h = (f_{m+1}(\mathbf{U}), f_{m+2}(\mathbf{U}), \ldots, f_n(\mathbf{U}))$ [41, 119].

Figure 1.28 shows the decomposition of the initial system (1.10), according to (1.11). As a result, the self-oscillating system takes a view of the ring structure, where subsystems form the united feedback loop.

After that, two identical systems obtained during the same decomposition, are separated. One of them remains in the form represented in Fig. 1.29, i.e. it is active (self-oscillating), "master" system. In the other system, the feedback loop is broken off, i.e. it becomes passive (non-oscillating), "slave" system [41]. We should note that the view the equations and figures take do not contradict to the fact that these two subsystems V and W have internal feedbacks, i.e. it does not exclude their oscillating properties. Nevertheless, the presence of internal feedback in them does not generally guarantee their transformation into the self-oscillating subsystems. For instance, the positive feedback factor in them can not exceed the threshold value. Another example: subsystems V and W can contain the elements with the hysteresis transfer function, the existence of which can be explained by the feedback presence. But, while interpreting the chaotic synchronous response concept, the authors of [41] limit themselves by the pointing out to the absence of internal feedbacks in subsystems V and W.

If the signal from the output of one subsystem (for example, the V subsystem), namely—$\mathbf{V}_1(t)$, of the master system U_1, is applied to the input of the other subsystem (the W subsystem) of the slave system U_2, as it is shown in Fig. 1.29, then, under definite conditions, the difference between the input $\mathbf{V}_1(t)$ and the output $\mathbf{V}_2(t)$ signals in the slave (open) system will be vanishing $|\mathbf{V}_1(t) - \mathbf{V}_2(t)| \rightarrow 0$ at $t \rightarrow \infty$. Thus, with the help of the decomposition, it is possible to form a pair of systems ("master–slave"), in which the slave system asymptotically repeats a behavior of the master system under the one-directional influence of the master system [41].

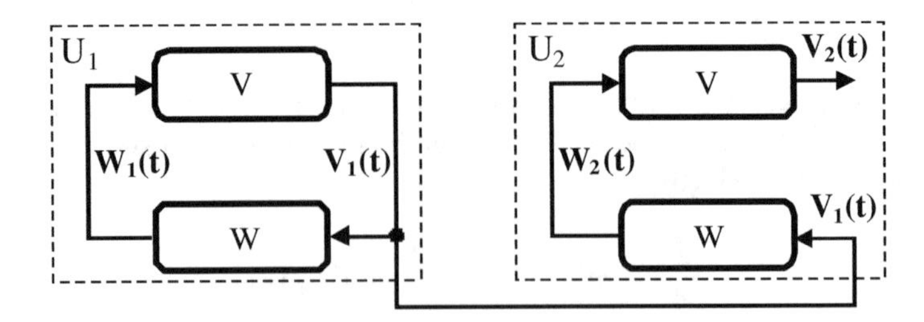

Fig. 1.29 Formation of a pair systems "master-slave" [41]

Peculiarities of the chaotic signals as the information carriers. The study of generation, synchronization and control peculiarities of the chaotic oscillations [30, 51, 136–139] reveals that the chaotic signals have some properties permitting their differentiation into the separate group from the point of view of application in the information transmission systems. Among such properties there are the following:

(1) chaotic signals possess high “plasticity”, i.e. there can be realized a variety of different types of oscillations with a wide spectrum in the same dynamic system;
(2) a choice of this or that chaotic mode is performed by means of very insignificant variations of dynamic system parameters;
(3) chaotic signals are not periodic in time, and they have the positive value of the Kolmogorov–Sinay entropy [31, p. 206], which radically differs them from the pseudo-random sequences (M-sequences etc.), as well as from different types of compound signals;
(4) chaotic signal can be easily obtained in the radio frequency, UHF and optical ranges with the help of simple structures (from the point of view of hardware implementation);
(5) there is a large variety of methods for information signal introduction into the chaotic carrier signal;
(6) the transmission is possible of the several information-bearing signals by means of the only one chaotic carrier;
(7) chaotic signals are wide-band, and, therefore, they have larger information-bearing capacity in comparison with the narrow-band signals (they are capable of carrying a large volume of useful information without significant variations in spectral and other physical characteristics);
(8) the wide frequency range of the chaotic signals leads to the high energy effectiveness and to the high rate of data transmission, and can increase the operation stability of the communication system in the presence of the disturbing factors;
(9) in the treatment of the chaotic signals there can be applied both the common processing methods suitable for the compound signals (for instance, correlation methods) and the specific methods based on the knowledge of the chaotic dynamic regulations;
(10) chaotic signals are adaptable for confidential communication, if the spectrum of the information signal can be concealed in the spectrum of the chaotic signal.

Thus, the dynamic chaos has some properties, which makes it attractive for the application in communication systems as the information carrier [134, 140].

1.4.1 General Schemes and Functioning Principles of the Confidential Communication Systems in the Mode of the Dynamic Chaos

Until now, the significant experience has been accumulated in the development of transmission schemes and information ciphering (coding) algorithms with the dynamic chaos employed [41, 51, 136–139, 141–144]. Growth of researchers' interest to these problems is illustrated by Fig. 1.30. We would like to emphasize that chaotic synchronization application constitutes the dominating idea in the nonlinear-dynamic cryptology, to which our book is devoted.

We now consider only some of the most common of them. We should note that there exists a large enough number of various versions of communication systems applying chaotic signal. It is associated with the fact that modulation and shift-keying of chaotic oscillations during their transmission and the methods of their selection at the reception are considerably more diverse compared to those of the traditional communication systems with the amplitude, frequency and phase modulation and shift-keying. For the same reason, the capabilities, the construction principles and the methods of implementation of the chaotic communication systems have not been fully examined yet.

The principle of the chaotic signal application for the information transmission is explained by the functional diagram shown in Fig. 1.31.

At the transmitted side, the information signal $s(t)$ somehow interacts with the chaotic oscillation $x(t)$, which is produced by the transmitter generator. The chaotic signal $y(t)$ obtained as a result of such interaction, propagates in some medium called the communication channel and, having undergone the inevitable distortions $n(t)$, comes to the receiver input in the form of the $z(t)$ chaotic signal. The receiver extracts the informational component from the $z(t)$ signal with the help of specific procedure. As a result, the $s'(t)$ signal, which is in the ideal case the exact copy of the initial information signal, comes to its output. Thus, the chaotic signal is used as the oscillation carrier.

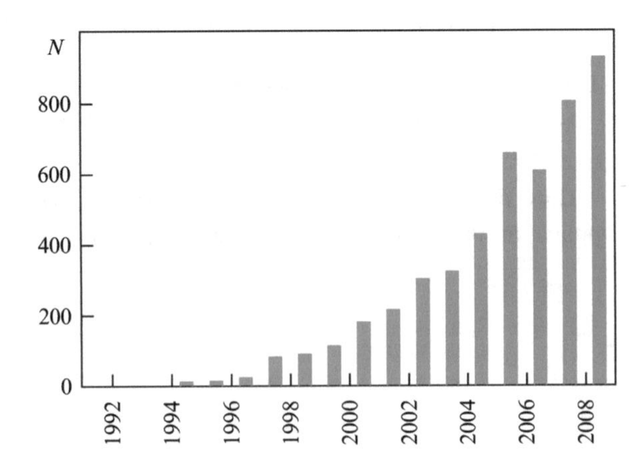

Fig. 1.30 Dynamics of the publication citing number (per one year in central scientific journal being reviewed), concerning application of chaotic synchronization in the communication systems (information from ISI Web of Knowledge) [144]

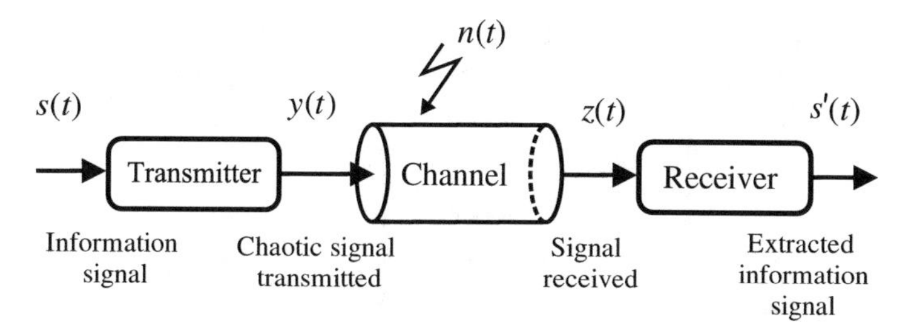

Fig. 1.31 The functional diagram of the information transmission system with chaotic carrier [137]

Various ways and means have been suggested of the information signal introduction into the chaotic signal at the transmitting side and its recovering at the receiving side. The second task is more complicated. At present, communication systems with the chaotic carrier use two types of receivers: coherent and non-coherent ones [41, 137].

Non-coherent receivers use statistical properties of the $z(t)$ signal, coming from the communication channel, for the information extraction. In this case, it is assumed that the method of introduction of information into the carrier (chaotic) signal is the only known parameter, and the other accurate parameters of the transmitter remain unknown. With such an approach, the traditional methods of compound signal processing can be used: correlation methods and matched filtering methods. It means that non-coherent receivers do not use any specific properties of the chaotic systems in their operation.

Application of non-coherent receivers allows obtaining the characteristics comparable to the characteristics produced during the application of the usual methods of information transmission. For instance the significant noise-immunity can be achieved. However, the transmitted information confidentiality cannot be provided.

Coherent receivers represent systems similar to the chaotic system generators in the transmitter part, or in some of their fragments. They are capable of synchronizing with a transmitter, and therefore, allow information extraction from the received chaotic signal. In other words, the essence of the coherent receiver operation is constituted by either the phenomenon of system synchronization—the chaotic synchronization [30], or by a response of the oscillating system fragment to the chaotic signal, which is known as the chaotic response.

The structure of the receiving-transmitting system here, as well as a totality of its parameters, can be considered as some cryptographic *key* allowing messages extraction, i.e. ensuring a confidentiality of information transmission. Besides, the chaotic subsystem performing as the transmitter, remains synchronized with the other subsystem—the receiver—during the entire communication session, and no other additional devices are required to achieve synchronization.

The main disadvantage of coherent receivers is their high sensitivity in the synchronous mode to noises in the communication channel and to parameters

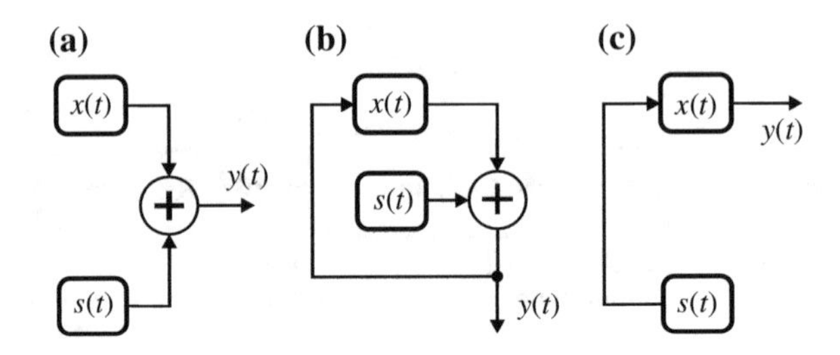

Fig. 1.32 Variants of insertion of the information signal into the chaotic carrier signal $x(t)$

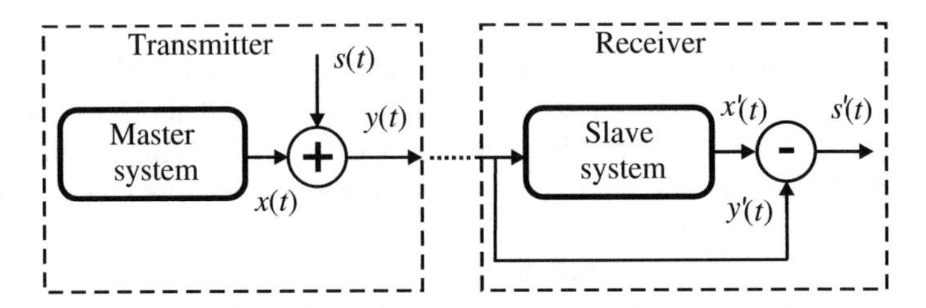

Fig. 1.33 The communication system with chaotic masking [137]

non-identity at transmitting and receiving sides. Nevertheless, to our mind, the second disadvantage becomes the advantage, if it is required to ensure a large number of keys.

The type of chaotic communication system is fundamentally specified by the way the information is introduced into the chaotic signal, as well as by the way of its extraction.

A variety of existing approaches to information signal overlapping can be reduced to the one of the three variants shown in Fig. 1.32.

The chaotic masking approach. The first variant (Fig. 1.32a) unifies systems with so-called *chaotic masking* (Fig. 1.33). In chaotic masking, the information signal $s(t)$ is summed with the input signal $x(t)$ of the chaotic signal (master system). The resulting signal $y(t) = s(t) + x(t)$ is transmitted into the communication channel. The receiver (slave system) represents the chaotic generator, which is identical to the generator of transmitting side. When the signal from the communication channel is applied to receiver input (in the absence of the information signal), the forced chaotic synchronization arises, in which signals at input and output of receiver's generator coincide. The information signal imposed onto the chaotic carrier, can be obtained at receiver output with the help of the subtraction device.

In spite of the fact that such a information transfer scheme can ensure effective information concealment in the spectrum of the transmitted chaotic signal (in the presence of a relatively high rate of data transmission), it is not used in practice. It is believed that, due to the inequivalence of information signal influence upon receiving and transmitting systems, the synchronization is only approximate. Besides, the information signal performs as a disturbing factor. Therefore, its power

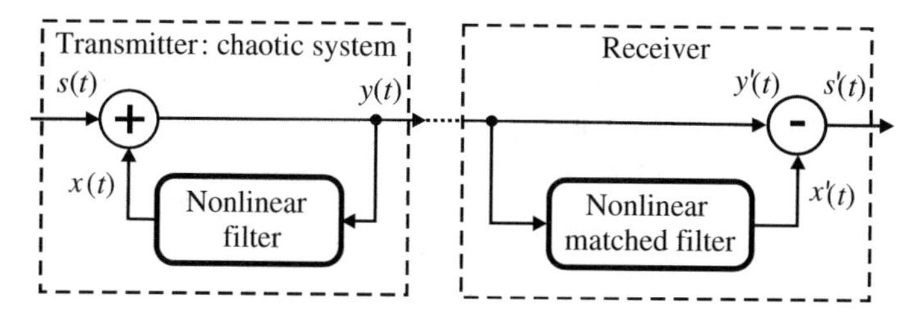

Fig. 1.34 The communication system with nonlinear mixing and synchronous chaotic response (following [134])

should be much less than the signal power in the communication channel (by 35–65 dB [41, 145]), which leads to its invisibility against the channel noise background and to the low signal/noise ratio at receiver output. For the reason mentioned, the energy efficiency of such transfer scheme is also low: the signal transmitted over the communication channel is mainly the one containing no information.

The nonlinear mixing approach. The direct introduction of information into the chaotic generator at the transmitting side is what unites the second group of methods (Fig. 1.32b)—the so-called *nonlinear mixing*. The term "direct introduction" means that the information signal—with the help of the auxiliary block—"is mixed" directly into the chaotic signal generated and circulated in the transmitter. Therefore, the information signal participates in the generation of the complicated chaotic behavior of the oscillating system. Its operation is not reduced to additive mixing, which is reflected by the adjective in the term "nonlinear mixing". Naturally, the information signal changes the form of the transmitter output signal.

The mentioned auxiliary block can perform any mathematical operation, for which the inverse operation exists, for example: addition and subtraction, addition and subtraction by modulus of any number, division and multiplication, etc. We should note that, as in the case of chaotic masking (Fig. 1.33), the replacement of addition—subtraction by the similar dual operation is possible.

The principle of operation of the device with nonlinear mixing performed in a transmitter by means of adding of information signal to the chaotic one is the following (Fig. 1.34). In the master system, the information signal $s(t)$ is mixed into the proper system signal $x(t)$: the signal $s(t)$ is summed up with $x(t)$. The information signal changes the operation mode of the transmitter, but, nevertheless, the mode remains chaotic. The formed mixture $F[s(t), x(t)]$ is transmitted via the communication channel. In order to extract information in the receiver, the element (for instance, *the nonlinear matched filter*) is used, which performs the same type of transformation as occurs in the transmitter (Fig. 1.34).

As a result, there occurs the chaotic synchronization of the receiving (slave) and the transmitting (master) systems. Or, in above-mentioned terms, the passive synchronization takes place, also known as the synchronous chaotic response of the element of the receiving system to the external chaotic influence. Recovering of the information signal $s(t)$ in the receiving device is ensured by the subtraction of the signal, which has passed through the filter, from the signal applied to the filter input.

It should be noted that in the system with nonlinear mixing, when there is identity of receiver and transmitter elements, and there is no interference in the communication channel, the information signal $s(t)$ is extracted at the receiver output perfectly accurately. Besides, there are no principal limitations on frequency ratio of information-bearing and chaotic signals, as well as on their power ratio. However, from the point of view of the transmission confidentiality, the frequency band of the information signal should not exceed the band of the chaotic signal, but should be completely "covered" by the last one, as in the case of chaotic masking. From the same point of view, the growth of the power level of the information signal, in relation to the chaotic signal (i.e. energy efficiency of communication), increases the data transmission quality, but decreases the communication security at the same time [41].

The above-discussed operation principle starts from the assumption that the information signal is capable of changing the oscillation character in the transmitter, but does not deprive it of its chaotic nature. For the purpose of ensuring the confidentiality of transmission, such a case is the most preferable one. However, we can face a situation, when the chaotic signal is not observed in the system, until the information signal is introduced into it. Such a mode saves the transmitter energy and does not pollute the ether without necessity. The situation opposite to the given one, when the intervention of the information signal switches the generator from the chaotic mode into the regular one, is fraught with communication confidentiality violation. In other words, if the choice of parameters of the master system is correct, this system constantly generates the chaotic signal, regardless of the presence, or absence of the external signal at the inputs.

Thus, let us list the advantages of the methods applying nonlinear information introduction and the synchronous chaotic response for the information extraction:

(1) the information signal in the absence of interference in the communication channel can be detected accurately, i.e. without distortions;
(2) since the information signal participates in the generation of the chaotic behavior of the transmitting system (and in the generation of the signal $y(t)$ spectrum as well), then it is difficult for the "external observer" to extract the information signal $s(t)$ by means of adaptive filtering;
(3) application of inverse circuits allows ensuring the high rate of information transmission, however, there should be taken into consideration the above discussed case of the overlapping of the spectrum of the information signal by the spectrum of the chaotic signal;
(4) for the compensation both of the losses in channel and the fluctuations of the transmitter parameters, it is possible to apply the self-correcting (adaptive) subsystems in the receiver. The parameters of these systems are adjusted during the reception process to match the characteristics of the received signal.

We mention the following disadvantages of the nonlinear mixing with synchronous chaotic response:

(1) the presence of noises, linear and nonlinear signal distortions in the communication channel may lead to the partial or full loss of the chaotic synchronization between transmitting and receiving systems;
(2) a power of the information signal should be much less than the chaotic signal power. The last requirement is caused by the fact that the information signal of, for example, the periodic form, may disrupt the chaotic generation mode and switch the transmitting device into the mode of regular oscillations, thus disturbing the communication confidentiality, as we have pointed out above.

One of the ways to overcome these shortcomings is to apply the communication system employing the dual transformation of the addition-subtraction type in modulo two [146]. For the potential user, it is necessary to take into consideration all the mentioned factors, in order to choose the optimal transmitted message level.

Parameter control of the transmitting system and chaotic mode switching. The third variant of information overlapping (Fig. 1.32c) combines the chaotic communication systems with the control of the parameters of the transmitting system (*chaos shift keying*) and switching of the chaotic modes (*chaotic switching*). The chaotic oscillating systems, as a rule, have several parameters, and we can introduce the information signal $s(t)$ by modulating any of these parameters. In turn, this causes modulation of the output signal $y(t)$.

At *chaotic modes switching*, the binary information signal $s(t)$ switches a certain parameter p of the chaotic generator in the transmitter from the state p_1 into the state p_2 (Fig. 1.35). The transmission of two different chaotic signals can be implemented for similar statistical properties. The similar chaotic generator with $p = p_1$ parameter is present in the receiver. In the case of coincidence of the p parameter values at the transmitting and receiving sides, the mode of chaotic synchronization is realized between generators of the transmitter and the receiver. Then, obviously, their signals will coincide by the shape and the differential signal of subtracter output will be absent ($e = 0$). In this case the signal "0" is transmitted ($s'(t) = 0$). Otherwise, an irregular error signal e, varying in time, occurs, by whose detection we can understand that the signal "1" ($s'(t) = 1$) is transmitted.

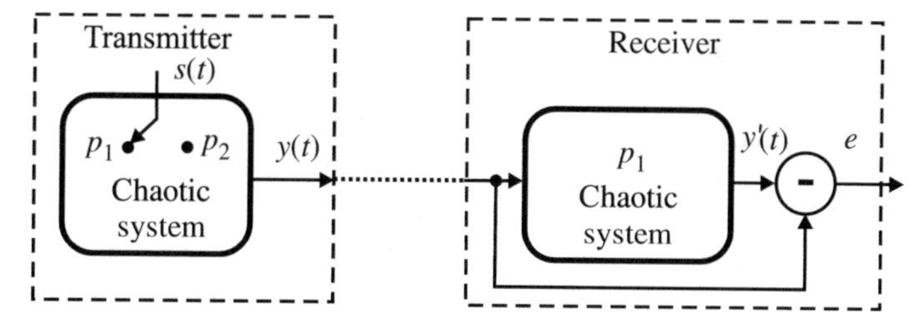

Fig. 1.35 The communication system with chaotic mode switching [137]

This version of the communication system can be simplified and we can increase its operation speed by constructing the receiver with the usage of the synchronous chaotic response, which is similar to the nonlinear mixing method.

In some variants of the communication systems with chaotic mode switching, a number of chaotic systems in the transmitter and in the receiver are doubled (Fig. 1.36). Thus, one may avoid the transient processes in the transmitter increasing the operation speed of the communication system. Nevertheless, if the special measures are not taken, the output signal will become discontinuous, and that can potentially decrease the communication immunity. The danger is that the enemy gets a chance to register the sharp jumps of the signal, recognizing the transition from "0" to "1" in it, et vice versa. On the other hand, such doubling of communication system components allows the permanent checking the authenticity of the transmitter in the receiver. You can see in this case that in any moment of time only one out of the two signals of the error e_i must be close to zero.

The information signal $s(t)$, controlling the parameter, can be not only binary, but the continuous one, as well. Thus, we approach the communication method with transmitting systems *parameters control*. From the signal obtained by such an approach, we can extract the usual information at the receiving side, tracking the evolution of the parameter adjusted in the adaptive manner.

As the advantages of the parameter control method and the chaotic mode switching method we understand:

(1) a possibility of simultaneous transmission of several messages via the same communication channel, when several parameters are modulated simultaneously;
(2) relative realization simplicity of the phase-locked loop systems in the radio-frequency range, with the chaotic oscillations spectrum directly extended (it refers to the method of parameter control);
(3) identical values for the parameters of the receiver and the transmitter are not required, and there are no strict recommendations concerning the noise level in the communication channel. This becomes possible due to the fact that for the information extraction it is sufficient for the receiver to determine whether the

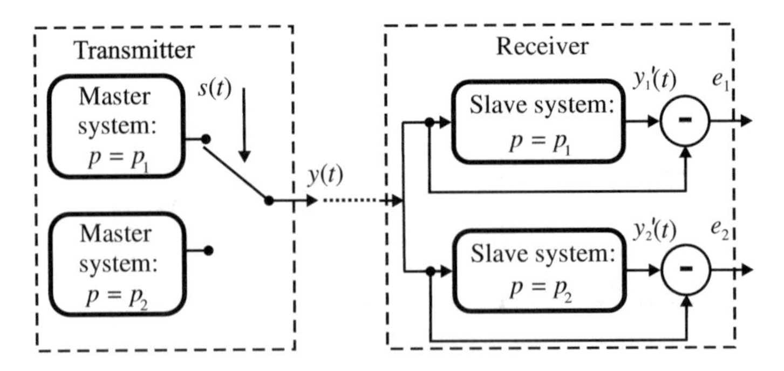

Fig. 1.36 The scheme of chaotic mode switching [137]

"native" or the "foreign" signal has been applied to its input (it refers to the method of chaotic mode switching); the system of adaptive parameter adjustment is capable of compensating these negative factors (it refers to the method of parameter control);
(4) the communication energy efficiency is high enough, because the entire chaotic signal transmitted is the information carrier.

Due to p. 3 and 4, such systems of data transmission are more stable to the noise influence in the communication channel, compared to the systems with chaotic masking and nonlinear mixing.

Disadvantages of these two methods are:

(1) the relatively low rate of information transmission, since the time is required to establish a synchronization mode of the inertial receiving system and (or) the system of adaptive parameter adjustment;
(2) the low crypto-resistance, since adaptive methods of the receiver parameter adjustment can be used by the enemy, who traces temporal realization of the signal transmitted [147]; besides, the enemy is capable of recognizing the mode switching. To increase the crypto-resistance, a careful selection of switched chaotic modes is required, to ensure that all these modes are close by their characteristics.

The double-channel systems of chaotic communication. In the first publications devoted to this problem [135, 148] the following principle of information transmission was suggested. The masking chaotic (or random) signal $\xi(t)$ is generated, and then this signal is divided into two channels. In the one channel the oscillation $\xi(t)$ passes through a functional transformer F and takes the form $\eta(t)$. This signal is summed up with the information signal $s(t)$, and the resulting signal $u(t) = s(t) + \eta(t)$ is proceeded into the *information communication channel*. The pure oscillation $\xi(t)$ passes into the *synchronization channel*. After having passed the channel, it passes through the similar transformer F and takes the form $\eta(t)$. The useful signal $s(t)$ is extracted into the subtracting device $s(t) = u(t) - \eta(t)$, as it is shown in Fig. 1.37.

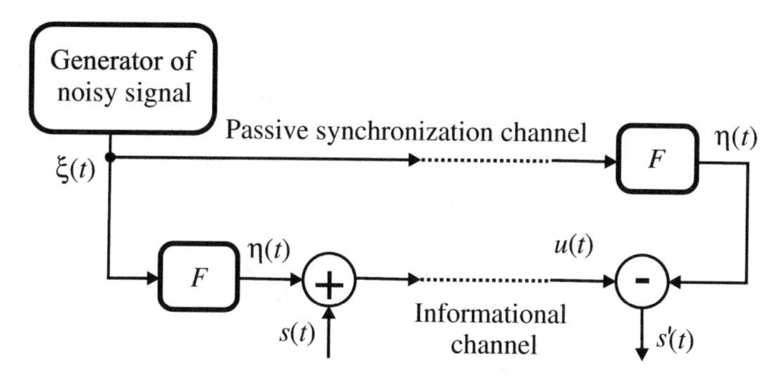

Fig. 1.37 The double-channel system of chaotic communication [135]

Efficiency of the considered communication system depends upon F that is a degree of identity of devices, which play the part of the secure key, and upon the identity of characteristics of communication channels. The presence of two communication channels is not always acceptable in practice, but when, for instance, the fiber-optical communication lines or the computer systems for information transmission, processing and storage are used, the mentioned difficulties are surmountable. Besides, the channels may be considered not only as physical communication channels, but also as the logical channels within the limits of one physical channel.

Evidently, in the double-channel system in Fig. 1.37 the functional converter F can be replaced by a dynamic system. Here the *active* synchronization of such systems is possible and in the transmitter and the receiver. As a result, the double-channel system is formed with active synchronization and with the "external" chaotic generator. An example of its implementation is described in the paper [149].

Reasoning by analogy, it is only natural to expect the circuit [149], in which the chaotic generator is both in the transmitter and in the receiver, and the information-bearing channel and the synchronization channel are divided between themselves (as in the system in Fig. 1.37). Obviously, the same system is freed from the disadvantage of communication system with chaotic masking (Fig. 1.33), i.e. the danger of the information signal breaking synchronization. In detail this situation is considered in Sect. 1.4.2.

The symbolization principle of system dynamics. There exists a still series of methods of information transmission based on the *symbolic dynamics*. In this case, sequences of positive peaks of the chaotic oscillation can be associated with "1", and negative peaks—with "0". As a result, sequences of zeros and units arise, which are unambiguously caused by initial conditions of the given chaotic systems. Starting the system start in various initial conditions, we can construct a coding function depending upon initial conditions. A set of chosen fragments of chaotic trajectories forms a "grammar" of the symbolic representation. Knowing the coding function, we can generate the required sequence from symbols "0" and "1". The main task for the receiver is to select and identify the received fragments of the chaotic trajectories [150].

Among the different variants of communication systems, which use the nonlinear dynamics, there are schemes with the *Poincare predictive control* and schemes using specific trajectories. There are also some other systems under development.

Predictive control by the Poincare section. Apparently, this method of *predictive Poincare control* [151] can be rightfully considered as a version of dynamics symbolization (Fig. 1.38). In the Poincare section of the analogous chaotic system, two or more separated areas are identified and the definite values are assigned to them for data coding (for instance, to one trajectory of the chaotic process there is conditionally assigned the value "0", and to the other—"1"). With the help of the appropriate way of control, the phase trajectory of the transmitting system is directed through one of the Poincare section area. Thereby, trajectories are realized

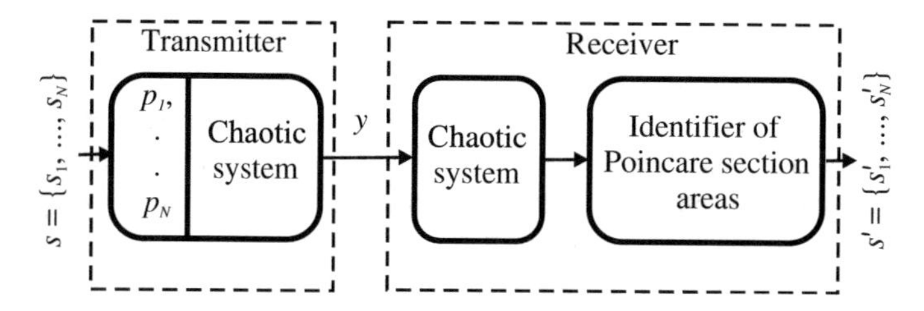

Fig. 1.38 The method of data transmission with predictive control by the Poincare section [151]

corresponding to the information signal. The control can be implemented, for example, by changing, in the insignificant limits, the parameter $p_i \in \{p_1, ..., p_N\}$ of the chaotic system is from the alphabet of the s message in accordance with the symbol transmitted—$s_i \in \{s_1, ..., s_N\}$.

At the receiving side, the chaotic system is approximately synchronized with the transmitting system. In the synchronized receiver the areas of the Poincare section are identified that the phase trajectory of receiver's chaotic system visits. Symbols from the message s alphabet correspond to these areas. Thereby, the information signal s' is extracted into the receiver.

The guarantee of confidentiality in such an approach is an agreement between the authorized subscribers about the parameters used in the chaotic system of the receiver, about the Poincare section and its fragmentation into areas, and about how much these areas correspond to the alphabet symbols. And in this case it is the responsibility of the transmitting side to ensure that the phase trajectory in the receiver passes through the appropriate area of the Poincare section. To achieve this, the transmitter should be able to predict the future behavior of the chaotic system of the receiver and to control it with the help of the output signal. But this presupposes that the transmitter can also control the behavior of its own chaotic system. For this purpose, the transmitter can use the high sensitivity of the both chaotic systems to the controlling influences s and y, relatively.

The directly chaotic approach to the information transmission. Authors of [140–142] introduce the term "*directly chaotic communication system*". These are systems, in which information is introduced into the chaotic signal directly generated in the frequency range of the signal transmitted through the communication channel [140–142, 152, 153]. In other words, information is "introduced" into the chaotic signal, which is generated directly in the radio-frequency, microwave or optical ranges that are used for airing. The signal can be introduced, for example, with the help of the transmitter parameters modulation. Or it can be performed through the modulation of the chaotic signal by the information signal after it has been already generated in the source. Or (if the spectra of the information-bearing and chaotic signals are close)—by mixing with linear, or nonlinear signals. Extraction of the information signals is also performed within the range of high or ultra-high frequencies by the coherent or non-coherent method.

Let us give an example of the communication system operating in the mode of the full chaotic synchronization (Fig. 1.39). The chaotic signal $y_1(t)$ is formed in the transmitter as the wide-band information carrier. Characteristics of y_1 are given in accordance with the frequency band of the communication channel and the information signal spectrum s_1. The modulation of the chaotic signal y_1 leads to formation of y_2 in the form of the sequence of chaotic pulses. Their duration and time intervals between them, as well as their mean-square amplitude can encode the "contents" of the information-bearing signal. The modulated signal y_2 passes through the communication channel to the receiver in the form of y_2'. The slave system in the receiver, matched (or partially matched) with the system in the transmitter, generates the chaotic signal y_1'. After subtraction the chaotic pulse sequence s_1' occurs. Demodulation of its parameters gives the useful signal. The power integration of the chaotic pulse is particularly possible within the limits of its duration.

Some notes on crypto-resistance and breaking communication systems employing the dynamic chaos. Different requirements can be made to the systems of confidential chaotic communication. The crypto-resistance is the one highly important (if not the main) property. In the pioneer publications devoted to the problem under investigation, it was assumed that the crypto-resistance of synchronous chaotic communication systems should be very high. Considerations about immunity of such communication systems were based on two circumstances:

(1) it is difficult to extract the information messages hidden in the chaos by methods of spectral analysis (due to wide-band nature and the continuous character both of the chaotic carrier signal and (as a rule) the information-bearing component);
(2) for correct recovering of the message transmitted, it is necessary to have the accurate knowledge of the receiver system parameters (or of the parameters of the chaotic system in the transmitter). The set of system parameters serves as the *family of keys*, which should not be accessible for the enemy.

However, further investigations showed that crypto-resistance of communication systems under consideration can be disturbed not by the access to the secret parameter set, but by the application of some other approaches. For communication schemes with the chaotic masking and the chaotic carrier modulation, it was shown that the hidden message can be extracted with the help of nonlinear dynamic forecast methods [154, 155]. It was assumed that the weakness of information protection lies in the low dimension of the attractor of the dynamic system (small number of the positive Lyapunov characteristic exponents), which generates the chaotic carrier signal. Therefore, it was suggested to use the hyper-chaotic systems presented as the coupled chaotic systems or the systems with time delay. However, a strategy of essential increase of crypto-resistance of communication systems under consideration is not yet developed. For example, in [156] it is shown that to recover the message, which is masked by the hyper-chaos of the six-dimensional dynamic system, you can use even the three-dimensional system.

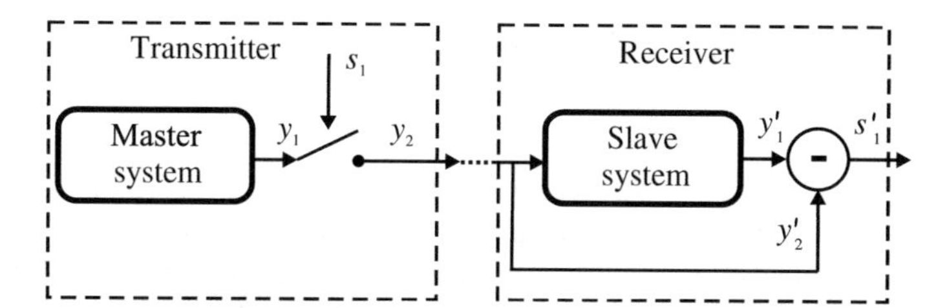

Fig. 1.39 Structure of direct chaotic communication with the help of full chaotic synchronization (coherent reception)

Studies on the protection of the communication systems with the chaotic modes switching and with the modulation of parameters and state variables of the transmission system [147, 157–164] show that the hidden messages can be extracted from the chaotic carrier signal. Both the global recovery of the chaotic system dynamics and the application of the specific return representations [157] are here effective. Nevertheless, if the transmitted signals are complicated enough, or when the differences between them are very thin, it is difficult to find a convenient return representation.

Although the results of investigation of communication systems with the dynamic chaos significantly decrease the confidence in the high level of their crypto-resistance, but they do not destroy it completely. Firstly, information deciphering even with the help of the modern supercomputers takes certain time, and that means that information confidentiality may have much less importance. Secondly, the serious training of the crypto-analytics is necessary and that requires financial, temporal and training resources. Thirdly, the transmission based on the chaotic carrier is compatible with the other approaches to the concealing and masking of information. Such combination of circumstances makes the deciphering of the intercepted message a rather uncertain and problematic task. Moreover, we can decrease the level of our expectations and choose the chaotic communication systems for their steganographic and (possibly) physical resistance.

Application of the chaotic signals as pseudo-random sequences extending the spectrum is also highly promising. This approach can add to the chaotic communication systems the properties of the systems with spectrum spreading, employing pseudo-random sequences. [165–167]. It is shown that correlation properties of the chaotic sequences can virtually coincide with the properties of the random process and be close to the characteristics of *M*-sequences. But, in contrast to the latter they may form the coded signals of arbitrary duration from the practically unlimited set of the codes. Thereby, there comes the perspective possibility for the application of the chaotic dynamics for the arrangement of the communication systems with the multi-user access [139].

Let us pass to the illustration of the above-described principles.

1.4.2 Examples of Radio Physical Systems for Information Protection

As far as we know, the first variants of the dynamic chaos application in the cryptosystems of radio-frequency range were implemented by authors of [134, 168–171] (more complete reference list on this problem is presented in the book [41] and in the review [144]). In cryptology context, the diversification problem of ciphering and deciphering methods is always relevant. Let us give several significant examples.

In the paper [134] the approach is suggested to the detection of information-bearing component of the chaotic signal. It is based on the principles of synchronous response of the slave oscillating system. The paper reveals the main results of the experimental investigation of the detection process. The generator of the ring type with 1.5 freedom degrees (Fig. 1.40) was used as a chaotic generator in the experiments. The transmitter operates in the mode of additive nonlinear mixing of the information-bearing component, and the receiver operates in the mode of the forced synchronous chaotic response. It is stated that there take place the asymptotic stability of dynamic coincidence of the chaotic generator and the forced synchronous chaotic response in the slave system.

As we already mentioned in Sect. 1.3.1, the Chua circuit serves as the common and widely-spread base for the chaotic generator. The publications [172, 173] are devoted to the analysis of the properties of the appropriate chaotic communication systems. Thus, M. Hasler in his classic review of the methods of the hidden transmission of the messages in the chaotic signal [173] assumes that the Chua chaotic generator forms a basis for the devices with different modes, such as additive mixing and active synchronization, the discrete manipulation of the chaotic generator parameter (switching) and the active synchronization.

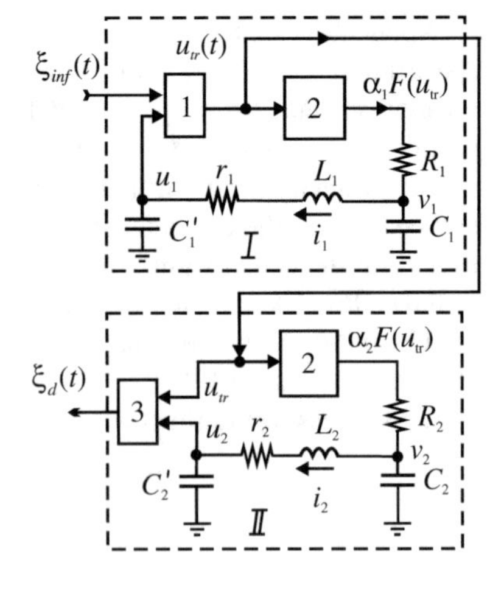

Fig. 1.40 The structural diagram of the experimental system of chaotic communication based on the mode of additive nonlinear mixing and the forced synchronous chaotic response [134]. (I) is the master system, (II) is the slave system, *1* is a summer, *2* is nonlinear converter, *3* is subtracting device

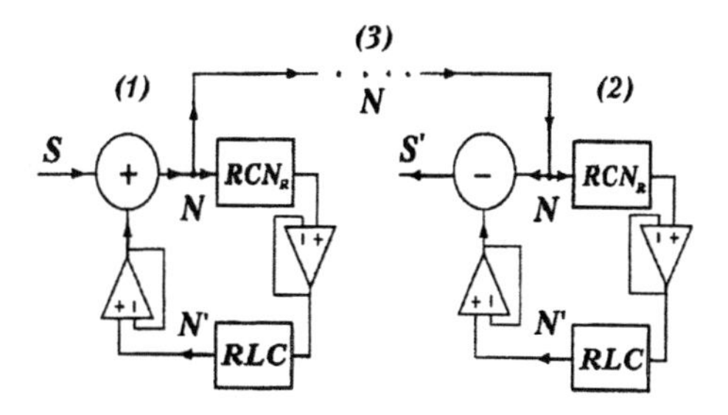

Fig. 1.41 The system of chaotic communication on the base of the Chua circuit subjected by decomposition [50]

For example, for the chaotic communication system based on the Chua circuit subjected to decomposition (Fig. 1.41), the authors [50, 51] numerically investigated the conditions of synchronization and message transmission (Fig. 1.11). They experimentally created the chaotic communication by means of the handheld miniature receiving-transmitting radio devices using an amplitude modulation of the carrier frequency of 27 MHz. Evidently, the cipherer here operates in the mode of nonlinear mixing, and the decipherer in the mode of the chaotic response.

In the paper [169] there are simulated the processes in the message transmission system shown in Fig. 1.42. The information-bearing digital signal in this system influences the parameter of the chaotic generator (transmitter) either ensuring its non-identity with the appropriate parameter of the receiver, or recovering its identity. This causes synchronization (or desynchronization) of the receiver creating (or eliminating, accordingly) the chaotic signal at the receiver output. Thus, we can evaluate the message content (Fig. 1.43), according to its presence and a level. It is clear that for synchronization (or desynchronization) some time is required, therefore, it is necessary that the informative interval length would exceed by several times the typical time of system oscillations. The theoretical analysis and experimental implementation of the similar circuit (but when the transmitter operates in the nonlinear mixing mode) are described in the paper [174].

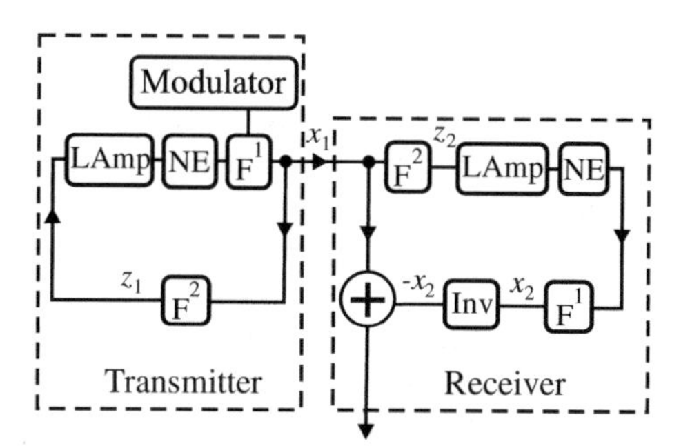

Fig. 1.42 The structural diagram of chaotic communication system based on the mode of discrete parameter manipulation of the chaotic oscillator (according to the information signal law) and the forced synchronous chaotic response [169]. LAmp is the linear amplifier, NE is a nonlinear element, F1 and F2 are low-frequency filters of the first and second order, Inv is an inverter, + is a summator

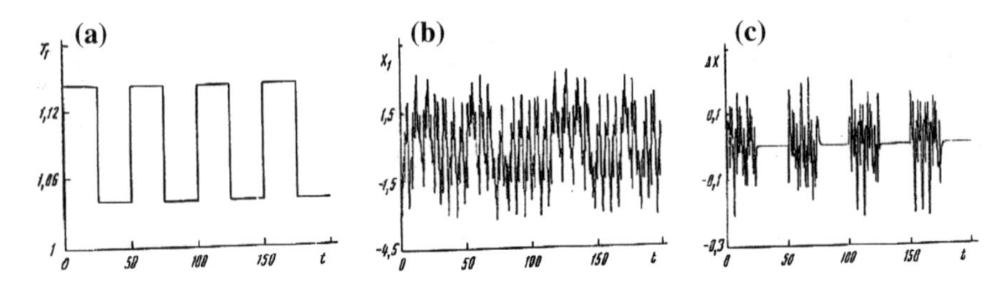

Fig. 1.43 Simulation of system in Fig. 1.42 operation: periodic variation of the time constant T_1 of the F_1 filter of the transmitter (**a**); signals on the transmitter outputs (**b**) and receiver output (**c**) [169]

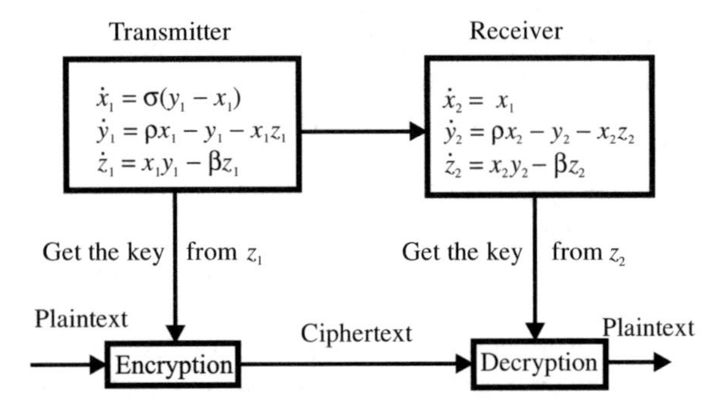

Fig. 1.44 The structural circuit for the key generation in the pair master-slave chaotic system [149]

By means of simulation, authors of [149] showed the operability of the pair of double-channel devices, representing the stages of the system development. In initial variant, the (active) synchronization of chaotic systems is used, which was suggested by L.M. Pecora and T.L. Carroll. The chaotic generator in the transmitter was implemented as the Lorenz model, and the one dynamic variable (x_2) of the Lorenz model in the receiver was replaced by the variable (x_1) transmitted through the synchronization channel (Fig. 1.44).

The modified system differs by the fact that the key is generated in the transmitter and in the receiver with the help of the two identical slave systems, which are synchronized by the signal from the common chaotic generator. The master system considerably differs from the slave systems (Fig. 1.45). We note that slave nonlinear dynamic systems have one freedom degree each, and, therefore, they are not the chaotic generators.

In [140, 141] some properties of the direct chaotic communication system (Fig. 1.46) are theoretically discussed and the experimental circuit is described performing the high-rate data transmission. The results obtained are also commented.

After acquaintance with the examples of cryptosystems realization in the radio-frequency range, we now address the optical devices similar in their purpose.

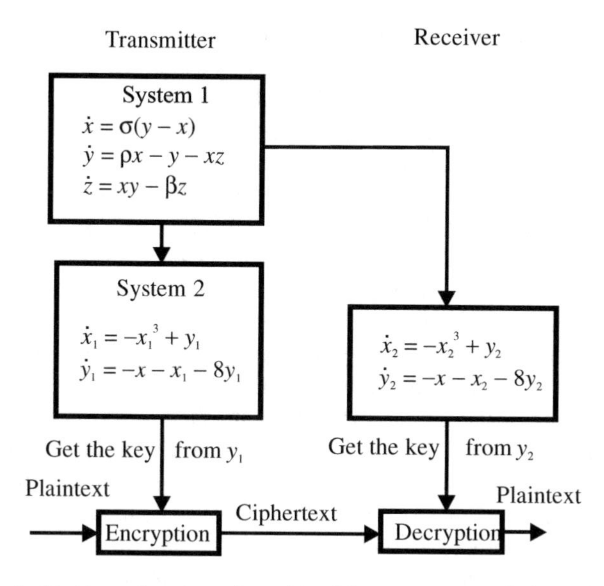

Fig. 1.45 The double-channel system of confidential message transmission using the synchronization of the dynamic systems in the transmitter and in the receiver by the signal from mutual chaotic oscillator [149]

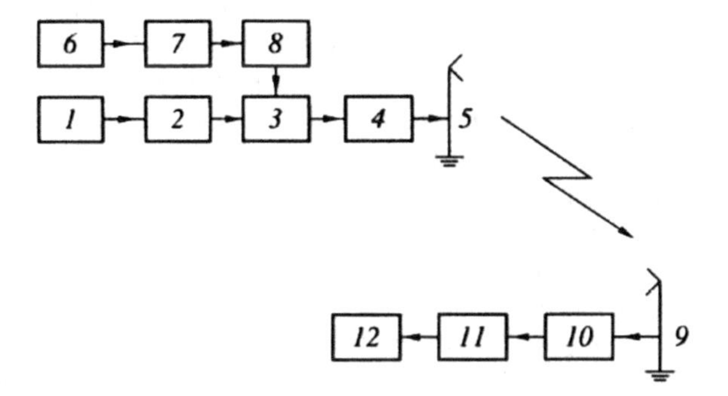

Fig. 1.46 An example of the structural circuit of the direct chaotic communication: *1* is the device for chaotic oscillator, *2* is control, *3* is the modulator of switching type, *4* and *11* are amplifiers, *5* and *9* are wide-band antennas, *6* is a source of the information signal, *7* and *8* are encoders of the source and the channel, *10* is a filter, *12* is the signal processing system [141]

1.4.3 Examples of the Application of Deterministic Chaos in Optical System of the Confidential Communication

Since the middle of 1990s, the researchers have been developing principles and devices for information protection in the radio-frequency range that are based on the properties of the deterministic chaos. The theme of secret information transmission in the optical range has been rather widely discussed (see, for instance,

[40, 86, 175–177], Chap. 5 in [90, 178]). However, the development of chaotic communication systems for the optical wavelength range is less investigated than the development of such systems for the radio-frequency and microwave ranges. Yet, the experimental verification of the theoretical conclusions confirming that the optical chaotic communication system can really be constructed has been obtained. For example, in [179, 180] we are informed about the fiber-optical communication over the distance of 35 km. Let us address some significant subjects.

In [181] the characteristics are studied of the chaotic solid-state laser with loss modulation for which the inverse maps on a plane—maxima of intensity and inter-peak intervals—are used. The simple relation between the intensity maximum and a pair of preceding and following inter-peal intervals is demonstrated by means of laboratory and numerical experiments. Authors of [181] performed the numerical investigation of the possibility of encoding a message on the output of the chaotic laser generator in the transmitter with subsequent decoding of this message by the same laser in the receiver. On basis of the relations between the intensity maxima and the inter-peak intervals, it is proved that the employed method of message coding is applicable in the information protection systems. In this paper, the alternative methods of message coding are analyzed: by modulation of the pumping level of the transmitter laser and by control over multi-frequency absorption modulation.

The structure of the optical information channel is suggested on the base of the two synchronous lasers, which operate in the chaotic mode with the periodic pumping of the active medium [40]. The dynamics of the laser system in the presence of the chaotic radiation injection, which is modulated by the external signal, is under investigation. The possibility of recovering information contained in the chaotic signal, which is transmitted through the channel, is shown, and an estimation of the acceptable spectrum width of the information signal is performed.

Somewhat later, this same research team [86] theoretically proved that the chaotic laser driven by the injection signal, which is fed from the similar chaotic master generator, is capable of operating even at the insignificant frequency shift and can function in the new synchronization mode. The definite relationship is revealed between average values of the laser fields, which are calculated approximately in the explicit form. The instantaneous values of fields are interpreted as dynamic variables. It is established that the system attractor is formed around a point defined by these average field values, with relatively small phase trajectories scattering near it. It is shown that such a configuration can be used for the confidential information transmission.

The next publication [177] of this research team is devoted to a problem of the influence of regular disturbances in the chaotic signal of the master generator on the synchronous chaos, when the master/slave generator is the two-sectional laser.

In [87], it is shown experimentally that synchronized diode lasers with the external feedback can be used for chaotic optical ciphering and deciphering. It is found out that effective deciphering requires that the signal intensity would exceed some value determined by the operation conditions.

Authors of [184] study a possibility of message ciphering at the output of the chaotic single-mode semiconductor laser with an external resonator. They analyzed

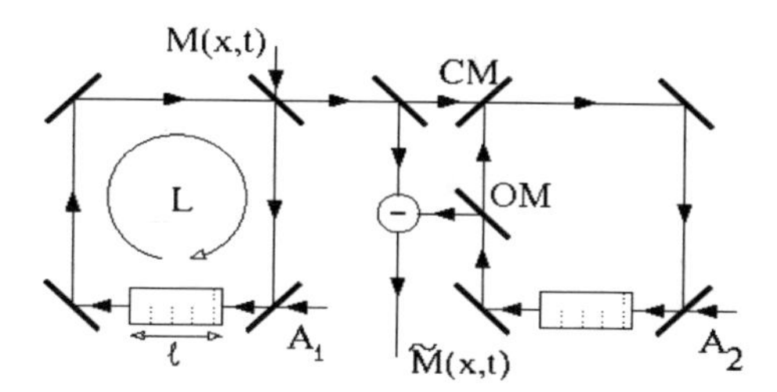

Fig. 1.47 The transmission scheme of spatial-temporal information signal using the optical chaos [100]. *CM* and *OM* are mirrors for connection and radiation output; A_1, A_2 are amplitudes of the plain waves, which are constantly applied into resonators; $M(x, t)$ and $\tilde{M}(x, t)$ are ciphering and deciphering signals; l is a length of the nonlinear element (saturable absorber); L is the optical length of interferometers

the message coding system with chaotic mode switching (*chaos shift keying*). It is shown numerically that a system with the single receiver ensures better characteristics compared to the circuit containing two receivers. The data transmission with reliable message deciphering (with the rate up to 3 Gbit/s) occurs at the frequency, which is lower than the relaxation frequency of the emitter (but with the frequency of the similar order).

In [100, 175], processes in the optical system for secrete information transmission are simulated on basis of a pair of four-mirror interferometers, which are generators of the deterministic spatial-temporal chaos (Fig. 1.47). Here the decipherer operates in the active synchronization mode; the diffraction is taken into consideration in the model; the saturated absorber forms the nonlinear medium; the deepness of the spatial modulation of the amplitude by the information signal is 0.005 apparently. The static image transmission is simulated, which, as a result of deciphering, becomes quite distinguishable, when the coupling coefficient is equal to 0.7, but still recovers with some distortions. The estimation of the influence of the synchronization process and the system efficiency upon the deciphering error is also presented.

We would like to note that the manifest disadvantage of the circuit in Fig. 1.47 is the requirement of coherence to the three laser beams: A_1, A_2, $M(x, t)$. Evidently, the requirement of small deepness of spatial light amplitude modulation by the information signal and the presence of non-zero deciphering error are also the disadvantages.

In [176] authors continue their researches and suggested two circuits for information transmission based on the ring nonlinear resonators. In the first circuit, the information signal is introduced according to the nonlinear mixing method (Fig. 1.47). In the second circuit, the chaotic generator parameter is modulated (intra-resonator modulation of the optical field). During deciphering we may use the chaotic response, i.e. the circuit is similar to Fig. 1.47, if the *OM* mirror is made as completely reflecting, and the *CM* mirror is eliminated. In order to determine the coding efficiency, it is offered to calculate the value of the mutual information between initial and transmitted messages. And to determine the coding effectiveness we can calculate the mutual information between the original and recovered

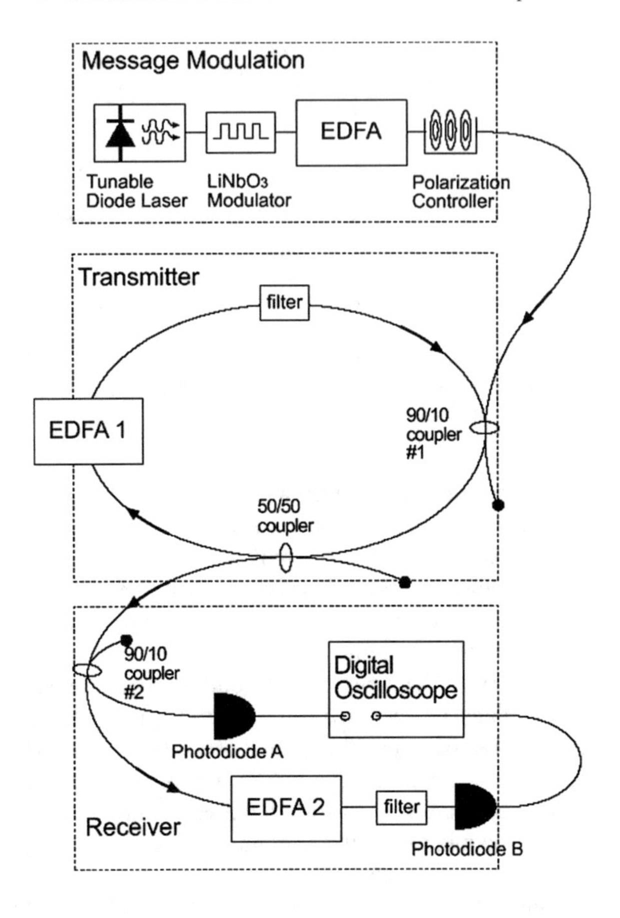

Fig. 1.48 Experimental diagram of the chaotic communication system with nonlinear mixing of the information signal [182]. EDFA is the amplifier of the base of fiber-optics doped by the erbium

messages. It was found out that message introduction according to the nonlinear mixing method has a larger ciphering efficiency while the intra-resonator modulation of the optical field is more preferable for deciphering.

Various circuits of the chaotic communication systems are suggested and experimentally tested in [179, 180, 182] terms of the fiber-optical ring laser systems. For instance, the single-circuit fiber-ring model of the chaotic synchronous communication system with nonlinear mixing (Fig. 1.48) is examined in detail in [182]. The case of the chaotic communication system with the control of the parameters of the transmitted system (Figs. 1.49 and 1.50) is described there as well.

For dynamic chaos generation it is suggested to use the ring laser on basis of fiber-optics doped by the erbium (EDFRL is the erbium doped fiber ring laser) [182]. A possibility for dynamic chaos excitation in EDFRL and the modes of synchronization between two EDFRLs is shown in [179, 181, 183, 184]. Erbium fiber-optical lasers radiate in the 1.53–1.55 μm wavelength, where losses in the fiber-optical elements are minimal.

To increase the potential secrecy of messages transmitted in the systems with nonlinear mixing, it is suggested to modify the above-mentioned circuit of

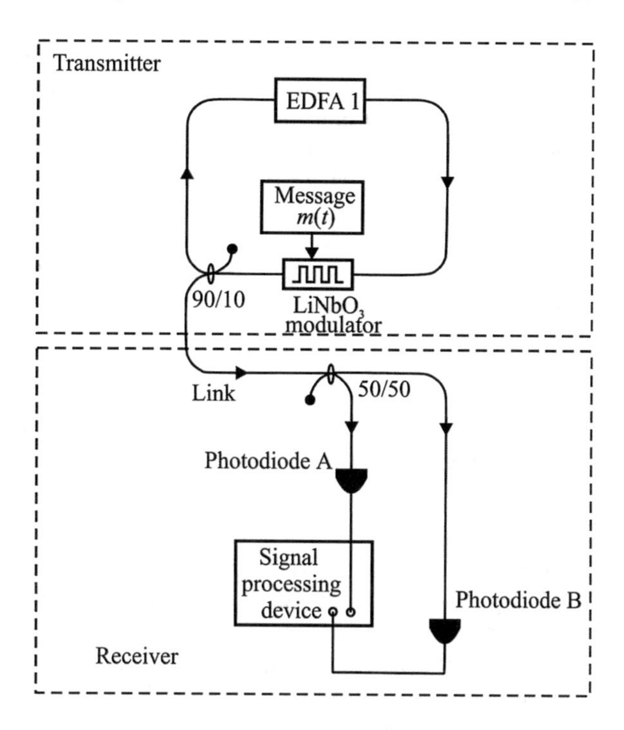

Fig. 1.49 Experimental diagram of the chaotic communication system with parameter control of the transmitting system [182]. EDFA1 is the amplifier of the base of fiber-optics doped by the erbium

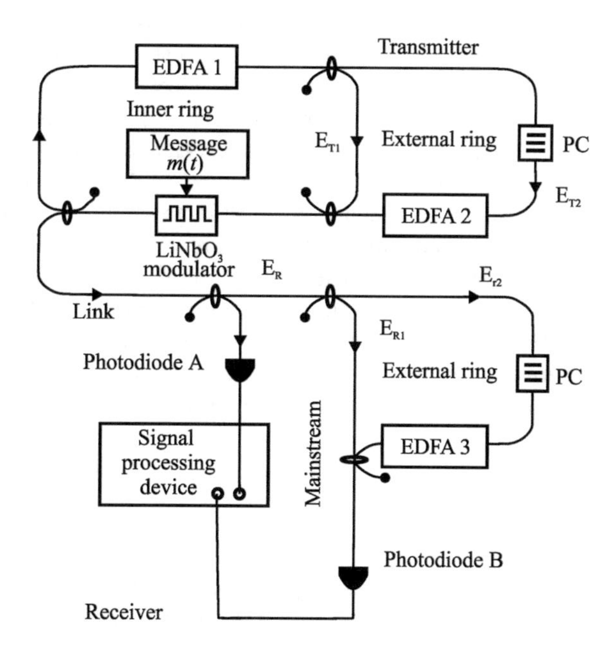

Fig. 1.50 Experimental diagram of the optical double-circuit chaotic communication system with parameter control of the transmitting system o the base of fiber-ring laser [180]

cipherer-decipherer and to add the second circuit in the transmitter and, hence, in the receiver (Fig. 1.50) [180, 182]. Thus, the larger is the number of system parameters having an influence on its dynamics, the larger is a number of cryptographic keys, if this system is used as a cipherer. This situation is confirmed by calculations in [182], where it is shown that reduction of delay time offset value between appropriate circuits of the transmitter and the receiver leads to the decrease in the deciphering information distortions. This proves in practice the importance of identity of parameters of the cipherer and the decipherer in the synchronization mode, as it effects the message deciphering quality.

1.4.4 Influence of Disturbing Factors on the Characteristics of the Data Transmission System

Among the factors [41] having an influence on the communication quality in the data transmission systems with the complete synchronization, we must note the following: non-coincidence of values of the element parameters of the transmitter and the receiver, irregularity in amplitude-frequency response of functional component characteristics, nonlinear distortions and noise in the communication channel. Let us consider the influence of these factors in the aspect of the *quality* of the *synchronous chaotic response* obtained at the receiver output.

Non-identity of elements. When the real devices are used, there is a mismatch of the element parameter values of the receiver and the transmitter. Even if only one transmitter element has deviation in parameters with the receiver, then this leads to signal mistiming appearance at the receiver output. As it was noted in [41], the non-identity of transmitter and receiver parameters may result in the arising of on-off alternation. This causes the appearance not only of the mismatching signal of small amplitude, which is proportional to the value of parameter non-coincidence, but it also causes irregular bursts of large amplitude, which are commensurable with the chaotic oscillation amplitude at receiver input. The average frequency of these irregular bursts grows with signal (noise) power increase and leads to sharp growth of the relative mistiming signal (noise) power η at the receiver output.

This non-identity may be quantitatively characterized by the absolute Δr_i and the relative δr_i value of i-th parameter non-coincidence in the receiver and the transmitter. The can be determined by formulas

$$\Delta r_i = r_{i,2} - r_{i,1}, \; \delta r_i = \frac{r_{i,2} - r_{i,1}}{r_{i,1}},$$

where $r_{i,1}$ and $r_{i,2}$ are values of i-th parameter of the transmitter and the receiver.

Irregularity of the amplitude-frequency characteristics of the communication channel and the auxiliary receiving-transmitting devices leads to the appearance of additional filtering of the chaotic signal and to the variation of its spectral properties. Most often, it is caused by a presence of the *RC*-circuit (low-frequency filter) between output of the chaotic generator, which forms the chaotic signal carrying information, and the input of the device in the receiver, where the chaotic signal arises. Therefore, while discussing the influence of amplitude-frequency characteristic irregularity upon the synchronous chaotic response quality, one usually assumes that this irregularity is caused by the presence of such a filter, and transmitter and receiver parameters are identical. Transformation of the signal by the filter is described by the following differential equation of the first order:

$$T\frac{\mathrm{d}x}{\mathrm{d}t}+x=y,$$

where y is the signal at transmitter output, x is the signal at receiver input. The time constant T defines the cutoff frequency f of the low-frequency filter $(f=T^{-1})$.

A quality (relevance) of the synchronous response is, naturally, determined by the ratio of the filter cutoff frequency f and upper boundary frequency F of the signal power spectrum at the transmitter output. At $f \gg F$ the presence of the filter does not disturb the system functioning. On the contrary, commensurability of values F and f degrades the synchronous response quality. Therefore, we can expect that at f growth the relative mistiming noise power η at the receiver output should decrease and the signal/noise ratio should increase. These regularities are illustrated by Fig. 1.51, where the μ ratio of power of information and chaotic signals in the transmitter is equal to 1 [41].

We note that in the analogous communication system it is accepted to use exactly the signal/noise ratio. In digital systems its role is played by the ratio of energy E_b to the noise power density N_0:

$$\frac{S}{N}=10\log_{10}\frac{E_b}{N_0},\ \mathrm{dB}.$$

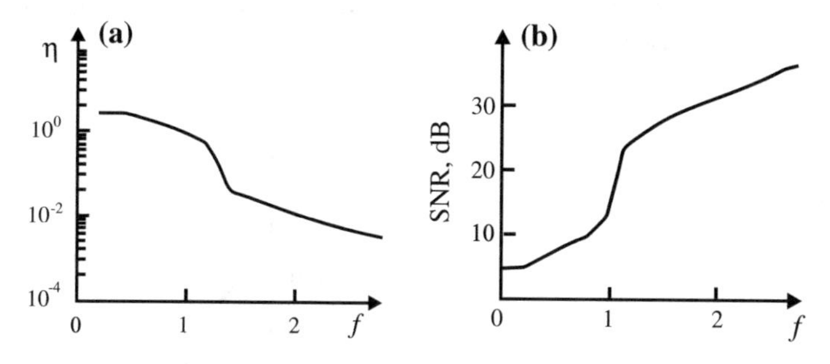

Fig. 1.51 Influence of the amplitude-frequency characteristic irregularity of the communication channel: **a** is the function of relative power of the mistiming noise η *versus* the cutoff frequency f of the low-frequency filter; **b** is the function of the signal/noise ratio (at $\mu=1$) versus the cutoff frequency f of the low-frequency filter [41]

Here:

- the signal energy E_b fitted per one bit of data transmitted and given by the product $E_b = P_s \cdot \Delta t$, where P_s is the signal power in the noise absence, Δt is time for one bit transmission;
- the spectral density of the noise power N_0 is a ratio of the noise power in the communication channel P_n to the channel pass-band width Δf, i.e. $N_0 = \frac{P_n}{\Delta f}$.

Nonlinear distortions in the communication channel. Research of nonlinear distortion influence in the communication channel [41] upon signal/noise ratio is performed in terms of the widely spread and practically important case of cubic nonlinearity

$$x = y(1 - \alpha y^2),$$

where y is a signal at the transmitter output, and x is a signal at receiver input. Then, the ratio

$$ND = 10 \log_{10} \frac{P_x}{P_y}, \text{ dB},$$

where P_x is a power on the transmitter output, P_y is the signal power on the receiver input, serves as the quantitative measure of nonlinear distortions ND in the communication channel.

The parameter α of the cubic distortions varies within the limits starting from 10^{-3}, which corresponds to the 1 %-level of the signal amplitude on the transmitter output, and up to 3×10^{-2} (30 %-level of the signal amplitude). The function of the signal/noise ratio (for $\mu = 1$) on the receiver output versus the nonlinear distortion order is presented at Fig. 1.52.

From this figure we can see that the function has the same tendency that in the case of amplitude-frequency characteristics. In particular, in both cases the maximal possible signal/noise ratio equals to 35–40 dB.

Influence of the additive normally-distributed noise in the communication channel. External noise influence on the communication system functioning is

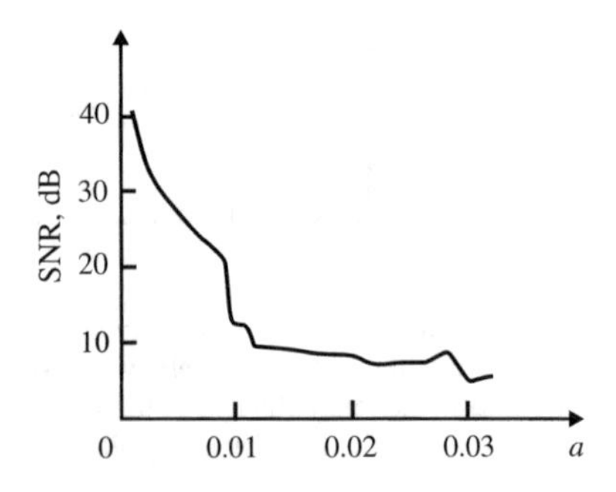

Fig. 1.52 Signal/noise ratio on receiver output at presence of nonlinear distortions in the communication channel (α is the cubic distortion parameter) [41]

manifested in systems, where there is a high level of the background noise power in the communication channel: for wireless data transmission in the optical communication channels, fiber-optical communication lines.

In numerical experiments [41] researchers study the additive normally-distributed noise influence in the channel on the quality of the synchronous chaotic response and the information signal at the receiver output. The noise frequency bandwidth is restricted by the bandwidth of the low-frequency filter of the first order, through which the noise passes. The noise power is measured on the filter output.

The functions of the signal/noise ratio (for $\mu = 1$) *versus* the cutoff frequency f for the noise levels –40, –50 and –60 dB are presented in Fig. 1.53. The noise presence in the frequency bandwidth of the carrier signal decreases the signal/noise ratio at the receiver output (in this case by 13–15 dB) compared with this ratio in the communication channel (at the received input). This can be observed at different levels of noise power and it is explained by loss in quality at extraction of the information signal from the mixture with the chaotic signal. The ambiguity zone *2* (shaded in Fig. 1.53) is caused by the finite length of the time samples and by irregularity of mistiming bursts [41].

Comparison of data transmission schemes. In the discussion above of the various principles of the secret data transmission, we have tried to attract your attention to the pioneer principles and schemes, which are applicable for information protection arrangement with the help of dynamic chaos. Some of them are fit for the digital signal transmission only, and the others—for the transmission of the analogous signals as well. The further chapters will be devoted to the discussion of the universally applicable systems. Among them, we prefer those, in which the nonlinear mixing of the information signal in the transmitter and the synchronous chaotic response in the receiver are used. Therefore, we have not yet considered a series of recently suggested circuits designed for the digital message transmission, for example, using the mode of the generalized synchronization [144].

Nevertheless, it is reasonable to consider in brief the destabilizing factors affecting the communication quality not only of the systems discussed above in this section, but of the other system as well. Accordingly, characteristics of some of the above-mentioned systems (lines 1–4) and new (mainly digital) systems (lines 5–9)

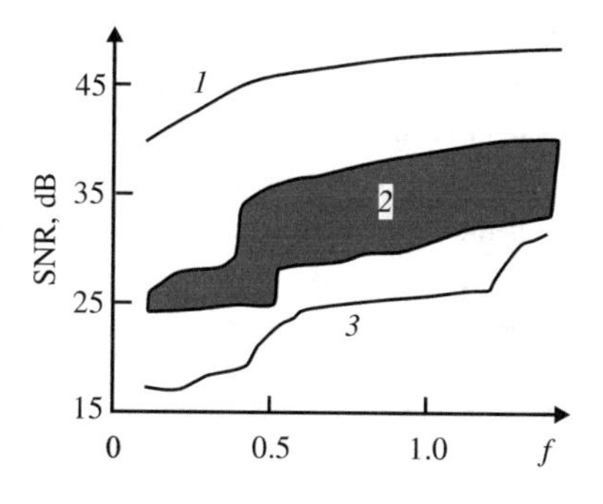

Fig. 1.53 The signal/noise ratio at receiver output: in the case of the additive noise presence in communication channel (f is the noise cutoff frequency). Curves *1*, *2*, *3* correspond to the noise levels –60, –50, –40 dB [41]

are compared in Table 1.2. This table is borrowed from the review [144]. The qualitative data presented in Table 1.2 are obtained by the simulation of the systems including the Rössler system (one of the standard systems of nonlinear dynamics) as the chaos source.

The critical levels (marked by subscript "c") of disturbing factors are systemized in Table 1.2:

S/N_C is a ratio of energy per bit to the noise spectral power S/N [185, 186], at which the data transmission scheme becomes disabled. In other words, in this case it is already impossible to recover the initial information signal $s(t)$ from $s'(t)$;

δr_c is a value of relative non-coincidence of receiver and transmitter parameters δr, at which the system becomes disabled;

ND_c is the nonlinear distortion level ND, at which the data transmission system is still efficient.

Table 1.2 confirms the high sensitivity of chaotic communication schemes based on the *complete synchronization* to parameter non-coincidence of the transmitter and the receiver, as well as to nonlinear distortions in the communication channel. On the one hand, this is a negative factor when using in the data transmission systems. But, on the other hand, it increases a number of cryptographic keys in the system of confident communication.

The scheme with the *generalized synchronization* demonstrates the lesser value of the critical signal/noise ratio and sensitivity to distortion in the communication channel. It does not also require the complete identity of control parameters of the transmitter and the receiver. But, its significant disadvantage is the presence of transients, which limits the transmission rate and also it assumes application the digital signals only.

The scheme using the *phase synchronization* mode demonstrates even the lesser critical value of the signal/noise ratio and sensitivity to distortions in the communication channel. The increased identity of the chaotic generators is not required in it as well. The given scheme is intended for the digital signal transmission. But, it is more complicated and requires additional chaotic generators.

Table 1.2 Comparison of critical levels of perturbing factors for different data transmission systems using a chaos [144]

Communication system type	S/N_C, dB	δr_c, %	ND_c, dB
Chaotic masking	56.48	0.30	1.03
Chaotic mode switching	30.76	2.0	23.3
Nonlinear mixing	64.99	0.3	0.26
Modulation of control parameters	30.76	2.0	23.3
Scheme on the base of phase synchronization mode	32.4	0.8	10.7
Scheme on the base of generalized synchronization	39.52	1.0	7.75
Scheme on the base of generalized and complete synchronization	39.24	0.5	4.83
Scheme with combined chaotic signal	61.47	0.2	2.63
Scheme ultrastable to noises	−10.01	2.0	27.2

1.4.5 Classification of Communication Systems Using the Dynamic Chaos

As it is known, in the modern science a taxonomic problem is felt more sharply the younger is the scientific school [187, pp. 123–130; 188]. The same conclusion, to some extent, can be attributed to the terminological problem. Rather young sphere of research on information protection in radio physical and optical communication systems is not the exception. The continuous flow of publications devoted to the practical aspects of nonlinear dynamics leads to the problem of classification of the known systems of chaotic communication.

The existing typology does not have clear and complete structure. It is far from complete, since it does not take into account several methods, which have independent theoretical and applied significance. It considerably affects the possibility of examining the chaotic communication system from some common point of view.

For example, in the paper [137], the author suggests distinguishing the following types of communication between master and slave systems.

(a) Communication with the imposition of the state variable of the master system to the slave system.
(b) Communication applying control response (based on the decomposition of the initial system into two separate systems).
(c) Communication with partial replacement of one variable.
(d) Communication applying the linear control circuit.

Coherent and non-coherent operation modes of the signal receivers can be used as another foundation for the classification.

The variety of systems and principles of confidential communication demonstrated in Sects. 1.4.1–1.4.3 allows suggesting the alternative classification of the known and yet unknown systems of chaotic communication, it is shown in Fig. 1.54 [135]. According to it, we must distinguish the following systems.

1. Asynchronous (i.e. non-coherent) systems, which do not require the mutual synchronization of transmitting and receiving sides. One uses the statistical signal properties only. In this class, besides traditional communication systems with wide-band signals, which are formed by the band spreading and/or a spectrum, we should also include systems, in which the deterministic chaotic signal may replace the pseudo-random and noise-like signals. As separate subsystems we distinguish the ones applying:

   1.1 Matching filtering (the optimal reception);
   1.2 Correlation reception;

2. Synchronous systems. They provide application of the specific properties of the systems, referring to the generation of chaotic oscillations: phenomena of the chaotic synchronization and the chaotic response, in particular. To this group the following systems belong:

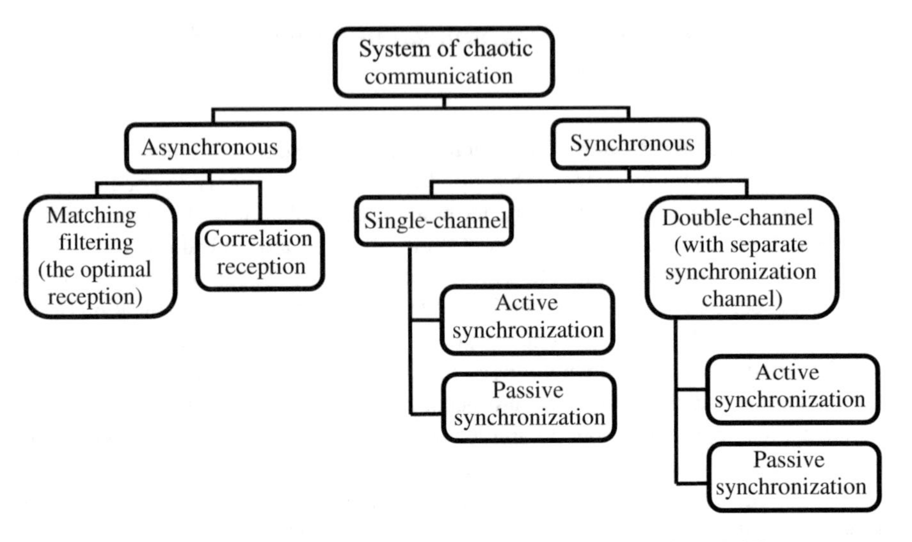

Fig. 1.54 Classification of communication systems using the dynamic chaos [135]

2.1. Single-channel systems, which are widely discussed in specialized literature. They are represented by the systems with chaotic masking, nonlinear mixing, controlled parameters, etc. [173].

    2.1.1. Systems with active synchronization. They are based on the phenomenon of (active) chaotic synchronization, which requires high identity of the chaotic generators of the transmitter and the receiver [100, 173].
    2.1.2. Systems with passive synchronization (with chaotic response). They are distinguished by the absence of the chaotic generator in the receiver. Devices described in [50, 134, 182] can serve as their examples.

2.2. The double-channel (with the separate synchronization channel) systems are not well studied, and they are not described in literature fully enough. They provide for the second communication channel to be added, specifically for the implementation of chaotic synchronization between the transmitting and receiving sides. Information hidden in the chaotic carrying signal is transmitted through the first channel. This improves a quality and stability (in terms of disturbances and distortions) of synchronization and ensures more accurate recovering of the transmitted message [189].

    2.2.1. The double-channel systems with active synchronization [149].
    2.2.2. The double-channel systems with passive synchronization.

On basis of this classification, the authors of [135] suggest a new double-channel system of the chaotic communication with passive synchronization of transmitting and receiving units. Later, independently, the authors of [148] offered the similar system. This approach allows considerable simplification of the communication system design as a whole, we also avoid decreasing its masking properties [135, 148].

It is clear that double-channel structure and arrangement of the separate synchronization channel are the mutually complementary classification features. Hence, additional resources for the classification development are present. In the General Conclusion we will reveal them.

1.5 Conclusions

From this brief literature review it becomes clear that in nonlinear-dynamic cryptography, which has been developing since the middle of 1990s as the new field for the research and engineering examination and testing, we can point out to a number of tendencies.

1. In the modern stage of development of the information protection methods their diversification is the tendency. Along the further development of the classical cryptography, there arise the quantum cryptography and the nonlinear-dynamic cryptography applying the phenomenon of the deterministic chaos. The evolution of the latter has produced, practically simultaneously, several types of systems for confidential communication that can be arranged in a tree-like scheme of classification.
   Besides, the steganographic and physical protection methods are still relevant. For them, in microwave, terahertz and optical wavelength ranges the beams are provided with the spiral dislocations of the wave front (as they say in singular optics, with optical vortices).
2. The development of principles and information protection systems applying the phenomenon of the deterministic chaos has been mainly carried out during the last decade, and it refers to the various radio electronic devices beginning with the Van der Pol generator and ending with the specially constructed devices. From the end of 1990s there have been attempts to devise the information protection devices in the optical range. However, a number of the variants of optical dynamic chaos generators is still not very large.
3. The analysis of structural and electric diagrams of information protection devices shows a variety of methods and engineering solutions. Evidently, it is associated with the early stage of nonlinear-dynamic cryptography evolution, when the study is focused on the diversification, in order to extend the possibilities. Each system variant would find its own "ecological niche", in which its self-reproduction should become rather stable, otherwise this system will perish. Nevertheless, the most promising operation principles are outlined for the information protection systems. But, up to now, a really practical, reliable and versatile system has not yet been developed.

4. Such kind of investigation and research, in turn, presuppose the search for the answer for the question what specific characteristics should the deterministic chaos and chaotic generators possess. Hence, we need to design the chaotic generators, which are capable of providing the required features (for instance, the wide continuous spectrum of the generated signal, a sufficient number and the value range of parameters, which are the keys of the nonlinear-dynamic cryptosystem).

The next chapter is devoted to the description of the dynamic chaos generators for radio-frequency and optical ranges.

References

1. Introduction to cryptography/under edition of V.V. Yashchenko. Sankt-Peterburg: Piter Publication; 2001. 288 pp. (in Russian).
2. Artamonov VA. Elements of cryptology. Soros Educ J. 2000; 6(5):123–127. (in Russian).
3. Nechaev VI. Elements of cryptography (fundamentals of theory): textbook. Moscow: Higher School Publication; 1999. 109 pp. (in Russian).
4. Allen L, Beijersbergen MW, Spreeuw RJC, Woerdman JP. Um of light and the transformation of Laguerre–Gaussian laser modes. Phys Rev A. 1992; 45:81–85.
5. Korolenko VP. Optical vortices. Soros Educ J.1998; 6:94–99. (in Russian).
6. Swartzlander GA Jr., Law CT. Optical vortex solitons observed in Kerr nonlinear media. Phys Rev Lett. 1992; 69(17):2503–2506.
7. Rozas D, Law CT, Swartzlander GA Jr. Propagation dynamics of optical vortices. J Opt Soc Am B. 1997; 14(11):3054–3065.
8. Aksenov VP, Banakh VA, Tikhomirova OV. Potential and vortex features of optical speckle field and visualization of wave-front singularities. Appl Opt.1998; 37(21):4536–4540.
9. Berry MV. Wave dislocation reactions in non-paraxial Gaussian beams. J Mod Opt. 1998; 45 (9):1845–1858.
10. Optical vortices horizons in world physics. In: Vasnetsov M, Staliunas K, editors. Braunschweig, Germany. 1999; 228. 218 pp.
11. Mansuripur M, Wrignt E. Linear optical vortices. Opt Photonics News. 1999; 10(2):40–44.
12. Weiss CO, Vaupel M, Staliunas K, Slekys G, Taranenko V.B. Solitons and vortices in lasers. Appl Phys B. 1999; 68:151–168.
13. Aksenov VP, Kolosov VV, Tartakovskiy VA, Fortes BV. Optical vortices in nonuniform media in nonuniform media. Atmos Oceanic Opt. 1999; 12(10):952–958. (in Russian).
14. Voliar AV, Zhilaitis VZ, Fadeeva TA. Optical vortices in small-mode fibers. III. Dislocation reactions, phase transitions and topological birefringence. Opt. Spectrosc. 2000; 88(3):446–455 (in Russian).
15. Masajada J. Optical vortices and their application to interferometry. Public Institute of Physics Wroclaw University of Technology, 2004 (Prace Naukowe Instytutu Fizyki Nr 36, Monografie Nr 25). 104 pp.
16. Progress in Optics, vol. 53. 2009. Elsevier. doi:10.1016/S0079-6638(08)00205-9.
17. Adaptive optics progress. Edited by Tyson RK, Publication: InTech; 2012. doi:10.5772/46199.
18. Aksenov VP, Izmailov IV, Poizner BN, Tikhomirova OV. Energy streamlines in conditions of optical vortices formation. In: Matvienko GG, Panchenko MV, eds. Proceedings of SPIE seven international symposium on atmospheric and ocean optics, vol. 4341, pp. 173–180. 16–19 July 2000. Tomsk, Russia. 2000. 8 pp.

19. Aksenov VP, Izmailov IV, Poizner BN, Tikhomirova OV. Spatial ray dynamics at forming of optical speckle-field. In: Soskin MS, VasnetsovMV (eds). Proceedings of SPIE second international conference on singular optics (optical Vortices): fundamentals and applications, pp. 108–114, vol. 4403. 2–6 October 2000, Alushta, Crimea, Republic of Ukraine. 2001. 7 pp.
20. Aksenov VP, Izmailov IV, Poizner BN, Tikhomirova OV. Wave front dislocations at laser beam propagation in the inhomogeneous medium. In: Corcoran VJ, Corcoran TA (eds). Proceedings of international conference on lasers'2000, pp. 76–82. Albuquerque, USA. 4–7 December 2000. USA: STS Press, 2001. 7 pp.
21. Aksenov VP, Izmailov IV, Poizner BN, Tikhomirova OV. Wave and beam spatial dynamics of the light field at arising, evolution and annihilation of phase dislocations. Opti Spectro. 2002; 92(3):452–461. (in Russian).
22. Aksenov VP, Izmailov IV, Poizner BN, Tikhomirova OV. Spatial dynamics of optical vortexes when Gauss–Laguerre beam propagates in the Kerr nonlinear medium. Ukr J Phys. 2004; 49(5):504–511.
23. Aksenov VP, Izmailov IB, Kanev FYu, Pogutza ChE. Properties of vortical laser beams propagating in the turbulent atmosphere. Atmospheric and oceanic optics. Physics of the atmosphere. In: Proceeding of xvii international symposium, pp. B195–B199. Tomsk: IOA SB RAS Publication. June 28th–July 1st 2011. (in Russian).
24. Molina-Terriza G, Torres JP, Torner L. Management of the angular momentum of light: preparation of photons in multidimensional vector states of angular momentum. Phys Rev Lett. 2002; 88:013601.
25. Gibson G, Courtial J, Padgett MJ, Vasnetsov M, Pas'ko V, Barnett SM, Franke-Arnold S. Free-space information transfer using light beams carrying orbital angular momentum. Opt Express. 2004; 12:5448–5454.
26. Rabinovich MI, Trubetzkov DI. Introduction to the theory of oscillations and waves. Dordrect, Boston, London: Kluwer Academic Publication; 1989. 440 pp.
27. Landa PS. Nonlinear oscillations and waves in dynamical systems. Dordrect, Boston, London: Kluwer Academic Publication; 1996. 538 pp.
28. Dmitriev AS, Kislov VYa. Oscillations in radiophysics and electronics. Moscow: Nauka Publication; 1989. 280 pp (in Russian).
29. Anishchenko VS, Astakhov VV, Vadivasova TE, Neiman AB, Strelkova GI, Shimanskiy-Gaer L. Nonlinear effects in chaotic and stochastic systems. Moscow–Izhevsk: Institute of Computer Researches Publication; 2003. 544 pp (in Russian).
30. Anishchenko VS, Vadivasova TE, Astakhov VV. Nonlinear dynamics of the chaotic and systems. Fundamentals and selected problems. Saratov, Saratov University Publication; 1999. 368 pp (in Russian).
31. Malinetskiy GG, Potapov AB. Modern problems of nonlinear dynamics. Moscow: Editorial URSS Publication; 2000. 336 pp. (in Russian).
32. Kuznetsov SP. Dynamic chaos (lecture course). Moscow: FizMatLit Publication; 2001. 296 p (in Russian).
33. Trubetskov DI. Introduction on synergetics: chaos and structures. Moscow: Editorial URSS Publication; 2004. 240 pp. (in Russian).
34. Vladimirov SN, Smolskiy SM. Non-traditional dynamics in electronics: theory and practice. Paceo Segovia Irvine (USA, CA): Scientific Research Publication Inc.; 2011. 260 pp.
35. Izmailov IV, Poizner BN. Axiomatic scheme of studying the dynamics: from criteria of their diversity to self-change. Tomsk: STT; 2011. 570 pp. (in Russian).
36. Dmitriev AS, Efremova EV Maksimov NA, Panas AI. Chaos generation. Moscow: Technosphere Publication; 2012. 424 pp. (in Russian).
37. Oraevskiy AN. Masers lasers and strange attractors. Quantum Electron. 1981;8(1):130–42 (in Russian).
38. Oraevskiy AN. Superlight wave in amplifying medium. Optical tachions. Soros Educ J. 1999; 10:75–50 (in Russian).

39. Khanin YaI. Fundamentals of laser dynamics. Moscow: Nauka Publication; 1999. 368 pp. (in Russian).
40. Napartovich AP, Sukharev AG. Information decoding in the chaotic laser circuit, which is controlled by chaotic signal. Quantum Electron. 1998; 25(1):85–88. (in Russian).
41. Dmitriev AS, Panas AI. Dynamic chaos: new information carrier for communication systems. Moscow: FizMatLit Publication; 2002. 252 pp. (in Russian).
42. Baranov SV, Kuznetsov SP. Hyperchaos in the system with delayed feedback on the base of van-der paul oscillator with modulated Q-factor. Izvestiya VUZ. Appl Nonlinear Dyn. 2010;18(4):101–110. (in Russian).
43. Kuznetsov SP, Sokha YuI. Hyperchaos in the model non-autonomous systems with cascade excitation transfer on spectrum. Izvestiya VUZ. Appl Nonlinear Dyn. 2010; 18(3):24–32. (in Russian).
44. Kuznetsov SP. Dynamic chaos and uniformly hyperbolic attractors: from mathematics to physics. Adv Phys Sci. 2011; 181(2):121–147. (in Russian).
45. Aidarova Yu S, Kuznetsov SP. Chaotic dynamics of the hunt model—the artificially constructed flow systems with hyperbolic attractor. Izvestiya VUZ. Appl Nonlinear Dyn. 2008; 16(3):176–196. (in Russian).
46. Arzhanukhina DS, Kuznetsov SP. Robust chaos in autonomous time-delay system. Izvestiya VUZ. Appl. Nonlinear Dyn. 2014; 22(2):36–49. (in Russian).
47. Kislov VYa, Zalogin NN, Myasin EA. Research of stochastic oscillators with delay. J Commun Technol Electron. 1979; 24(6):1118–1130. (in Russian).
48. Dmitriev AS, Дмитриев АС. Dynamic chaos in the ring oscillator systems with the nonlinear filter. Izvestiya VUZ. Radiophys. 1985; 28(4):429–439. (in Russian).
49. Dmitriev AS, Kislov VYa. Oscillations in the oscillator with inertial delay of the first order. Radiotekhnika i Electronika. 1984; 29(12):2389–2398. (in Russian).
50. Dmitriev AS, Panas AI, Starkov SO, Kuzmin LV. Experiments on RF band communications using chaos. Int J Bifurcat Chaos. 1997; 7(11):2511–2527.
51. Dmitriev AS, Panas AI, Starkov SO. Dynamic chaos as paradigm of the modern communication systems. Foreign radio electronics. Achiev. Mod Radio Electron. 1997; 10:4–26. (in Russian).
52. Astakhov V, Shabunin A, Anishchenko V. Synchronization of self-oscillations by parametric excitation. Int J Bifurcat Chaos. – 1998. – V. 8, № 7. – P. 1605–1612.
53. Panas AI, Yang T, Chua LO. Experimental results of impulsive synchronization between two Chua's circuits. Int J Bifurcat Chaos. 1998; 8(3):639–644.
54. Dmitriev AS, Efremova EV, Nikishov AYu, Panas AI. Generation of microwave chaotic oscillations in CMOS structure. Nonlinear Dyn. 2010; 6(1):159–167. (in Russian).
55. Dmitriev AS, Efremova EV, Nikishov AYu. Generation of dynamic chaos in microwave range in oscillator system on the base of SiGe. Lett J Tech Phys 2/19. 2009; 35(23):40–46. (in Russian).
56. Nikishov AYu, Panas AI. Ultra-wide-band UHF chaotic oscillator of the ring structure on amplifying micro-assemblies. Achiev Mod Radio Electron. 2008; 1:54–62. (in Russian).
57. Anisimova Yu V, Dmitriev AS, Zalogin NN, Kalinin VI, Kislov V Ya., Panas AI. About one mechanism of transition to chaos in the system "electronic beam—electromagnetic wave". Lett JETF. 1983; 37(8):387–390. (in Russian).
58. Kislov V Ya. Theoretical analysis of noisy oscillations in electronic-wave systems. Radiotekhnika i Elektronika. 1980; 25(8):1683–1690. (in Russian).
59. Feldbaum AA, Butovskiy AG. Methods of automatic control systems. Moscow: Nauka; 1971. 743 pp. (in Russian).
60. Vladimirov SN, Zolotov SV, Negrul' VV, Perfiliev VI. Deterministic chaos in UHF electronics. Electron Ind. 2002; 2:87–90. (in Russian).
61. Vladimirov SN. Dynamic instabilities of flows and maps. A glance from radio physicist. Tomsk: Tomsk University Publication; 2008. 352 pp. (in Russian).
62. Vladimirov SN, Negrul VV. On autoparametric route leading to chaos in dynamical systems. Int J Bifurcat Chaos. 2002; 12(4):819–826.

63. Zalogin NN, Kalinin VI, Kislov V Ya. Nonlinear resonance and stochasticity in the oscillator system with delay. Radiotekhnica i Electron. 1983; 28(10):2001–2007. (in Russian).
64. Man'kin IA, Shkolnikov VG. Theoretical Analysis on Interaction of lengthy electronic flow with the field of the wide-band stochastic signal. Radiotekhnika i Elektronika. 1981; 26 (9):1932–1938. (in Russian).
65. Myasin EA, Panas AI. On issue of steady-state of the UHF oscillator of wide-band stochastic oscillations. Radiotekhnica i Electron. 1983; 28(12):2423–2429. (in Russian).
66. Bezruchko BP, Kuznetsov SP, Trubetskov DI. Experimental observation of stochastic oscillations in the dynamic system the Dynamic System the Electronic flow – reverse electromagnetic wave. J Exp Theor Phys Lett. 1979; 29(3):180–184. (in Russian).
67. Bezruchko BP, Bulgakova LV, Kuznetsov SP, Trubetskov DI. Stochastic oscillations and instability in the backward wave tube. Radiotekhnika i Elektronika. 1983; 28(6):1136–1139. (in Russian).
68. Ginzburg NS, Kuznetsov SP. Periodic and stochastic auto-modulation modes in electronic generators with distributed interaction. In the book: Relativistic High-Frequency Electronics. Gorkiy: IPF AN SSSR Publication; 1981. pp. 101–144. (in Russian).
69. Dmitrieva TV, Ryskin NM, Shigaev AM. Complex dynamics of simple models of distributed self-oscillating delayed feedback systems. Nonlinear Phenom Complex Syst. 2001; 5(4):376–382.
70. Filatov RA, Hramov AE, Koronovskii AA. Chaotic in coupled spatially extended beam-plasma systems. Phys Lett A. 2006;358(4):301–8.
71. Filatov RA, Hramov AE, Bliokh YP, Koronovskii AA, Felsteiner J. Influence of background gas ionization on oscillations in a virtual cathode with a retarding potential. Phys. Plasmas. 2009; 16 C 033106.
72. Arefiev AA, Baskakov EN, Stepanova LN. Radio engineering devices on transistor equivalents of p-n-p-n-structures. Moscow: Radio i Sviaz Publication; 1982. 104 pp. (in Russian).
73. Romanov IV, Izmailov IV, Kohanenko AP, Poizner BN. The chaos in experiments with radio oscillator having a delay and nonlinearity formed by two Λ-diodes. In: Proceeding of the synergy in natural science: fifth anniversary Kurdiumov interdiscipline conference, pp. 106–109. 15–18th April 2009. Tver: Tver University Publication. 2009 (in Russian).
74. Chua LO, Yu JB, Yu YY. Bipolar-JFET-MOSFET negative resistance devices. IEEE Trans Circuits Syst. 1985; 32(1):46–61.
75. Chua LO, Yu JB, Yu YY. Negative resistance devices. Int J Circuits Theory Appl. 1983; 11:161–186.
76. Porter JA. JEFT transistor yields devices with negative resistance. IEEE Trans Electron Devices. 1976; 23(9):1098–1099.
77. Kumar U. A complication of negative resistance circuits generated by two novel algorithms. Act Passive Electron Comp. 2002; 25:211–214.
78. Chua LO, Desoer CA, Kuh EA. Linear and nonlinear circuits. New York: McGraw-Hill; 1987. 839 pp.
79. Tadic N, Gobovic DA. Floating, negative-resistance voltage-controlled resistor. In: Proceedings of the 18th IEEE on instrumentation and measurement technology conference (IMTC 2001), vol. 1, pp. 437–442. Budapest. 2001.
80. Rulkov NF. Images of synchronized chaos. CHAOS. 1996; V 6(3):262–279.
81. Zhong GQ, Ko KT, Man KF, Tang KS. A systematic procedure for synthesizing two-terminal devices with polynomial non-linearity. Int J Circ Theor Appl. 2001; 29:241–249.
82. Zeiger SG, klimontovich Yu L, Landa PS, et al. Wave and fluctuation processes in lasers. Moscow: Nauka Publication; 1974. 256 pp. (in Russian).
83. Papoff F, D'Alessandro G, Oppo G-L, et al. Local and global effects of boundaries on optical-pattern formation in Kerr media. Phys Rev A. 1994; 48(1):634–641.
84. Mel'nikov LA, Koniukhov AI, Riabinina MV. Dynamics of transverse polarization field structure in lasers. Izvestiya VUZ. Appl Nonlinear Dyn. 1996; 4(6):33–53. (in Russian).

85. Rosanov NN. Optical bistability and in distributed nonlinear systems. Moscow: Nauka Publication; 1997. 336 pp. (in Russian).
86. Napartovich AP, Sukharev AG. Synchronizing a chaotic laser by injecting a chaotic signal with a frequency offset. J Exp Theoret Phys. 1999; 88(5):875–881.
87. Sivaprakasam S, Shore KA. Critical signal strength for effective decoding in diode laser chaotic optical communications. Phys Rev E. 2000; 61(5):5997–5999.
88. Mel'nikov LA, Konukhov AI, Veshneva IV, et al. Nonlinear dynamics of spatial and temporal patterns in lasers and atom optics: Kerr-lens mode-locked laser, Zeeman laser and Bose-Einstein atomic condensate. Izvestiya VUZ. Appl nonlinear Dyn. 2002;10(3):40–62. (in Russian).
89. Smirnov E, Stepin M, Shandarov V, Kip D. Pattern formation by spatially incoherent light in a nonlinear ring cavity. Appl Phys B. 2006; 85:135–141.
90. Agrawal GP. Application of nonlinear fiber optics. New York: Academic Press; 2008. 624 pp.
91. Bjorkholm JE, Smith PW, Tomlinson WJ, Kaplan AE. Optical bistability based on self-focusing. Opt Lett. 1981; 6(7):345–347.
92. Akhmanov SA, Vorontsov MA, Ivanov V Yu, Larichev AV, Zhelezykh NI. Controlling transverse-wave interaction of spatiotemporal structures. Opt Soc Am B. 1992; 9(1): P. 78–90.
93. D'Alessandro G, Firth WJ. Hexagonal spatial patterns for a Kerr slice with a feedback mirror. Phys Rev A. 1992; 46(1):537–548.
94. Adachihara H, Faid H. Two-dimensional nonlinear-interferometer pattern analysis and decay of spirals. Opt Soc Am. 1993; 10(7):1242–1253.
95. Vorontsov MA, Firth WJ. Pattern formation and competition in nonlinear optical systems with two-dimensional feedback. Phys Rev A. 1994; 49(4):2891–2905.
96. Arshinov AI, Mudarisov RR, Poizner BN. Mechanisms of formation of the simplest optical structures in the nonlinear Fizeau interferometer. Russ Phys J. 1995; 6:77–81.
97. Larichev AV, Nikolaev IP, Shmalgausen VI. Optical dissipative structures with controlled spatial period in the nonlinear system with fourier filter in the feedback loop. Quant Electron. 1996; 10:894–898. (in Russian).
98. Vorontsov MA, Karpov AYu. Pattern formation due to interballon spatial mode coupling. Opt Soc Am B. 1997; 1.
99. Martin R, Oppo G-L, Harkness GK, et al. Controlling pattern formation and spatio-temporal disorder in nonlinear optics. Opt Express. 1997; 1(1):39–43.
100. Garcia-Ojalvo J, Roy R. Spatiotemporal communication with synchronized optical chaos. 2000. 4 pp. http://xxx.lanl.gov/abs/nlin.CD/0011012. Accessed 6 Nov 2000.
101. Chesnokov SS, Rybak AA. Spatiotemporal chaotic behavor of time-delayed nonlinear optical systems. Laser Phys. 2000;10(3):P. 1–8.
102. Vorontsov MA, Carhart GW, Dou R. Spontaneous optical pattern formation in a large array of optoelectronic feedback circuits. J Opt Soc Am B. 2000;17(2):266–274.
103. Baliakin AA, Ryskin NM. Transition to the chaos in nonlinear ring resonator at excitation by the multi-frequency signal. Bull Russ Acad Sci Phys. 2001;65(12):1741–1744. (in Russian).
104. Baliakin AA. Investigation of chaotic dynamic of the ring nonlinear resonator at double-frequency external impact. Izvestiya VUZ. Appl Nonlinear Dyn.2003;11(4):3–15. (in Russian).
105. Ryskin NM, Khavroshin OS. Chaos Control in the Ikeda system: simplified model in the form of point representastion. Izvestiya VUZ. Appl Nonlinear Dyn. 2009;17(2):66–78. (in Russian).
106. Ryskin NM, Khasvroshin OS. Chaos control in the Ikeda system: spatial-temporal model. Izvestiya VUZ. Appl. Nonlinear Dyn. 2009; 17(2):87–100. (in Russian).
107. IkedaK. Multiple-valued stationary state and its instability of the transmitted light by ring cavity system. Opt. Comm. 1979;30(2):257–260.
108. Akhmanov SA, Voromtsov MA. Nonlinear waves: dynamics and evolution: collection of papers. Moscow: Nauka Publication; 1989.pp. 228–237. (in Russian).

109. New physical principles of optical information processing: collection of papers. Under edition of Akhmanov, SA, Vorontsov, MV. Moscow: Nauka Publication; 1990. pp. 263–326. (in Russian).
110. Vorontsov MA. Nonlinear wave spatial dynamics of light fields. Bull Russ Acad Sci Phys. 1992; 56(4):7–15. (in Russian).
111. Arshinov AI, Mudarisov RR, Poizner BN. The mechanism of optical structures formation in the nonlinear fizeau interferometer at mirror shift and beam parameter variation. Russ Phys J. 1997; 7:67–72. (in Russian).
112. Rybak AA, Chesnokov SS. Distributed nonlinear-optical systems with the delay line as generators of artificial optical turbulence "Optics-99" In: proceedings of internation conference of young researchers and experts, p. 73. Sankt-Peterburg, ITMO Publication, 19–21 Oct. 1999 (in Russian).
113. Beliakov VA, Sonin AS. Optics of cholesterol liquid crystals. Moscow: Nauka Publication. 360 pp. (in Russian).
114. Pestriakov VB, Afanasiev VP, Gurvich VL et al. Noise-like sugnals in systems of information transmission. Moscow: Sovetskoe Radio Publication; 1973. 424 pp. (in Russian).
115. Varakin I.E. Communication systems with noise-like signals. Moscow: Radio i Sviaz; 1985. 384 pp. (in Russian).
116. Harmuth HF Nonsinusoidal waves for radar and radio communications. Toronto: Academic Press; 1981. 362 pp.
117. Fujisaka H, Yamada T. Stability theory of synchronized motion in coupled systems. Prog. Theor. Phys. 1983; 69:32–46.
118. Afraimovich VS, Verichev NN, Rabinovich MI. Synchronization of oscillations in the dissipative systems. Izvestiya VUZ. Radiophys. 1986; 29:1050–1060. (in Russian).
119. Pecora LM, Carroll TL. Synchronization in chaotic systems. Phys Rev Lett. 1990; 64:821–824.
120. Kuznetsov AP, Kuznetsov SP, Ryskin NM. Nonlinear oscillations: textbook for universities. Moscow: FizMatLit Publication; 2002. 292 pp. (in Russian).
121. Pecora LM, Carrol TL. Driving systems with chaotic signals. Phys Rev A. 1991; 4 (44):2374–2384.
122. Rosenblum VG, Pikovsky AS, Kurths J. Phase synchronization of chaotic oscillators. Phys Rev Lett. 1996; 76(11):1804–1807.
123. Chen JY, Wong KW, Cheng LM, Shuai JW. A secure communication scheme based on the phase of chaotic systems. Chaos. 2003; 13(2):508–514.
124. Rulkov NF, Sushchik MM, Tsimring LS, Abarbanel HI. Generalized of chaos in directionally coupled chaotic systems. Phys Rev E. 1995; 51(2):980–994.
125. Terry JR, VanWiggeren GD. Chaotic communication using generalized synchronization. Chaos Solitons Fractals. 2001; 12(1):145–152.
126. Rosenblum M.G., Pikovsky A.S., Kurths J. From phase to lag synchronization in coupled chaotic oscillators. Phys Rev Lett. 1997; 78(22):4193–4196.
127. Taherion S, Lai Y-C. Observability of lag synchronization of coupled chaotic oscillators. Phys Rev E.1999; 59(6):6247–6250.
128. Fahy S, Hamann DR. Transition from chaotic to nonchaotic behavior in randomly driven systems. Phys Rev Lett. 1992; 69(5):761–764.
129. Martian A, Banavar JR. Chaos, noise, and synchronization. Phys Rev Lett. 1994; 72 (10):1451–1454.
130. Kaulakys B, Vektaris G. Transition to nonchaotic behavior in a Brownian-type motion. Phys Rev E. 1995; 52(2):2091–2094.
131. Chen Y-Y. Why do chaotic orbits converge under a random velocity reset? Phys. Rev. Lett. 1996; 77(21):4318–4321.
132. Kaulakys B, Ivanauskas F, Mekauskas T. Synchronization of chaotic systems driven by identical noise. Int J Bifurcat Chaos. 1999; 9(3):533–539.

133. Toral R, Hernandez-Garcia E, Piro O, Mirasso CR. Analytical and numerical studies of noise-induced synchronization of chaotic systems. Chaos. 2001; 11(3):665–673.
134. Volkovskiy AR, Rul'kov NF. Synchronous chaotic response of the nonlinear oscillating system as a principle of detection of the chaos informative component. Tech Phys Lett. 1993; 19(3):71–75. (in Russian).
135. Vladimirov SN, Negrul VV. Communication systems with passive chaotic synchronization. In: Proceedings of 5-th International Conference (Relevant Problems of Electronic Instrumentation APEP-2000. Novosibirsk, vol. 7. pp. 39–41. 26–29th Sept 2000 (in Russian).
136. Shalfeev VD, Osipov GV, Kozlov AK, Volkovskiy AV. Chaotic oscillations: generation, synchronization, control. Foreign radio electronics. Achiev. Mod. Radio Electron. 1997; 10:27–49. (in Russian).
137. Hasler M. Achievements in the field of information transmission with chaos utilization. Foreign radio electronics. Achiev Mod Radio Electron. 1998; 11:33–43. (in Russian).
138. Andreev Yu V, Dmitriev AS, Emez SV, Panas AI, Starkov SO, Balabin AM, Dmitriev AA, Kishik VB, Kuzmin LV, Borisenko AG. Strategies of dynamic chaos utilization in communication systems and computing networks. Separation of chaotic coder and channel coder. Foreign radio electronics. Achiev Mod Radio Electron. 2000; 11:4–26. (in Russian).
139. Starkov SO, Schwarz V, Abel A. Multi-user communication systems with utilization of the dynamic chaos. Foreign radio electronics. Achiev Mod Radio Electron. 2000; 11:34–47. (in Russian).
140. Dmitriev AS, Starkov SO, Kuzmin LV. Abstracts of international conference "Progress in nonlinear science", p. 204. Nizhny Novgorod. 2–6 July 2000.
141. Dmitriev AS, Kriaginskiy BE, Panas AI, Starkov SO. Dynamic Chaos in Radiophysics and electronics. Direct-chaotic schemes of information transmission in Ultra-High-Frequency Range. Radiotekhnika i Elektronika. 2001; 46(2):224–233. (in Russian).
142. Dmitriev AS, Kiarginskiy BE, Maksimov NA, Panas AI, Starkov SO. Perspectives of development of the direct-chaotic communication systems in radio and UHF ranges. Radio Eng. 2000; 3:9–20. (in Russian).
143. Dmitriev AS. Dynamic Chaos as the information carrier. News in synergy: a glance to third millennium. Moscow: Nauka; 2002. pp. 82–122. (in Russian).
144. Koronovskii AA, Moskalenko OI, Hramov AE. On the use of chaotic for secure communication. Uspekhi Fizicheskikh Nauk. 2009;179(12):1281–310. doi:10.3367/UFNr.0179.200912c.1281.
145. Downes P. Secure communication using chaotic synchronization. SPEE Chaos Commun. 1993:227–233.
146. Dmitriev AS, Kuzmin LV, Panas AI. Communication system with summation on modulo of chaotic and information signals. Radiotekhnika i Elektronika. 1999;44(8):988–96 (in Russian).
147. Yang T. Recovery of digital signals from chaotic. Int J Circu Theor Appl. 1995;23(6):611–5.
148. Kalianov EV, Grigoryants VV. Information transmission with usage of masking chaotic oscillations. J Tech Phys Lett. 2001; 27(6):71–76. (in Russian).
149. He R, Vaidya PG. Implementation of chaotic with chaotic. Phys. Rev. E. 1998; 57(2):1532–1535.
150. Hayes S, Grebogi C, Ott E. Communicating with chaos. Phys. Rev. Lett. 1993; 70(20):3031–3034.
151. Schweizer J, Kennedy M. Predictive poincare control modulation: a new method for modulating digital information onto a chaotic signal. In: Proceedings of Irish DSP and control colloquium, pp. 125–132. 1994.
152. Dmitriev AS, Starkov SO. Transmission using the chaos and classical information theory. Foreign radio electronics. Achiev. Mod. Radio Electron. 1998; 11:4–31. (in Russian).
153. Dmitriev AS, Panas AI, Starkov SO et al. The method of information transmission using chaotic signals: Russian patent no 2185032. 27.07.2000.
154. Short KM. Step toward unmasking secure communication. Int J chaos. 1994;4(4):959–77.

155. Short KM. Unmasking a modulated chaotic communication scheme. Int J chaos. 1996;6 (2):367–75.
156. Short KM, Parcer AT. Unmasking a hyperchaotic communication scheme. Phys. Rev. E. 1998; 58(1):1159–1162.
157. Perez G, Cerdeira HA. Extracting messages masked by chaos. Phys. Rev. Lett. 1995; 74 (11):1970–1974.
158. Changsong Z, Lai C-H. Decoding information by following parameter modulation with parameter adaptive control. Phys. Rev. E. 1999; 59(6):1014–1019.
159. Wu CW, Yang T, Chua LO. On adaptive synchronization and control of nonlinear dynamical systems. Int J chaos. 1996;6(3):455–71.
160. Kaplanov MV, Kuminov VN, Larionov AG, Udalov AG. Properties of information transmission systems with shift-keying of parameters and initial conditions of chaotic oscillation generators. Foreign radio electron. Achiev Mod Radio Electron. 2000; 11:48–60. (in Russian).
161. Anishchenko VS, Pavlov AN, Yanson NB. Reconstruction of dynamic systems in application to problems solutions of information protection. J Tech Phys. 1998;68(12):1–8 (in Russian).
162. Pecora LM, Carrol TL, Jonson J, Marr D. Volume-preserving and volume expanding synchronized chaotic system. Phys Rev E. 1997;56(5):5090–7.
163. Baker GL, Golub JP, Blackbum JA. Inverting chaos: extracting system parameters from experimental data. Chaos. 1996;6(4):528–33.
164. Mathiazhagan C. Secure chaotic communication. http://xxx.lanl.gov/abs/chao-dyn/9905001. Accessed 3 May 1999.
165. Kislov VYa, Kislov VV. New type of signals for information transmission. Wide-band chaotic signals. J. Commun. Technol. Electron. 1997; 42(9):962–973. (in Russian).
166. Kislov VYa, Kalmykov VV, Beliaev RV, Vorontsov GM. Correlation properties of noise-like signals generated by systems with dynamic chaos. J Commun Technol Electron. 1997; 42(11):1341–1349. (in Russian).
167. Beliaev RV, Vorontsov GM, Kolesov VV. Random sequences formed by nonlinear algorithm with delay. J. Commun. Technol. Electron. 2000; 45(8):954–960. (in Russian).
168. Kocarev LJ, Halle KS, Eckert K, Chua LO, Partlitz U. Experimental demonstration of secure communication via chaotic. Int J Chaos. 1992; 2(3):709–713.
169. Belskiy Yu L, Dmitriev AS. Information transmission using deterministic chaos. J. Commun. Technol. Electron. 1993; 37(7):1310–1315 (in Russian).
170. DedieuH, Kennedy MP, Hasler M. Chaos shift keying: modulation and demodulation of a chaotic using self-synchronizing Chua's circuits. Procc IEEE Trans Circuits Syst. 1993; 40 (2):634–642.
171. Halle KS, Wu CW, Itoh M, Chua LO. Spread spectrum communication through modulation of chaos. Int J chaos. 1993; 3:469–477.
172. Alexeyev AA, Green MM. Secure communications based on variable topology of chaotic circuits. Int J Chaos. 1997; 7(12):2861–2869.
173. Hasler M. Synchronization of chaotic systems and transmission of information. Int J Chaos. 1998; 8(4):647–659.
174. Dmitriev AS, Kuzmin LV, Panas AI. Scheme of information transmission on the base of synchronous chaotic response at filtering presence in the communication channel. Radio Eng. 1999; 4:75–80. (in Russian).
175. Garcia-Ojalvo J., Roy R. Parallel communication with optical spatiotemporal chaos. IEEE Trans circuits syst. I: Fundam Theory Appl. 2001; 48(12):1491–1497.
176. Garcia-Ojalvo J, Roy R. Communicating with optical spatiotemporal chaos. In: Proceedings of SPIE, pp. 1–8. 2002.
177. Napartovich AP, Sukharev AG. Influence of chaotic regular perturbations in master oscillator signal on synchronized chaos in a iscillators each being two-element array. In: Advance program of the 17th International Conference on Coherent and Nonlinear Optics "ICONO'2001". Belarus, Minsk, 26 June–1 July 2001. P. 69.

178. Progress in optics. 2005; vol. 48. Elsevier. doi:10.1016/S0079-6638(08)00205-9.
179. VanWriggeren GD, Roy R. Communication with chaotic lasers. Science. 1998; 279:1198–1200.
180. VanWriggeren GD, Roy R. Optical communication with chaotic waveforms. Phys Rev Lett. 1998; 81(16):3547–3550.
181. Lasing PM, Gavrielides A, Kovanis V, Roy R, Thornburn KS. Encoding and decoding messages with chaotic lasers. Phys Rev E. 1997; 56(6):6302–6310.
182. Mirasso CR, Mulet J, Masoller C. Chaos shift-keying in chaotic external-cavity semiconductor lasers using a scheme. IEEE Photonics Technol Lett. 2002; 14(4):456–458.
183. VanWriggeren GD, Roy R. Chaotic communication using time-delayed optical system. Int J Chaos. 1999; 9(11):2129–2156.
184. Goedgebuer J-P, Larger L, Porte H. Optical based on synchronization of hyperchaos generated by a tunable laser diode. Phys Rev Lett. 1998; 80(10):2249–2252.
185. Poberezhskiy ES. Digital radio receiving devices. Moscow: Radio i Sviaz Publication; 1987. 184 pp. (in Russian).
186. Sklar B. Digital communications: fundamentals and applications.New York: Prentice Hall; 2001. 1079 pp.
187. Rozova SS. Classification problem in the modern science. Novosibirsk: Nauka Publication; 1986. 224 pp. (in Russian).
188. Pokrovskiy MP. Classiology as a System. Philos Probl. 2006; 7:108–117. (in Russian).
189. Vladimirov SN, Negrul' VV Robastic variant of chaotic communication system. In: Proceedings of the International Symposium on Antennas and Propagation. vol. 3, pp. 1403–1406. Fukuoka, Japan. 21–25 August 2000.

Chapter 2
Radiophysical and Optical Chaotic Oscillators Applicable for Information Protection

Let us proceed to the presentation of the original results of our studies. We begin from the description of the block-diagram of the chaotic oscillator of the radio-frequency range, introducing its mathematical model, simulation results and laboratory experiments. After that, we shall proceed to the discussion of the models and the properties of the nonlinear ring interferometers as the chaotic oscillators. We would like to draw the reader attention to their advantages owing to the two-dimensional delayed feedback. Let us consider the versions of the interferometer with one or two of such feedback loop circuits. At first, we will discuss the feasibility of the development of the nonlinear interferometer based on the optical fiber. We will "refract" the optical chaotic oscillators through the "prism" of the hypothesis suggested by S.R. Hameroff and R. Penrose considering the microtubule of the cytoskeleton as an intracellular calculator.

We now begin from the discussion of the suggested radio-electronic chaotic source.

2.1 The Radio-Electronic Oscillator of the Deterministic Chaos with Nonlinearity in the Form of Parabola Compositions

2.1.1 The Structure and the Mathematical Model of the Oscillator

Let us examine the self-oscillating system of the ring type with delay, its structure shown in Fig. 2.1. The oscillator consists of the following elements sequentially closed into the ring: a linear adjustable amplifier, a circuit of the DC components of the signal, a nonlinear element (NE), a low-frequency filter of the first order (LPF), a high frequency filter of the first order (HPF), a delay line. The delay line provides

I. Izmailov et al., *Cryptology Transmitted Message Protection*,
Signals and Communication Technology, DOI 10.1007/978-3-319-30125-9_2

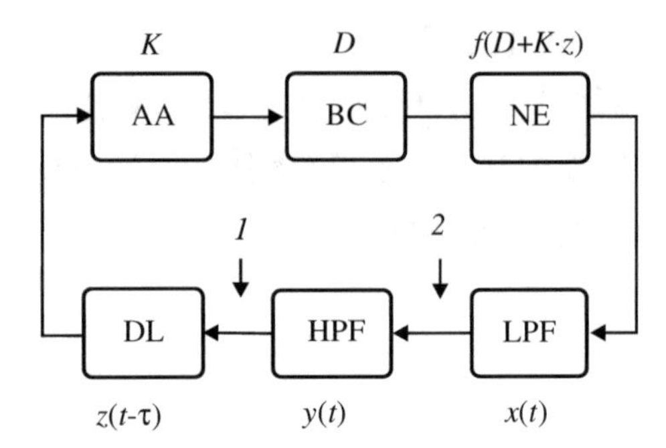

Fig. 2.1 The structural diagram of the deterministic chaos oscillator. Accepted abbreviations: *AA* is a linear adjustable amplifier, *BC* is the bias circuit for DC signal component, *NE* is a nonlinear element, *LPF* is a low-pass filter of the first order, *HPF* is a high-pass filter of the first order, *DL* is a delay line

the high dimensionality of the phase space for the dynamic system (see discourse on the influence of the equivalence of the delay line and the inertial element upon dynamics at the end of Sect. 1.3.1). Due to this, the necessary condition for the chaotic mode implementation is fulfilled in the chaotic oscillator (as in case of the second-order (or higher order) filter performance, for example, in [1, 2]). In the aggregate, the adjustable amplifier, the bias circuit and the NE form the nonlinear amplifier.

In practice, nonlinear element is characterized by the certain persistence. But, we will assume that the nonlinear element is ideal, i.e. it immediately converts the input signal into the output one. Such an assumption is true when the characteristic time t_c of the signal transformation in the nonlinear element is much less than the time constant T_1 of LPF ($t_c \ll T_1$).

The low-frequency filter is meant for the limiting of the maximal oscillation frequency in the chaotic oscillator. The high-frequency filter limits the oscillating spectrum in the low-frequency region. Besides, since development of the deterministic chaos oscillator is directed to its application in the communication systems, then it is desired that the signal at transmitter outlet is set centralized (i.e. contains no DC component). This condition can also be achieved owing to this HPF. The signal centralization simplifies a circuit solution on the reception side, saving from the necessity of recovering the DC component.

Let us consider the signal passing through the feedback loop of the chaotic oscillator. The signal $z(t)$ (in the point 1 in Fig. 2.1) passes through the delay line and lags by τ time. The obtained signal $z(t-\tau)$ arrives at the nonlinear amplifier input. Its first block is the adjustable linear amplifier and it performs the signal scaling transformation. The bias circuit defines the operation point D at the transfer characteristic of the nonlinear element. The signal $D + Kz(t-\tau)$ passes to NE, where it is subject to nonlinear transformation of $f[D + Kz(t-\tau)]$. At LPF output (in the point 2 in Fig. 2.1) a certain signal $x(t)$ is produced. It passes to the input of HPF, at which output (in the point 1 in Fig. 2.1) the signal $z(t) = x(t) - y(t)$ is formed, $y(t)$ being the voltage drop in the HPF capacitor. The circle is closed.

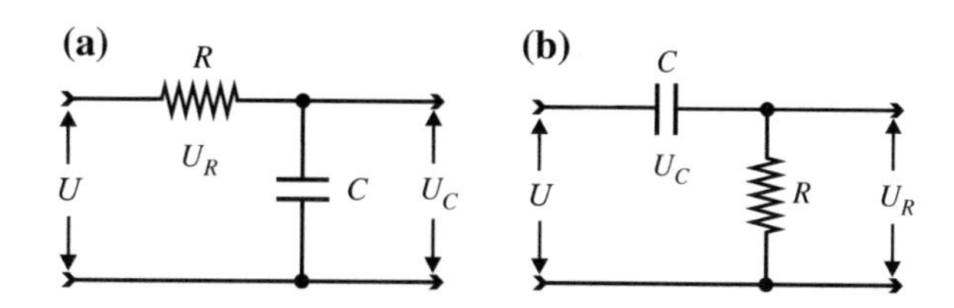

Fig. 2.2 Electrical circuit **a** of the LPF, **b** of the HPF of the first order

Let us proceed to the construction of the mathematical model of the chaotic oscillator, whose variants and development stages were approved at the Conferences [3–7] and described in papers [8–10].

The model of the deterministic chaos oscillator with delay. Let us establish the connection between input and output voltages of each elements of the oscillator and the whole circuit. Electrical circuits of the differential and integration RC—networks, which are LPF and HPF, are shown in Fig. 2.2. The dependence between the input voltage U and the LPF output voltage U_C can be obtained from the equality of currents flowing through the resistor R and the capacitor C:

$$I = I_C = I_R = C\frac{\mathrm{d}U_C}{\mathrm{d}t} = \frac{U_R}{R}.$$

From the equality $U = U_R + U_C$, we obtain the equation for LPF

$$\frac{\mathrm{d}U_C}{\mathrm{d}t} = \frac{U - U_C}{RC} \tag{2.1}$$

The dependence between the voltage at the input U and at the output U_R in HPF can be obtained from the equality of the currents flowing through R and C:

$$I = I_C = I_R = C\frac{\mathrm{d}U_C}{\mathrm{d}t} = \frac{U_R}{R}.$$

Taking into account the equalities $U = U_C + U_R$, we get the equation

$$\frac{\mathrm{d}U_R}{\mathrm{d}t} = \frac{\mathrm{d}U}{\mathrm{d}t} - \frac{U_R}{RC},$$

where $RC = T$ is a time constant of the filter. The HPF equation can be more conveniently written relating to the variable $U_C = U - U_R$

$$\frac{\mathrm{d}U_C}{\mathrm{d}t} = \frac{U - U_C}{RC}. \tag{2.2}$$

The account of delay can be performed by the introduction of the deviating argument:

$$U_n = U(t-\tau), \tag{2.3}$$

where U and U_n are input and output voltages of the delay line. Operations of scaling and determination of the operation point are carried out by the linear amplifier and the bias circuit and are described by the expression

$$U_k(t) = D + KU_n(t), \tag{2.4}$$

where U_n and U_k are input and output voltages of the adjustable amplifier, D—is the bias voltage, K is the gain coefficient. The nonlinear element (Fig. 2.1) transforms the signal according to the law of given transfer characteristic $f(U)$:

$$U_f = f(U_k), \tag{2.5}$$

where U_k and U_f are input and output voltages of NE. Now we do not dwell on the NE structure, as it will be discussed in detail later.

Equations (2.1)–(2.5) allow obtaining the system of the two differential equations describing the dynamics of the chaotic oscillator, which is formed by the series of elements connected according to the circuit in Fig. 2.2 [9]:

$$\begin{aligned} \frac{\mathrm{d}x(t)}{\mathrm{d}t} &= \frac{f(D+K(x(t-\tau)-y(t-\tau)))-x(t)}{T_1}, \\ \frac{\mathrm{d}y(t)}{\mathrm{d}t} &= \frac{x(t)-y(t)}{T_2}, \end{aligned} \tag{2.6}$$

where T_1 and T_2 are the time constants of LPF and HPF. We would like to remind that here $x(t)$ is the value of voltage at LPF output of the first order (the point 2 in Fig. 2.1); $y(t)$ is the voltage on the HPF capacitor; $z(t) = x(t) - y(t)$ is the voltage at the HPF output (the point 1); D is the bias DC voltage; K is the gain coefficient of adjustable amplifier; τ is the signal delay time in the delay lines.

Evidently, the obtained mathematical model describes the nonlinear dynamic system with the infinite-dimension phase space. It allows assumption that dynamics in this oscillator is chaotic. We note that the delay line (Fig. 2.1) is not the only way to increase the phase space dimension of the simple dynamic systems. We know [11–13] that the condition for the chaotic dynamics arising is the phase space dimension equal to three and more. In order to perform the transition to the finite analog of the chaotic oscillator, we may use the way of delay line replacement by the one or several transit-time elements. In this case, the dynamic system dimension is not large but equal to three or more, i.e. it is sufficient for the emergence of the chaotic dynamics in the oscillator. We have already mentioned (see Sect. 1.3.1) the parallels between the "hereditary" systems and the systems with inertia elements.

2.1.2 *The Nonlinear Element: A Structure, a Mathematical Description*

In addition to the literature review presented in Sect. 1.3.2, we have to mention several approaches to the synthesis of the radio-electronic nonlinear elements for the chaotic oscillators. The first approach consists in the integration of nonlinearity and amplitude limitation functions in the one separate device (or in the set of active and passive elements), whose volt-ampere characteristic has a segment with negative differential resistance. The tunnel diode, the IMPATT diode, the electronic tube (dynatrone effect), the transistor (bipolar, single-junction, field effect etc.) can be attributed to such devices and elements. In the most cases, this segment is not large compared to the whole volt-ampere characteristic. The significant disadvantage (in the aspect of the data transmission application) of the discrete nonlinear elements is the parameter dispersion of different specimens of semiconductor devices, a variation of volt-ampere characteristic shape with a change of temperature during the period of device operation. Moreover, high resistance values, together with the spurious distributed capacitors, limit the operating frequency band.

Under the second approach the functions of nonlinearity and limitations (at the ends of the transfer characteristics) are, on the contrary, separated. For instance, the nonlinear transfer function is formed by sequential connection of the amplifier, operating at the nonlinear segment, and the amplifier set in the limitation mode. This allows growing the operating frequency band and reducing requirements to the parameters of the nonlinear element parts (at the expense of the task sharing between the NE parts). However, the descending segment of the transfer characteristic of such an element is absent, or it is weakly expressed [14–16].

The third approach is similar to the second one. According to it, the nonlinear transfer characteristic can be created by the combination of the piecewise-linear and the nonlinear circuits performing the mathematical operations: addition, subtraction, involution, taking the logarithm, exponentiation [17–20]. Application of the operational amplifiers allows realization of the controllable transfer characteristics in the wide frequency bands. Moreover, it is expedient to apply the digital signal processors for the generation of the transfer characteristic. But, there are difficulties as well, for example, a high cost value, as well as the delay during signal processing.

Mentioned approaches (together with a variety of semiconductor discrete elements and the integrated circuits) allow synthesis of the desired transfer characteristic in the frequency band from the units of hertz to the tens of gigahertz.

In the context of chaotic oscillator creation, the choice is important of the transfer characteristic. Its shape (a local slope, a swing, location of the singular points along the curve) mainly determines the dynamic modes and scenarios of the transition to the chaos in the oscillator that can be realized. For instance, the value of the local slope and the transfer characteristic swing affect the conditions of the stability loss in the oscillator. The more a slope is, the less value of gain is required (in the oscillator) to excite the dynamic mode. When the deterministic chaos

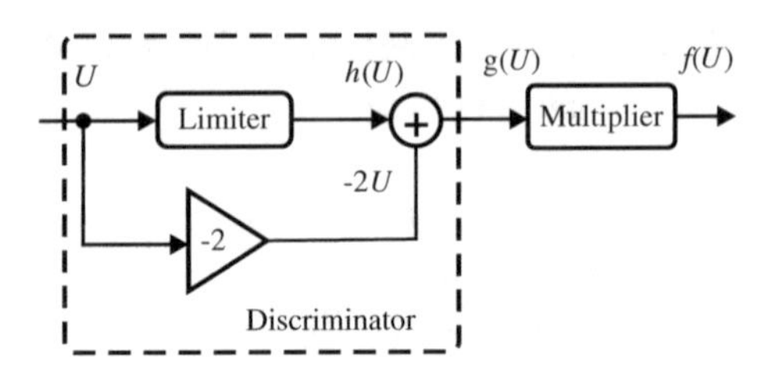

Fig. 2.3 Structural circuit of the nonlinear element

oscillator is used in the structure of communication systems, the reduction of gain leads to the signal/noise ratio growth at the expense of the decrease of the additional noise influence. In turn, in the singular points vicinities the transfer characteristic changes its local slope. This allows the control of the oscillator dynamic mode. But growth of the singular point number, i.e. the presence of descending and ascending segments, increases a number and a variety of dynamic modes.

Therefore, we offered the deterministic chaotic oscillator, in which NE has a transfer characteristic in the form of parabola composition. In our case, the characteristic has five singular points: two maxima and three minima. The local slope of segments and a position of singular points can be controlled by NE parameters.

Figure 2.3 shows the circuit of the nonlinear element consisting of discriminator and multiplier. This discriminator realizes the inversed *N*-like characteristic. Indeed, having arrived to its input, the U signal divides and passes through the voltage limiter into the discriminator arms. In the first arm the signal passes through the voltage limiter thus obtaining the shape of $h(U)$(Fig. 2.4a). Propagating along the second arm of the discriminator, the U signal undergoes the linear transformation and the signal of $-2U$ is obtained. After that, both signals are added, and the resulting signal $g(U)$ come out of the discriminator (Fig. 2.4b). The signal $g(U)$ from the discriminator passes to the both inputs of multiplier and the signal $g^2(U)$ arrives at its output (Fig. 2.4c). So, the NE transfer characteristic $f(U)$ is formed.

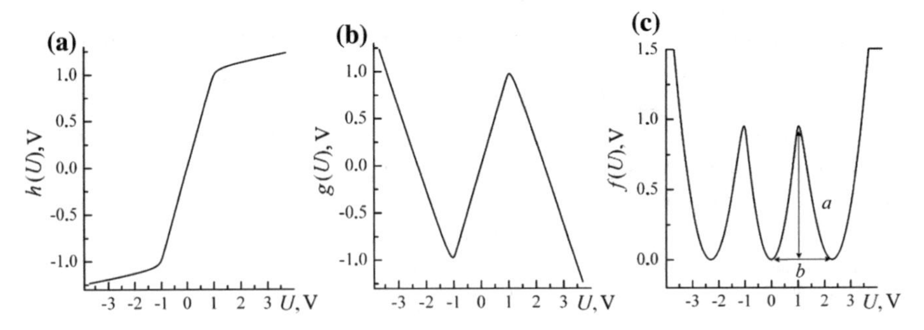

Fig. 2.4 Transfer characteristics: **a** of voltage limiter $h(U)$ in NE; **b** of the discriminator $g(U)$ in NE; **c** of the nonlinear element as a whole $f(U)$: $a = 0.95$ V, $b = 2.3$ V

The transfer characteristic of the limiter $h(U)$ (Fig. 2.4a) can be approximated by the exponential function. Then, using Fig. 2.3, we obtain

$$\begin{aligned} h(U) &= \operatorname{sgn}(U) \cdot M \cdot (1 - e^{\frac{-|U|}{V_t}}), \\ g(U) &= h(U) - 2 \cdot U, \\ f(U) &= g^2(U), \end{aligned} \tag{2.7}$$

where V_t, M are coefficients of approximation. The function (configuration of the chain of transposition 2.7) in the first approximation with accepted accuracy describes the transfer characteristic of the limiter and the nonlinear element as a whole [8].

Further, we shall call the $f(U)$ function as a transfer (volt-volt) characteristic, which has a shape of the parabola composition (Fig. 2.4c). The distance between maxima and minima of the transfer characteristic along the ordinate axis will be called as a swing a. The distance between the nearest minima along the abscissa axis will be called as a period b.

2.1.3 *Analysis of Equilibrium State Stability in the Model of the Deterministic Chaos Oscillator*

It is known that in mathematics there is a concept of the system roughness, or the structural stability. The roughness means that under the small variation of the system parameter the behavior of this system can change only in details, but it will not undergo the quantitative (structural) mode variation. For rough systems, the transition through the bifurcation point means the changing of the one structurally stable mode by another mode. In the very bifurcation point, the system ceases to be rough, i.e. the small change of the parameter leads to the replacement of the one mode by another [13, 21].

In the case of bifurcations of the equilibrium states, the points of bifurcation often correspond to the boundary areas of stability and instability of these states. The necessary condition for the existence of such boundaries is the equality to zero of the real part ($\mathrm{Re}\ \lambda = 0$) of the λ eigenvalues of the linearization matrix of the dynamic system. The last condition determines the points from the parameter space, in which the solution bifurcation occurs [22, 23].

Let us examine the equilibrium states stability analysis in the mathematical model (2.6) of the deterministic chaotic oscillator [7]. We linearize (2.6) by finding the partial derivatives with respect to the variables x, y and fulfilling the replacement $U = D + K[x(t-\tau) - y(t-\tau)]$:

$$\frac{\partial}{\partial x}\left(T_1\frac{\mathrm{d}x}{\mathrm{d}t}\right) = K\frac{\partial f(U)}{\partial U}\frac{\mathrm{d}x(t-\tau)}{\mathrm{d}x(t)} - 1, \quad \frac{\partial}{\partial y}\left(T_1\frac{\mathrm{d}x}{\mathrm{d}t}\right) = K\frac{\partial f(U)}{\partial U}\frac{\mathrm{d}y(t-\tau)}{\mathrm{d}y(t)},$$
$$\frac{\partial}{\partial x}\left(T_2\frac{\mathrm{d}y}{\mathrm{d}t}\right) = 1, \quad \frac{\partial}{\partial y}\left(T_2\frac{\mathrm{d}y}{\mathrm{d}t}\right) = -1.$$

Let us construct the equation for eigenvalues of λ

$$\begin{vmatrix} K\frac{\partial f(U)}{\partial U}\frac{\mathrm{d}x(t-\tau)}{\mathrm{d}x(t)} - 1 - \lambda & K\frac{\partial f(U)}{\partial U}\frac{\mathrm{d}y(t-\tau)}{\mathrm{d}y(t)} \\ 1 & -1-\lambda \end{vmatrix} = 0.$$

Expanding the determinant, we obtain the characteristic equation

$$\left(\frac{\partial f(U)}{\partial U}K\frac{\mathrm{d}x(t-\tau)}{\mathrm{d}x(t)} - 1 - \lambda\right)(-1-\lambda) - \frac{\partial f(U)}{\partial U}\left(K\frac{\mathrm{d}y(t-\tau)}{\mathrm{d}y(t)}\right) = 0.$$

In designations

$$\begin{aligned} S &= 2 - K\frac{\partial f(U)}{\partial U}\frac{\mathrm{d}x(t-\tau)}{\mathrm{d}x(t)}, \\ J &= 1 - K\frac{\partial f(U)}{\partial U}\left(\frac{\mathrm{d}x(t-\tau)}{\mathrm{d}x(t)} + \frac{\mathrm{d}y(t-\tau)}{\mathrm{d}y(t)}\right). \end{aligned} \tag{2.8}$$

The characteristic equation takes a form

$$\lambda^2 + S\cdot\lambda + J = 0. \tag{2.9}$$

It is known that in the small vicinity of solutions $x(t)$ and $y(t)$, the perturbed solution $X(t)$ and $Y(t)$ can be represented as $X(t) = x(t) + \exp(\lambda t/T_1)$ and $Y(t) = y(t) + \exp(\lambda t/T_2)$.

Then, for the *static state (state of equilibrium)*, differentials from $X(t)$ and $Y(t)$ are

$$\mathrm{d}X(t) = 0 + \frac{\lambda}{T_1}\exp\left(\lambda\frac{t}{T_1}\right)\mathrm{d}t, \ \mathrm{d}X(t-\tau) = 0 + \frac{\lambda}{T_1}\exp\left(\lambda\frac{t-\tau}{T_1}\right)\mathrm{d}t,$$

$$\mathrm{d}Y(t) = 0 + \frac{\lambda}{T_2}\exp\left(\lambda\frac{t}{T_2}\right)\mathrm{d}t, \ \mathrm{d}Y(t-\tau) = 0 + \frac{\lambda}{T_2}\exp\left(\lambda\frac{t-\tau}{T_2}\right)\mathrm{d}t,$$

and ratios in (2.8) take the form of

$$\frac{\mathrm{d}X(t-\tau)}{\mathrm{d}X(t)} = \frac{\frac{\lambda}{T_1}\exp\left(\lambda\frac{t-\tau}{T_1}\right)\mathrm{d}t}{\frac{\lambda}{T_1}\exp\left(\lambda\frac{t}{T_1}\right)\mathrm{d}t} = \exp\left(-\lambda\frac{\tau}{T_1}\right),$$

$$\frac{\mathrm{d}Y(t-\tau)}{\mathrm{d}Y(t)} = \frac{\frac{\lambda}{T_2}\exp\left(\lambda\frac{t-\tau}{T_2}\right)\mathrm{d}t}{\frac{\lambda}{T_2}\exp\left(\lambda\frac{t}{T_2}\right)\mathrm{d}t} = \exp\left(-\lambda\frac{\tau}{T_2}\right).$$

We substitute the expressions obtained into (2.8) and, in the final form, write the characteristic equation (2.9) for equilibrium states as

$$\begin{aligned} &\lambda^2 + S\lambda + J = 0, \\ &S = 2 - K\frac{\partial f(U)}{\partial U}\exp\left(-\lambda\frac{\tau}{T_1}\right), \\ &J = 1 - K\frac{\partial f(U)}{\partial U}\left(\exp\left(-\lambda\frac{\tau}{T_1}\right) + \exp\left(-\lambda\frac{\tau}{T_2}\right)\right). \end{aligned} \tag{2.10}$$

Now we find the equilibrium states by equating the right parts of the equation system (2.6) to zero

$$\begin{aligned} &\frac{\partial x(t)}{\partial t} = \frac{f(D + K(x(t-\tau) - y(t-\tau))) - x(t)}{T_1} = 0, \\ &\frac{\partial y(t)}{\partial t} = \frac{x(t) - y(t)}{T_2} = 0. \end{aligned}$$

Hence,

$$f(D + K(x(t-\tau) - y(t-\tau))) - x(t) = 0, \;\; x(t) - y(t) = 0.$$

Due to the time isotropy, i.e. the arbitrariness of t value, from the equation $x(t) - y(t) = 0$ follows the equation $x(t-\tau) - y(t-\tau) = 0$. The same result will be obtained, if to take into consideration that we consider the equilibrium states, i.e. $x(t) = x(t-\tau)$ and $y(t) = y(t-\tau)$. Therefore,

$$x(t) = y(t) = f(D).$$

We may conclude that in the equilibrium state $U = D$. The characteristic equation (2.10) does not contain the values of dynamic variables x, y that would vary independently. But this Eq. (2.10) contains U as their image, strictly equal to the D parameter. Therefore, if *several* equilibrium states take place in the system, they are all either stable or instable simultaneously, which is impossible. It means that the equilibrium state is *unique*. The relation $x = y = f(D)$ can serve as another proof of this conclusion.

2.2 Simulation of Static and Dynamic Modes of the Deterministic Chaos Oscillator

2.2.1 Stability of Equilibrium States

We know that under the loss of stability of the equilibrium states oscillations arise in the systems. Therefore, we can judge about the stability of these oscillations not only by referring to (2.10), but also by analyzing the conditions for the oscillation existence. Let consider them in detail. For this, it is necessary to provide the fulfillment of the conditions for the phase and amplitude balance. In the simplest case of our ring system, the amplitude balance means that the product of gain coefficient and the losses in the feedback loop should be equal to 1. Coincidence of phases at input and output of the linear amplifier is called the phase balance. To excite the sinusoidal oscillations, it is sufficient to fulfill the conditions for the amplitude and phase balance on the same frequency. On the contrary, to excite the multi-frequency oscillations, we must fulfill these conditions for each of the frequencies. The frequency selectivity in the chaotic oscillator can be performed by the given bandwidth of the pass-band filter of the first order (with characteristic times T_1, T_2), the delay time τ, the gain coefficient K and the value of the local slope of the transfer characteristic of the nonlinear element $df(U)/dU$.

Equations for the amplitude-frequency characteristic (AFC) and the phase-frequency characteristic of the pass-band filter, taking into account the delay line (τ), are represented by the following equations

$$\begin{aligned} k(\omega) &= \frac{1}{\sqrt{1+(T_1\omega)^2}} \times \frac{T_2\omega}{\sqrt{1+(T_2\omega)^2}}, \\ \varphi(\omega) &= \arctan\left(\frac{1}{T_2\omega}\right) - \arctan(T_1\omega) - \tau\omega. \end{aligned} \tag{2.11}$$

Based on (2.11), as for the neglecting of the phase delay in the nonlinear element and adjustable amplifier, we can draw a series of plots for AFC and the phase-frequency characteristics (Fig. 2.5). As it is known, with the help of the phase-frequency characteristic, it is rather convenient to graphically determine for which frequencies (and for what number of them) the condition for the phase balance is satisfied. And after that, referring to AFC, one can estimate the required gain coefficient K and the slope value $df(U)/dU$ of the NE transfer characteristic, necessary for the oscillation excitement at these frequencies in the oscillator. Let us give several examples of such an analysis.

Thus, in absence of delay ($\tau = 0$), when the operating point is on the ascending part of the NE transfer characteristic, it is necessary to use the curve for $\varphi(\omega)/\pi$ in Fig. 2.5a. According to its shape, we can easily conclude that the phase balance condition is satisfied on the radian frequency $\omega_0 = 0.45$ rad/s. Moreover, we can see from the AFC shape that this frequency ω_0 is near the AFC maximum, and

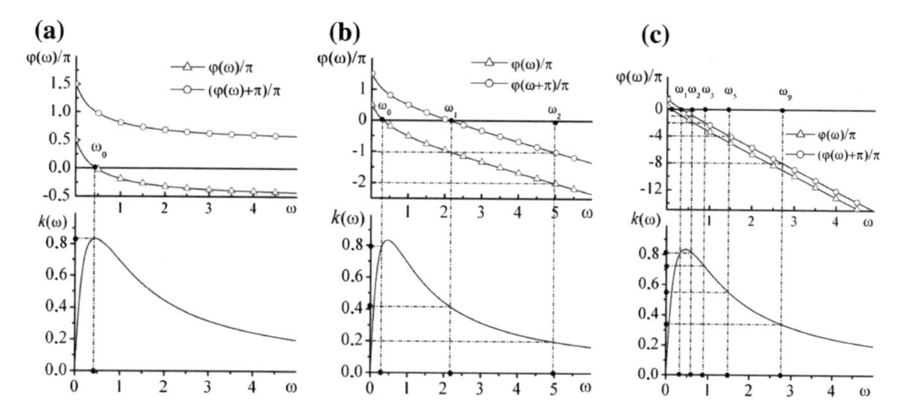

Fig. 2.5 The phase-frequency and amplitude-frequency characteristics of the pass-band filter of the first order with delay $\tau = 0$ (**a**), $\tau = 1$ (**b**) and $\tau = 10$ (**c**), calculated according formulas (2.11) at $T_1 = 1$ (LPF) and x(HPF). *Curves* $\varphi(\omega)/\pi$ correspond to oscillator operation on the ascending part of the NE transfer characteristics, and *curves* $[\varphi(\omega) + \pi]/\pi$—on the descending part

therefore, the amplitude balance condition fulfillment becomes easier. Let it now the operating point be on the descending part of the transfer characteristic (Fig. 2.5a, the curve $[\varphi(\omega) + \pi]/\pi$). Then, evidently, the oscillation of the arbitrary frequency arisen, for instance, at the output of the adjustable amplifier will return to its output at the same phase, which will lead to the decreasing of its amplitude and, as a consequence, to damping. In this case arising of the dynamic modes is impossible.

If $\tau \neq 0$, the phase balance condition will be satisfied for some countable frequency set, independent of the operating point position on the NE transfer characteristic. So, at $\tau = 1$ for the operating point, which is on the ascending part of the transfer characteristic (Fig. 2.5b, the curve $\varphi(\omega)/\pi$), this condition is satisfied not only for ω_0, but for higher frequencies ω_{2n}, where $n \in 0, 1, \ldots, N$. If the operating point is on the descending part (Fig. 2.5b, the curve $[\varphi(\omega) + \pi]/\pi$, the phase balance condition is satisfied for the frequency set with the odd indices ω_{2n+1}, where $n \in 0, 1, \ldots, N$.

The amplitude balance condition requires that the total oscillation gain in the oscillator would be not less than 1. The filter AFC irregularity $k(\omega)$ for LPF and HPF for the case corresponding to Fig. 2.5b, when $k(\omega_0) = 0.79$, $k(\omega_1) = 0.41$, leads to the fact that for oscillation arising at the frequency ω_0 it is enough to have almost two times less gain than it is required for the frequency ω_1. It is clear that we assume here that for chosen operating points on the transfer characteristic its modulus of the local slope $|df(U)/dU|$ is the same. In other words, in this example, oscillations at operating point position at the ascending part of the characteristic can be excited easier.

The growth of the delay time τ increases the number of frequencies, for which the phase and amplitude balance conditions are satisfied for the given gain K and $|df(U)/dU|$ (compare Fig. 2.5b, with Fig. 2.5c).

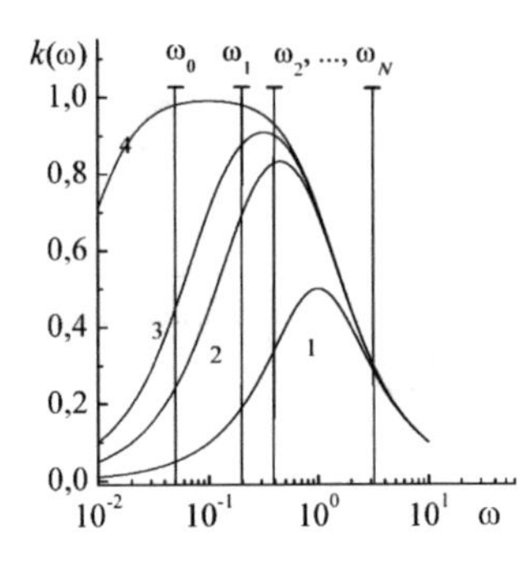

Fig. 2.6 The amplitude-frequency characteristic of the pass-band filter of 1st order and positions of spectral components $\omega_0, \ldots, \omega_N$ at $T_1 = 1$ and different values of T_2 : *1*—$T_2 = 1$; *2*—$T_2 = 5$; *3*—$T_2 = 10$; *4*—$T_2 = 100$

Figure 2.6 shows that, with the growth of HPF time constant value T_2 from 1 to 100, the AFC maximum increases from 0.5 to 1.0, and the pass-band filter bandwidth also increases. Hence, the greater is the T_2 time constant, the smaller values of the gain K and the local slope $|df(U)/dU|$ are enough to satisfy the amplitude balance condition (for frequencies $\omega_0, \ldots, \omega_N$).

Let us return to the analysis of stability of equilibrium states on the basis of (2.10). It is known that, if the Lyapunov characteristic exponent is lower than zero (Re $\lambda < 0$), then such a state of the dynamic system is considered as stable, that is any arisen oscillations in the chaotic oscillator will disappear in time. If Re $\lambda > 0$, than the smallest fluctuation can turn the oscillator into the dynamic mode. Equation (2.10) is transcendental, therefore, the general (analytical) investigation of it can be labor-intensive. In the specific case, of course, we can get a solution analytically. For instance, at $K = 0$, or when $df(U)/dU = 0$, there is the single solution $\mathrm{Re}\lambda = \lambda = -1$.

Figure 2.7 presents the calculation results for $\mathrm{Re}\lambda$ on the plane of parameters $D - K$. The stable states ($\mathrm{Re}\lambda < 0$) are marked by white color. Values of $\mathrm{Re}\lambda$, which are greater or lower than zero, are marked by grey color tins. Maps in Fig. 2.7 are drawn in the order of delay time τ increase in the feedback loop. Figure 2.7a illustrates the case when $\tau = 0$. On it the areas corresponding to the three descending parts of the NE transfer characteristic $(df(U)/dU < 0)$ have a shape of the horizontal white bands, i.e. equilibrium states at such bias D are stable. Their stability loss (dark bands in the figure) is possible only for bias D corresponding to the ascending parts $(df(U)/dU > 0)$ with the proper combination of values K and D. These conclusions agree with the results of phase and amplitude balance condition analysis for the case 2.5, *a*.

The delay τ in the feedback loop τ leads to the cardinal change in the structure of instability regions (as well as of stability regions) of equilibrium states. These dark areas grow with τ increase, and their asymmetry with respect to horizontal lines corresponding to values of $df(U)/dU = 0$ decreases (Fig. 2.7b) and practically disappears (Fig. 2.7c, d).

So, for delay $\tau = 0.05$ (Fig. 2.7b) the instability regions fill the larger part of the plane $D - K$. Nevertheless, areas of both dark and white color remain asymmetric with respect to horizontal lines corresponding to values $df(U)/dU = 0$. Regions of

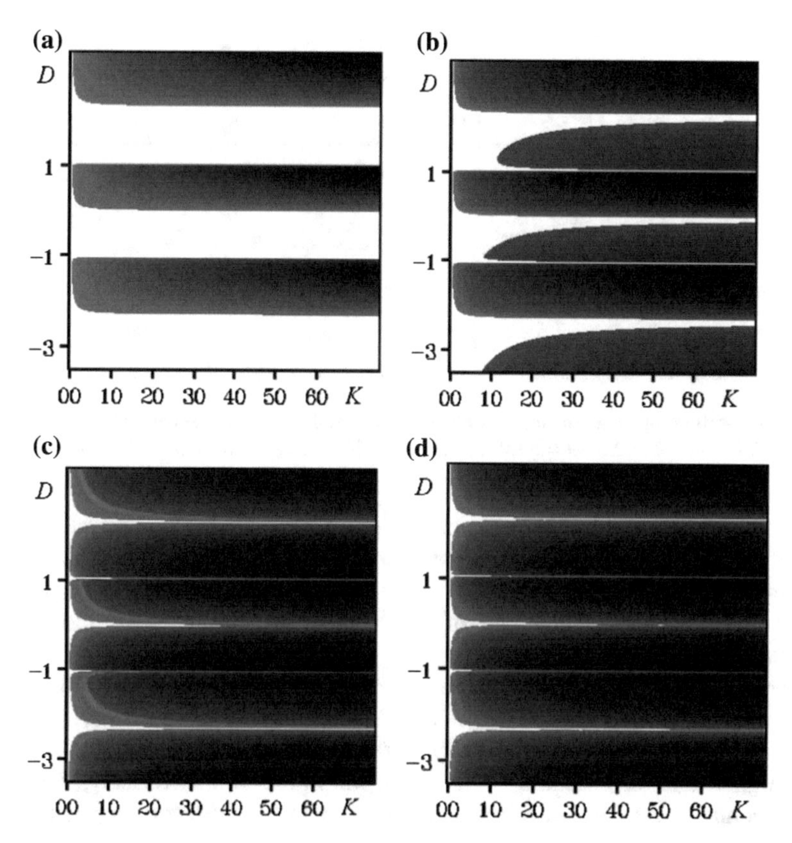

Fig. 2.7 Maps of Reλ distribution on the plane D—K for equilibrium states in the model of chaotic oscillator (2.6) for $T_1 = 1$ (LPF), $T_2 = 7$ (HPF), a—$\tau = 0.05$, b—$\tau = 0.5$, c—$\tau = 30$; d—$\tau = 1000$

stability corresponding to operating points on the descending parts of the NE transfer characteristic have greater area than areas of operation on the ascending part.

This agrees with the results of analysis of the phase and amplitude balance conditions (Fig. 2.5b). Indeed, in the first case the phase balance conditions are satisfied for frequencies $\omega_1, \omega_3, .., \omega_{2n-1}$, but the amplitude balance conditions are achieved only at the rather great values of the gain ($K > 10$). And in the second case, even for small values of K the amplitude balance condition is fulfilled for ω_0, and for this frequency the phase balance condition is already set (Fig. 2.5b). It is clear, that with gain K growth the amplitude balance condition is satisfied for greater number of frequencies, and thus, the multi-frequency oscillations are possible.

These calculations conform with experimental results. As an example, we introduce Fig. 2.8, presenting the signal for the output of the chaotic oscillator, where the delay time is by one order less than the time constant of HPF ($\tau \ll T_2$),

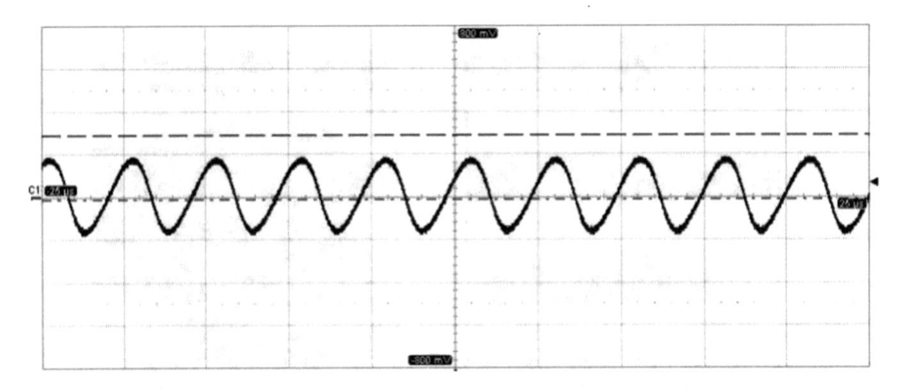

Fig. 2.8 Oscilloscope patterns for oscillations in the oscillator implemented on the base of circuit in Fig. 2.1, when $K = 2.5$, $D = -1.2$ V, $T_1 = 2.7 \cdot 10^{-8}$ s, $T_2 = 1.2 \cdot 10^{-4}$ s, $\tau = 1.5 \cdot 10^{-8}$ s

the bias voltage $D = 1.2V$ corresponds to operating point on the ascending part of the NE transfer characteristic, and the gain $K = 2.5$. The main oscillation occurs at the cyclic frequency $\omega_0 = 2\pi \cdot 196$ kHz = $1.23 \cdot 10^6$ rad/s. When the operating point is on the descending part of the NE transfer characteristic, there are no oscillations in chaotic oscillator.

Figures 2.9 and 2.10 show calculation results for Reλ on the plane of parameters $T_2 - \tau$ for cases when $df(U)/dD > 0$ and $df(U)/dD < 0$, respectively. Inside the maps Re$\lambda(T_2, \tau)$ we can conditionally extract three zones. The zone *1* is on the left and higher than the ascending diagonal of the figure. More precisely, the zone *1* corresponds to the case when the signal delay τ in the feedback loop is lower, or

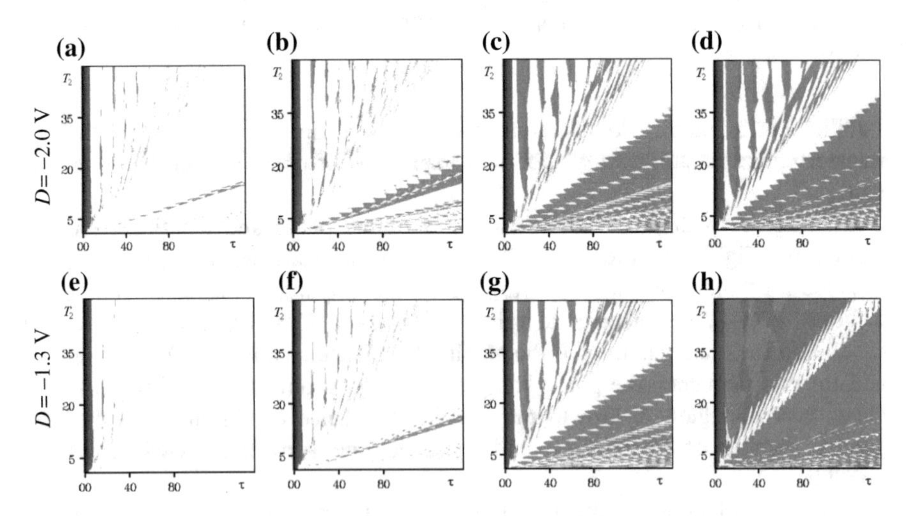

Fig. 2.9 Maps of Reλ distribution on the plane $T_2 - \tau$ for equilibrium states in the chaotic oscillator model (2.6) for $T_1 = 1$; **a** $K = 1.3$; **b** $K = 1.5$; **c** $K = 1.7$; **d** $K = 2.1$; **e** $K = 0.4$; **f** $K = 0.5$; **g** $K = 0.7$; **h** $K = 1.0$. $D = -2.0$ V (**a**–**d**) and $D = -1.3$ V (**e**–**h**)

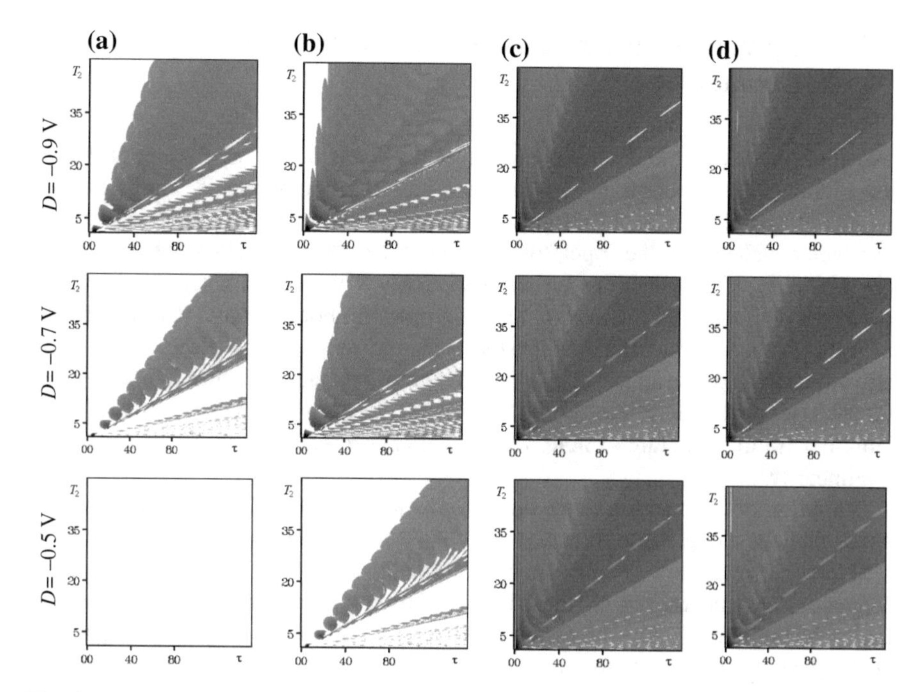

Fig. 2.10 Maps of Reλ distribution on the plane $T_2 - \tau$ for equilibrium states in the chaotic oscillator model (2.6) for T_1 = 1; **a** K = 0.5; **b** K = 0.7; **c** K = 1.9; **d** K = 2.3

equal to the doubled time constant T_2 of HPF ($\tau/T_2 < 2$). Regions of stability and instability of equilibrium states in the zone *1* are oriented vertically: instability regions have look like irregular spots of different sizes and shapes, but they are elongated along the $0T_2$ axis. In this zone, for all values of T_2, the presence of the instability region of the equilibrium states for $\tau < 5$ (Fig. 2.9) is typical, and that agrees with the map structures on the plane $D - K$ (Fig. 2.7a, b). Instability regions increase in sizes with the gain K growth, while stability regions decrease (Fig. 2.9a–d, e–g). When the value of K is great enough, only the instable equilibrium states correspond to zone *1* (Re$\lambda > 0$). See, for example, Fig. 2.9g.

The zone *2* is situated more right-hand and below of the ascending diagonal (Fig. 2.9). It corresponds to the case, at which the signal delay value τ in the feedback loop is more or equal to four value of the time constant T_2 of HPF ($\tau/T_2 > 4$). In contrast to the zone *1*, areas of stability and instability of equilibrium states in the zone 2 are situated along the beams of various slope, which comes from the origin.

The zone *3* belongs to the diagonal in the area of parameters $2 < \tau/T_2 < 4$ (Fig. 2.9). In it the region of stable equilibrium states has a shape of a triangle with the apex near the origin. The stability region becomes narrower with the growth of the gain K.

2.2.2 Operating Modes in the Deterministic Chaos Oscillator

Dynamic modes in the deterministic chaos oscillator breadboard are specified by the oscillator construction and by the properties of its components. They are set in its model by the form of the evolution operator of the dynamic system and by the parameter values in the operator, in our case by the values of the vector (D, K, T_1, T_2, τ). Depending on the variation of its controlling parameters, the oscillator demonstrates different dynamic properties. For example, with the growth of gain K the complexification of the dynamic mode takes place. The simple attractors in the phase space change into the complex ones, and that can finally lead to the rise of the deterministic chaos. In this regard, the gain K of the linear amplifier (in the structure of nonlinear amplifier) becomes a measure of the system nonlinearity.

Variation of the position of the operating point (D, K, T_1, T_2, τ) in the parameter space and set along some direction allows observing the appropriate sequence of bifurcations resulting in the chaotic attractor formation. Typical bifurcation sequences are generalized by the concept of bifurcation mechanisms, i.e. the scenarios of the transition to the chaos (development of chaos) [24]. In practice, it is convenient to represent the chaos development scenarios in the bifurcation diagrams.

For our oscillator model (2.6) with NE having the transfer characteristic, shown in Fig. 2.4c, the phase portraits are presented in Fig. 2.11 for different values of controlling parameters. On this figure, along the abscissa axis the values of the dynamic variable x are marked, and along the ordinate axis—the values of the variable y, while the maps are drawn in the order of gain K increase within the range from 1 to 25. To simplify their understanding, the figures are drawn in grey color gradation technique. The color of the phase portrait area reflects the frequency (duration of stay) of the dynamic system in the appropriate condition: darker areas correspond to the long-term and the light areas—to the short-term system stay in the given area.

Under $K = 1$ we have the stable limit cycle. With K growth up to the value of 1.25, there develops the aggravation of anharmonicity of dynamic variable oscillations, and it leads to the growth in the number of maxima during the period of these oscillations. At the bifurcation diagram (Fig. 2.12) this event *looks like* the first act of period doubling. We can easily distinguish the typical divarication of the previously existed branches; moreover, pairs of branches can arise from the single point.

However, the classical bifurcation of the period doubling takes place during the further growth of K: from 1.25 to 1.29 (Figs. 2.11 and 2.12a). A cascade of such bifurcations in the value range $K \in (1.29; 1.31$ leads to the deterministic chaos mode (Fig. 2.11, $K = 1.31$)) at some critical value of K_{cr}. The bifurcation diagrams (Fig. 2.12a, b) graphically demonstrate the transition from the limit cycle to the chaotic attractor. In the specialized literature, this scenario is called the transition to

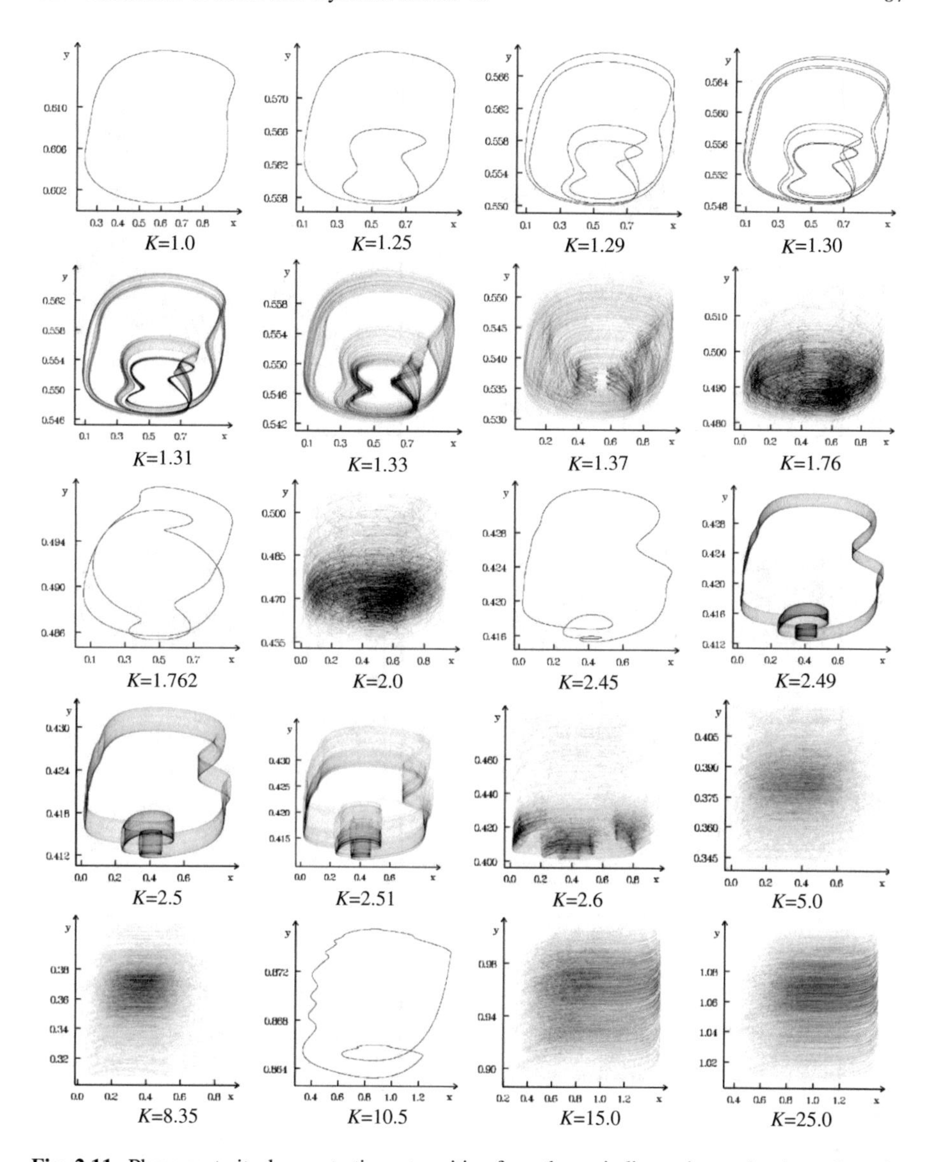

Fig. 2.11 Phase portraits demonstrating a transition from the periodic motion to the chaos through the period doubling bifurcation, as well as through an intermittency in the model (2.6) for gain K changing from 1 to 25, when $D = -0.8$ V, $T_1 = 1$, $T_2 = 100$, $\tau = 5$; $a = 0.95$ V, $b = 2.3$ V

the chaos through a sequence of period doubling bifurcations (by sub-harmonic cascade). We know that the deterministic chaos mode is aperiodic (i.e. the period is infinite) and unstable according to Lyapunov, and is also characterized by the continuous frequency spectrum. The strange attractor in the system phase space corresponds to these oscillations. Its properties are clearly seen in Fig. 2.12 for $K = 1.37$ and $K = 1.76$.

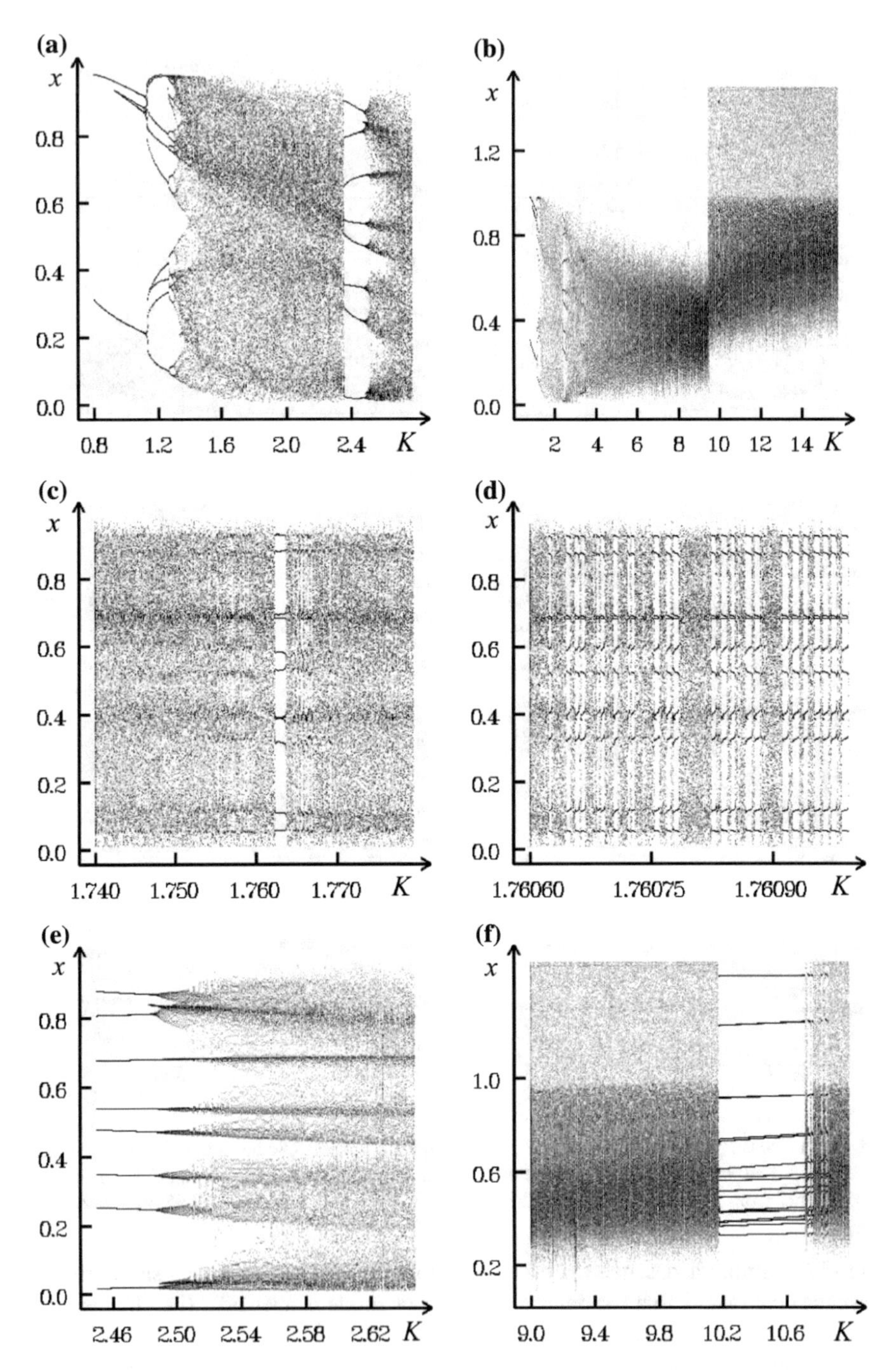

Fig. 2.12 Bifurcation diagrams in the chaotic oscillator model (2.6) in the plane: gain K—extreme values of x voltage at LPF output, when the bias $D = -0.8$ V, $T_1 = 1$, $T_2 = 100$, $\tau = 5$, $a = 0.95$ V, $b = 2.3$ V. Diagrams **a**, **c**–**e** are zoomed fragments of Figure **b**

Phase portraits for $K = 1.762$, $K = 2.45$, $K = 10.5$ in Fig. 2.11, and also the structure of the bifurcation diagrams in Fig. 2.12c, a, e respectively, justify the existence of "periodicity windows". In the literature, they call so the areas of the parameter space corresponding to periodic motions in the system, if, with a change of the certain parameter, these areas are surrounded by the areas of such parameters, under which the chaos arises. Periodicity windows in Fig. 2.12a, e (for $K \in [2.39;\ 2.48]$) can be explained by a presence of slightly sloping part near the extreme of the NE transfer characteristic (the same can be observed in logistic representation). If this characteristic were, for instance, piecewise-linear (say, of the triangle, or of the pyramid type), then periodicity windows would disappear, as well as the sequence of the period doubling bifurcation [24].

"Bandy" shape of the phase portraits in Fig. 2.11 for $K = 2.5$ and 2.51 corresponds to the scenario of transition to the chaos through the collapse of the double-frequency oscillating mode (collapse of the two-dimensional torus) [13, p. 205–217; 25, 26, p. 247–256; 27, p. 209–216]. The similar attractor has been observed in the our model of the modified radio-frequency prototype of the nonlinear ring interferometer [28, 29] (under the title "the chaos of the displaced limit cycle"). In the last model, the intermittency is observed as well. Both aspects of this similarity are probably caused by the fact that there are common circumstances in all these cases:

(1) the presence of LPF and HPF in circuits of both chaotic oscillators, at that, $T_2 \gg T_1$;
(2) the (quasi)-periodic shape of the nonlinear transfer characteristic (the harmonic function and a composition of the type Fig. 2.4c);
(3) the presence of delay.

It is known that the transition from the periodic oscillations may happen by a jump resulting from the only one bifurcation. Such a mechanism is called rigid. It is accompanied by the *intermittency* phenomenon. Intermittency is the alternation mode of the almost-regular oscillations (the laminar phase) with intervals of the chaotic behavior (the turbulent phase), which is observed just after the threshold of the chaos arising. Exactly such a structure is typical for the signal (Fig. 2.13) at LPF output in the model of the chaotic oscillator. The intermittence mode corresponds to the phase portraits in Fig. 2.11 for $K = 1.76$, $K = 2$ and $K = 2.6$. Figure 2.12c demonstrates the rigid mechanism: "periodicity window" ($K \in [1.760;\ 1.765]$) has the sharp boundaries.

Properties of intermittency are distinguishable at the bifurcation diagrams (Fig. 2.12c, d) due to the grey gradation and to the peculiarities of the figure drawing approach. This approach differs by the fact that this diagram contains the extremes only of the dynamic variables that can be met at the time segment several times less than the characteristic duration of intermittence stages. Because of the random distribution (on the time axis t) of these stages, they are represented by the same random manner in the structure of the bifurcation diagram with K growth (Fig. 2.12d). Therefore, as a whole, the diagram looks like as a set of alternating small-scaling areas, whose view is typical for the diagram depicting the chaotic or periodic mode.

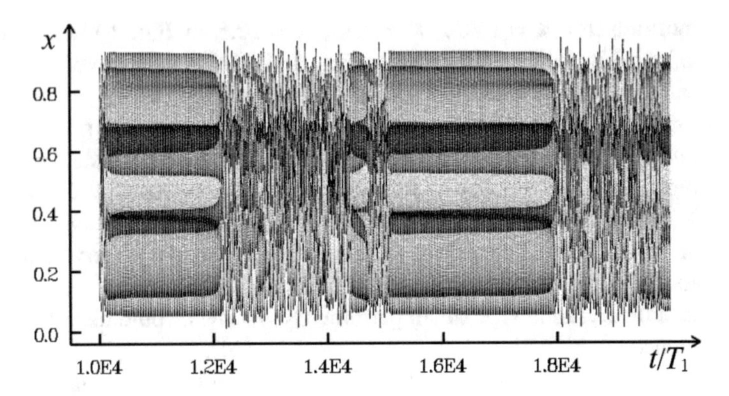

Fig. 2.13 A signal at LPF output in the model (2.6) for $K = 1.76075$, $D = -0.8$ V, $T_1 = 1$, $T_2 = 100$, $\tau = 5$, $a = 0.95$ V, $b = 2.3$ V

We must note in Fig. 2.12b two typical areas separated by the straight line $K \approx 9.5$. For the first area ($K < 9.5$) the dynamic of the x variable is typical, which is limited by the values 0.0 V and 1.0 V. The second area ($K > 9.5$) is characterized by dynamics of x variable within the limits from 0.2 V to 1.5 V.

For a series of bifurcation diagrams in Fig. 2.14, the bias voltage D is a bifurcation parameter. A series is drawn for different values of gain K: 1.3 (a, b); 5.0 (c, d);

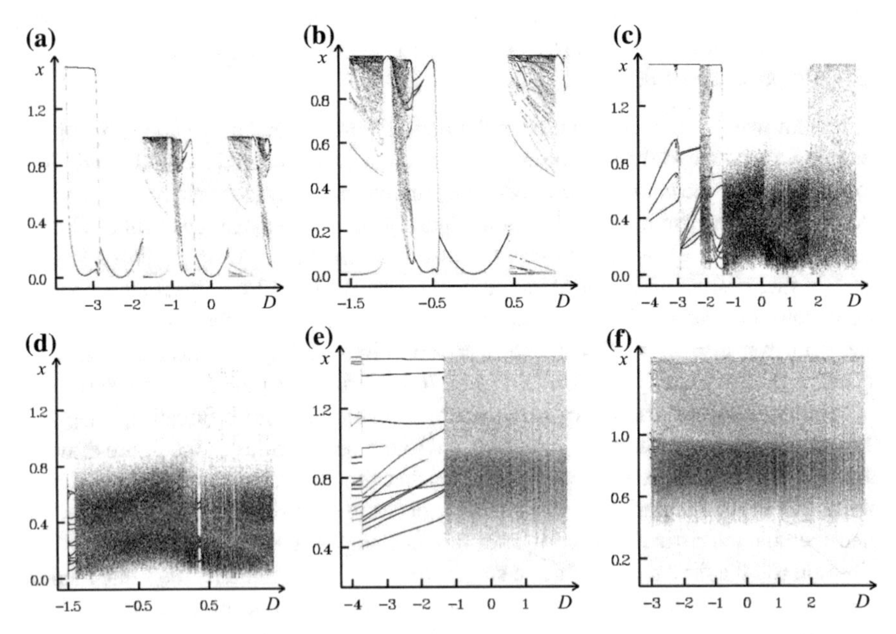

Fig. 2.14 Bifurcation diagrams in the model of the chaotic oscillator (2.6) in the plane xOD, when $T_1 = 1$, $T_2 = 100$, $\tau = 5$, $a = 0.95$ V, $b = 2.3$ V: $K = 1.3$ (**a**, **b**); $K = 5.0$ (**c**, **d**); $K = 25.0$ (**e**, **f**). Figure 2.14**e** versus **f** is drawn at decrease of the value D

25 (*e*, *f*). Figure 2.14b, d are enlargement of the area $D \in [-1.5\ \text{V};\ 1.5\ \text{V}]$ on Fig. 2.14a, c, but these diagrams are drawn for another initial conditions than the diagrams in Fig. 2.14a, c. The diagram in Fig. 2.14g, in contrast to the other diagrams (primarily, to Fig. 2.14e), is drawn for decrease of the value D.

According to the view of Fig. 2.14a, b, for small values of K ($K = 1.3$) the static and periodic modes are dominating. The chaotic mode is realized near maxima of the transfer characteristic ($D = \pm 1.2$ V) only in narrow areas ($D \in [-1.5\ \text{V};\ -0.7\ \text{V}]$ and $D \in [0.7\ \text{V};\ 1.5\ \text{V}]$), dissected by even narrower intervals of regular dynamics near the same maxima. At gain K growth (Fig. 2.14c, d, e), areas of chaotic dynamics widen. On bifurcation diagrams in Fig. 2.14c, d ($K = 5$) areas of periodic and chaotic dynamics are easily distinguished. We would like to note two types of chaotic modes: the first one arises (Fig. 2.14d) in the area $D \in [-1.5\ \text{V};\ 1.5\ \text{V}]$ and is characterized by variations of x within the limits [0.0; 1.0]; the second area takes place for $D \approx -2.0$ V and $D \in [1.5\ \text{V};\ 4.0\ \text{V}]$ (Fig. 2.14c) and differs by larger range of values $x \in [0.0;\ 1.5]$. The same behavior is observed in Fig. 2.12b.

At higher gain ($K = 25$) the periodic mode and the chaotic one of the second type (Fig. 2.12e, f) are observed in the chaotic oscillator mode. Besides it should be noted that, in the last case, x varies within the limits $x \in [0.4;\ 1.5]$, and it realizes in the area $D \in [-1.5\ \text{V};\ 4.0\ \text{V}]$ for quasi-stationary increase of D from −4.0 V to 4.0 V, and in the area $D \in [-3.0\ \text{V};\ 4.0\ \text{V}]$ at decrease of D from 4.0 V to −4.0 V.

The area of stable dynamics on the diagram (Fig. 2.14e) is located within the boundaries $D \in [-4.0\ \text{V};\ -1.5\ \text{V}]$ or $D \in [-4.0\ \text{V};\ -3.0\ \text{V}]$ for quasi-stationary increase, or decrease of the bifurcation parameter D. Mentioned differences depending on the D variation direction can be explained by the hysteresis presence (multi-stability). The essential width of its "loop" corresponds to the noticeable difference between these boundaries on D. Due to hysteresis, we should observe the dependence of the diagram shape upon the initial conditions in the chaotic oscillator (in the moment of diagram construction beginning). It would principally become apparent as the differences in the diagram structure in Fig. 2.14a, c and b, d.

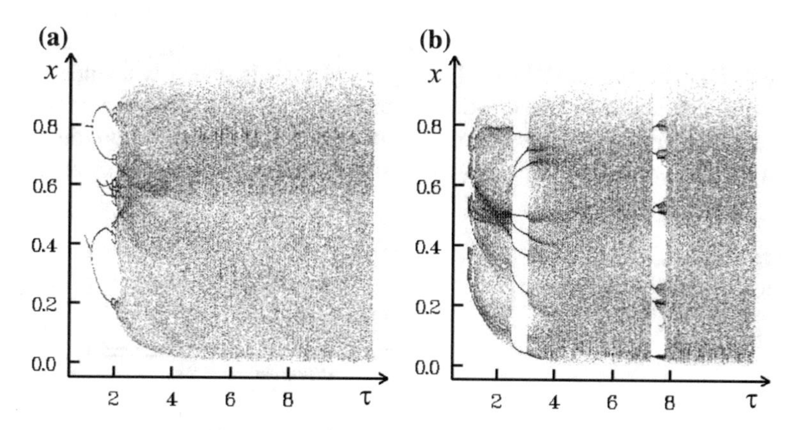

Fig. 2.15 Bifurcation diagrams in the chaotic oscillator model (2.6) in the plane $xO\tau$, when $D = -0.8$ V, $T_1 = 1$, $T_2 = 100$, $a = 0.95$ V, $b = 2.3$ V: $K = 2.0$ (**a**); $K = 3.0$ (**b**)

Figure 2.15 represents bifurcation diagrams for the case when the controlling parameter is the delay time. We have clearly distinguishable areas of periodicity (stability) in this Figure, and there are two such areas in Fig. 2.15b.

Let us be limited the analysis of calculation experiments by this.

2.3 Modes and Scenarios of Transitions to Chaotic Oscillations in the Radio-Frequency Oscillator of Deterministic Chaos

Let us proceed to the discussion of the processes in the chaotic oscillator breadboard realized in accordance with the model (2.6) and the structural circuit in Fig. 2.1. It is developed and designed for experimental research and comparison with mathematical simulation results. But firstly we must briefly describe the breadboard.

2.3.1 The Breadboard of Deterministic Chaos Oscillator

The structural circuit of the breadboard is shown in Fig. 2.16.

Design of the chaotic oscillator breadboard has been carried out with reference to the requirement of the module implementation. Together with the application of the unified units, this allows maximal simplification of the development of the data transmission system breadboard.

The module *1* (Fig. 2.16) ensures matching and commutation of signals in the chaotic oscillator, as well as the power supply of its units. It consists of the four main sub-modules. The sub-module *1.1* (a power source) provides the regulated voltage for all modules *1* and *2*. The sub-module *1.2* (an adjustable amplifier) performs linear transformations of signals (amplification with gain K and bias by the value D). The sub-module *1.3* (analogous electronic gates) is intended for the signal commutation and specification of the initial conditions in the breadboard. The module *2* ensures nonlinear transformation of signals passing from the module *1*, with their subsequent return. It includes sub-modules: *2.1* (the nonlinear element)

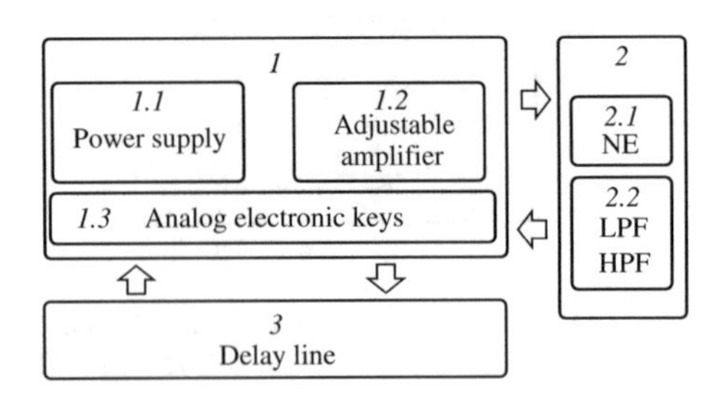

Fig. 2.16 The structural diagram of the laboratory breadboard of the chaotic oscillator

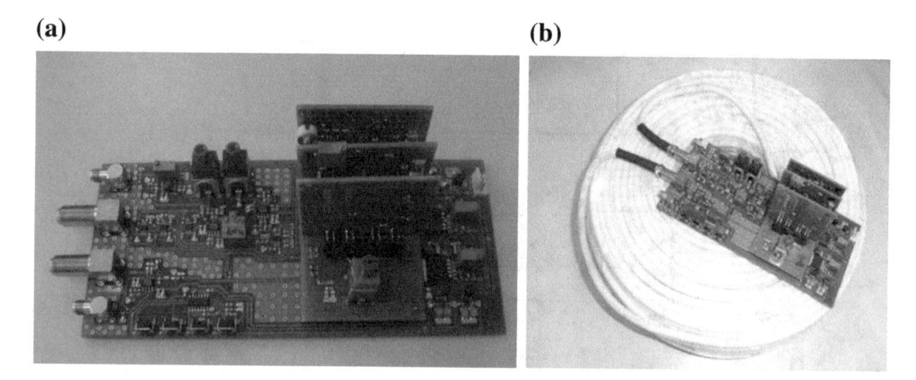

Fig. 2.17 The external view of the chaotic oscillator without the delay

and *2.2* (filters of high and low frequencies of the first order). The module *3* (the delay line) provides signal delay, thus, increases the dimension of the phase space of the dynamic system (the chaotic oscillator).

Figure 2.17 shows the laboratory breadboard of the deterministic chaos oscillator without the delay line (*a*) and with it (*b*). The delay line *9* represents the segment of the coaxial cable (RG-6) with wave impedance 75 Ω. The cable with a length from 0.1 to 100 m in this case performs the lag in the time range from 15 to 300 ns (the minimal unavoidable lag arises as the signal passes via electrical circuits of the oscillator). The time lag increase due to the coaxial cable application is limited mainly by the bulkiness of the design.

If the experiments require longer signal time lags, one can apply the delay line based on the fiber optics (Fig. 2.18). To match the electrical and optical sections of the breadboard, we can use the optical-electronic transceiver. In our case, it allows operation even when the length of the optical cable is up to 3 km.

Application of the waveguide elements as the delay lines is justified by the following circumstances. Due to small damping, AFC regularity and linearity of the

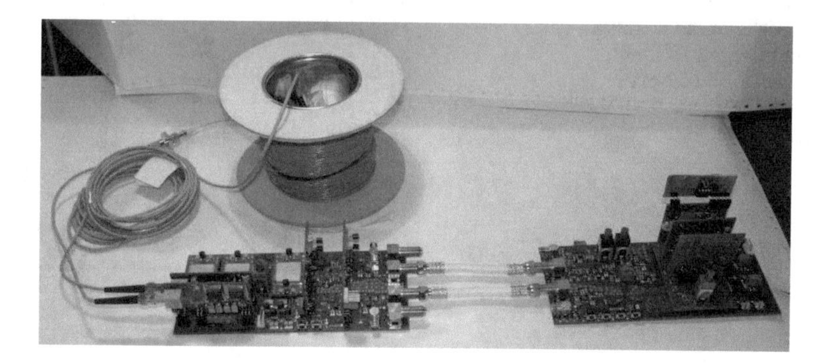

Fig. 2.18 The laboratory breadboard of the deterministic chaos oscillator with the fiber-optical delay line

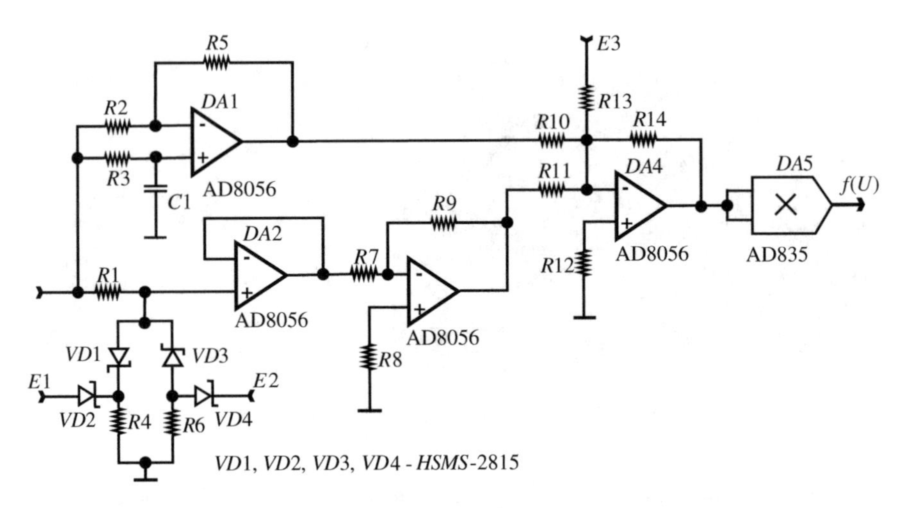

Fig. 2.19 The electrical circuit of the nonlinear element

phase-frequency characteristic of the optical fiber in the band from 0 to 100 MHz, the lag time up to 15 μs is achievable. The further increase of the length of the fiber-optical line is limited by the necessity of multiple connections of the fiber hanks. This attenuates the signal and leads to the low signal/noise ratio at the output of the transceiver.

From all the developed modules and sub-modules of the deterministic chaotic oscillator breadboard (Fig. 2.16) we will now present only the electrical circuit of the nonlinear element (Fig. 2.19). Its transfer characteristic has a form of parabola composition (Fig. 2.20).

Proceeding to the discussion of the processes in the laboratory breadboard of the chaotic oscillator, we shall structure description according scenario types of transition to the chaos.

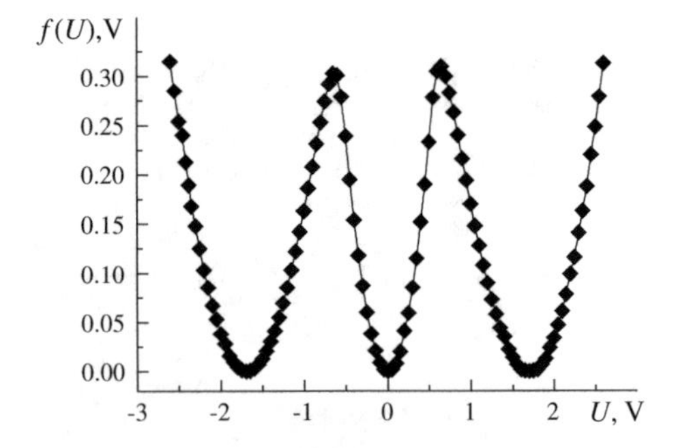

Fig. 2.20 The transfer characteristic of the nonlinear element in Fig. 2.19 for parameters $E_1 = 667$ mV, $E_2 = -667$ mV: $a = 0.31$ V, $b = 1.7$ V

2.3.2 Transition to the Chaos Through the Period Doubling Bifurcation

As we know, the graphic representation of the dynamic modes and their direct analysis during the experiment is realized through the construction of the phase portraits. Figure 2.21 shows a series of phase portraits plotted for the case of the controlling parameter K growth, the voltage value at the delay line input is used for

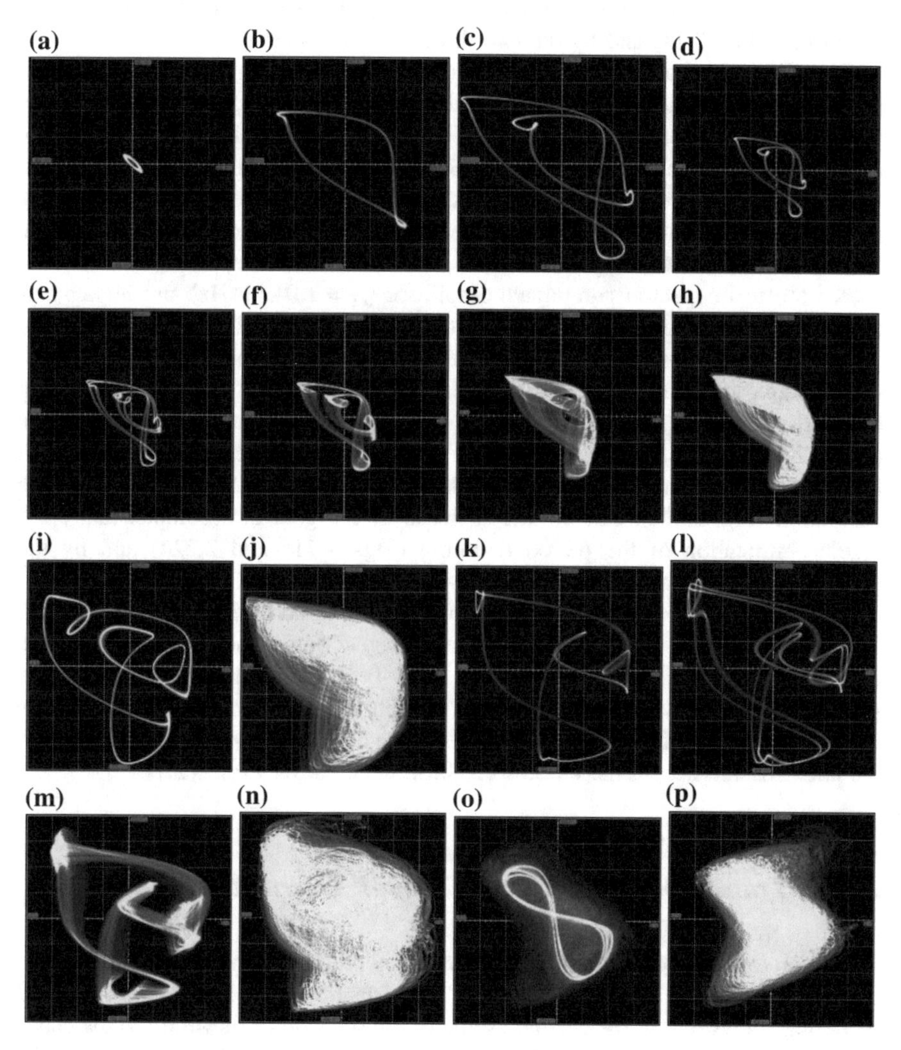

Fig. 2.21 Phase portraits obtained in the experiment with the chaotic oscillator: the signal $z(t)$ is on the abscissa axis, the signal $z(t{-}\tau)$ is on the ordinate axis. The gain K increases from 1 (**a**) to 8 (**p**), the bias voltage D = −0.57 V

ordinate axis, and for the abscissa axis—the output delay line voltage is used. There are modes in the chaotic oscillator, for which the dynamics performs near the frequency, given by the inversed time delay ($f_1 = 1/\tau$). Therefore, representation of signals on the phase portraits in such coordinates simplifies distinguishing between the dynamic modes and the scenarios of transition to them.

Figure 2.21 is plotted in the afterglow mode of the oscilloscope screen: the brighter are the separate places of the screen, the oftener the signal "falls" into these places. In other words, such method of the portrait plotting allows evaluating the frequency of the "visits" the dynamic system pays to a certain area of its phase space. The similar possibility is provided by the phase portraits in the Sect. 2.2 (for instance, in Fig. 2.11) and by the two-dimension histograms.

The ellipse in the center of the oscilloscope screen in Fig. 2.21a, confirms the presence of the periodic oscillation mode in the chaotic oscillator. There is one dominating component in the power spectrum of these oscillations, and there are also components, which are less by 40 dBm (Fig. 2.22a). In this case, the phase and amplitude balance conditions in the oscillator are satisfied at the frequency $f_1 = 1.036$ MHz. With K gain growth, the stable periodic process is realized in the chaotic oscillator with the period $T(K) = 0.959$ μs (the limit cycle C). At the same time, both main spectral component amplitude ($f_1 = 1.043$ MHz) and higher harmonics increase. In the power spectrum there are both even (2nd, 4th, 6th) and odd (3rd, 5th, 7th) harmonics present, whose values are commensurable with the level of the main oscillation (Fig. 2.22b). With further increase of K, the period doubling bifurcation occurs, leading to the emerging of the stable limit cycle $2C$ with the period $2T(K) = 1.918$ μs (Fig. 2.21c, d). This is accompanied by the signal with half-frequency harmonics appearing in the spectrum $f_{1/2} = 0.52$ MHz of the main oscillation (Fig. 2.22c). The further increase of the gain is accompanied by the regular bifurcation of the period doubling (Figs. 2.21e and 2.22d) and by the appearance of the stable limit cycle $4C$ with the period $4T(K) = 3.836$ μs.

The similar set of period doubling bifurcations (caused by K growth) gives birth to sub-harmonics $f_{1/2}, f_{1/4}, f_{1/8} \ldots f_{1/(2n)}$ in the signal power spectrum leading to the dynamic chaos arising (Fig. 2.21f–i and 2.22f–j). The uniform attractor filling the phase portraits, as well as the continuous spectrum of the chaotic signal in some frequency band, confirms the applicability of the deterministic chaotic oscillator for the data transmission in the mode with the chaotic carrier (Table 2.1).

2.3.3 *Transition to the Chaos Through Intermittency*

Transition to the chaos through the intermittency phenomenon was investigated for the first time in the studies by Pomeau and Manneville [24, 30]. Mathematical modeling shows that in the chaotic oscillator the scenario of transition to the chaos is possible by the exactly mentioned way. Thus, it can be distinguished on the bifurcation diagrams (Figs. 2.12c, d and 2.13) for the chaotic oscillators with the operating point on the descending part of the NE transfer characteristic in the

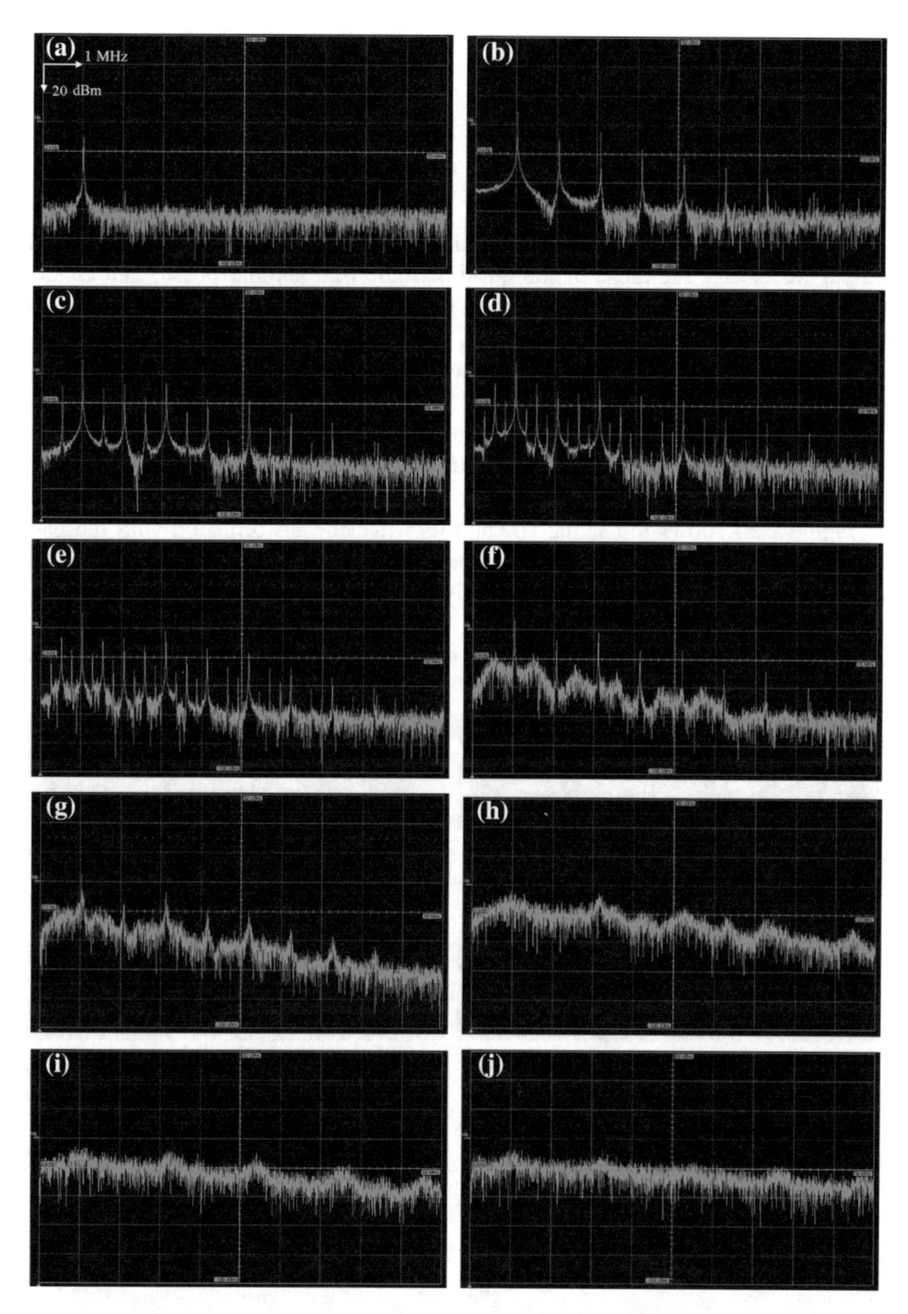

Fig. 2.22 Variation of the power spectrum of the $z(t)$ signal on the HPF output in the experiment with the chaotic oscillator at gain K increase from 1 (**a**) to 8 (**j**), the bias voltage D = −0.57 V. One cell of the axes is 1 MHz and 20 DB. Correspondence of the power spectra in Fig. 2.22 to the phase portraits in Fig. 2.21 see in Table 2.1

Table 2.1 Correspondence of the power spectra in Fig. 2.22 to phase portraits in Fig. 2.21

Figure 2.22	*a*	*b*	*c*	*d*	*e*
Figure 2.21	*a*	*b*	*c*, *d*	*d*	*e*
Figure 2.22	*f*	*g*	*H*	*h*	*j*
Figure 2.21	*i*	*i*	*J*	*n*	*p*

vicinity of the stable area. At the photo picture from the oscilloscope screen at HPF output we clearly see areas of the quasi-periodic (the laminar phase) and chaotic (the turbulent phase) dynamics. In the experiment with the chaotic oscillator, for parameters $D = -570$ mV and $K = 3.62$, the periodic more is realized (Fig. 2.23a). In the vicinity of this point on the parameter plane *DOK* the transition to the chaos occurs through intermittency: (1) with K growth to the critical value $K = 3.64$, the regular oscillations begin to be interrupted by chaotic bursts (Fig. 2.23b); (2) the bursts duration grows (Fig. 2.23c, d) with growth of controlling parameter K until oscillations become completely irregular.

Figure 2.24 shows the signal $z(t)$ and its wavelet-transformation with the base function Morlet wavelet (Gabor wavelet). Time realizations are obtained on the digital oscilloscope Lecroy Wave Surfer × 62 with embedded ADC with discretization period $\Delta t = 1$ ns. However, to create the wavelet-spectrograms the samples have been resampled down: each 50th sample is used, i.e. for the

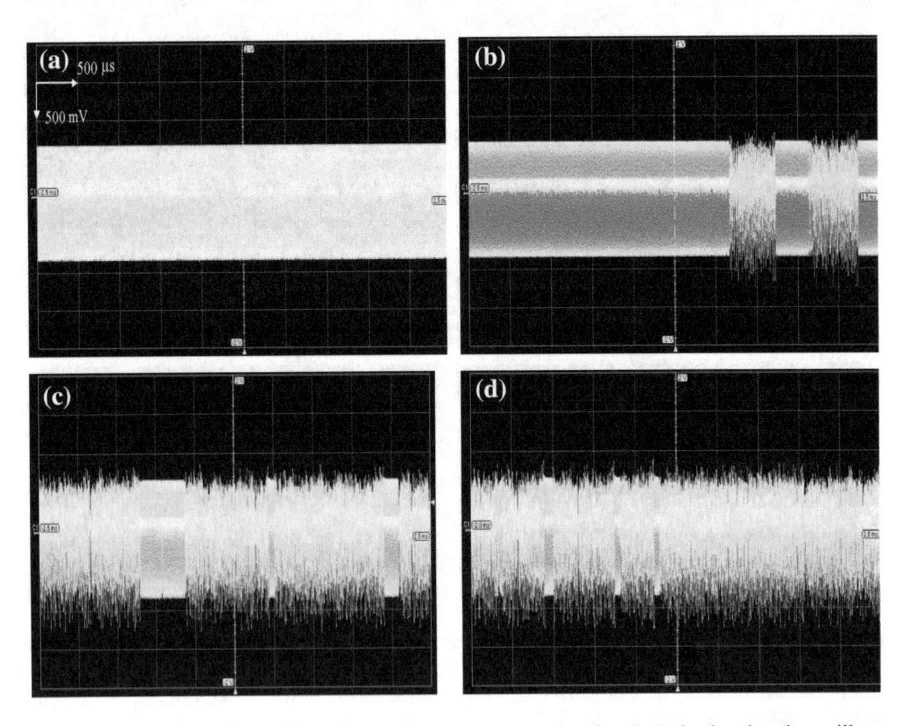

Fig. 2.23 The photo picture form the oscilloscope screen: the signal $z(t)$ in the chaotic oscillator for $D = -0.57$ B: **a** $K = 3.62$; **b** 3.64; **c** 3.7; **d** 3.71

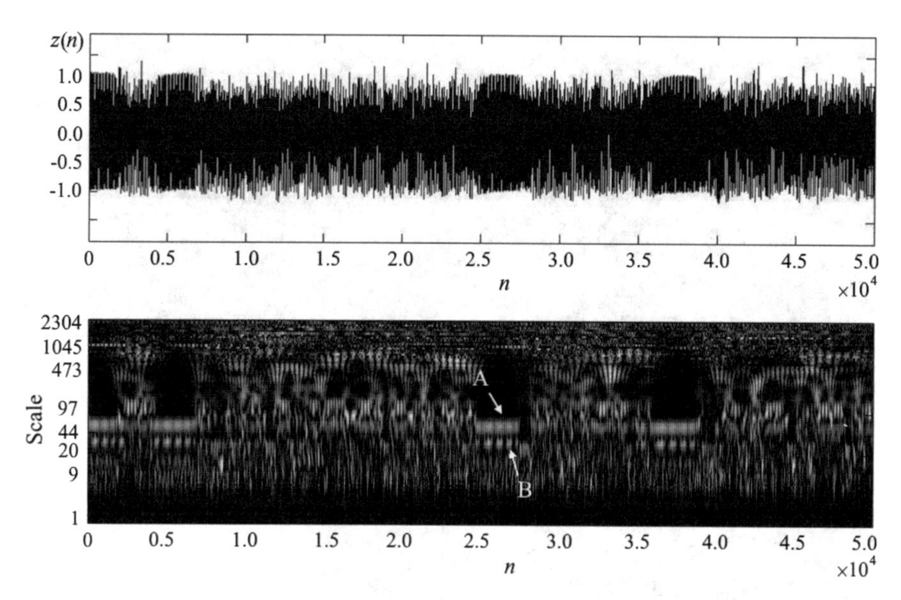

Fig. 2.24 Time realization of $z(t)$ in the chaotic oscillator breadboard and its wavelet-spectrogram in the intermittence mode for $D = -0.572$ V

wavelet-spectrograms $\Delta t = 50$ nsec. The typical structures, corresponding to the laminar (regular) and turbulent phases, are clearly seen on the wavelet plane. The regular phase is mainly characterized by two global maxima on the scales $1/f_1 \approx 60$, $1/f_2 \approx 30$. Two light areas A and B (oriented parallel to time axis) consisting of alternating light and dark bands, correspond to them. The presence of these areas and its almost periodic structure relates, first of all, to the shape of the regular signal (with its Fourier spectrum).

After the dynamics in the chaotic oscillator transfers to the turbulent phase, the shape of the wavelet-spectrum changes: the scale of the splitting observed is $1/f_1$, $1/f_2$, and there is the "burst" of oscillating phenomena with different time scaling. The excitement (irregular-alternative), as well as the oscillation suppression in the chaotic oscillator with the various scaling, occurs during the whole turbulent phase. It is known [30], that in such cases the main oscillation energy fit into this very scaling range, within which the two light areas are located before the "burst". Nevertheless, the energy scale distribution considerably changes during the turbulent phase.

2.3.4 *Transition to the Chaos Through a Collapse of Two-Frequency Oscillating Mode*

For the value of the parameter $D = -570$ mV, $K = 5.56$ the periodic mode (the limit cycle in Fig. 2.25a) is realized in the laboratory breadboard of the chaotic oscillator.

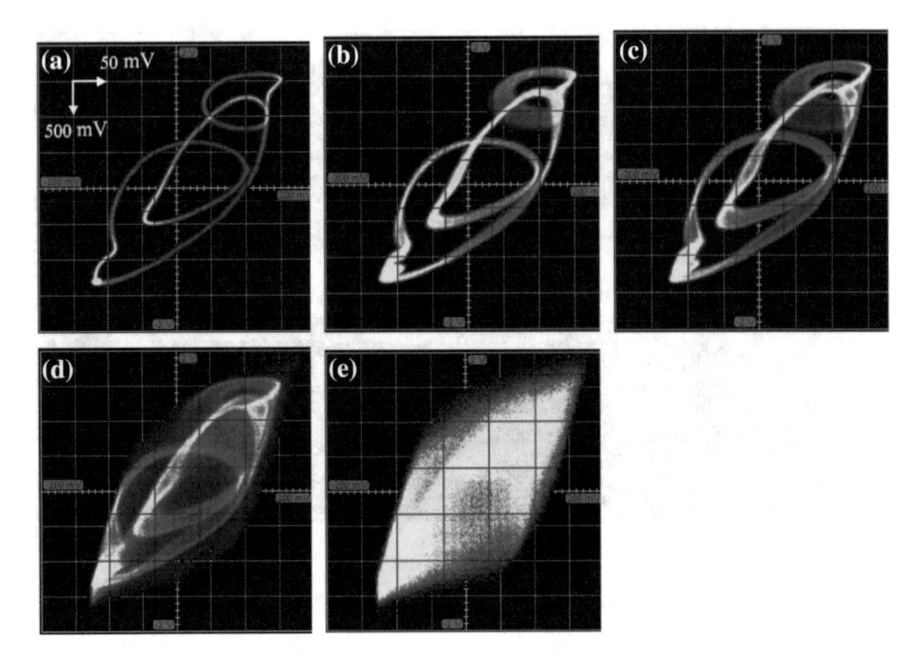

Fig. 2.25 Phase portraits obtained in the experiment: signals $x(t)$ and $y(t)$ are marked on the axes of abscissa and ordinate. Here the bias $D = -570$ mV, the gain K has values: **a** 5.56; **b** 5.63; **c** 5.71; **d** 5.74; **e** 5.83

If you start increasing the value of the K parameter, the attractor shape changes (Fig. 2.25a–e): the line thickness increases and the trajectories on the attractor are dispersed (dithered), which means the chaos arising.

Figure 2.26a–e shows the power spectra corresponding to this transition from the periodic mode to the chaos with K growth. The limit cycle (Fig. 2.25a, $K = 5.56$) is characterized by the presence of harmonics $f_2 = 1.316$ MHz, $f_3 = 1/\tau = 1.974$ MHz etc. ($f_n = n \times 658$ kHz) together with the main frequency $f_1 = 658$ kHz. Growth of the gain to values $K = 5.63$ leads to appearance the components on frequencies $f_s = 110$ kHz and $2\,f_s = 220$ kHz in the power spectrum. In contrast to the Feigenbaum scenario, new harmonics mf_s appear at frequencies aliquant to the half of the fundamental frequency f_1 (Fig. 2.26b). Nonlinear interaction of the last harmonics and the harmonics f_n of the "main series" leads to appearance of oscillations on the frequencies $f_{n1} = (n \times 658) \pm 110$ kHz in the signal spectrum. The presence of the low-frequency components leads to the dithering of the limit cycle trajectories (Fig. 2.25b).

The further growth of K enriches the spectrum of the new low-frequency components. Their nonlinear interaction becomes a reason for the appearance of the new harmonics and sub-harmonics in the high-frequency part of the power spectrum. The further growth of K leads to the increase in the number of the components

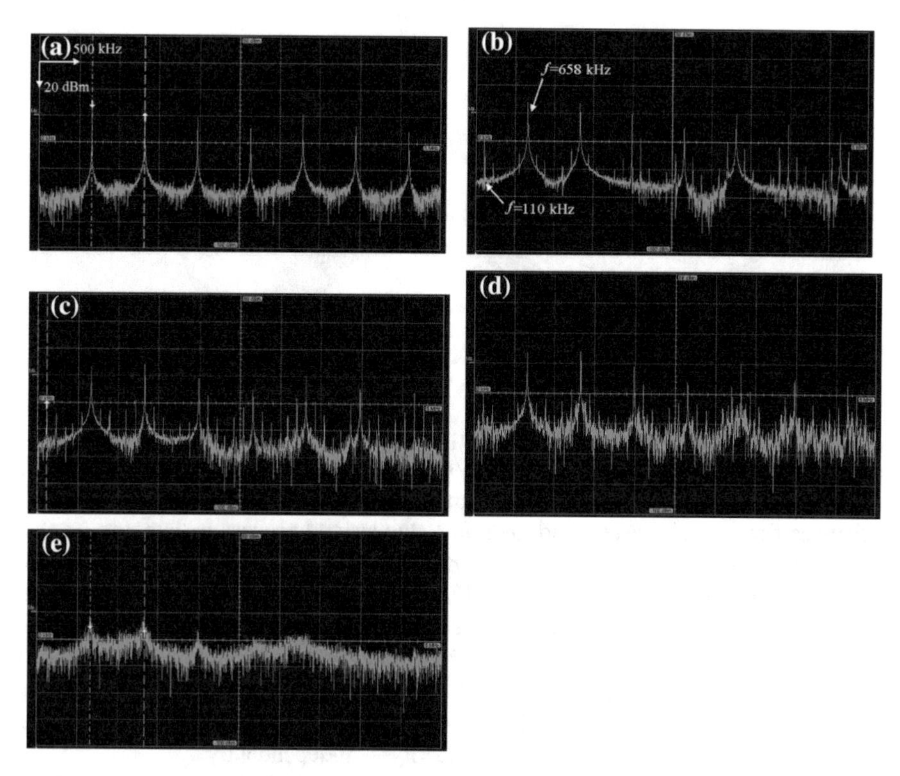

Fig. 2.26 Variation of the Fourier spectrum of the $y(t)$ signal at HPF output in the experimental set up at the gain K growth

in the power spectrum (Fig. 2.26c, d). After exceeding some threshold value K one recognizes the oscillation mode as chaotic (Fig. 2.26e).

2.3.5 *Transition to the Chaos Through a "Semi-Torus" Collapse*

In [2, 31] there is described the transition to the chaotic motion type developing in accordance with the auto-parametric scenario in the autonomous dissipative non-linear dynamic system. Let its state be instable, owing to which the system performs the periodic motion $\varepsilon(t)$ with the frequency ω_1. Let the system nonlinearity be described by the function $f[\varepsilon(t)]$. At $\varepsilon(t)$ evolution, the local slope of nonlinearity $df[\varepsilon(t)]/d\varepsilon(t)$ changes influencing to stability conditions of this or that possible states of the system under consideration. Let properties of the given system be such that with the exceeding by the intensity of the process $\varepsilon(t)$ of some threshold ξ the conditions for self-excitation of the second process $\theta(t)$ with the frequency

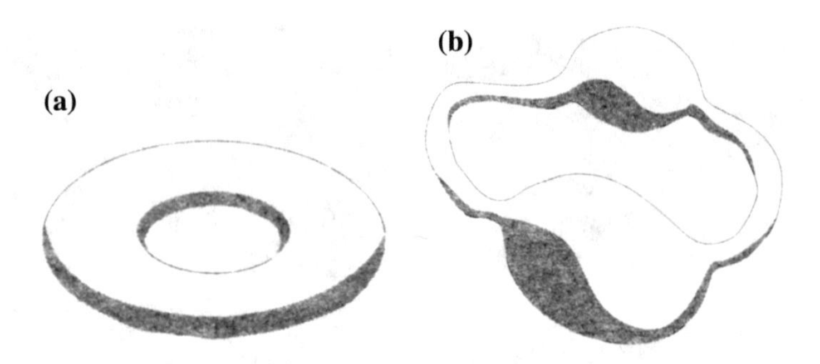

Fig. 2.27 An example of 2D-torus (**a**) and of a set diffeomorphic to 2D-torus (**b**) [21]

$\omega_2 > \omega_1$ are created. The second (fast) process exists only during the part of the period of the first process $\Delta t \in T_\varepsilon = 2\pi/\omega_1$, for which the intensity of the process $\varepsilon(t)$ exceeds ξ. During the remaining part of the period $T_\varepsilon - \Delta t$, the conditions of existence for $\theta(t)$ are absent, and, therefore, the second process is damping. In this case in the phase space there is a *semi-torus*—object representing the limit cycle, in whose jumps the torus fragments arise [2, 31]. Because of the presence of internal fluctuations, the torus fragments appear each time in various initial conditions. Therefore, the transition to chaos is possible either by the collapse of half-torus as a whole, or by the "winding" of the phase trajectory over it. The transition to chaos has been called *auto-parametric*, since in this case the system changes its parameter —the local slope $df[\varepsilon(t)]/d\varepsilon(t)$ of nonlinearity—"independently".

Let us try to consider the semi-torus (as an object in the phase space involved into the complex behavior of the dynamic systems) in a more *general* way. To do so, it is useful, firstly, to take into account that very fact that in nonlinear dynamics and in the context of the attractor typology it is normal to speak about attractors of the torus type and about motion on the torus, etc. Under two-dimensional torus in the three-dimension phase space we here assume not only the ideal circumference rotating around some axis, but not intersecting it (the torus in its traditional understanding—Fig. 2.27a), but also the so-called set, which is diffeomorphic in relation to the 2D-torus (Fig. 2.27b).

Less strictly speaking, we may say that a radius of the rotation circle can vary greatly during rotation, but the period of this variation may be equal to the rotation period. Moreover, it is the arbitrary closed line that may rotate instead the circle! Simultaneously, we can change the distance between this closed curve and the rotation axis. In other words, the attractor of the torus type (or simply—torus) is similar to the highly deformed (inflated or flattened here and there) hollow hoop, but without breaks on the hoop surface.

Under such representation of these attractors, the half-torus looks like as the limiting case of the torus, when some of its parts in their transverse size are compressed to the size of a line. We put it as "looks like", since when the intensity of the process $\varepsilon(t)$ does not exceed the threshold ξ and the conditions of existence for $\theta(t)$ are absent, then it means that earlier existed fast motion $\theta(t)$ is damping and, sooner or later, this damping will go asymptotically and exponentially. In other words, the motion $\theta(t)$ does not disappear completely and, therefore, the part of the torus does not degenerate into a line. Hence, the inaccuracy of the interpretation of the half-torus as a limit cycle, in whose "jumps" there are fragments of the torus, is obvious.

Secondly, the slow process $\varepsilon(t)$ controlling the excitation conditions for the fast motion $\theta(t)$ can be a product of the external system. That is, $\varepsilon(t)$ can be external modulation influence. Then, due to the specific choice of the dynamic system, its parameters and impact characteristics, a certain half-torus can be realized in the phase space [32]. Such an approach requires the presence of the external system, but, owing to this, it is more flexible concerning the possibility of generating attractors with arbitrary number of the sectors of the fast motion and of the types of this fast motion. Besides, it is expedient to draw the bifurcation and stability diagrams of the static states of the controllable system and to interpret them as the modulation characteristics of this modulator system [33].

Let us return to the consideration of the modes of our radio-frequency chaotic oscillator. Now we are to find out, if this scenario of the transition to chaos could be realized. Figure 2.28 presents phase portraits, time realizations and signal power spectra in the deterministic chaotic oscillator, when its operating point lays on the *ascending* part of the NE transfer characteristic (Fig. 2.20) in contrast with the above discussed case. The attractor shape corresponds to the shape of the half-torus, as we can see, for instance, from [2, p. 71].

Probably, we have not observed the semi-torus earlier, due to the choice of the operating point on the *descending* part. This assumption, evidently, could be substantiated by involving the maps of stability of equilibrium states (Figs. 2.7, 2.9 and 2.10) and by referring to the amplitude-phase conditions for oscillation excitement (Fig. 2.5). We must remind that they confirm the essential differences in excitement conditions for the chaotic oscillator—depending on at which part of the NE transfer characteristic the operating point is situated—on ascending, or on descending one.

The transition to the deterministic chaos mode will now be considered in detail. Under the value of parameter $K = 1.76$, the periodic oscillation (the limit cycle) with the fundamental frequency $f = 97.1$ kHz arises in the chaotic oscillator. With K growth, the period of this oscillation in the chaotic oscillator increases (under $K = 2.27$ the fundamental oscillation occurs at the frequency $f = 87.2$ kHz), due to the dependence of the amplitude-phase condition upon K, as well as to the increasing role of the nonlinearity in the oscillator. Under $K = 3.37$ the higher frequency oscillations appear (the torus) at the time periods corresponding to the extremes of the dynamic variable $x(t)$ (that also corresponds to the signal extremes in the input of the nonlinear element). The further growth of gain ($K = 4.43$) leads to the increase of the high-frequency oscillation amplitude.

After that, the phase trajectory intermixing (destruction) takes place and the strange attractor appears in the phase space ($K = 4.48$ and $K = 4.69$). To such a behavior in the oscillator the continuous Fourier spectrum explicably corresponds (Fig. 2.28c).

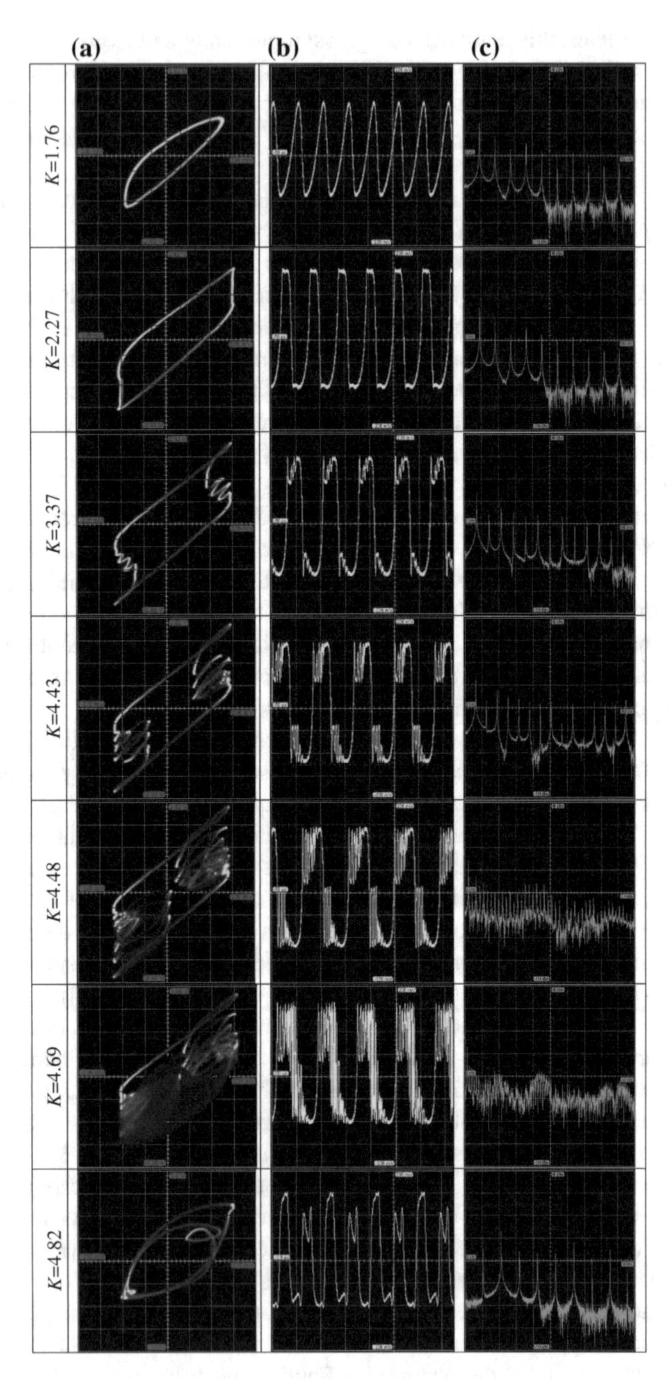

Fig. 2.28 The phase portraits (**a**), the signals (**b**), the power spectra (**c**) for D = 300 mV

If to continue growing the gain, then the destruction of the strange attractor occurs, and it is replaced by the limit cycle with a shape different from the earlier existed one and the fundamental frequency turns to be $f = 1.061$ MHz ($K = 4.82$, Fig. 2.28a). The phase portrait has a shape of the doubled loop typical for the attractor that has gone through the period doubling bifurcation. This allows suspecting that here are the presence of a *hysteresis* and the possibility (with K reduction) of observing "the single" limit cycle with the fundamental frequency $f = 2.122$ MHz. Then, in Fig. 2.28a ($K = 4.82$), we can consider the spectrum component with $f = 1.061$ MHz as a sub-harmonic for the frequency 2.122 MHz. The fact that the further growth of K leads to the period doubling bifurcation cascade and to the appearance of the chaotic mode confirms this assumption.

In addition to the phase portraits and spectra, we are now to handle the bifurcation diagrams. They will also allow revealing both static and dynamic modes studied in the experiment.

2.3.6 *Bifurcation Diagrams*

Let us draw the bifurcation diagrams for the case when the bias voltage D (Fig. 2.29a), or the nonlinearity coefficient K (Fig. 2.30), serves as the bifurcation parameter. The variation of bias voltage D affects the local slope of the transfer characteristic (Fig. 2.29c), accompanied by the mode changes during both the physical experiment (Fig. 2.29a) and the modeling phase (Fig. 2.29b). The values of K have been chosen in order to ensure the dynamic mode operability under the maximal number of D bias values, varying in the range from –1.9 V to 1.9 V with a step of 0.05 V. It has made 76 measurements possible.

Because of the noise in experimental set up, the false extremes of the signals appear, and therefore, their representation on the bifurcation diagram (Fig. 2.29a) lead to larger blur compared to the calculated one. Nevertheless, comparison of

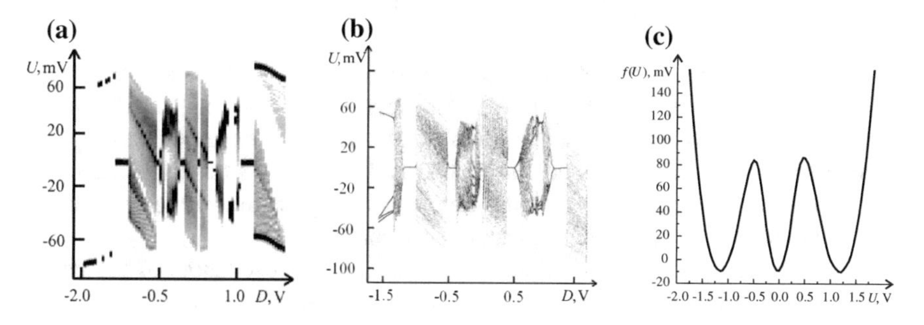

Fig. 2.29 The bifurcation diagram with the D bifurcation parameter for $K = 11.3$: **a** experiment, **b** calculation; **c** experimental NE transfer characteristic ($a = 86$ mV, $b = 1.2$ V), included in the model (2.6) structure. Darkening degree on the diagrams correspond to the falling frequency of the signal local extreme in the given point

Figs. 2.29a, b confirms qualitative (structural and morphological) and even quantitative similarity of calculation and laboratory results. We can clearly distinguish zones with periodic and chaotic dynamics, as well as static modes. Lack of coincidence in the amplitude values is the result of the presence of the amplifier at the output of the chaotic oscillator.

The static mode, i.e. the absence of oscillation in the chaotic oscillator is represented on the bifurcation diagram (Fig. 2.29) by the single line. Static modes correspond to the slightly sloping parts on the transfer characteristic, where the derivative $df(U)/dU$ is small or equal to zero. The static mode by itself is not suitable for the data transmission by the chaotic carrier, but the knowledge of parameter areas in the chaotic oscillator corresponding to the static modes is necessary for the reliable implementation of dynamic modes in the transmitter.

Periodic modes are interpreted on the bifurcation diagram as pairs of lines (or several pairs of lines), and areas with the chaotic mode are represented by the continuum of points having different tints of grey color.

During the plotting of the experimental bifurcation diagrams with bifurcation parameter K (Fig. 2.30) ten values of the D controlling parameter from –0.8 V to 0.8 V have been chosen within the range of D, the ones corresponding to the different values of the local slope of NE transfer characteristic in Fig. 2.29c.

These experiments demonstrate various scenarios of transition to chaos. The sequence of period doubling bifurcations is observed both on the descending and the ascending parts of the transfer characteristic (Fig. 2.30a–g, j). The transition from the limit cycle mode to chaos through the collapse (by jump as a result of the only one bifurcation) in operating point ($D = 0.8$ V) on the descending part (Fig. 2.30j). Moreover, the collapse is observed at $D = -0.5$ V (Fig. 2.30c) and $D = 0.5$ V (Fig. 2.30i). The mentioned operating points correspond to the extremes on the NE transfer characteristic (Fig. 2.29c). Initially, the dynamic system is in the equilibrium state, but with the K growth up to the critical value ($K \approx 13$) it abruptly transmits to the deterministic chaos mode. Near the value $K = 12$ on the bifurcation diagrams (Fig. 2.30b, d, f, j) there are stability areas present. According to the modeling results (Fig. 2.12), with parameter K growth and while approaching the right boundary of the area, the scenario of transition to chaos is realized through the collapse of the two-frequency self-oscillations mode.

Thus, we get the mathematical model of the chaotic oscillator and the appropriate radio-electronic breadboard. We also have investigated the conditions for both theoretical and experimental researches, allowing the dynamic chaos appearance. Besides we have revealed the similarity between the results of computer modeling and laboratory experiments. This allows application of the oscillator in the confidential communication systems using chaos.

Now we are to proceed to the analysis of the chaotic oscillators of the optical range, which are applicable for the information protection also [34].

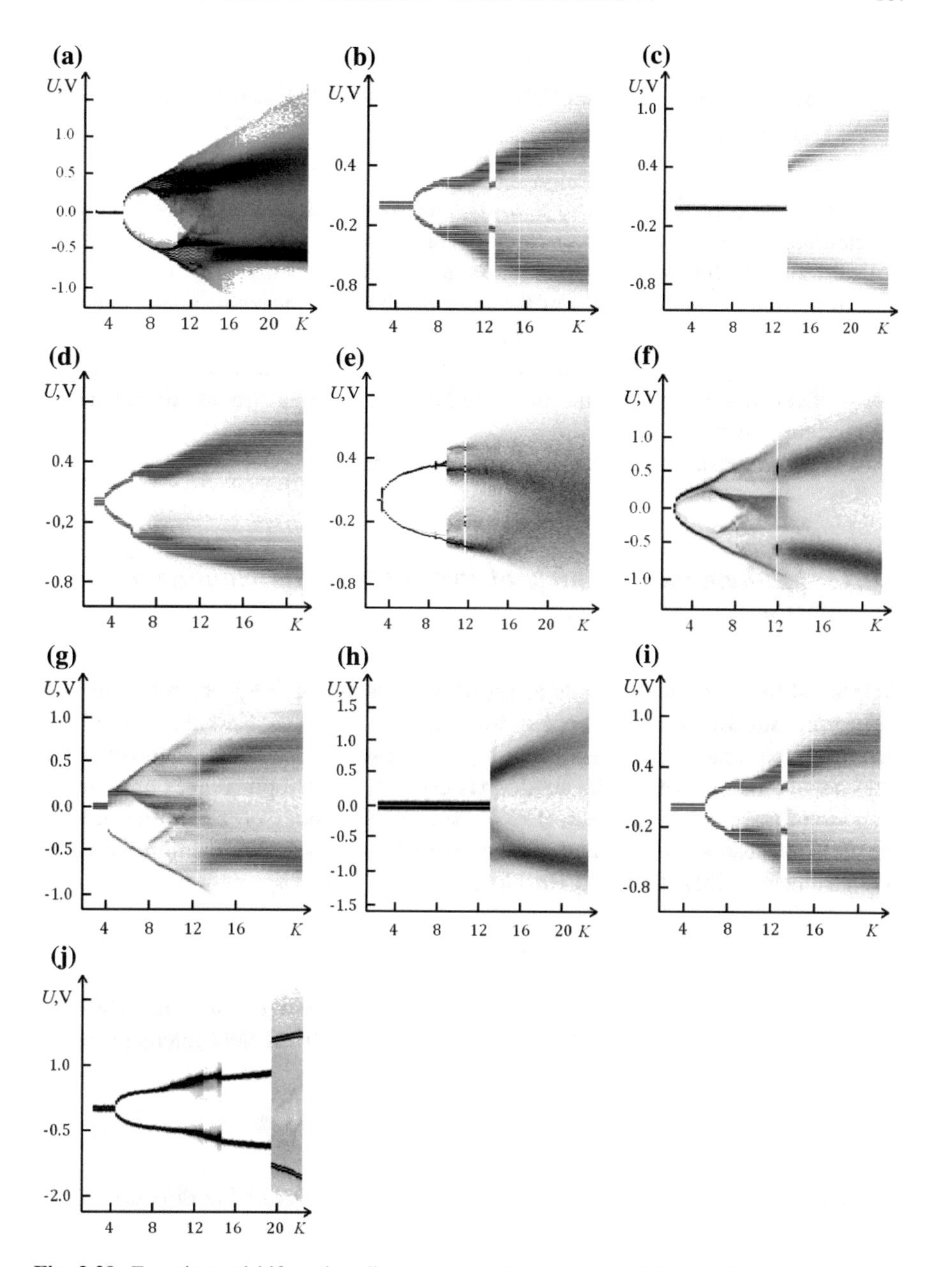

Fig. 2.30 Experimental bifurcation diagrams upon the parameter K for a = 86 mV, b = 1.2 V: **a** $D = -0.8$ V; **b** $D = -0.65$ V; **c** $D = -0.5$ V; **d** $D = -0.4$ V; **e** $D = -0.3$ V; **f** $D = 0.3$ V; **g** $D = 0.4$ V; **h** $D = 0.5$ V; **i** $D = 0.65$ V; **j** $D = 0.8$ V

2.4 The Ring Interferometer with the Kerr Nonlinear Medium and Its Modifications as the Deterministic Chaos Oscillators

We rely on ideas of K. Ikeda and S.A. Akhmanov and M.A. Vorontsov and choose the nonlinear ring interferometer as an oscillator of the deterministic chaos in the optical range. It differs from the Ikeda system by a possibility of two-dimensional transformation of the light field of the laser beam in its transverse plane, and from the Akhmanov and Vorontsov interferometer by a presence of many bypasses of radiation through the feedback loop. Moreover, we will offer two-circuit fiber-optical modification of the interferometer, as well as structurally related set of interferometers.

Naturally, we start with producing the mathematical model [34–38].

2.4.1 *Mathematical Models of Processes in the Nonlinear Ring Interferometer*

Relationship between the fields at input and output of the Kerr medium with diffusion, and the linear element in the interferometer. G. Kerr showed [39–42] that many isotropic substances (gases, liquids, glasses) placed in the static electrical field behave themselves as the single-axis crystals whose optical axes are directed along the force lines of the field. If n_0 is the refraction index of the substance when the field is absent, and $n_{\|}$ and $n_{\perp}$ are refraction indices for extraordinary and ordinary beams, then Kerr and Havelock laws are true:

$$\left(n_{\|} - n_{\perp}\right) = \lambda B \mathbf{E}^2, \ \left(n_{\|} - n_{\perp}\right) = 2(n_{\perp} - n_0). \tag{2.12}$$

Here B is the Kerr constant, λ is the wavelength of the light wave, for which the refraction index is measured, $\mathbf{E}$ is a vector of the electrical field intensity, which causes anisotropy. From (2.12) it is easy to obtain:

$$n_{\perp}(\lambda) = n_0 + n_2 \mathbf{E}^2, \ n_{\|}(\lambda) = n_0 + 2n_2 \mathbf{E}^2, \tag{2.13}$$

where $n_2 = \lambda_0 B$. If the field alternate in time and with invariable direction of the electrical vector intensity act upon the medium, then the uniform molecule orientation, and, hence, refraction indices $n_{\|}$ and $n_{\perp}$, will be established asymptotically. We designate the appropriate alternative refraction indices in the given time moment as $n_{\|}(t)$ and $n_{\perp}(t)$. Let us suppose that the asymptotic transient process is carried out in accordance with the time constant τ_n. Then, for alternative fields the following relations will act:

$$\frac{dn_{\perp}(t)}{dt} = \frac{[n_{\perp} - n_{\perp}(t)]}{\tau_n}; \frac{dn_{||}(t)}{dt} = \frac{[n_{||} - n_{||}(t)]}{\tau_n}.$$

Substituting (2.13) into this equation, we now generalize these two equations into the form:

$$\frac{dn(t)}{dt} = \frac{n_0 + an \cdot n_2 \mathbf{E}^2(t) - n(t)}{\tau_n}. \tag{2.14}$$

Here $an = 1$, if $n = n_{\perp}$; $an = 2$, if $n = n_{||}$. The refraction index of the nonlinear medium (NM) varies due to the diffusion (with coefficient D) of its molecules or the charge carriers. This leads to the appearance of local (small-scale) transverse coupling of the fields. The diffusion effect can be taken into account phenomenologically [43, 44] by introducing the Laplace operator into Eq. (2.14). In approximation of the thin medium (when there is no diffusion along z—axis) and after designation $\mathbf{r} \equiv (x, y)$ we obtain:

$$\frac{\partial n(\mathbf{r},t)}{\partial t} = \frac{n_0(\mathbf{r}) + an \cdot n_2(\mathbf{r})\mathbf{E}^2(\mathbf{r},t) - n(\mathbf{r},t)}{\tau_n(\mathbf{r})} + D\Delta[n(\mathbf{r},t) - n_0(\mathbf{r})]. \tag{2.15}$$

Let us introduce the effective diffusion coefficient that is the coefficient normalized in accordance with relaxation time for NM and the square of some typical size r_0: $D_e(\mathbf{r}) = \frac{\tau_n(\mathbf{r})D}{r_0^2}$. Then, from (2.15) and designating $\mathbf{r} := \mathbf{r}/r_0$, we obtain the equation describing the evolution of the refraction index of the thin (along z axis) spatially irregular (in the plane xOy) Kerr medium under the impact of the spatially non-uniform (in the plane xOy) and time-varying electric field. The specific condition here is that the diffusion of NM molecules should be taken into account:

$$\begin{aligned} \tau_n(\mathbf{r})\frac{\partial n(\mathbf{r},t)}{\partial t} &= an \cdot (\mathbf{r})\mathbf{E}(\mathbf{r},t) - [n(\mathbf{r},t) - n_0(\mathbf{r})] \\ &+ D_e(\mathbf{r})\Delta[n(\mathbf{r},t) - n_0(\mathbf{r})] \end{aligned} \tag{2.16}$$

Let the NM expansion l (Fig. 2.31) along the propagation direction (Oz axis) be so small that we can neglect the beams curvature during their passing through the medium and operate with some l-average value of the refraction index. Then we can also neglect the variation of the Δt_n time field propagation in the medium t_n for the time interval t_n, i.e. $\omega \Delta t_n \ll 2\pi$. In this case, having considered energy transformation effects due to the NM polarization and relaxation, as well as dispersion and diffraction, negligible, we obtain the relationship between electrical field intensity (or its some component) at input and output of the nonlinear medium:

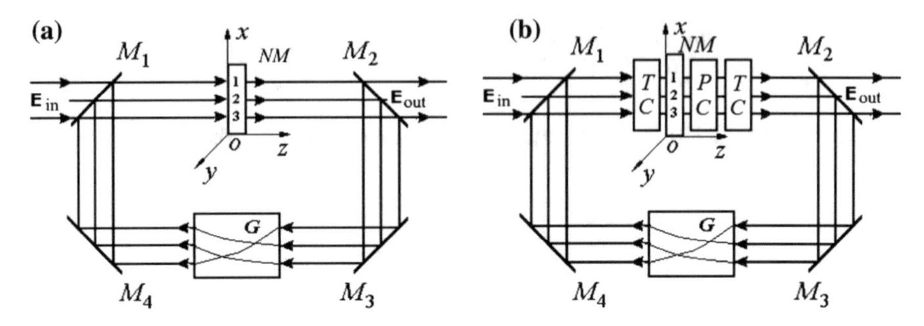

Fig. 2.31 Schematic diagram of two versions of the nonlinear ring interferometer. The beam course is shown for the case of the beam rotation (by the element G) in the plane xOy by 120°. We accept the following designations: NM is the nonlinear medium with the length l; G is a linear element performing large-scale field transformation; mirrors M_1, M_2 have the reflection factor $R_i[\Theta(\mathbf{r},t)]$ in intensity, and M_3, M_4 have the factor equaled to 1, TC are the transparent conductors, on which the DC voltage is applied; PC is the photo conductor (its conductance depends on light intensity, which passes through it)

$$\mathbf{E}_{out,nm}(\mathbf{r},t) = [\alpha(\mathbf{r},t)]^{0,5} C_n(\mathbf{r})\mathbf{E}_{in,nm}[\mathbf{r}, t - t_n(\mathbf{r},t)], \tag{2.17}$$

where $C_n(\mathbf{r})$ are losses in the nonlinear medium, for instance the Bouguer $[\exp(-\alpha_b l/2)]$; $t_n(\mathbf{r},t) = n(\mathbf{r},t)l/c$ is the total time of propagation of the light wave (or its component) that, by the time moment t, arrives to the point $\mathbf{r}$ of the NM output plane from the point $\mathbf{r}$ of the NM input plane; c is the light speed; $n(\mathbf{r},t)$ is the refraction index for this wave (component); $\alpha(\mathbf{r},t) = 1/\{1 + dt_n(\mathbf{r},t)/dt\}$ is a coefficient showing how many times (at first approximation) the volumetric density of the wave energy will increase as a result of the variation of the NM optical length [35]. From inequality $\omega\Delta t_n \ll 2\pi$ and with the assumption that $\Delta t_n \approx t_n(\mathbf{r},t)\frac{dt(\mathbf{r},t)}{dt}$, we obtain another formulation of the above-mentioned approximation: $\frac{dt_n(\mathbf{r},t)}{dt} \ll \frac{\lambda}{n(\mathbf{r},t)l} < 1$. It means that $\alpha(\mathbf{r},t) \approx 1$ and in (2.16) the variation speed of the refraction index is limited: $\frac{d[n(\mathbf{r},t)l]}{cdt} \ll \frac{2\pi c}{\omega n(\mathbf{r},t)l}$ or $\frac{dn(\mathbf{r},t)}{dt} \ll \frac{2\pi c^2}{\omega n(\mathbf{r},t)l^2}$.

The impact of the linear optical element (G in Fig. 2.31) upon the large-scale field transformation is that the beam having arrived to the point $\mathbf{r}' \equiv (x', y')$ of the transverse section of the laser beam then transfers to the point $\mathbf{r} \equiv (x, y)$. As a result, the beam is capable of undergoing a series of consecutive transformations:

(1) M_x (or M_y) is its mirror representation with respect to the straight line parallel to Ox axis (or Oy) passing through the beam center (optical axis of the interferometer);
(2) it is rotated by the Δ angle;
(3) stretching of its transverse section by σ_x times along Ox axis (or by σ_y times along the Oy axis);

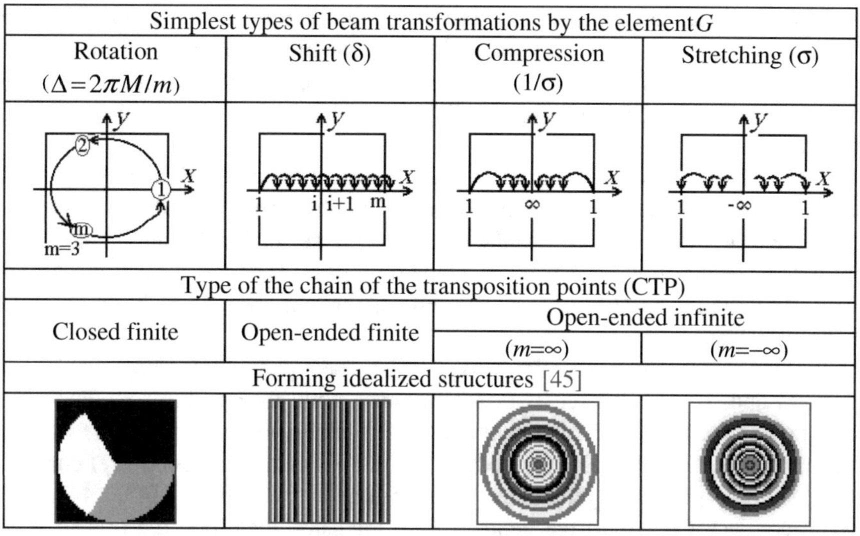

<table>
<tr><td colspan="4">Simplest types of beam transformations by the element G</td></tr>
<tr><td>Rotation ($\Delta=2\pi M/m$)</td><td>Shift (δ)</td><td>Compression (1/σ)</td><td>Stretching (σ)</td></tr>
<tr><td></td><td></td><td></td><td></td></tr>
<tr><td colspan="4">Type of the chain of the transposition points (CTP)</td></tr>
<tr><td rowspan="2">Closed finite</td><td rowspan="2">Open-ended finite</td><td colspan="2">Open-ended infinite</td></tr>
<tr><td>($m=\infty$)</td><td>($m=-\infty$)</td></tr>
<tr><td colspan="4">Forming idealized structures [45]</td></tr>
<tr><td></td><td></td><td></td><td></td></tr>
</table>

Fig. 2.32 The relation between simplest types of transformations of the laser beam in the NRI feedback loop (which gives the configuration of the chain of transposition points) and the view of formed optical structure $U(\mathbf{r}, t)$ in the transverse beam section

(4) it is shifted by the value δ_x along the Ox axis, or by δ_y along the Oy axis (Fig. 2.32). The beam shift $|\mathbf{r} - \mathbf{r}'|$ can then be congruent to the size of the light beam.

We assume that the vector **E** direction does not change here, as well as the input beam center coincides with the origin of the plane xOy. Then we can calculate the result of the field transformation by the element of the large-scale transfiguration in the case when light diffraction and dispersion can be set as negligible by the application of the equations system:

$$\begin{aligned} \mathbf{E}_{out}(\mathbf{r}, t) &= \kappa[\mathbf{r}', t, \Theta(\mathbf{r}', t)] \frac{\mathbf{E}_{in}\{\mathbf{r}', t - t_G[\Theta(\mathbf{r}', t)]\}}{(\sigma_x \sigma_y)^{0.5}}, \\ x' &= M_y \left[(x - \delta_x) \cos \frac{(\Delta)}{\sigma_x} + (y - \delta_y) \sin \frac{(\Delta)}{\sigma_y} \right], \\ y' &= M_x [(y - \delta_y) \cos \frac{(\Delta)}{\sigma_y} - (x - \delta_x) \sin \frac{(\Delta)}{\sigma_x}], \end{aligned} \tag{2.18}$$

where $M_x = 1$ when mirror image is absent with respect to the straight line parallel to the Ox axis; $M_x = -1$ when mirror image is present; similarly, $M_y = 1$ when mirror image is absent with respect to the straight line parallel to the Oy axis, and $M_y = -1$ when mirror image is present; $\Theta(\mathbf{r}', t)$ is the angle between the Ox axis and the vector $\mathbf{E}(\mathbf{r}', t)$; $t_G[\Theta(\mathbf{r}', t)]$ is a time for field propagation in the element G; $\kappa[\mathbf{r}', t, \Theta(\mathbf{r}', t)]$ is the loss coefficient in amplitude [36].

The model of processes in the nonlinear ring interferometer. Equations (2.16), (2.17), (2.18) allow obtaining the mathematical description of the processes in the nonlinear ring interferometers (NRI), which are realized according to the various circuits (for example, Fig. 2.31a, b). Let us discuss in detail their differences.

In the first of them (Fig. 2.31a) the high-intensive optical field in the NM causes the anisotropy of the Kerr medium. Due to this, the refraction index changes. As (2.16) does not take into account the feature of the vector **E** orientation change, we must assume that this light wave is plane-polarized. Then the refraction index is defined by the Eq. (2.16) for $an = 2$, $n = n_{||}$, as the NM optical axis is codirectional with the vector **E**.

In the second interferometer (Fig. 2.31b), the optical field intensity is low enough, and we can neglect its direct influence upon the nonlinear medium. The light arriving in the photo-conductor changes its conductivity, and then the value of electrical (not optical) field (applied to NM and causing the anisotropy) changes. The similar physical principles were used in researches [43, 44]. The photo-conductor reacts to the incident light intensity, therefore, the optical wave polarization can be arbitrary. The refraction index obeys to the Eq. (2.16) for $an = 1$, $n = n_{-}$ (the optical axis of NM is co-directional with the direction of light propagation). Here under $\mathbf{E}(\mathbf{r}, t)$ we should understand the alternate component of the intensity of the non-optical field that is applied to NM, and under D_e—the equivalent diffusion coefficient taking into consideration the diffusion both of the NM molecules and of the charge carriers of the photo-conductor. For simplicity, we assume that the equality $\Delta\,\mathbf{E}^2_{nonoptic} = \mathbf{E}^2_{optic}$ is true.

Let us reveal some relationships referring to the linear-polarized quasi-monochrome component of the field propagating in the interferometer. In this case, we presuppose that there is neither rotation of the polarization plane, no changing of its view. We use index "p" meaning the presence of the appropriate dependence of the certain value upon the orientation of the vector of the electric field intensity. We also use index "w" meaning frequency dependence.

Let the "delay time" $t_{0p}(\mathbf{r}', t)$ be the name for the time of the propagation of the field arriving from the point $\mathbf{r}'$ of the NM output plane (through the feedback loop) to the point $\mathbf{r}$ of the NM input plane by the time moment t. The time $t_{0p}(\mathbf{r}', t)$ also includes time t_{Gp}. Then, evidently,

$$
\begin{aligned}
\mathbf{E}_{in,nm,pw}(\mathbf{r}, t) &= \mathbf{E}_{fb,pw}(\mathbf{r}, t) + \left[1 - R_{1p}\right]^{0.5}\mathbf{E}_{in,pw}(\mathbf{r}, t),\\
\mathbf{E}_{fb,pw}(\mathbf{r}, t) &= R_{1p}^{0.5}\mathbf{E}_{out,G,pw}(\mathbf{r}, t) = R_{1p}^{0.5}\kappa_p(\mathbf{r}', t)E_{in,G,pw}\left(r', t - t_{Gp}\right)/\sigma,\\
\mathbf{E}_{in,G,pw}\left(\mathbf{r}', t - t_{Gp}\right) &= \left\{\frac{R_{2p}}{1 + \frac{dt_{0p}(r', t)}{dt}}\right\}^{0.5} E_{out,nm,pw}\left[r', t - t_{0p}(r', t)\right],\\
\mathbf{E}_{out,nm,pw}\left[\mathbf{r}', t - t_{0p}(\mathbf{r}', t)\right] &= \left\{\alpha_w\left[r', t - t_{op}(r', t)\right]\right\}^{0.5}\\
&\quad\times C_n(\mathbf{r}')\mathbf{E}_{in,nm,pw}\left\{\mathbf{r}', t - t_{nw}\left[\mathbf{r}', t - t_{0p}(\mathbf{r}', t)\right] - t_{0p}(\mathbf{r}', t)\right\}.
\end{aligned}
$$

Here $\mathbf{E}_{fb,pw}$ is the field having arrived to the NM input from the feedback loop (FBL); $\mathbf{E}_{in,pw}$ is the field at the interferometer input; $t_{nm}(\mathbf{r},t) = l \cdot n_w(\mathbf{r},t)/c$ is the light wave propagation time with the wavelength λ_w in the medium; $n_w(\mathbf{r},t) = n_0 + \frac{[n(\mathbf{r},t)-n_o]\lambda_w}{\lambda_0}$; $\kappa(\mathbf{r}',t)$ are the losses in FBL of NRI, i.e. the amplitude transfer coefficient of FBL in contract to the losses $\kappa[\mathbf{r}',t,\Theta(\mathbf{r}',t)]$ in the element G in (2.18). Hence,

$$\begin{aligned}\mathbf{E}_{in,nm.pw}(\mathbf{r},t) &= \left(1-R_{1p}\right)^{0.5}\mathbf{E}_{in,pw}(\mathbf{r},t)\\ &\quad + C_n(\mathbf{r}')\left(R_{1p}R_{2p}\right)^{0.5}\frac{\kappa_p(\mathbf{r}',t)}{\sigma}\eta_{pw}(\mathbf{r}',t)\\ &\quad \times \mathbf{E}_{in,nm,pw}\left[r',t-t_{nw}\left(r',t-t_{0p}(r',t)\right)-t_{0p}(r',t)\right],\end{aligned}$$

where $n_{pw}(\mathbf{r}',t) = \frac{1}{\left[1+\frac{d\left[t_{0p}(\mathbf{r}',t)+t_{nw}\left\{\mathbf{r}',t-t_{0p}(\mathbf{r}',t)\right\}\right]}{dt}\right]^{0.5}}$ (implicitly η^2_{pw} is similar to α_w).

Similarly to [35, 43, 44], we introduce the loss characteristic $\gamma_p(r',t) \equiv 2\kappa_p(r',t)C_n(r')\left(R_{2p}R_{1p}\right)^{0.5}$. Obviously, $\gamma \in [0;2]$, but for the real systems, without limitation of generality, we can assume that $\gamma \in (0;2)$. Keeping in mind that the field $\mathbf{E}$ can be represented as the sum of the two components $\mathbf{E}_x, \mathbf{E}_y$, each of which in turn is the sum of the set of the quasi-monochrome components $\mathbf{E}_{xw}, \mathbf{E}_{yw}$ with the frequency w, we obtain the model of the processes in the interferometer:

$$\begin{aligned}\tau_n(\mathbf{r})\frac{\partial n(\mathbf{r},t)}{\partial t} &= an\cdot n_2(\mathbf{r})\left[\mathbf{E}^2_{in,nm,x}(\mathbf{r},t)+\mathbf{E}^2_{in,nm,y}(\mathbf{r},t)\right]\\ &\quad -[n(\mathbf{r},t)-n_0(\mathbf{r})]+D_e(\mathbf{r})\Delta[n(\mathbf{r},t)-n_0(\mathbf{r})],\\ \mathbf{E}_{in,nm.pw}(\mathbf{r},t) &= \left[1-R_{1p}\right]^{0.5}\mathbf{E}_{in,pw}(\mathbf{r},t)+\frac{0.5\gamma_p(\mathbf{r}',t)}{\sigma}\eta_{pw}(\mathbf{r}',t)\\ &\quad \times \mathbf{E}_{in,nm,pw}[\mathbf{r}',t-t_{nw}\left[\mathbf{r}',t-t_{0p}(\mathbf{r}',t)\right]-t_{op}(\mathbf{r}',t)].\end{aligned} \tag{2.19}$$

Here $\mathbf{E}_{in,nm,x}(\mathbf{r},t) = 0$, p = "y" or $\mathbf{E}_{in,nm,y}(\mathbf{r},t) = 0$, p = "x", $an = 2$ for the optical scheme in Fig. 2.31a; $an = 1$, $p \in \{$"x", "y"$\}$ for the scheme in Fig. 2.31b.

Due to the equivalence of the influence of the components of the refraction index n and the delay time t_0 upon the dynamics of the processes in the interferometer [35], we merge $t_{0p}(\mathbf{r},t)$ and $n_0(\mathbf{r})l/c$ into the single variable $t_{ep}(\mathbf{r},t) \equiv n_0(\mathbf{r})l/c + t_{op}(\mathbf{r},t)$. The value $t_{ep}(\mathbf{r},t)$ means the fraction of the time of the light field propagation. It is conditioned by the presence of the linear part $n_0(\mathbf{r})$ of the NM refraction index, as well as the time $t_{0p}(\mathbf{r},t)$, while $t_{ep}(\mathbf{r},t)$ can be referred to as the equivalent delay time in NRI. We are to use the value of delay time conditioned by the non-linear part of the refraction index $t_u(\mathbf{r},t) = l[n(\mathbf{r},t)-n_0(\mathbf{r})]/c$. Then, the processes modeling allowing for the many field bypasses in NRI (2.19) can be rewritten as:

$$
\begin{aligned}
\tau_n(\mathbf{r})\frac{\partial t_u(\mathbf{r},t)}{\partial t} &= an\cdot n_2(\mathbf{r})\frac{l}{c}\left[\mathbf{E}^2_{in,nm,x}(\mathbf{r},t)+\mathbf{E}^2_{in,nm,y}(\mathbf{r},t)\right] \\
&\quad - t_u(\mathbf{r},t)+D_e(\mathbf{r})\Delta t_u(\mathbf{r},t), \\
\mathbf{E}_{in,nm,pw}(\mathbf{r},t) &= \left(1-R_{1p}\right)^{0.5}\mathbf{E}_{in,pw}(\mathbf{r},t) \\
&\quad + \frac{0.5\gamma_p(r',t)}{\sigma}\eta_{pw}(r',t)E_{in,nm.pw}\left(r',t-\tau_{pw}\right).
\end{aligned}
\tag{2.20}
$$

Here $\tau_p \equiv \tau_p(r',t) \equiv t_{ep}(r',t)+t_u\left[r',t-t_{op}(r',t)\right] \approx t_{ep}(r',t)+t_u\left[r',t-t_{ep}(r',t)\right]$ is the propagation time of the p component of the light field with a wavelength λ_0 that, by the time moment t, has arrived (through the feedback loop) to the point $\mathbf{r}$ of the NM input plane from the point $\mathbf{r}'$ of the same plane (full time delay, a time of the complete bypass of the interferometer);

$$
\begin{aligned}
\tau_{pw} \equiv \tau_{pw}(r',t) &= t_{ep}(r',t)+t_{uw}\left[r',t-t_{0p}(r',t)\right], \\
t_{uw}(\mathbf{r},t) &\equiv \frac{\lambda_w t_u(\mathbf{r},t)}{\lambda_0} = \frac{\omega_0 t_u(\mathbf{r},t)}{\omega_w}; \\
\eta_{pw}(r',t) &= \frac{1}{\left[1+\frac{d\tau_{pw}(r',t)}{dt}\right]^{0.5}}.
\end{aligned}
$$

The mathematical model (2.20) is true for the field of arbitrary type at the input of the interferometer (excluding the polarization type limitations in the system in Fig. 2.31a). Therefore, it can serve as a foundation for further developments producing new different variants of the models.

The process model in approximation of the slowly-varying amplitudes, phases, modulations of the polarization plane position, delay time and electrical field losses. In Eq. (2.20) the electrical field intensity appears, which can vary with a period much less than the variation time of the optical field intensity and the value of the refraction index. Therefore, it is reasonable to transform it in approximation of slowly-varying amplitudes, phases, modulations of the polarization plane position, delay time and field energy losses. For this, it is necessary to specify the field type $\mathbf{E}(\mathbf{r},t)$ at the NRI input.

In literature we can find the cases, when $\mathbf{E}(\mathbf{r},t)$ is the linear-polarized monochrome non-modulated field [43–48]. It is interesting for us to study a more general case. We would like to note that optimization of the interference interaction of elliptically-polarized beams requires the homothetic polarization ellipses [49]. The circular and linear polarization types are universal, as far as they satisfy the mentioned condition.

Let the field in the NRI input consists of the two components with circular polarization (Fig. 2.33):

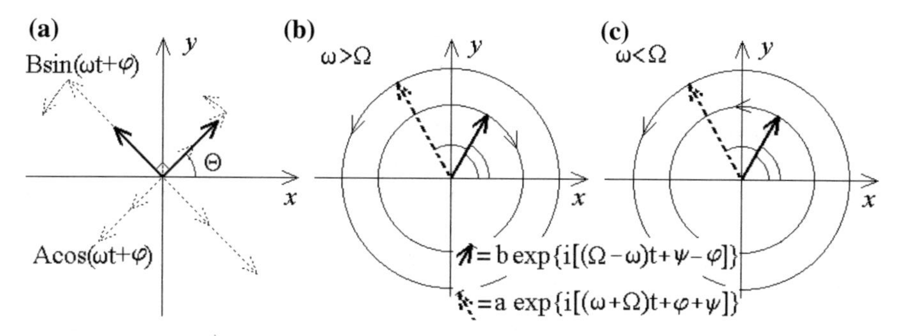

Fig. 2.33 The structure of bi-chromatic optical radiation $\mathbf{E}(\mathbf{r},t)$: *a*—for writing in the form (2.21), **b**, **c**—for writing in the form (2.22). *Thick lines* in figure *a* correspond to instantaneous state of the intensity vectors for $\omega t+\varphi(\mathbf{r},t)=30°$, $\theta(\mathbf{r},t)=45°$, $A/B=2/3$, the *dashed lines* illustrate possible states of these vectors at arbitrary $\omega t+\varphi(\mathbf{r},t)$, and its rotation direction is shown for $\Omega>0$. In figures **b**, **c** the *thick lines* corresponding to instantaneous state of intensity vectors for $\omega t+\varphi(\mathbf{r},t)=30°$, $\theta(\mathbf{r},t)=90°$, $A/B=1/5$, rotation directions (opposite to **b** when $\omega>\Omega$ and equal to **c** when $\omega<\Omega$) are shown for $\omega>0$, $\Omega>0$. The *dashed line* corresponds to the field component with the frequency $\omega+\Omega$ and the amplitude $a(\mathbf{r},t)$, and the *solid thick lines*—to the component with $\omega-\Omega$ and $b(\mathbf{r},t)$

$$\begin{aligned}\mathbf{E}(\mathbf{r},t)&=\mathbf{e}[\Theta(\mathbf{r},t)]A(\mathbf{r},t)\cos[\omega t+\varphi(\mathbf{r},t)]\\&+\mathbf{e}\left[\Theta(\mathbf{r},t)+\frac{\pi}{2}\right]B(\mathbf{r},t)\sin[\omega t+\varphi(\mathbf{r},t)],\end{aligned}\tag{2.21}$$

where ω is the fundamental frequency of the light field, $\Theta(\mathbf{r},t)=\psi(\mathbf{r},t)+\Omega t$ is the angle between the $\mathbf{e}(\Theta)$ vector, which gives the polarization direction, and the Ox axis that lies in the (xOy) plane of the beam transverse section (Ω can be commensurable with ω), Ω is the frequency of synchronous rotation of the polarization vectors $\mathbf{e}$ lying in the plane that we call the polarization plane; $A(\mathbf{r},t)$, $B(\mathbf{r},t)$, $\varphi(\mathbf{r},t)$, $\psi(\mathbf{r},t)$ are the amplitude, phases, the polarization plane position of the light field insignificantly varying during the time $T=2\pi/\omega$. In contrast to [35], there is here the second member, i.e. $B\neq 0$. The sign Ω characterizes the rotation direction of the polarization vectors $\mathbf{e}$.

If to introduce designations $a(\mathbf{r},t)\equiv\frac{A(\mathbf{r},t)+B(\mathbf{r},t)}{2}$, $b(\mathbf{r},t)\equiv\frac{A(\mathbf{r},t)-B(\mathbf{r},t)}{2}$, then (2.21) will be expressed via projections of $\mathbf{E}(\mathbf{r},t)$:

$$\begin{aligned}E_x(\mathbf{r},t)&=a(\mathbf{r},t)\cos[(\omega+\Omega)t+\varphi(\mathbf{r},t)+\psi(\mathbf{r},t)]\\&\quad+b(\mathbf{r},t)\cos[(\omega-\Omega)t+\varphi(\mathbf{r},t)-\psi(\mathbf{r},t)],\\E_y(r,t)&=a(\mathbf{r},t)\sin[(\omega+\Omega)t+\varphi(\mathbf{r},t)+\psi(\mathbf{r},t)]\\&\quad-b(r,t)\sin[(\omega-\Omega)t+\varphi(r,t)-\psi(r,t)].\end{aligned}\tag{2.22}$$

Thus, at interferometer input there arrives a sum of quasi-monochrome fields with amplitudes $a(\mathbf{r},t)$, $b(\mathbf{r},t)$ and with frequencies $\omega+\Omega$ of the circular polarizations of the different (for $\omega>\Omega$), or of the identical (for $\omega<\Omega$), rotation directions (Fig. 2.33b, c). Here ω (or Ω for $\omega < \Omega$) means average frequency, and $2\Omega(2\omega$ for $\omega < \Omega)$ is the frequency interval between field components. To reflect the specificity of the optical field under consideration, we deal with $q \equiv \Omega/\omega$ that is bi-chromaticity parameter, or non-monochromaticity parameter, according to [36].

We assume that using the specific engineering approaches we have achieved the fulfillment of the conditions: $t_{ex}(\mathbf{r},t) = t_{ey}(\mathbf{r},t) = t_e(\mathbf{r},t)$, $R_{1x} = R_{1y} = R_{2x} = R_{2y} = R$, $\gamma_x(\mathbf{r},t) = \gamma_y(\mathbf{r},t) = \gamma(\mathbf{r},t)$, $\eta_{xw}(\mathbf{r},t) = \eta_{yw}(\mathbf{r},t) = \eta_w(\mathbf{r},t)$, $\gamma(\mathbf{r},t) \equiv 2R\kappa(\mathbf{r},t)C_n(\mathbf{r})$. Let us agree that we do not take into account the dependence of the field propagation time in NM upon frequency (in this case we may designate $\eta_w(\mathbf{r},t) = \eta(\mathbf{r},t)$, $t_{uw}(\mathbf{r},t) = t_u(\mathbf{r},t)$ etc., and $\tau_p \equiv \tau_p(\mathbf{r}',t) = t_{ep}(\mathbf{r}',t) + t_u[\mathbf{r}', t - t_{ep}(\mathbf{r}',t)] = \tau$, $\tau_{pw} \equiv \tau_{pw}(\mathbf{r}',t) \equiv t_{ep}(\mathbf{r}',t) + t_{uw}\left[\mathbf{r}', t - t_{0p}(\mathbf{r},t)\right] = \tau$, where $\tau \equiv \tau(\mathbf{r}',t) \equiv t_e(\mathbf{r}',t) + t_u[\mathbf{r}', t - t_e(\mathbf{r}',t)]$. These approximations essentially simplify examination because all fields $\mathbf{E}$ in (2.20) become presentable in the form (2.21).

In typical case, the relaxation time of the nonlinear part of the refraction index $\tau_n(\mathbf{r}) \gg T$ (or, more strictly,—$\omega T \frac{dt_u}{dt} \ll 2\pi$ and $\Omega T \frac{dt_u}{dt} \ll 2\pi$), where $T \equiv \frac{2\pi}{\omega}$ is a period of the light oscillations. Taking into account that during the time T the values A, B and φ vary by values much less than A, B and 2π, we carry out averaging of the Kerr medium equation (in the model (2.20)) over the time interval $[t; t+T]$:

$$\begin{aligned} \tau_n(\mathbf{r})\frac{\partial t_u(\mathbf{r},t)}{\partial t} &= D_e(\mathbf{r})\Delta t_u(\mathbf{r},t) - t_u(\mathbf{r},t) \\ &\quad + an \cdot n_2(\mathbf{r})\frac{l}{c}\frac{1}{T}\int\limits_{t}^{t+T}\left\{\mathbf{E}^2_{in,nm.p'}(\mathbf{r},t') + \mathbf{E}^2_{in,nm,p}(\mathbf{r},t')\right\}dt \\ &= D_e(\mathbf{r})\Delta t_u(\mathbf{r},t) - t_u(\mathbf{r},t) + an \cdot n_2(\mathbf{r})\frac{l}{c}\left[a^2_{in,nm}(\mathbf{r},t) + b^2_{in.nm}(\mathbf{r},t)\right]. \end{aligned} \tag{2.23}$$

This equation describes the dynamics of the medium refraction index dependence [through $t_u(\mathbf{r},t)$] upon the amplitude square of the incident field, i.e. the high-frequency Kerr effect [39]. The similar procedure of "averaging" is equivalent to the action of the low-frequency filter, which excludes the fast varying processes (not interesting for us) from the consideration.

Let us estimate the magnitude $\eta(\mathbf{r}',t) = \frac{1}{\left\{1+\frac{d\tau(\mathbf{r}',t)}{dt}\right\}^{0.5}}$. Proceeding from the accepted approximation $\omega\Delta\tau \ll 2\pi$ and assuming that $\Delta\tau \approx T\frac{d\tau(\mathbf{r}',t)}{dt}$, we obtain the inequality $\frac{d\tau(\mathbf{r}',t)}{dt} \ll \frac{2\pi}{T\omega} = 1$, from which we get $\eta(\mathbf{r}',t) \approx 1$. Further omitting the index "in", and replacing the index "in.nm" by "nm" and using (2.23) from (2.20), we get the model for the processes in NRI:

$$\tau_n(\mathbf{r})\frac{\partial U(\mathbf{r},t)}{\partial t} = D_e(\mathbf{r})\Delta U(\mathbf{r},t) - U(\mathbf{r},t) + f(\mathbf{r},t),$$
$$f(r,t) = n_2(\mathbf{r})lk \cdot an\langle \mathbf{E}^2_{nm}(\mathbf{r},t)\rangle_T = an \cdot n_2(\mathbf{r})lk\left[a^2_{nm}(\mathbf{r},t) + b^2_{nm}(\mathbf{r},t)\right], \quad (2.24)$$
$$\mathbf{E}_{nm}(\mathbf{r},t) = (1-R)^{0.5}\mathbf{E}(\mathbf{r},t) + \frac{\gamma(\mathbf{r}',t)}{2\sigma}\mathbf{E}_{nm}(\mathbf{r}',t-\tau).$$

Here $k = \omega/c$; $U(\mathbf{r},t) \equiv \omega t_u(\mathbf{r},t)$ is a nonlinear phase shift; $\tau \equiv \tau(\mathbf{r}',t) = t_e(\mathbf{r}',t) + \frac{U[\mathbf{r}',t-t_e(\mathbf{r}',t)]}{\omega}$; $\gamma(\mathbf{r}',t) \equiv 2R\kappa(\mathbf{r}',t)C_n(\mathbf{r}')$; $an = 1$ (Fig. 2.31b); $an = 2$, $\Omega = 0$, $\psi = const$ (Fig. 2.31a).

Here the formulas for $a(\mathbf{r},t)$ and $\varphi(\mathbf{r},t)$ are useful:
if

$$a(\mathbf{r},t)\exp\{j[\omega t + \varphi(\mathbf{r},t)]\} = a_1(\mathbf{r},t)\exp\{j[\omega t + \varphi_1(\mathbf{r},t)]\} + a_2(\mathbf{r},t)\exp\{j[\omega t + \varphi_2(\mathbf{r},t)]\},$$

then $a(\mathbf{r},t) = (Ac^2 + As^2)^{0.5} = \left[a_1^2 + a_2^2 + 2a_1a_2\cos(\varphi_1 - \varphi_2)\right]^{0.5} \geq 0$,

$$\varphi(\mathbf{r},t) = \arg(Ac,As) \equiv \begin{cases} \frac{\pi}{2} - \arctan\frac{Ac}{As}, & As > 0 \\ \frac{3\pi}{2} - \arctan\frac{Ac}{As}, & As < 0 \\ 0 & , As = 0, Ac \geq 0 \\ \pi & , As = 0, Ac < 0 \end{cases} \in [0, 2\pi), \quad (2.25)$$

where $Ac \equiv a_1\cos\varphi_1 + a_2\cos\varphi_2$, $As = a_1\sin\varphi_1 + a_2\sin\varphi_2$.

We would like to remind that the field $\mathbf{E}(\mathbf{r},t)$ has a view (2.22). Then (2.25) should be used independently for the components with frequencies $(1+q)\omega$ and $(1-q)\omega$. Hence,

$$a_{nm}(\mathbf{r},t) = \left(Ac^2 + As^2\right)^{0.5}, \quad \varphi_{nm}(\mathbf{r},t) + \psi_{nm}(\mathbf{r},t) = \arg(Ac,As), \quad (2.26)$$

where

$$Ac = (1-R)^{0,5}a(\mathbf{r},t)\cos[\varphi(\mathbf{r},t) + \psi(\mathbf{r},t)] + \frac{\gamma(\mathbf{r}',t)}{2\sigma}a_{nm}(\mathbf{r}',t-\tau)\cos[\varphi_{nm}(\mathbf{r}',t-\tau) + \psi_{nm}(\mathbf{r}',t-\tau) - (1+q)\omega\tau],$$

$$As = (1-R)^{0.5}a(\mathbf{r},t)\sin[\varphi(\mathbf{r},t) + \psi(\mathbf{r},t)] + \frac{\gamma(\mathbf{r}',t)}{2\sigma}a_{nm}(\mathbf{r}',t-\tau)\sin[\varphi_{nm}(\mathbf{r}',t-\tau) + \psi_{nm}(\mathbf{r}',t-\tau) - (1+q)\omega\tau],$$

$$q \equiv \frac{\Omega}{\omega}, \quad \tau \equiv \tau(\mathbf{r}',t) = t_e(\mathbf{r}',t) + \frac{U[\mathbf{r}',t-t_e(\mathbf{r}',t)]}{\omega}.$$

Similarly,

$$b_{nm}(\mathbf{r},t) = \left(Ac^2 + As^2\right)^{0.5},\ \varphi_{nm}(\mathbf{r},t) - \psi_{nm}(\mathbf{r},t) = \arg(Ac, As) \tag{2.27}$$

where

$$\begin{aligned} Ac =& (1-R)^{0.5} b(\mathbf{r},t)\cos[\varphi(\mathbf{r},t) - \psi(\mathbf{r},t)] \\ &+ \frac{\gamma(\mathbf{r}',t)}{2\sigma} b_{nm}(\mathbf{r}',t-\tau)\cos[\varphi_{nm}(\mathbf{r}',t-\tau) - \psi_{nm}(\mathbf{r}',t) - (1-q)\omega\tau], \end{aligned}$$

$$\begin{aligned} As =& (1-R)^{0.5} b(\mathbf{r}',t)\sin[\varphi(\mathbf{r},t) - \psi(\mathbf{r},t)] \\ &+ \frac{\gamma(\mathbf{r}',t)}{2\sigma} b_{nm}(\mathbf{r}',t-\tau)\sin[\varphi_{nm}(\mathbf{r}',t-\tau) - (1-q)\omega\tau]. \end{aligned}$$

Let us rewrite (2.24) in the form

$$\begin{aligned} \tau_n(\mathbf{r})\frac{\partial U(\mathbf{r},t)}{\partial t} =& D_e(\mathbf{r})\Delta U(\mathbf{r},t) - U(\mathbf{r},t) + f(\mathbf{r},t), \\ f(\mathbf{r},t) =& an \cdot n_2(\mathbf{r}) lk\left[a_{nm}^2(\mathbf{r},t) + b_{nm}^2(\mathbf{r},t)\right]. \end{aligned} \tag{2.28}$$

Formulas (2.26)–(2.28) form the mathematical model of processes in NRI, the one taking into account many bypasses of the field equivalent to (2.24), but of a more understandable form. Using (2.26) and (2.27), we can prove that the following relation is true:

$$\begin{aligned} a_{nm}^2(\mathbf{r},t) + b_{nm}^2(\mathbf{r},t) =& (1-R)\left[a^2(\mathbf{r},t) + b^2(\mathbf{r},t)\right] \\ &+ \left[\frac{\gamma(\mathbf{r}',t)}{2\sigma}\right]^2 \left[a_{nm}^2(\mathbf{r}',t-\tau) + b_{nm}^2(\mathbf{r}',t-\tau)\right] \\ &+ \frac{(1-R)^{0.5}\gamma(\mathbf{r}',t)}{\sigma} \times \{a(\mathbf{r},t)a_{nm}(\mathbf{r}',t-\tau) \\ &\times \cos[(1+q)\omega\tau + \varphi(\mathbf{r},t) - \varphi_{nm}(\mathbf{r}',t-\tau) + \psi(r,t) - \psi_{nm}(\mathbf{r}',t-\tau)] \\ &+ b(\mathbf{r},t)b_{nm}(\mathbf{r}',t-\tau) \\ &\times \cos[(1-q)\omega\tau + \varphi(\mathbf{r},t)] - \varphi_{nm}(\mathbf{r}',t-\tau) - \psi(r,t) + \psi_{nm}(\mathbf{r}',t-\tau)\}. \end{aligned} \tag{2.29}$$

Let us introduce nonlinearity coefficients corresponding to the usual scheme. For this, we rewrite $f(\mathbf{r},t)$ using (2.29) into the form

$$
\begin{aligned}
f(\mathbf{r},t) =& an\cdot n_2(\mathbf{r})lk(1-R)\left[a^2(\mathbf{r},t)+b^2(\mathbf{r},t)\right]\\
&+an\cdot n_2(\mathbf{r})lk(1-R)\left[\frac{\gamma(\mathbf{r},t)}{2\sigma}\right]^2\frac{a_{nm}^2(\mathbf{r}',t-\tau)+b_{nm}^2(\mathbf{r}',t-\tau)}{1-R}\\
&+an\cdot n_2(\mathbf{r})lk(1-R)^{0.5}\left[(1-R)^{0.5}\frac{\gamma(\mathbf{r}',t)}{\sigma}\right]\\
&\times\{a(\mathbf{r},t)a(\mathbf{r}',t-\tau)\frac{a_{nm}(\mathbf{r}',t-\tau)}{a(\mathbf{r}',t-\tau)(1-R)^{0.5}}\\
&\times\cos[(1+q)\omega\tau+\varphi(\mathbf{r},t)-\varphi_{nm}(\mathbf{r}',t-\tau)+\psi(\mathbf{r},t)-\psi_{nm}(\mathbf{r}',t-\tau)]\\
&+b(\mathbf{r},t)b(\mathbf{r}',t-\tau)\frac{b_{nm}(\mathbf{r}'t-\tau)}{b(\mathbf{r}',t-\tau)(1-R)^{0.5}}\\
&\times\cos[(1-q)\omega\tau+\varphi(\mathbf{r}',t)-\varphi_{nm}(\mathbf{r}',t-\tau)-\psi(\mathbf{r},t)+\psi_{nm}(\mathbf{r}',t-\tau)]\}.
\end{aligned}
$$

Taking into consideration that the second summand can be transformed as

$$
\begin{aligned}
&an\cdot n_2(\mathbf{r})lk(1-R)[\frac{\gamma(\mathbf{r}',t)}{2\sigma}]^2\frac{a_{nm}^2(\mathbf{r}',t-\tau)+b_{nm}^2(\mathbf{r}',t-\tau)}{1-R}\\
&\quad=an\cdot n_2(\mathbf{r})lk(1-R)\left[\frac{\gamma(\mathbf{r}',t)}{2\sigma}\right]^2\frac{a^2(\mathbf{r}',t-\tau)a_{nm}^2(\mathbf{r}',t-\tau)}{a^2(\mathbf{r}',t-\tau)(1-R)}\\
&\quad+an\cdot n_2(\mathbf{r})lk(1-R)\left[\frac{\gamma(\mathbf{r}',t)}{2\sigma}\right]^2\frac{b^2(\mathbf{r}',t-\tau)b_{nm}^2(\mathbf{r}',t-\tau)}{b^2(\mathbf{r}',t-\tau)(1-R)}\\
&\quad=\left[\frac{\gamma(\mathbf{r}',t)}{2\sigma}\right]^2 Ka(\mathbf{r},\mathbf{r}',t-\tau,\mathbf{r}',t-\tau)a_{nm,r}^2(\mathbf{r}'.t-\tau)\\
&\quad+\left[\frac{\gamma(\mathbf{r}',t)}{2\sigma}\right]^2 Kb(\mathbf{r},\mathbf{r}',t-\tau,\mathbf{r}',t-\tau)b_{nm,r}^2(\mathbf{r}'.t-\tau),
\end{aligned}
$$

where

$$
\begin{aligned}
a_{nm,r}(\mathbf{r},t)&\equiv\frac{a_{nm}(\mathbf{r},t)}{(1-R)^{0.5}a(\mathbf{r},t)},\\
b_{nm}(\mathbf{r},t)&\equiv\frac{b_{nm}(\mathbf{r},t)}{(1-R)^{0.5}b(\mathbf{r},t)},\\
Ka(\mathbf{r}_n,\mathbf{r},t,\mathbf{r}',t-\tau)&\equiv(1-R)an\cdot n_2(\mathbf{r}_n)lka(\mathbf{r},t)a(\mathbf{r}',t-\tau),\\
Kb(\mathbf{r}_n,\mathbf{r},t,\mathbf{r}',t-\tau)&\equiv(1-R)an\cdot n_2(\mathbf{r}_n)lkb(\mathbf{r},t)b(\mathbf{r}',t-\tau),
\end{aligned}
\tag{2.30}
$$

we get

$$\begin{aligned}
\tau_n(\mathbf{r})\frac{\partial U(\mathbf{r},t)}{\partial t} =& D_e(\mathbf{r})\Delta U(\mathbf{r},t) - U(\mathbf{r},t) + f(\mathbf{r},t),\\
f(\mathbf{r},t) =& Kab(\mathbf{r},t,\mathbf{r}) + [\frac{\gamma(\mathbf{r}',t)}{2\sigma}]^2 Ka(\mathbf{r},\mathbf{r}',t-\tau,\mathbf{r}',t-\tau)a^2_{nm,r}(\mathbf{r}',t-\tau)\\
&+ [\frac{\gamma(\mathbf{r}',t)}{2\sigma}]^2 Kb(\mathbf{r},\mathbf{r}',t-\tau,\mathbf{r}',t-\tau)b^2_{nm,r}(\mathbf{r}',t-\tau)\\
&+ \frac{\gamma(\mathbf{r}',t)}{\sigma} \times \{Ka(\mathbf{r},\mathbf{r},t,\mathbf{r}',t-\tau)a_{nm,r}(\mathbf{r}',t-\tau)\\
&\times \cos[(1+q)\omega\tau + \varphi(\mathbf{r},t) - \varphi_{nm}(\mathbf{r}',t-\tau) + \psi(\mathbf{r},t) - \psi_{nm}(\mathbf{r}',t-\tau)]\\
&+ Kb(\mathbf{r},\mathbf{r},t,\mathbf{r}',t-\tau)b_{nm,r}(\mathbf{r}',t-\tau)\\
&\times \cos[(1-q)\omega\tau + \varphi(\mathbf{r},t) - \varphi_{nm}(\mathbf{r}',t-\tau) - \psi(\mathbf{r},t) + \psi_{nm}(\mathbf{r}',t-\tau)]\}.
\end{aligned} \tag{2.31}$$

Here $Kab(\mathbf{r},t,\mathbf{r}_n) \equiv (1-R)an \cdot n_2(\mathbf{r}_n)lk[a^2(\mathbf{r},t) + b^2(\mathbf{r},t)]$, $\tau \equiv \tau(\mathbf{r}',t) = t_e(\mathbf{r}',t) + U[\mathbf{r}',t - t_e(\mathbf{r}',t)/\omega$. Formulas (2.26), (2.27), (2.30) and (2.31) form the mathematical model of the processes in the multipass NRI. This model is equivalent to the formulas (2.26)–(2.28), but they also contain the nonlinearity coefficients.

If to use approximation of the single field pass in NRI $\{a_{nm}(\mathbf{r}',t-\tau) = (1-R)^{0.5}a(\mathbf{r}',t-\tau),\quad b_{nm}(\mathbf{r}',t-\tau) = (1-R)^{0.5}b(\mathbf{r}',t-\tau),$ $\varphi_{nm}(\mathbf{r}',t-\tau) = \varphi(\mathbf{r}',t-\tau),\ \psi_{nm}(\mathbf{r}',t-\tau) = \psi(\mathbf{r}',t-\tau)\}$, or of large losses $\{[\frac{\gamma(r',t)}{2\sigma}]^2 \approx 0,\ \varphi_{nm}(\mathbf{r}',t-\tau) = \varphi(\mathbf{r}',t-\tau),\ \psi_{nm}(\mathbf{r}',t-\tau) = \psi(\mathbf{r}',t-\tau)\}$, and if to accept: $B_{in} = 0,\ A_{in} = const,\ \psi = const,\ \Omega = 0,\ p = 0,\ an = 2,\ \sigma = 1,\ t_e = 0,$ $\gamma(\mathbf{r}',t) = \gamma,\ n_2(\mathbf{r}) = n_2,\ I_0 = A^2_{in}$, then we obtain from (2.31) the result coincident with the mathematical model described in [43, 44, 46] and presented in Chap. 1. As this fact demonstrates the absence of contradictions between (2.31) and the model in [43, 44, 46], we can evaluate it as a verification.

The point model of processes in the interferometer. We have described above the optical density of the nonlinear medium by the differential equation in the partial derivatives, supposing that medium molecules or the charge carriers are included into diffusion motion. Examining such systems for solution stability and for bifurcation arising is a more complicated problem than the study of the systems describing by the ordinary differential equations. Therefore, in (2.31) we can neglect diffusion ($D_e = 0$) and, thus, to undergo to the so-called point model.

The name "point model" is justified by the fact that the total set of transverse section points of the light beam (depending on the type of large-scale field transformation by the G element in the feedback loop) is divided into the infinite number of subsets, each independent from one another. Under independence of subsets we understand that the processes (nonlinear and interference) in *different* subsets will run independently. In other words, between the field and the parts of the nonlinear medium (as well as between fields themselves), the physical interaction is absent, if they correspond to the *different* subsets.

Vice versa, such interactions exist, if fields and medium parts correspond to the points of the *same* subset. These subsets represent the chains of points, “coupled” by the sequential displacement of the light beam in the transverse beam plane (due to the action of the G element (Fig. 2.31)).

In other words, the light beam passing through the nonlinear medium and the feedback loop of NRI in the point i (for instance, in Fig. 2.31 i = 1, 2, 3) acquires the phase shift U_i and feels the time delay t_{e1}. Because of the presence of the G element, the beam gets to the point i + 1. Here, “adding” with the one of the input beams of the interferometer, it, in accordance with the model (2.24), affects the rate of variation of the nonlinear phase shift value U_{i+1}. We note that due to the beam closing in Fig. 2.31, the index value $i + 1 = 4$ should be accepted as equal to $i + 1=1$. Exactly so the U_i shift in the point i influences on the U_{i+1} shift in the point i + 1. If the point number in these subsets is finite and equal to m, then we speak about the **degenerated** two-dimensional feedback of m-order [44]. The given types of points, according to terminology of [47], are called the transposition points, and m is the order of transposition (from Latin *trans*—through + *positio*—position). Accordingly, we shall speak about chains of the transposition points (CTP). The illustration to the CTP concept is given in Fig. 2.32. With such feedback structure, the **trajectory** of the beam is closed after m passes of NRI. It is easy to see that, according to the accepted approach of numeration of the transposition points, we imply the operation $(i \bmod m) + 1$ under writing i + 1, where the symbol $i \bmod m$ means the reminder after division of i by m. Physically it means that the beam from the m-th point will get to the 1st one.

The point approximation, understood in the mentioned way, makes it possible that from the model (2.26), (2.27), (2.30) and (2.31), we can obtain the system of ordinary differential equations (ODE), which describes dynamics of the nonlinear phase shift for the single CTP

$$\begin{aligned}
\tau_{ni}\frac{dU_i(t)}{dt} &= -\,U_i(t) + f_i,\\
f_i \equiv f_i(t) &= an\cdot n_{2i}lk\left[a_{nm}^2(t) + b_{nm}^2(t)\right] = Kab_{i,i}(t)\\
&\quad + [\frac{\gamma_{i-1}(t)}{2\sigma}]^2 Ka_{i,i-1,i-1}(t-\tau, t-\tau)a_{nm,r,i-1}^2(t-\tau)\\
&\quad + [\frac{\gamma_{i-1}(t)}{\sigma}]^2 Kb_{i,i-1,i-1}(t-\tau, t-\tau)b_{nm,r,i-1}^2(t-\tau)\\
&\quad + \frac{\gamma_{i-1}(t)}{2\sigma}\{Ka_{i,i,i-1}(t, t-\tau)a_{nm,r,i-1}(t-\tau)\\
&\quad \times \cos\left[(1+q)\omega\tau + \varphi_i(t) - \varphi_{nm,i-1}(t-\tau) + \psi_i(t) - \psi_{nm,i-1}(t-\tau)\right]\\
&\quad + Kb_{i,i,i-1}(t, t-\tau)b_{nm,r,i-1}(t-\tau)\\
&\quad \times \cos\left[(1-q)\omega\tau + \varphi_i(t) - \varphi_{nm,i-1}(t-\tau) - \psi_i(t) + \psi_{nm,i-1}(t-\tau)\right]\}.
\end{aligned} \tag{2.32}$$

Here

$$
\begin{aligned}
a_{nm,r,i}(t) &\equiv \frac{a_{nm,i}(t)}{(1-R)^{0.5}a_i(t)},\\
b_{nm,r,i}(t) &\equiv \frac{b_{nm,i}(t)}{(1-R)^{0.5}b_i(t)},\\
Ka_{i,j,k}(t,t-\tau) &\equiv (1-R)an \cdot n_{2i}lka_j(t)a_k(t),\\
Kb_{i,j,k}(t,t-\tau) &\equiv (1-R)an \cdot n_{2i}lkb_j(t)b_k(t),\\
Kab_{i,j}(t) &\equiv (1-R)an \cdot n_{2j}lk\left[a_i^2(t)+b_i^2(t)\right],\\
\tau &\equiv \tau_{i-1}(t) = t_{e,i-1}(t) + U_{i-1}[t - t_{e,i-1}(t)]/\omega,\\
\gamma_{i-1}(t) &\equiv 2R\kappa_{i-1}(t)C_{n,i-1},\\
a_{nm,i}(t) &= \left(Ac_a^2 + As_a^2\right)^{0.5},\ \varphi_{nm,i}(t) + \psi_{nm,i}(t) = \arg(Ac_a, As_a),\\
b_{nm,i}(t) &= \left(Ac_b^2 + As_b^2\right)^{0.5},\ \varphi_{nm,i}(t) - \psi_{nm,i}(t) = \arg(Ac_b, As_b),
\end{aligned}
$$

$$
\begin{aligned}
Ac_a =& (1-R)^{0.5}a_i(t)\cos[\varphi_i(t)+\psi_i(t)]\\
&+ \frac{\gamma_{i-1}(t)}{2\sigma}a_{nm,i-1}(t-\tau)\cos\left[\varphi_{nm,i-1}(t-\tau)+\psi_{nm,i-1}(t-\tau)-(1+q)\omega\tau\right],\\
As_a =& (1-R)^{0.5}a_i(t)\sin[\varphi_i(t)+\psi_i(t)]\\
&+ \frac{\gamma_{i-1}(t)}{2\sigma}a_{nm,i-1}(t-\tau)\sin\left[\varphi_{nm,i-1}(t-\tau)+\psi_{nm,i-1}(t-\tau)-(1+q)\omega\tau\right],\\
Ac_b =& (1-R)^{0.5}b_i(t)\cos[\varphi_i(t)-\psi_i(t)]\\
&+ \frac{\gamma_{i-1}(t)}{2\sigma}b_{nm,i-1}(t-\tau)\cos\left[\varphi_{nm,i-1}(t-\tau)-\psi_{nm,i-1}(t-\tau)-(1-q)\omega\tau\right],\\
As_b =& (1-R)^{0.5}b_i(t)\sin[\varphi_i(t)-\psi_i(t)]\\
&+ \frac{\gamma_{i-1}(t)}{2\sigma}b_{nm,i-1}(t-\tau)\sin\left[\varphi_{nm,i-1}(t-\tau)-\psi_{nm,i-1}(t-\tau)-(1-q)\omega\tau\right].
\end{aligned}
$$

Depending on the type of the field transformation into the G range, the variation of i range is different. Under beam compression $i \in [0, +\infty]$; under stretching $i \in [-\infty, 0]$; with rotation by the angle $\Delta \neq 2\pi N/m$ $i \in (-\infty, +\infty)$; with rotation by the angle $\Delta = 2\pi N/m$ $i \in [1, m]$; when there is a shift by the δ value $i \in [1, r_b/\delta]$ (m, N are the mutually prime integer numbers, r_b is the beam radius). Under rotation by the angle $\Delta = 2\pi N/m$, the chain of transposition points is closed, i.e. the field from the last CTP point m after passing the feedback loop arrives into the first one (U_m affects on U_1). During the other field transformations, the unclosed CTPs are probably formed. In case of the closed chain of the transposition points, the

m parameter defines a number of independent dynamic variables, i.e. the phase space dimension of the dynamic system equals to 4 m. This gives exclusiveness to this parameter: variation of its value performs the transition to another dynamic system.

Mathematical models (2.24), (2.32) of the processes are universal, with respect to the view of spatial-temporal functions, whose laws determine the modulation of the field characteristics and interferometer parameters (excluding limitation on maximal rate of modulation), and, therefore, they represent wide opportunities for the development of the theoretical and applied aspects. Molecules diffusion of the nonlinear medium and the optical field diffraction occurring in laboratory conditions do not allow us to consider the processes taking place in the near points of the beam's transverse section plane (in neighbor chains of the transposition points) as independents ones. This circumstance applies restriction on the point model (2.33) application.

Together with the ODE point models, the discrete mapping models are widely used in nonlinear dynamics. Their advantages are the simplification of calculations and the calculation time reduction with keeping the main regulations in system behavior intact. Let us construct such a model and, continuing our selected strategy, reveal point model of the processes in the interferometer. We note that the ODE point model transfers into this model when we use the approximation of the instantaneous response, or if we use the static mode of the NRI operation, i.e. in the absence of variations in the field amplitude time and in phases in NRI and, hence, without the phase shift ($dU/dt = 0$).

The model of optical field dynamics in the nonlinear ring interferometer (in terms of discrete mapping). In NRI static operation mode, the point model (2.32) is reduced to the recurrent relation. We assume that the static mode takes place when the values a_i, b_i, φ_i, ψ_i, γ_i, t_{e1} are constant in time. Then, from (2.32) we have:

$$
\begin{aligned}
U_i =& an \cdot n_{2i} lk\left[a^2_{nm,i} + b^2_{nm,i}\right] = Kab_{i,i} + [\frac{\gamma_{i-1}}{2\sigma}]^2 \\
&\times \left[Ka_{i,i-1,i-1}a^2_{nm,t,i-1} + Kb_{i,i-1,i-1}b^2_{nm,r,i-1}\right] \\
&+ \frac{\gamma_{i-1}}{\sigma}\{Ka_{i,i,i-1}a_{nm,r,i-1} \\
&\times \cos\left[(1+q)(\Phi_{i-1} + U_{i-1}) + \varphi_i + \psi_i - \varphi_{nm,i-1} - \psi_{nm.i-1}\right] \\
&+ Kb_{i,i,i-1}b_{nm,r,i-1} \\
&\times \cos\left[(1-q)(\Phi_{i-1} + U_{i-1}) + \varphi_i - \psi_i - \varphi_{nm,i-1} + \psi_{nm.i-1}\right]\}.
\end{aligned}
\tag{2.33}
$$

Here

$$\begin{aligned}
a_{nm,r,i} &\equiv \frac{a_{nm,i}}{(1-R)^{0.5}a_i},\\
b_{nm,r,i} &\equiv \frac{b_{nm,i}}{(1-R)^{0.5}b_i},\\
Ka_{i,j,k} &\equiv (1-R)an\cdot n_{2i}lka_ja_k,\\
Kb_{i,j,k} &\equiv (1-R)an\cdot n_{2i}lkb_jb_k,\\
Kab_{i,j} &\equiv (1-R)an\cdot n_{2j}lk\left[a_i^2+b_i^2\right],\\
\gamma_{i-1} &\equiv 2R\kappa_{i-1}C_{n,i-1},\ \Phi_i\equiv\omega t_{e,i},\\
a_{nm,i} &= \left(Ac_a^2+As_a^2\right)^{0.5},\ \varphi_{nm,i}+\psi_{nm,i}=\arg(Ac_a,As_a),\\
b_{nm,i} &= \left(Ac_b^2+As_b^2\right)^{0.5},\ \varphi_{nm,i}-\psi_{nm,i}=\arg(Ac_b,As_b),
\end{aligned}$$

$$\begin{aligned}
Ac_a &= (1-R)^{0.5}a_i\cos(\varphi_i+\psi_i)+\frac{\gamma_{i-1}}{2\sigma}a_{nm,i-1}\\
&\quad\times\cos\left[\varphi_{nm,i-1}+\psi_{nm,i-1}-(1+q)(\Phi_{i-1}+U_{i-1})\right],\\
As_a &= (1-R)^{0.5}a_i\sin(\varphi_i+\psi_i)+\frac{\gamma_{i-1}}{2\sigma}a_{nm,i-1}\\
&\quad\times\sin\left[\varphi_{nm,i-1}+\psi_{nm,i-1}-(1+q)(\Phi_{i-1}+U_{i-1})\right],\\
Ac_b &= (1-R)^{0.5}b_i\cos(\varphi_i-\psi_i)+\frac{\gamma_{i-1}}{2\sigma}b_{nm,i-1}\\
&\quad\times\cos\left[\varphi_{nm,i-1}-\psi_{nm,i-1}-(1-q)(\Phi_{i-1}+U_{i-1})\right],\\
As_b &= (1-R)^{0.5}b_i\sin(\varphi_i-\psi_i)+\frac{\gamma_{i-1}}{2\sigma}b_{nm,i-1}\\
&\quad\times\sin\left[\varphi_{nm,i-1}-\psi_{nm,i-1}-(1-q)(\Phi_{i-1}+U_{i-1})\right].
\end{aligned}$$

In case of the uniform (within the limits of the single CTP) optical properties of NM in NRI ($n_2=n_{2j}$) and input field amplitudes ($a=a_i$, $b=b_i$), the following relations are true: $Kab=Ka+Kb$, where $Ka\equiv(1-R)an\cdot n_2lk\cdot a^2$, $Kb=(1-R)an\cdot n_2lk\cdot b^2$, $Kab\equiv(1-R)an\cdot n_2lk(a^2+b^2)$. It is convenient to introduce the total nonlinearity parameter K and a part Q_a of intensity of the component with a the frequency $(1+q)\omega$ according to the rule: $K\equiv Kab=(Ka+Kb)$, $Q_a\equiv Ka/K$. Then $Ka=KQ_a$, $K_b=K(1-Q_a)$. In case of the uniformity of the other optical properties of NRI ($\Phi=\Phi_i$, $\gamma=\gamma_i$) and the input field ($\psi_i=0$, $\varphi_i=0$), it is easy to obtain the discrete map (DM) from (2.33)

$$\begin{aligned}
U_i &= an\cdot n_2lk\left(a_{nm,i}^2+b_{nm,i}^2\right)=K\{1+(\frac{\gamma}{2\sigma})^2\left[Q_aa_{nm,r,i-1}^2+(1-Q_a)b_{nm,r,i-1}^2\right]\\
&\quad+\frac{\gamma}{\sigma}\{Q_aa_{nm,r,i-1}\cos\left[(1+q)(\Phi+U_{i-1})+\Psi-\Psi_{nm,i-1}\right]\\
&\quad+(1-Q_a)b_{nm,r,i-1}\cos\left[(1-q)(\Phi+U_{i-1})+\Theta-\Theta_{nm,i-1}\right]\}\}.
\end{aligned}\tag{2.34}$$

Here

$$
\begin{aligned}
a_{nm,r,i} &\equiv \frac{a_{nm,i}}{(1-R)^{0.5}a}, \\
b_{nm,r,i} &\equiv \frac{b_{nm,i}}{(1-R)^{0.5}b}, \\
\gamma &\equiv 2R\kappa C_n, \Phi \equiv \omega t_e, \\
a_{nm,i} &= \left(Ac_a^2 + As_a^2\right)^{0.5}, \\
\Psi_{nm,i} &= \arg(Ac_a, As_a), \\
b_{nm,i} &= \left(Ac_b^2 + As_b^2\right)^{0.5}, \\
\Theta_{nm,i} &= \arg(Ac_b, As_b),
\end{aligned}
$$

$$
\begin{aligned}
Ac_a &= (1-R)^{0.5}a\cos\Psi + \frac{\gamma}{2\sigma}a_{nm,i-1}\cos\left[\Psi_{nm,i-1} - (1+q)(\Phi + U_{i-1})\right], \\
As_a &= (1-R)^{0.5}a\sin\Psi + \frac{\gamma}{2\sigma}a_{nm,i-1}\sin\left[\Psi_{nm,i-1} - (1+q)(\Phi + U_{i-1})\right], \\
Ac_b &= (1-R)^{0.5}b\cos\Theta + \frac{\gamma}{2\sigma}b_{nm,i-1}\cos\left[\Theta_{nm,i-1} - (1-q)(\Phi + U_{i-1})\right], \\
As_b &= (1-R)^{0.5}b\sin\Theta + \frac{\gamma}{2\sigma}b_{nm,i-1}\sin\left[\Theta_{nm,i-1} - (1-q)(\Phi + U_{i-1})\right],
\end{aligned}
$$

and for the simplification of formula writing, the total and differential phases are introduced:

$$
\Psi = \varphi + \psi, \Theta = \varphi - \psi, \Psi_{nmi} = \varphi_{nm,i} + \psi_{nm,i}, \Theta_{nm,i} = \varphi_{nm,i} - \psi_{nm,i}.
$$

Obviously, $U_i = an \cdot n_2 lk \cdot \left(a_{nm,i}^2 + b_{nm,i}^2\right) = K\left[Q_a a_{nm,r,i}^2 + (1-Q_a)b_{nm,r,i}^2\right]$. Then we can rewrite (2.34) as

$$
\begin{aligned}
a_{nm,i} &= \left(Ac_{a,i}^2 + As_{a,i}^2\right), \Psi_{nm,i} = \arg\left(Ac_{a,i}, As_{a,i}\right), \\
b_{nm,i} &= \left(Ac_{b,i}^2 + As_{b,i}^2\right), \Theta_{nm,i} = \arg\left(Ac_{b,i}, As_{b,i}\right),
\end{aligned}
\tag{2.35}
$$

where

$$
\begin{aligned}
Ac_{a,i} &= (1-R)^{0.5}a\cos\Psi + \frac{\gamma}{2\sigma}a_{nm,i-1}\cos\left[\Psi_{nm,i-1} - (1+q)(\Phi + U_{i-1})\right], \\
As_{a,i} &= (1-R)^{0.5}a\sin\Psi + \frac{\gamma}{2\sigma}a_{nm,i-1}\sin\left[\Psi_{nm,i-1} - (1+q)(\Phi + U_{i-1})\right], \\
Ac_{b,i} &= (1-R)^{0.5}b\cos\Theta + \frac{\gamma}{2\sigma}b_{nm,i-1}\cos\left[\Theta_{nm,i-1} - (1-q)(\Phi + U_{i-1})\right], \\
As_{b,i} &= (1-R)^{0.5}b\sin\Theta + \frac{\gamma}{2\sigma}b_{nm,i-1}\sin\left[\Theta_{nm,i-1} - (1-q)(\Phi + U_{i-1})\right]
\end{aligned}
$$

$$
\begin{aligned}
U_{i-1} &= K\left[Q_a a_{nm,r,i-1}^2 + (1-Q_a) b_{nm,r,i-1}^2\right] \\
&= \frac{K}{1-R}\left\{Q_a\left[\frac{a_{nm,i-1}}{a}\right]^2 + (1-Q_a)\left[\frac{b_{nm,i-1}}{b}\right]^2\right\},
\end{aligned}
$$

$$
a_{nm,r,i} \equiv \frac{a_{nm,i}}{(1-R)^{0.5} a}, \quad b_{nm,r,i} \equiv \frac{b_{nm,i}}{(1-R)^{0.5} b}, \quad \gamma \equiv 2\mathrm{R}\kappa \mathrm{C}_n, \quad \Phi \equiv \omega t_e.
$$

We divide $a_{nm,i}$, $Ac_{a,i}$, $As_{a,i}$ by $\left[(1-R)^{0.5} a\right]$, and $b_{nm,i}$, $Ac_{b,i}$, $As_{b,i}$—by $\left[(1-R)^{0.5} b\right]$, then (2.35) will transform to the form

$$
\begin{aligned}
a_{nm,r,i} &= \left(Ac_{a,r,i}^2 + As_{a,r,i}^2\right)^{0.5}, \quad \Psi_{nm,i} = \arg\left(Ac_{a,i}, As_{a,i}\right), \\
b_{nm,r,i} &= \left(Ac_{b,r,i}^2 + As_{b,r,i}^2\right)^{0.5}, \quad \Theta_{nm,i} = \arg\left(Ac_{b,i}, As_{b,i}\right),
\end{aligned} \tag{2.36}
$$

where

$$
\begin{aligned}
Ac_{a,r,i} &= \cos\Psi + \frac{\gamma}{2\sigma} a_{nm,r,i-1} \cos\left[\Psi_{nm,i-1} - (1+q)(\Phi + U_{i-1})\right], \\
As_{a,r,i} &= \sin\Psi + \frac{\gamma}{2\sigma} a_{nm,r,i-1} \sin\left[\Psi_{nm,i-1} - (1+q)(\Phi + U_{i-1})\right], \\
Ac_{b,r,i} &= \cos\Theta + \frac{\gamma}{2\sigma} b_{nm,r,i-1} \cos\left[\Theta_{nm,i-1} - (1-q)(\Phi + U_{i-1})\right], \\
As_{b,r,i} &= \sin\Theta + \frac{\gamma}{2\sigma} b_{nm,r,i-1} \sin\left[\Theta_{nm,i-1} - (1-q)(\Phi + U_{i-1})\right], \\
U_{i-1} &= K\left[Q_a a_{nm,r,i-1}^2 + (1-Q_a) b_{nm,r,i-1}^2\right].
\end{aligned}
$$

As we know, the maximal dimension of the phase space corresponding to the dynamic system is determined by the minimal number of independent variables necessary for the complete description of the system behavior. Since the model (2.36) is composed with respect to the four variables: $a_{nm,r,i}$, $b_{nm,r,i}$, $\varphi_{nm,i,}$, $\psi_{nm,i}$, the maximal value of the phase space dimension equals to four.

We note that having introduced the change $\mathbf{a}_{nm,r,i} = a_{nm,r,i} \exp\left(j\Psi_{nm,i}\right)$, $\mathbf{b}_{nm,r,i} = b_{nm,r,i} \exp\left(j\Theta_{nm,i}\right)$, $\mathbf{a} = \exp(j\Psi)$, $\mathbf{b} = \exp(j\Theta)$, we can rewrite (2.36) in the more compact complex form:

$$
\begin{aligned}
\mathbf{a}_{nm,r,i} &= \mathbf{a} + \frac{\gamma}{2\sigma} \mathbf{a}_{nm,r,i-1} \exp[-j(1+q)(\Phi + U_{i-1})], \\
\mathbf{b}_{nm,r,i} &= b + \frac{\gamma}{2\sigma} \mathbf{b}_{nm,r,i-1} \exp[-j(1-q)(\Phi + U_{i-1})],
\end{aligned} \tag{2.37}
$$

where $U_{i-1} = K\left\{Q_a\left|\mathbf{a}_{nm,r,i-1}\right|^2 + (1-Q_a)\left|\mathbf{b}_{nm,r,i-1}\right|^2\right\}$.

Let us proceed to the two-circuit modification of the nonlinear ring interferometer. It has structurally more complicated (sometimes, even—"elaborate") configuration of chains of the transposition points. And for this very reason, this modification is more promising for the construction of the optical systems for confidential communication.

2.4.2 *Double-Circuit Nonlinear Ring Interferometer and Models of Processes in It*

As in the case of single-circuit ring interferometer, we construct a hierarchy of processes description in the double-circuit device starting from the model composed of equations in the partial derivatives and finishing by description in terms of discrete mapping.

The diagram of the double-circuit nonlinear ring interferometer and the dynamic model of the optical field in it (equation in partial derivatives). We have already mentioned above that the nonlinear ring interferometer (Fig. 2.31) is an example of the system capable of generating both regular optical structures and the deterministic chaos. It is natural to think that traditional NRI construction is the specific case of ring optical systems. Say, such systems can have several feedback loops (FBL), as well as the unequal nonlinear media. Developing this assumption, we examine (as the first stage) the double-circuit ring interferometer (DNRI), shown in Fig. 2.34. It is equipped with the additional FBL, in which one more large-scale transformation (the same or different) of the light field is produced. We can expect that increase in the number of device parameters makes it more difficult for the enemy to "crack" the cryptosystem, which may have DNRI as its basis, owing to the key length growth (see Sect. 1.3).

In approximations used during the construction of the NRI model (2.24), the model of processes in DNRI, evidently, takes a view

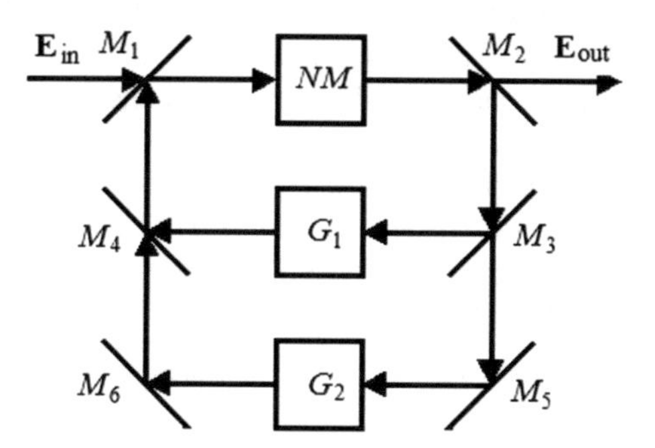

Fig. 2.34 The diagram of the double-circuit nonlinear ring interferometer (2.38)

$$\tau_n(\mathbf{r})\frac{\partial U(\mathbf{r},t)}{\partial t} = D_e(\mathbf{r})\Delta U(\mathbf{r},t) - U(\mathbf{r},t) + f(\mathbf{r},t),$$
$$f(\mathbf{r},t) = n_2(\mathbf{r})lkan\langle \mathbf{E}^2_{nm}(\mathbf{r},t)\rangle_T = an\cdot n_2(\mathbf{r})lk\left[a^2_{nm}(\mathbf{r},t) + b^2_{nm}(\mathbf{r},t)\right], \tag{2.38}$$
$$\mathbf{E}_{nm}(\mathbf{r},t) = (1-R_1)^{0.5}\mathbf{E}(\mathbf{r},t) + \mathbf{E}_{FBL}(\mathbf{r},t).$$

Here $\mathbf{E}_{nm}(\mathbf{r},t)$ is the field at nonlinear medium input, $\mathbf{E}(\mathbf{r},t)$ is the field at DNRI input, $\mathbf{E}_{fbl}(\mathbf{r},t)$ is the field component arriving at NM input from the "united" feedback loop. It is easy to show that

$$\mathbf{E}_{FBL}(\mathbf{r},t) = \frac{\gamma_1(\mathbf{r}',t)}{2\sigma_1}\mathbf{E}_{nm}\left(\mathbf{r}'_1,t-\tau_1\right) + \frac{\gamma_2\left(\mathbf{r}'_2,t\right)}{2\sigma_2}E_{nm}\left(\mathbf{r}'_2,t-\tau_2\right), \tag{2.39}$$

where $\gamma_1(\mathbf{r}',t) \equiv 2C_n(\mathbf{r}'_1)\kappa(\mathbf{r}'_1,t)R_I$ and $\gamma_2(\mathbf{r}'_2,t) \equiv 2C_n(\mathbf{r}'_2)\kappa_2(\mathbf{r}'_2,t)R_{II}$ is doubled coefficient of loss/transfer in the first and second FBL; $R_I \equiv (R_2R_3R_4R_1)^{0.5}$, $R_{II} \equiv [R_2(1-R_3)(1-R_4)R_1]^{0.5}$; R_i are reflection coefficients of the appropriate mirrors; $C_n(\mathbf{r})$ are losses in the nonlinear medium; $\kappa_i(\mathbf{r}'_i,t)$ are losses in the elements (excluding mirrors) of i-th FBL of DNRI; σ_i is a stretching beam coefficient in i-th FBL; $\tau_i \equiv \tau_i(\mathbf{r}'_i,t) = t_{ei}(\mathbf{r}'_i,t) + U\left[\mathbf{r}',t-t_{e,i}(\mathbf{r}'_1,t)\right]/\omega$ is the propagation time for the optical field component arrived (through i-th FBL) to the time moment t in the point $\mathbf{r}$ of the NM input plane from the point $\mathbf{r}'_i$ of the same plane (the total delay time, the total time of the interferometer bypass through i-th FBL). Expressions (2.38) and (2.39) form the process model in DNRI.

Let us remind that if the amplitude $a_i(\mathbf{r},t)$ and phase $\varphi_i(\mathbf{r},t)$ of the interfering components are known, the amplitude $a(\mathbf{r},t)$ and phase $\varphi(\mathbf{r},t)$ of the total field $\mathbf{i}a(\mathbf{r},t)\cos[\omega t+\varphi(\mathbf{r},t)] = \mathbf{i}\sum_i a_i(\mathbf{r},t)\cos[\omega t+\varphi_i(\mathbf{r},t)]$ can be found according to the formulas

$$a(\mathbf{r},t) = \left(Ac^2 + As^2\right)^{0.5};\ \varphi(\mathbf{r},t) = \arg(Ac,As), \tag{2.40}$$

where $Ac \equiv a(\mathbf{r},t)\cos[\varphi(r,t)] = \sum_i a_i(\mathbf{r},t)\cos[\varphi_i(\mathbf{r},t)]$, $As \equiv a(\mathbf{r},t)\sin[\varphi(r,t)] = \sum_i a_i(\mathbf{r},t)\sin[\varphi_i(\mathbf{r},t)]$. Besides, let us remind that the field of intensity varies in time according to the law (2.22) and is characterized by amplitudes $a(\mathbf{r},t)$, $b(\mathbf{r},t)$, phases $\Psi(\mathbf{r},t) = \varphi(\mathbf{r},t) + \psi(\mathbf{r},t)$, $\Theta(\mathbf{r},t) = \varphi(\mathbf{r},t) - \psi(\mathbf{r},t)$, of two components with frequencies $\omega+\Omega$, $\omega-\Omega$.

Hence, to calculate $\mathbf{E}_{nm}(\mathbf{r},t)$ in the case of DNRI $i = 1,\ldots, 3$, and we must use formulas (2.40) for $a(\mathbf{r},t)$ and $\varphi(\mathbf{r},t)$ independently for components with frequency $\omega+\Omega$ and $\omega-\Omega$. In other words, for the description of the component with frequency $\omega+\Omega$ we should accept $a_1 = (1-R_1)^{0.5}a(\mathbf{r},t)$, $a_2 = \frac{\gamma_1\left(\mathbf{r}'_1,t\right)a_{nm}\left[\mathbf{r}'_1,t-\tau_1\right]}{2\sigma_1}$,

$a_3 = \frac{\gamma_2(\mathbf{r}'_2,t)a_{nm}[\mathbf{r}'_2,t-\tau_2]}{2\sigma_2}$, $\varphi_1 = \Psi(\mathbf{r},t)$, $\varphi_2 = \Psi_{nm}[\mathbf{r}'_1,t-\tau_1] - (\omega+\Omega)\tau_1$, $\varphi_3 = \Psi_{nm}[\mathbf{r}'_2,t-\tau_2] - (\omega+\Omega\tau_2)$, and for the components with frequency $\omega - \Omega$, similarly, $a_1 = (1-R_1)^{0.5}b(\mathbf{r},t)$, $a_2 = \frac{\gamma_1(\mathbf{r}'_1,t)b_{nm}[\mathbf{r}'_1,t-\tau_1]}{2\sigma_1}$, $a_3 = \frac{\gamma_2(\mathbf{r}'_2,t)b_{nm}[\mathbf{r}'_2,t-\tau_2]}{2\sigma_2}$, $\varphi_1 = \Theta(\mathbf{r},t)$, $\varphi_2 = \Theta_{nm}[\mathbf{r}'_1,t-\tau_1] - (\omega-\Omega)\tau_1$, $\varphi_3 = \Theta_{nm}[\mathbf{r}'_2,t-\tau_2] - (\omega-\Omega\tau_2)$. Then, the model (2.38), (2.39) of processes in DNRI can be rewrite in the form

$$
\begin{aligned}
&\tau_n(\mathbf{r})\frac{\partial U(r,t)}{\partial t} = D_e(\mathbf{r})\Delta U(\mathbf{r},t) - U(\mathbf{r},t) + f(\mathbf{r},t),\\
&f(\mathbf{r},t) = n_2(\mathbf{r})lkan\cdot\left\langle \mathbf{E}^2_{nm}(\mathbf{r},t)\right\rangle_T = an\cdot n_2(\mathbf{r})lk\left[a^2_{nm}(\mathbf{r},t) + b^2_{nm}(\mathbf{r},t)\right],\\
&\mathbf{E}_{nm}(\mathbf{r},t) = (1-R_1)^{0.5}\mathbf{E}(\mathbf{r},t) + \frac{\gamma_1(\mathbf{r}'_1,t)}{2\sigma_1}\mathbf{E}_{nm}(\mathbf{r}'_1,t-\tau_1)\\
&+\frac{\gamma_2(\mathbf{r}'_2,t)}{2\sigma_2}\mathbf{E}_{nm}(\mathbf{r}'_2,t-\tau_2),
\end{aligned}
$$

$$
\begin{aligned}
a_{nm}(\mathbf{r},t) &= \left(Ac_a^2 + As_a^2\right)^{0.5}, \quad \Psi_{nm}(\mathbf{r},t) = Arg(Ac_a, As_a),\\
b_{nm}(\mathbf{r},t) &= \left(Ac_b^2 + As_b^2\right)^{0.5}, \quad \Theta_{nm}(\mathbf{r},t) = Arg(Ac_b, As_b),
\end{aligned}
$$

$$
\begin{aligned}
Ac_a &= (1-R_1)^{0.5}a(\mathbf{r},t)\cos[\Psi(\mathbf{r},t)]\\
&+\frac{\gamma_1(\mathbf{r}'_1,t)}{2\sigma_1}a_{nm}(\mathbf{r}'_1,t-\tau_1)\cos\left[\Psi_{nm}(\mathbf{r}'_1,t-\tau_1) - (1+q)\omega\tau_1\right]\\
&+\frac{\gamma_2(\mathbf{r}'_2,t)}{2\sigma_2}a_{nm}(\mathbf{r}'_2,t-\tau_2)\cos\left[\Psi_{nm}(\mathbf{r}'_2,t-\tau_2) - (1+q)\omega\tau_2\right],\\
As_a &= (1-R_1)^{0.5}a(\mathbf{r},t)\sin[\Psi(\mathbf{r},t)]\\
&+\frac{\gamma_1(\mathbf{r}'_1,t)}{2\sigma_1}a_{nm}(\mathbf{r}'_1,t-\tau_1)\sin\left[\Psi_{nm}(\mathbf{r}'_1,t-\tau_1) - (1+q)\omega\tau_1\right]\\
&+\frac{\gamma_2(\mathbf{r}'_2,t)}{2\sigma_2}a_{nm}(\mathbf{r}'_2,t-\tau_2)\sin\left[\Psi_{nm}(\mathbf{r}'_2,t-\tau_2) - (1+q)\omega\tau_2\right],
\end{aligned}
$$

$$
\begin{aligned}
Ac_b &= (1-R_1)^{0.5}b(\mathbf{r},t)\cos[\Theta(\mathbf{r},t)]\\
&+\frac{\gamma_1(\mathbf{r}'_1,t)}{2\sigma_1}b_{nm}(\mathbf{r}'_1,t-\tau_1)\cos\left[\Theta_{nm}(\mathbf{r}'_1,t-\tau_1) - (1-q)\omega\tau_1\right]\\
&+\frac{\gamma_2(\mathbf{r}'_2,t)}{2\sigma_2}b_{nm}(\mathbf{r}'_2,t-\tau_2)\cos\left[\Theta_{nm}(\mathbf{r}'_2,t-\tau_2) - (1-q)\omega\tau_2\right],\\
As_b &= (1-R_1)^{0.5}b(\mathbf{r},t)\sin[\Theta(\mathbf{r},t)]\\
&+\frac{\gamma_1(\mathbf{r}'_1,t)}{2\sigma_1}b_{nm}(\mathbf{r}'_1,t-\tau_1)\sin\left[\Theta_{nm}(\mathbf{r}'_1,t-\tau_1) - (1-q)\omega\tau_1\right]\\
&+\frac{\gamma_2(\mathbf{r}'_2,t)}{2\sigma_2}b_{nm}(\mathbf{r}'_2,t-\tau_2)\sin\left[\Theta_{nm}(\mathbf{r}'_2,t-\tau_2) - (1-q)\omega\tau_2\right],
\end{aligned}
\tag{2.41}
$$

where $\tau_i \equiv \tau_i(r'_i,t) = t_{ei}(r'_i,t) + U\left[r'_i, t - t_{ei}(r'_i,t)\right]/\omega$.

Let us introduce designations

$$\begin{aligned}
a_{nm,r}(\mathbf{r},t) &\equiv \frac{a_{nm}(\mathbf{r},t)}{(1-R_1)^{0.5}a(\mathbf{r},t)},\\
b_{nm,r}(\mathbf{r},t) &\equiv \frac{b_{nm}(\mathbf{r},t)}{(1-R_1)^{0.5}b(\mathbf{r},t)},\\
Kab(\mathbf{r},t,\mathbf{r}_n) &\equiv (1-R_1)an\cdot n_2(r_n)lk[a^2(r,t)+b^2(r,t)],\\
Ka(\mathbf{r}_n,\mathbf{r},t,\mathbf{r}',t-\tau) &\equiv (1-R_1)an\cdot n_2(\mathbf{r}_n)lka(\mathbf{r},t)a(\mathbf{r}',t-\tau),\\
Kb(\mathbf{r}_n,\mathbf{r},t,\mathbf{r}',t-\tau) &\equiv (1-R_1)an\cdot n_2(\mathbf{r}_n)lkb(\mathbf{r},t)b(\mathbf{r}',t-\tau).
\end{aligned} \tag{2.42}$$

Than, the model (2.41) can be rewritten as

$$\begin{aligned}
&\tau_n(\mathbf{r})\frac{\partial U(\mathbf{r},t)}{\partial t} = D_e(\mathbf{r})\Delta U(\mathbf{r},t) - U(\mathbf{r},t) + f(\mathbf{r},t),\\
&f(\mathbf{r},t) = Kab(\mathbf{r},t,\mathbf{r})\\
&+\frac{\gamma_1^2(\mathbf{r}_1',t)}{4\sigma_1^2}[Ka[\mathbf{r},\mathbf{r}_1',t-\tau_1,\mathbf{r}_1',t-\tau_1]a_{nm,r}^2(r_1',t-\tau_1)\\
&+Kb[\mathbf{r},\mathbf{r}_1',t-\tau_1,\mathbf{r}_1',t-\tau_1]b_{nm,r}^2(r_1',t-\tau_1)]\\
&+\frac{\gamma_2^2(\mathbf{r}_2',t)}{4\sigma_2^2}[Ka[\mathbf{r},\mathbf{r}_2',t-\tau_2,\mathbf{r}_2',t-\tau_2]a_{nm,r}^2(r_2',t-\tau_2)\\
&+Kb[\mathbf{r},\mathbf{r}_2',t-\tau_2,\mathbf{r}_2',t-\tau_2]b_{nm,r}^2(r_2',t-\tau_2)]\\
&+\frac{\gamma_1(\mathbf{r}_1',t)}{\sigma_1}Ka[\mathbf{r},\mathbf{r},t,\mathbf{r}_1',t-\tau_1]a_{nm,r}(\mathbf{r}_1',t-\tau_1)\\
&\times\cos[\Psi(\mathbf{r},t)-\Psi_{nm}(\mathbf{r}_1',t-\tau_1)+(1+q)\omega\tau_1]\\
&+\frac{\gamma_2(\mathbf{r}_2',t)}{\sigma_2}Ka[\mathbf{r},\mathbf{r},t,\mathbf{r}_2',t-\tau_1]a_{nm,r}(\mathbf{r}_2',t-\tau_1)\\
&\times\cos[\Psi(\mathbf{r},t)-\Psi_{nm}(\mathbf{r}_2',t-\tau_2)+(1+q)\omega\tau_2]\\
&+\gamma_1(\mathbf{r}_1',t)Kb(\mathbf{r}_n,\mathbf{r},t,\mathbf{r}_1',t-\tau_1)b_{nmr}(\mathbf{r}_1',t-\tau_1)\\
&\times\cos[\Theta(\mathbf{r},t)-\Theta_{nm}(r_1',t-\tau_1)+(1-q)\omega\tau_1]/\sigma_1\\
&+\gamma_2(\mathbf{r}_2',t)Kb(\mathbf{r}_n,\mathbf{r},t,\mathbf{r}_2',t-\tau_2)b_{nmr}(\mathbf{r}_2',t-\tau_2)\\
&\times\cos[\Theta(\mathbf{r},t)-\Theta_{nm}(r_2',t-\tau_2)+(1-q)\omega\tau_2]/\sigma_2
\end{aligned}$$

$$
\begin{aligned}
&+\frac{1}{2}\gamma_1(\mathbf{r}_1',t)\gamma_2(r_2',t)Ka(\mathbf{r}_n,\mathbf{r}_1',t-\tau_1,\mathbf{r}_2',t-\tau_2)\\
&\times a_{nm,r}(\mathbf{r}_1',t-\tau_1)a_{nm,r}(\mathbf{r}_2',t-\tau_2)\\
&\times \cos\left[(1+q)(\tau_2-\tau_1)\omega+\Psi_{nm}(\mathbf{r}_1',t-\tau_1)-\Psi_{nm}(\mathbf{r}_2',t-\tau_2)\right]/(\sigma_1\sigma_2)\\
&+\frac{1}{2}\gamma_1(\mathbf{r}_1',t)\gamma_2(r_2',t)Kb(\mathbf{r}_n,\mathbf{r}_1',t-\tau_1,\mathbf{r}_2',t-\tau_2)\\
&\times b_{nm,r}(\mathbf{r}_1',t-\tau_1)b_{nm,r}(\mathbf{r}_2',t-\tau_2)\\
&\times \cos\left[(1-q)(\tau_2-\tau_1)\omega+\Theta_{nm}(\mathbf{r}_1',t-\tau_1)-\Theta_{nm}(\mathbf{r}_2',t-\tau_2)\right]/(\sigma_1\sigma_2),
\end{aligned}
$$

$$
\begin{aligned}
a_{nm}(\mathbf{r},t) &= \left(Ac_a^2+As_a^2\right)^{0.5}, \quad \Psi_{nm}(\mathbf{r},t)=Arg(Ac_a,As_a),\\
b_{nm}(\mathbf{r},t) &= \left(Ac_b^2+As_b^2\right)^{0.5}, \quad \Theta_{nm}(\mathbf{r},t)=Arg(Ac_a,As_a),
\end{aligned}
$$

$$
\begin{aligned}
Ac_a &= (1-R_1)^{0.5}a(\mathbf{r},t)\cos[\Psi(\mathbf{r},t)]\\
&+\frac{\gamma_1(\mathbf{r}_1',t)}{2\sigma_1}a_{nm}(\mathbf{r}_1',t-\tau_1)\cos\left[\Psi_{nm}(\mathbf{r}_1',t-\tau_1)-(1+q)\omega\tau_1\right]\\
&+\frac{\gamma_2(\mathbf{r}_2',t)}{2\sigma_2}a_{nm}(\mathbf{r}_2',t-\tau_2)\cos\left[\Psi_{nm}(\mathbf{r}_2',t-\tau_2)-(1-q)\omega\tau_2\right],\\
As_a &= (1-R_1)^{0.5}a(\mathbf{r},t)\sin[\Psi(\mathbf{r},t)]\\
&+\frac{\gamma_1(\mathbf{r}_1',t)}{2\sigma_1}a_{nm}(\mathbf{r}_1',t-\tau_1)\sin\left[\Psi_{nm}(\mathbf{r}_1',t-\tau_1)-(1+q)\omega\tau_1\right]\\
&+\frac{\gamma_2(\mathbf{r}_2',t)}{2\sigma_2}a_{nm}(\mathbf{r}_2',t-\tau_2)\sin\left[\Psi_{nm}(\mathbf{r}_2',t-\tau_2)-(1-q)\omega\tau_2\right],\\
Ac_b &= (1-R_1)^{0.5}b(\mathbf{r},t)\cos[\Theta(\mathbf{r},t)]\\
&+\frac{\gamma_1(\mathbf{r}_1',t)}{2\sigma_1}b_{nm}(\mathbf{r}_1',t-\tau_1)\cos\left[\Theta_{nm}(\mathbf{r}_1',t-\tau_1)-(1-q)\omega\tau_1\right]\\
&+\frac{\gamma_2(\mathbf{r}_2',t)}{2\sigma_2}b_{nm}(\mathbf{r}_2',t-\tau_2)\cos\left[\Theta_{nm}(\mathbf{r}_2',t-\tau_2)-(1-q)\omega\tau_2\right],\\
As_b &= (1-R_1)^{0.5}b(\mathbf{r},t)\sin[\Theta(\mathbf{r},t)]\\
&+\frac{\gamma_1(\mathbf{r}_1',t)}{2\sigma_1}b_{nm}(\mathbf{r}_1',t-\tau_1)\sin\left[\Theta_{nm}(\mathbf{r}_1',t-\tau_1)-(1-q)\omega\tau_1\right]\\
&+\frac{\gamma_2(\mathbf{r}_2',t)}{2\sigma_2}b_{nm}(\mathbf{r}_2',t-\tau_2)\sin\left[\Theta_{nm}(\mathbf{r}_2',t-\tau_2)-(1-q)\omega\tau_2\right].
\end{aligned}
\tag{2.43}
$$

In practical modeling, instead of (2.42), it is more convenient to use the designation:

$$
\begin{aligned}
a_{nm,n}(\mathbf{r},t) &= \frac{a_{nm}(\mathbf{r},t)}{(1-R_1)^{0.5}a_{in,\max\{\mathbf{r},t\}}},\\
b_{nm,n}(\mathbf{r},t) &= \frac{b_{nm}(\mathbf{r},t)}{(1-R_1)^{0.5}b_{in,\max\{\mathbf{r},t\}}},\\
a_n(\mathbf{r},t) &= \frac{a(\mathbf{r},t)}{a_{in,\max\{\mathbf{r},t\}}},\\
b_n(\mathbf{r},t) &= \frac{b(\mathbf{r},t)}{b_{in,\max\{\mathbf{r},t\}}},\\
n_{2n}(\mathbf{r}) &= \frac{n_2(\mathbf{r})}{n_{2\max\{\mathbf{r}\}}},\\
K_a &= (1-R_1)an\cdot n_{2\max\{\mathbf{r}\}}lk\left(a_{in,\max\{\mathbf{r},t\}}\right)^2,\\
K_b &= (1-R_1)an\cdot n_{2\max\{\mathbf{r}\}}lk\left(b_{in,\max\{\mathbf{r},t\}}\right)^2,\\
K &= K_a+K_b,\ Q_a = K_a/K,
\end{aligned}
$$

where $a_{in\max\{\mathbf{r},t\}}$, $b_{in\max\{\mathbf{r},t\}}$ are maximal values of the input field amplitude, $n_{2\max\{\mathbf{r}\}}$ is the maximal value of the nonlinear refraction parameter. Then, instead of (2.43) we finally obtain from the model (2.38)

$$
\begin{aligned}
\tau_n(\mathbf{r})\frac{\partial U(r,t)}{\partial t} =& D_e(\mathbf{r})\Delta U(\mathbf{r},t) - U(\mathbf{r},t) + f(\mathbf{r},t),\\
f(\mathbf{r},t) =& Q_aKn_{2n}(\mathbf{r})a^2_{nm,n}(\mathbf{r},t) + K(1-Q_a)n_{2n}(\mathbf{r})b^2_{nm.n}(\mathbf{r},t),\\
a_{nm,n}(\mathbf{r},t) =& \left(Ac_a^2+As_a^2\right)^{0.5},\ \Psi_{nm}(\mathbf{r},t) = Arg(Ac_a,As_a),\\
b_{nm,n}(\mathbf{r},t) =& \left(Ac_b^2+As_b^2\right)^{0.5},\ \Theta_{nm}(\mathbf{r},t) = Arg(Ac_b,As_b),
\end{aligned}
$$

$$
\begin{aligned}
Ac_a =& a_n(\mathbf{r},t)\cos[\Psi(\mathbf{r},t)]\\
&+\frac{\gamma_1(\mathbf{r}'_1,t)}{2\sigma_1}a_{nm,n}(\mathbf{r}'_1,t-\tau_1)\cos\left[\Psi_{nm}(\mathbf{r}'_1,t-\tau_1)-(1+q)\omega\tau_1\right]\\
&+\frac{\gamma_2(\mathbf{r}'_2,t)}{2\sigma_2}a_{nm,n}(\mathbf{r}'_2,t-\tau_2)\cos\left[\Psi_{nm}(\mathbf{r}'_2,t-\tau_2)-(1+q)\omega\tau_2\right],\\
As_a =& a_n(\mathbf{r},t)\sin[\Psi(\mathbf{r},t)]\\
&+\frac{\gamma_1(\mathbf{r}'_1,t)}{2\sigma_1}a_{nm,n}(\mathbf{r}'_1,t-\tau_1)\sin\left[\Psi_{nm}(\mathbf{r}'_1,t-\tau_1)-(1+q)\omega\tau_1\right]\\
&+\frac{\gamma_2(\mathbf{r}'_2,t)}{2\sigma_2}a_{nm,n}(\mathbf{r}'_2,t-\tau_2)\sin\left[\Psi_{nm}(\mathbf{r}'_2,t-\tau_2)-(1+q)\omega\tau_2\right],
\end{aligned}
$$

$$
\begin{aligned}
Ac_b =& b_n(\mathbf{r},t)\cos[\Theta(\mathbf{r},t)] \\
&+\frac{\gamma_1(\mathbf{r}_1',t)}{2\sigma_1}b_{nm,n}(\mathbf{r}_1',t-\tau_1)\cos\left[\Theta_{nm}(\mathbf{r}_1',t-\tau_1)-(1-q)\omega\tau_1\right] \\
&+\frac{\gamma_2(\mathbf{r}_2',t)}{2\sigma_2}b_{nm,n}(\mathbf{r}_2',t-\tau_2)\cos\left[\Theta_{nm}(\mathbf{r}_2',t-\tau_2)-(1-q)\omega\tau_2\right], \\
As_b =& b_n(\mathbf{r},t)\sin[\Theta(\mathbf{r},t)] \\
&+\frac{\gamma_1(\mathrm{r}_1',t)}{2\sigma_1}b_{nm,n}(\mathbf{r}_1',t-\tau_1)\sin\left[\Theta_{nm}(\mathbf{r}_1',t-\tau_1)-(1-q)\omega\tau_1\right] \\
&+\frac{\gamma_2(\mathrm{r}_2',t)}{2\sigma_2}b_{nm,n}(\mathbf{r}_2',t-\tau_2)\sin\left[\Theta_{nm}(\mathbf{r}_2',t-\tau_2)-(1-q)\omega\tau_2\right].
\end{aligned}
\tag{2.44}
$$

Similar to this, as it was made above for NRI, let us perform the transition from (2.44) to the DNRI point model (i.e. those in which $D_e = 0$).

The model of optical field dynamics in the double-circuit nonlinear ring interferometer ("point" model, in terms of discrete mapping). Under rotation of the light beam by the angle $\Delta_j = 2\pi M_j/m$ (or for the beam shift by $\delta_j = M_j\delta$ or when its compression is by $\sigma_j = M_j\sigma$), the point model composed for single CPT has a form

$$
\begin{aligned}
\tau_{ni}\frac{dU(t)}{dt} =& -U_i(t)+Q_aKn_{2n,i}a^2_{nm,n,i}(t)+K(1-Q_a)n_{2n,i}b^2_{nm,n,i}(t), \\
a_{nm,n,i}(t) =& \left(Ac^2_{ai}+As^2_{ai}\right)^{0.5},\ \Psi_{nm,i}(t)Arg(Ac_{ai},As_{ai}), \\
b_{nm,n,i}(t) =& \left(Ac^2_{bi}+As^2_{bi}\right)^{0.5},\ \Theta_{nm,i}(t)Arg(Ac_{bi},As_{bi}), \\
Ac_{ai} =& a_{ni}(t)\cos[\Psi_i(t)] \\
&+\frac{\gamma_{1,i-M_1}(t)}{2\sigma_1}a_{nm,n,i-M_1}(t-\tau_1)\cos\left[\Psi_{nm,i-M_1}(t-\tau_1)-(1+q)\omega\tau_1\right] \\
&+\frac{\gamma_{2,i-M_2}(t)}{2\sigma_2}a_{nm,n,i-M_2}(t-\tau_2)\cos\left[\Psi_{nm,i-M_2}(t-\tau_2)-(1+q)\omega\tau_2\right],
\end{aligned}
$$

$$\begin{aligned}
As_{ai} =& a_{ni}(t)\sin[\Psi_i(t)] \\
&+ \frac{\gamma_{1,i-M_1}(t)}{2\sigma_1} a_{nm,n,i-M_1}(t-\tau_1)\sin\left[\Psi_{nm,i-M_1}(t-\tau_1)-(1+q)\omega\tau_1\right] \\
&+ \frac{\gamma_{2,i-M_2}(t)}{2\sigma_2} a_{nm,n,i-M_2}(t-\tau_2)\sin\left[\Psi_{nm,i-M_2}(t-\tau_2)-(1+q)\omega\tau_2\right], \\
Ac_{bi} =& b_{ni}(t)\cos[\Theta_i(t)] \\
&+ \frac{\gamma_{1,i-M_1}(t)}{2\sigma_1} b_{nm,n,i-M_1}(t-\tau_1)\cos\left[\Theta_{nm,i-M_1}(t-\tau_1)-(1-q)\omega\tau_1\right] \\
&+ \frac{\gamma_{2,i-M_2}(t)}{2\sigma_2} b_{nm,n,i-M_2}(t-\tau_2)\cos\left[\Theta_{nm,i-M_2}(t-\tau_2)-(1-q)\omega\tau_2\right], \\
As_{bi} =& b_{ni}(t)\sin[\Theta_i(t)] \\
&+ \frac{\gamma_{1,i-M_1}(t)}{2\sigma_1} b_{nm,n,i-M_1}(t-\tau_1)\sin\left[\Theta_{nm,i-M_1}(t-\tau_1)-(1-q)\omega\tau_1\right] \\
&+ \frac{\gamma_{2,i-M_2}(t)}{2\sigma_2} b_{nm,n,i-M_2}(t-\tau_2)\sin\left[\Theta_{nm,i-M_2}(t-\tau_2)-(1-q)\omega\tau_2\right],
\end{aligned} \tag{2.45}$$

where $\tau_1 \equiv \tau_{1,i-M_1}(t) = t_{e1,i-M_1}(t) + U_{i-M_1}\left[t - t_{e1,i-M_1}(t)\right]/\omega$, $\tau_2 \equiv \tau_{2,i-M_2}(t) = t_{e2,i-M_2}(t) + U_{i-M_2}\left[t - t_{e2,i-M_2}(t)\right]/\omega$.

Let us ask the question: how great is the number of points in CTP (what is the range for the variation of the point numbers)? It is clear that this number is conditioned by the CTP structure type. This type, in turn, is determined by the character of the large-scale spatial transformation of the light field. In contrast with the NRI case, in DNRI such transformation is performed by two linear elements $G_i()$ and not by only one of them. We remind that the main types of CTP for NRI are demonstrated in Fig. 2.32. They are the following: single-type beam transformations in two FBL $i \in [0, +\infty]$ with the bean compression; $i \in [-\infty, 0]$ with stretching; $i \in [1, r_b/\delta]$ when there is a shift by the value $\delta_j = M_j\delta$ (r_b is the beam radius); $i \in [1, m]$ under rotation by angles $\Delta_j = 2\pi M_j/m$ (M_1 and m, M_2 and m are mutually prime integer numbers). Here, (in terms of CTP) M_i is a value of displacement step within CTP. For instance, when there is a field rotation, M_i is a numerical expression (in units of $2\pi/m$) of the field rotation angle Δ_i. We shall speak later in detail about CTP types for DNRI.

As we already mentioned in the preamble to the dynamics U model construction in terms of discrete mapping (2.36) and (2.37), together with the ODE point models, it is expedient to research these mapping procedures. To reveal them in the case of DNRI, we assume that the latter operates in the static mode, i.e. values

$a_{nm,n,i}$, $b_{nm,n.i}$, $a_{n,i}$, $b_{n,i}$, $\Psi_{nm,i}$, $\Theta_{nm,i}$, Ψ_i, Θ_i, $\gamma_{j,i}$, $t_{e,j,i}$ are constants in time. Then, instead of (2.45), we have:

$$\begin{aligned}
a_{nm,n,i} =& \left(Ac_{a,i}^2 + As_{a,i}^2\right)^{0.5}, \quad \Psi_{nm,i} = Arg\left(Ac_{a,i}, As_{a,i}\right), \\
b_{nm,n,i} =& \left(Ac_{b,i}^2 + As_{b,i}^2\right)^{0.5}, \quad \Theta_{nm,i} = Arg\left(Ac_{a,i}, As_{a,i}\right), \\
Ac_{ai} =& a_{ni}(t)\cos[\Psi_i] \\
&+ \frac{\gamma_{1,i-M_1}}{2\sigma_1} a_{nm,n,i-M_1} \cos\left[\Psi_{nm,i-M_1} - (1+q)\omega\tau_1\right] \\
&+ \frac{\gamma_{2,i-M_2}}{2\sigma_2} a_{nm,n,i-M_2} \cos\left[\Psi_{nm,i-M_2} - (1+q)\omega\tau_2\right], \\
As_{ai} =& a_{ni}\sin[\Psi_i] \\
&+ \frac{\gamma_{1,i-M_1}}{2\sigma_1} a_{nm,n,i-M_1} \sin\left[\Psi_{nm,i-M_1} - (1+q)\omega\tau_1\right] \\
&+ \frac{\gamma_{2,i-M_2}}{2\sigma_2} a_{nm,n,i-M_2} \sin\left[\Psi_{nm,i-M_2} - (1+q)\omega\tau_2\right], \\
Ac_{bi} =& b_{ni}\cos[\Theta_i] \\
&+ \frac{\gamma_{1,i-M_1}}{2\sigma_1} b_{nm,n,i-M_1} \cos\left[\Theta_{nm,i-M_1} - (1-q)\omega\tau_1\right] \\
&+ \frac{\gamma_{2,i-M_2}}{2\sigma_2} b_{nm,n,i-M_2} \cos\left[\Theta_{nm,i-M_2} - (1-q)\omega\tau_2\right], \\
As_{bi} =& b_{ni}\sin[\Theta_i] \\
&+ \frac{\gamma_{1,i-M_1}}{2\sigma_1} b_{nm,n,i-M_1} \sin\left[\Theta_{nm,i-M_1} - (1-q)\omega\tau_1\right] \\
&+ \frac{\gamma_{2,i-M_2}}{2\sigma_2} b_{nm,n,i-M_2} \sin\left[\Theta_{nm,i-M_2} - (1-q)\omega\tau_2\right],
\end{aligned} \tag{2.46}$$

where i is a point number in the chain of the transposition points (in the transverse plane of the laser beam), $\tau_1 \equiv \tau_{1,i-M_1} = t_{e1,i-M_1} + U_{i-M_1}/\omega$, $\tau_2 \equiv \tau_{2,i-M_2} = t_{e2,i-M_2} + U_{i-M_2}/\omega$, $U_i = Q_a K n_{2n,i} a_{nm,n,i}^2 + K(1-Q_a) n_{2n,i} b_{nm,n,i}^2$.

Evidently, obtained Eq. (2.46) can be interpreted as the vector function of the vector argument of type $\mathbf{E}_i = \mathbf{F}(\mathbf{E}_{i-M_1}, \mathbf{E}_{i-M_2})$, where the designation $\mathbf{E}_i \equiv \left(a_{nm,n,i}, \Psi_{nm,i}, b_{nm,n,i}, \Theta_{nm,i}\right)$ is used. Let $M_1 \geq M_2$. Then, DM serves in DNRI as a model of the *spatial* variation of amplitudes $a_{nm,n,i}$, $b_{nm,n,i}$ and phases $\Theta_{nm,i}$ of fields, and it is composed of the M_1 of vector equalities through the function $\mathbf{E}_i$ (here DM has a dimension $4M_1 \leq 4m$):

$$\mathbf{E}_i = \mathbf{F}(\mathbf{E}_{i-M_1}, \mathbf{E}_{i-M_2}),$$
$$\mathbf{E}_{i+1} = \mathbf{F}(\mathbf{E}_{i+1-M_1}, \mathbf{E}_{i+1-M_2}),$$
$$\ldots\ldots\ldots\ldots\ldots\ldots\ldots\ldots\ldots\ldots$$
$$\mathbf{E}_{i+M_2} = \mathbf{F}[\mathbf{E}_{i+M_2-M_1}, \mathbf{F}(\mathbf{E}_{i+M_2-M_2+M_1}, E_{i+M_2-M_2-M_2})],$$
$$\mathbf{E}_{i+M_2+1} = \mathbf{F}[\mathbf{E}_{i+M_2+1-M_1}, \mathbf{F}(\mathbf{E}_{i+M_2+1-M_2-M_1}, \boldsymbol{E}_{i+M_2+1-M_2-M_2})],$$
$$\ldots\ldots\ldots\ldots\ldots\ldots\ldots\ldots\ldots\ldots\ldots\ldots\ldots\ldots$$
$$\mathbf{E}_{i+2M_2} = \mathbf{F}\{\mathbf{E}_{i+2M_2-M_1}, \mathbf{F}[\mathbf{E}_{i+2M_2-M_2-M_1}, \mathbf{F}(\mathbf{E}_{i+2M_2-2M_2-M_1}, \mathbf{E}_{i+2M_2-3M_2})]\},$$
$$\mathbf{E}_{i+2M_2+1} = \mathbf{F}\{\mathbf{E}_{i+2M_2+1-M_1}, \mathbf{F}[\mathbf{E}_{i+2M_2+1-M_2-M_1},$$
$$\mathbf{F}(\mathbf{E}_{i+2M_2+1-2M_2-M_1}, \mathbf{E}_{i+2M_2+1-3M_2})]\},$$
$$\ldots\ldots\ldots\ldots\ldots\ldots\ldots\ldots\ldots\ldots\ldots\ldots$$
$$\mathbf{E}_{i+M_1-1} = \ldots$$

Obtained DM should be rewritten into the traditional view after extracting two indices from the only one: the first one i responding to the iteration number, the second l—to the equation number in DM, where $l = nM_2 + j,\ l = [1; M_1]$; $j = [1; M_2 - 1]$:

$$\mathbf{E}_{i+1,l} = \mathbf{F}\Big\{\mathbf{E}_{i,l-0M_2}, \mathbf{F}\Big[\mathbf{E}_{i,l-1M_2}, \mathbf{F}(\mathbf{E}_{i-2M_2}, \ldots \mathbf{F}\big(\mathbf{E}_{i,l-nM_2}, \mathbf{E}_{i,M_1-M_2+j}\big)\ldots)\Big]\Big\}. \tag{2.47}$$

The essential DNRI peculiarity is the rich variety of values of spatial field transformation angles in i-th FBL and, to a considerable degree, by the variety of their combinations. This gives birth to the variety of the geometric structures of CTP, which requires separate consideration.

Features of the field transformation in DNRI feedback loops and a concept of isodynamics. We have already mentioned above the availability of DNRI application for the problem of information protection in the optical range. So, even within the framework of the point model, for the case of optical field rotation in feedback loops by angles $\Delta_i = 2\pi M_i/m$ ($i \in \{1, 2\}$), a plenty of combinations of integer numbers M_1, M_2, m arises. This indicates the increase of the power of the key set compared to the NRI case. In its turn, the question arises about the presence of *equivalent* (isodynamic) combinations of the mentioned numbers. Let us specify in what terms we must understand here the term *isodynamics*. The idea of isodynamics can be illustrated with the help of the Ohm law. If some specific values of currents i_i and resistances r_i satisfy the relations $i_1 r_1 = i_2 r_2$, $i_1 \neq i_{2\ 2}$, $r_1 \neq r_2$, then the values (i_1, r_1) and (i_2, r_2) are isodynamic in terms of the influence on the voltage u at each of the two different conductors, since $u = i_1 r_1 = i_2 r_2$.

Now we generalize these interpretations for the case of the two abstract dynamic systems with dynamical variables (variables of system states) $U_{ij}(\mathbf{r}, t) \in \mathbf{U}_i(\mathbf{r}, t) \equiv \{U_{ij}(\mathbf{r}, t)\}$ and parameters $\mathbf{p}_i(\mathbf{r}, t) \equiv \{p_{i,k}\}$, where $i \in \{1; 2\}$ responds to the system number; $j \in \{1; \ldots; m\}$, $k \in \{1; \ldots; N_i\}$; m_i and N_1 is a number of dynamic variables and parameters for the i-th system. We will state that dynamics (evolutions) of the two system copies are isodynamic in terms of $\mathbf{F}$, if the relation of isodynamics of evolutions is satisfied

$$\mathbf{F}[\mathbf{r}, t, \mathbf{U}_1(\mathbf{r}, t), \mathbf{U}_1(\mathbf{r}, 0), \mathbf{U}_2(\mathbf{r}, t), \mathbf{U}_2(\mathbf{r}, 0), \mathbf{p}_1(\mathbf{r}, t), \mathbf{p}_2(\mathbf{r}, t)] \approx 0, \qquad (2.48)$$

where $\mathbf{F}[\ldots] \equiv \{F_1[\ldots]; \ldots; F_N[\ldots]\}$ is the some vector-function; $\mathbf{U}_1(\mathbf{r}, 0)$, $\mathbf{U}_2(\mathbf{r}, 0)$ are initial conditions. In the general case, in the relation (2.48) under arguments $\{\mathbf{U}_1(\mathbf{r}, t),\ \mathbf{U}_1(\mathbf{r}, 0),\ \mathbf{U}_2(\mathbf{r}, t),\ \mathbf{U}_2(\mathbf{r}, 0),\ \mathbf{p}_1(\mathbf{r}, t),\ \mathbf{p}_2(\mathbf{r}, t)\}$ we should understand the complete spatial-temporal realization of the functions, but not their separate values $\mathbf{U}_1, \mathbf{U}_2, \mathbf{p}_1, \mathbf{p}_2$ in the single point $(\mathbf{r}, t)$. The relation (2.48) in the specific case can take a meaning of the relation of identity evolutions: $\mathbf{U}_1(\mathbf{r}, t) - U_2(\mathbf{r}, t) = 0$. We will put that values $[\mathbf{U}_1(\mathbf{r}, 0),\ \mathbf{p}_1(\mathbf{r}, t)]$ and $[\mathbf{U}_2(\mathbf{r}, 0),$ $\mathbf{p}_2(\mathbf{r}, t)]$ are equivalent in terms of the chosen relation $\mathbf{F}[\ldots] \approx 0$, if the condition (2.48) is satisfied [50, 51]. Naturally, the ascertainment of isodynamic properties is relevant for the solution of the problem about unicity of the cryptographic key [50, 51].

Let us return to the analysis of the structure of the chain of the transposition points in the transverse section of the laser beam. For the simplest *types* of field transformations in the element G of the single-circuit NRI, the CTP *types* are shown in Fig. 2.32. It its turn, the CTP type determines the structure of the mutual dependence of dynamic variable values, i.e. the dependences of U_{i+1} from U_i, in combination with the characteristics of closeness and non-closeness and finiteness (infiniteness).

In the case of DNRI, the complication of CTP structure is possible not only due to the inevitable extension of typology (because of increase in the number of combinations of the field transformation types), but because of the complication of CTP structure inside each of its type. As before, we are limited by rotation of the light field in i-th FBL by the angle $\Delta_i = 2\pi M_i/m$. Then, depending on combination of numbers M_i and m values, the CTP structure essentially changes. For three numbers (three indices): m, M_1, M_2 we can consider as the natural "coding" of the structure. In Tables 2.2, 2.3 and 2.4, where various configurations of transitions between points in CTP (various structures of chains of the transposition points), and in further description we use exactly this coding. In figures and tables, the lines represent transitions between points in CTP: the dashed lines—transitions in FBL_1, the solid ones—in FBL_2.

From Tables 2.2, 2.3 and 2.4 we clearly see how manifold the CTP structure begins to complicate with the number m growth at the expense of the second FBL presence. In case of the single FBL, each of the Tables 2.3 and 2.4 would contain only the one cell corresponding to the three numbers $m11$. But the representation way used in Tables 2.3 and 2.4 is redundant. For instance, CTP with indices 221 is *isodynamic* to CTP with indices 212, since DNRI with CTP 221 and DNRI 212 are isodynamic (in terms of a variety of the possible types of modes in the interferometer [51]). Moreover, CTP 222 is divided into a pair of independent CTP 111, i.e., strictly speaking, CTP 222 is not the chain of the transposition points. Similarly, CTP 333 is divided into three CTP 111. In the last two cases, the DNRI with CTP mmm is identical to the DNRI with CTP 111 [52].

Table 2.2 Transition configuration between points in CTP in case of the "single-point" (m = 1) process model in DNRI—double-circuit nonlinear ring interferometer (DNRI)

m=1		M_2
		1
M_1	1	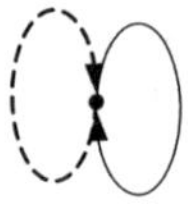 111

Table 2.3 Transition configuration between point in CTP in case of the "double-point" (m = 2) process model in DNRI—double-circuit nonlinear ring interferometer (DNRI)

m=2		M_2	
		1	2
M_1	1	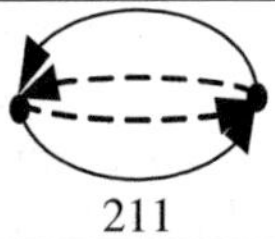 211	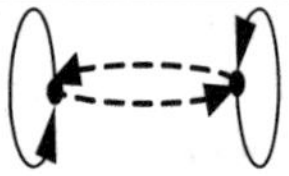 212
	2	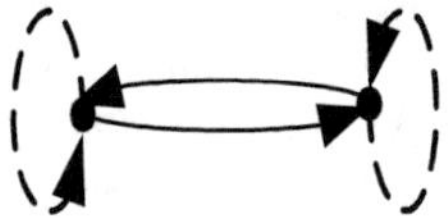 221=212	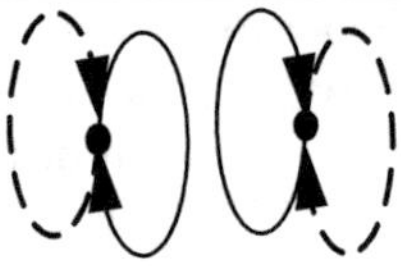 222=2(111)

If m is a prime number, the only first line is relevant in Tables 2.2, 2.3 and 2.4, all other transition configurations can be represented through configurations, which are present in it. Otherwise, one should choose values of triple indices in tables according to the following rule: $m = M_1N_1$, $m = M_2N_2$, where M_1 and M_2 are mutually prime numbers.

For example, for CTP composed of six points, the $M_1 = 3$ copies of $N_1 = 2$ (double) point structures and $M_2 = 2$ copies of $N_2 = 3$ (triple) point structures are realized. This case is represented in Fig. 2.35.

The case presented in Fig. 2.35, can be interpreted as a system of *three* oscillators (double-point NRI) related to six communication channels, but we can interpret it as a system of *two* oscillators (three-point NRI) related to the same six communication channels. On the contrary, the case CTP 511 and others, seemingly, should be interpreted as an appearance of the system of five additional connections in the one oscillator (five-point NRI), due to the arrangement of the second FBL [52].

Table 2.4 Transition configuration between points in CTP in case of the "triple-point" ($m = 3$) process model in DNRI—double-circuit nonlinear ring interferometer (DNRI)

m=3		M_2		
		1	2	3
M_1	1	311	312	313
	2	321=312	322=311	323=313
	3	331=313	332=323	333=3(111)

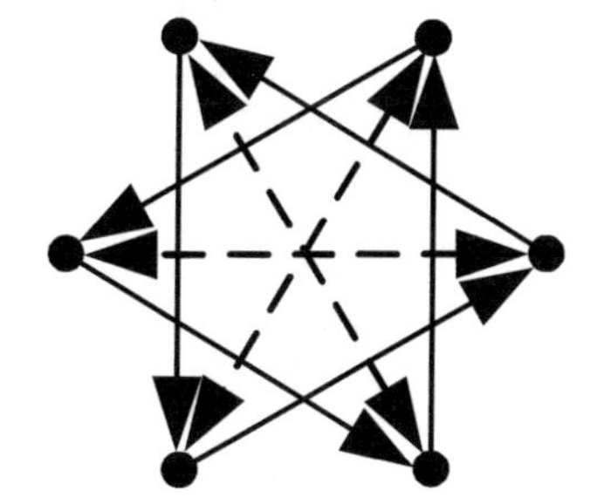

Fig. 2.35 Possible structures of transitions between six points forming CTP. *Dashed lines* represent transition between CTP points in the first FBL, *solid lines*—in the second FBL

Now we have mathematical models for dynamics of amplitude and phase of the light fields in ordinary and modified NRIs. Besides, for DNRI, the connection of CTP structure with possible spatial field transformations is discovered with the help of linear elements G_i in FBL, and the isodynamic combinations of these transformations are also revealed. We do not aim at the systematic of all the aspects of nonlinear ring interferometers behavior, we, however, now proceed to the demonstration of their properties relevant for the cryptologic applications specifically. Involving the numerical methods for this purpose, we are to show that NRI and DNRI are capable of functioning in the chaotic mode.

2.4.3 Dynamics in the Ring Interferometer Models

Regular and chaotic behavior in the NRI point model and in NRI DM model. A concept of the spatial deterministic chaos. For instance, the computing experiment, whose results Figs. 2.36 and 2.37 demonstrate, indicates the complicated dynamics in the point model (2.32) of NRI. These figures show, for cases of absence and presence of amplitude modulation of the input field for two different nonlinearity coefficients, the time realization, phase portraits and Fourier spectra of the wave amplitude A_i at the output, where i is a number of transposition points in CTP. The figures allow us to see: the modulation action changes the amplitude A_i spectrum so greatly that under regular (especially, chaotic) behavior of the model the problem of recovery of the modulation type at NRI input is non-trivial.

Let us proceed to the discrete mapping (DM) model. We remind that DM (2.36) and (2.47) are obtained in approximation of the static operation mode of NRI. We

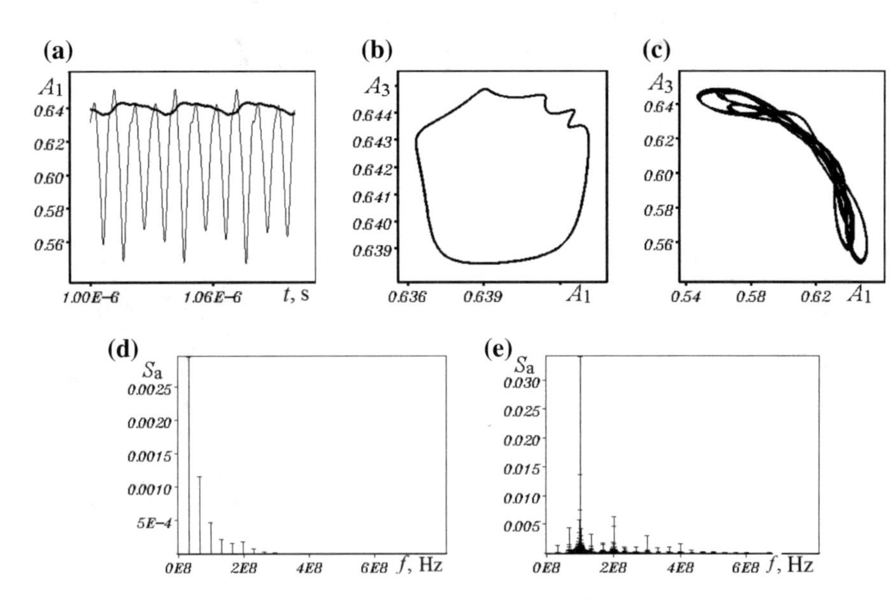

Fig. 2.36 Time realization A_1 (*a*), phase portraits (**b**, **c**), Fourier spectra S_a of output amplitude A_i of the wave (**d**, **e**) in the case of autonomous mode [*a* (*thick curve*), **b**, **d**] and the amplitude modulation mode [**a** (*this curve*), **c**, **d**]. $A_1 = A(\mathbf{r}, t)$, where $\mathbf{r} =(0{,}5;0)$; $A_3 = A(\mathbf{r}, t)$, where $\mathbf{r} = (0; 0.5)$. $K = 4.55$. The modulation law: $A_{in}(r, t) = [1 + 0.01\cos(2\pi f_1 t)/1.01]$, where $f_1 = \frac{1}{30.618 \cdot 10^{-9}} \approx 0.3266 \cdot 10^8$ corresponds to the frequency of the Fourier spectrum harmonic, which has the maximal amplitude (**e**)

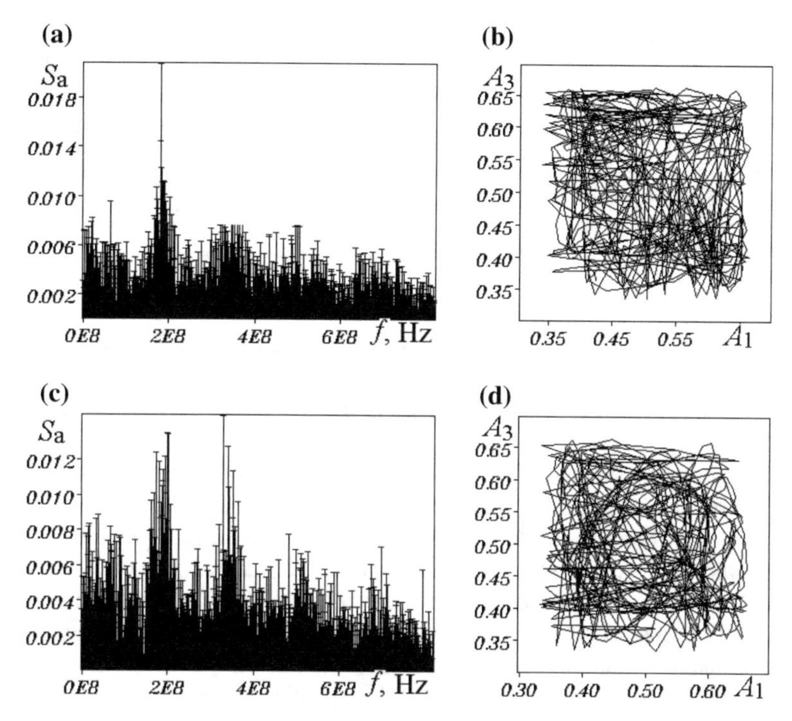

Fig. 2.37 Fourier spectra S_a of the output wave power A_1 (**a**, **c**) and phase portraits (**b**, **d**) in the case of the autonomous mode (*a*, *b*) and the amplitude modulation mode (**b**, **d**), which are the same as in Fig. 2.36. $K = 10$

can understand that this circumstance may cause confusion, as the question arises: what non-trivial behavior is possible in DM describing the NRI in its *static* mode? Here two answers suggest themselves. Firstly, the ODE point model transfers into the DM model, if we use the approximation of the instantaneous response. Secondly, the static mode, while excluding all time dynamics, by no means exclude the complex spatial distributions of the system dynamic variables. In the last case it is rightful to speak about the "spatial" deterministic chaos [53, 54]. Then, the variation of the discrete evolution variable i in DM formally corresponds to the examination of the space points (according to some algorithm) by the *observer*, who registers, for example, the values of $a_{nm,r,i}$, $b_{nm,r,i}$, $\varphi_{nm,i}$, $\psi_{nm,i}$, revealed in (2.36).

Here the term "deterministic" indicates that the disarray obeys some regulation built into the mathematical model and not caused by some random factor. Such a word usage is intended to specify the existing similarity/contrast with the widely known temporal, i.e. dynamic (deterministic), chaos in the single-dimensional system models, and with spatial-temporal chaos, or the turbulence, in the multi-dimensional models. We would like to note that systems, in which the spatial deterministic chaos is possible, *are not practically studied.*

For instance, Fig. 2.38 shows the complicated spatial dependence of the nonlinear phase shift from the number of the transposition point. The DM $U_{i+1} =$

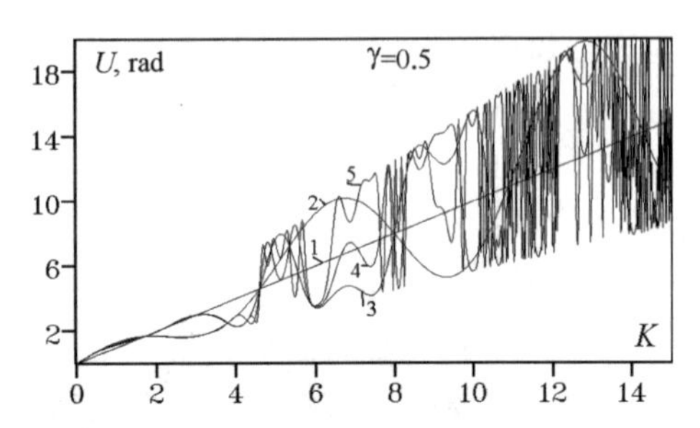

Fig. 2.38 Dependence of values of nonlinear phase shift U upon the nonlinearity parameter K in the first points of non-closed CTP for the static NRI mode at shift by the element G of the optical field (along the x-axis in the plane xOy). Numbers are the point numbers in CTP

$K[1+\gamma\cos(U_i+\Phi)]$ serves as its base, describing the spatial variation U_i in NRI in great losses approximation [55]. Then the complex distribution of the nonlinear phase shift $U(\mathbf{r})$ (see Fig. 2.32) is formed in the transverse plane of the laser beam in NRI.

Taking into consideration the cryptologic orientation of investigations, the phase portraits (Figs. 2.39 and 2.40) are constructed and the fractal dimension (capacity) of D_0 attractor is calculated for different sets of the DM model parameters (2.36). The example of attractor in four-dimensional dynamic system is shown in Fig. 2.40, where it is projected on the plane and is drawn in the three-dimensional space. As expected, increase of the fractal dimension is caused by the complication of the attractor structure and, accordingly, by the complication of NRI mode functioning. Usually, this is caused by the chaotic dynamics, necessary for the development of the confidential communication systems.

More complete picture of the influence of the parameters of the model upon the character of dynamics it demonstrates is achieved by the construction of the so-called Lyapunov characteristic exponents Λ (Fig. 2.41) and the maps of the fractal dimension on the plane of parameters under consideration (Fig. 2.42). Among them the most important are the nonlinearity coefficient K, the loss characteristic γ and the fraction Q_a of the component intensity with the frequency $\omega(1+q)$, since their set determines the nonlinearity degree in the system. The map $\Lambda(K,\gamma)$ in Fig. 2.41 reveals the possibility of both the regular ($\Lambda \leq 0$) and the chaotic dynamics ($\Lambda > 0$) in the model. A variety of the fractal dimension map structures $D_0(K,\gamma)$, depending on the parameters of input radiation spectrum, is illustrated by the series in Fig. 2.42. Maps are represented by grey color tins. The white color corresponds to the minimal value of the fractal dimension ($D_0 = 0$), the black color – to the maximal one. Since the attractor of zero dimension corresponds to the chaotic attractor, the structure of maps $\Lambda(K,\gamma)$ and $D_0(K,\gamma)$ should be similar. Indeed, maps in Figs. 2.41 and 2.42, *a* are similar in the structure, in spite of some difference in spectral characteristics of radiation q, Q_a.

As we know, there are no methodical recommendations in the specialized literature referring to the comparable analysis of maps of such a type. Therefore, it is difficult to present the impartial estimation of the variation degree in the map structure. In order to simplify the formation of the integral representation of the dynamic system properties, it is expedient to construct the plots of dimension density for these maps (Fig. 2.43), together with the maps of the fractal dimension

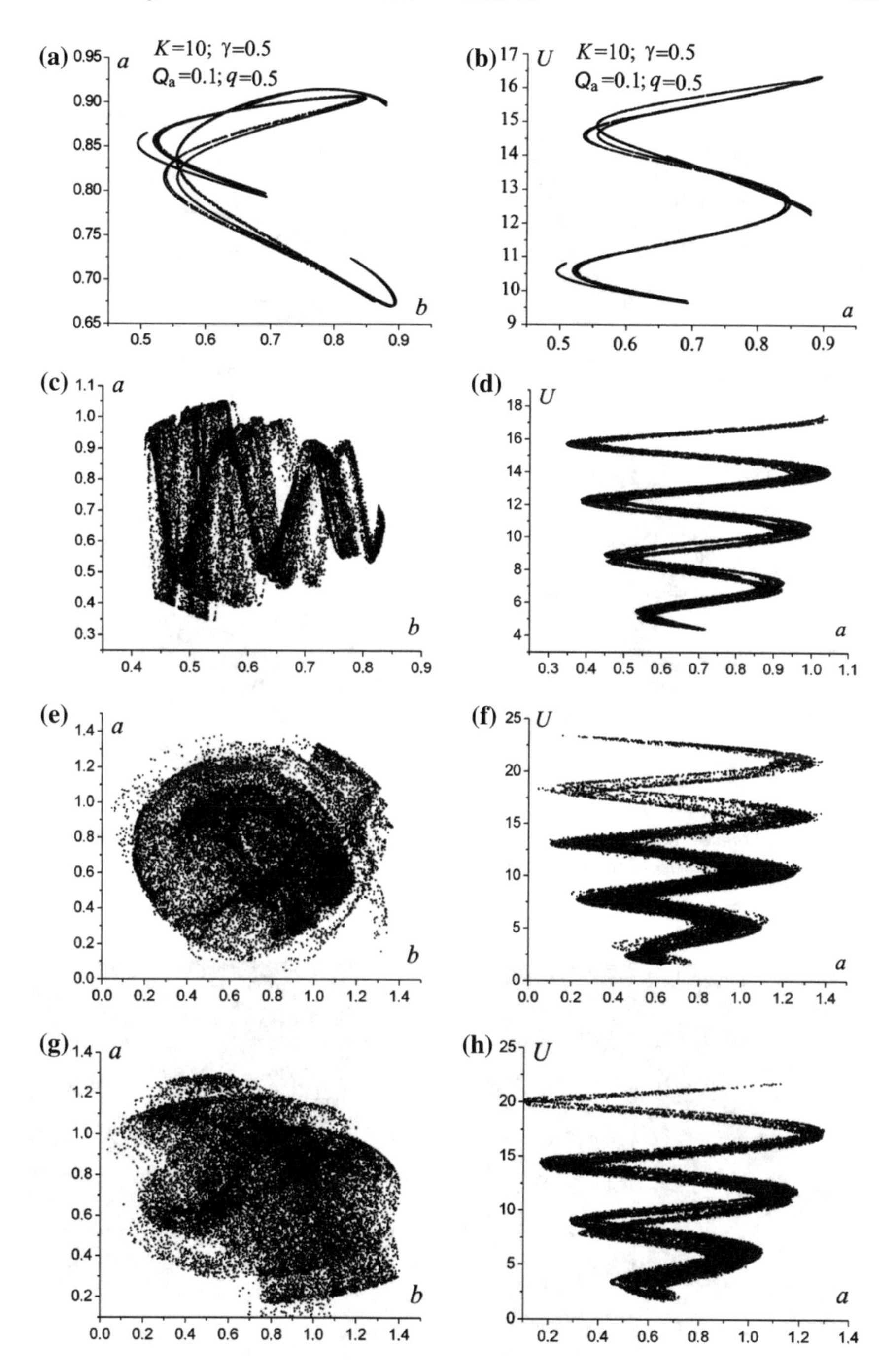

Fig. 2.39 Phase portraits of attractors with fractal dimension: $D_0 = 1.18496$ (**a**, **b**); $D_0 = 1.62791$ (**c**, **d**); $D_0 = 2.03342$ (**e**, **f**); $D_0 = 2.13637$ (**g**, **h**)

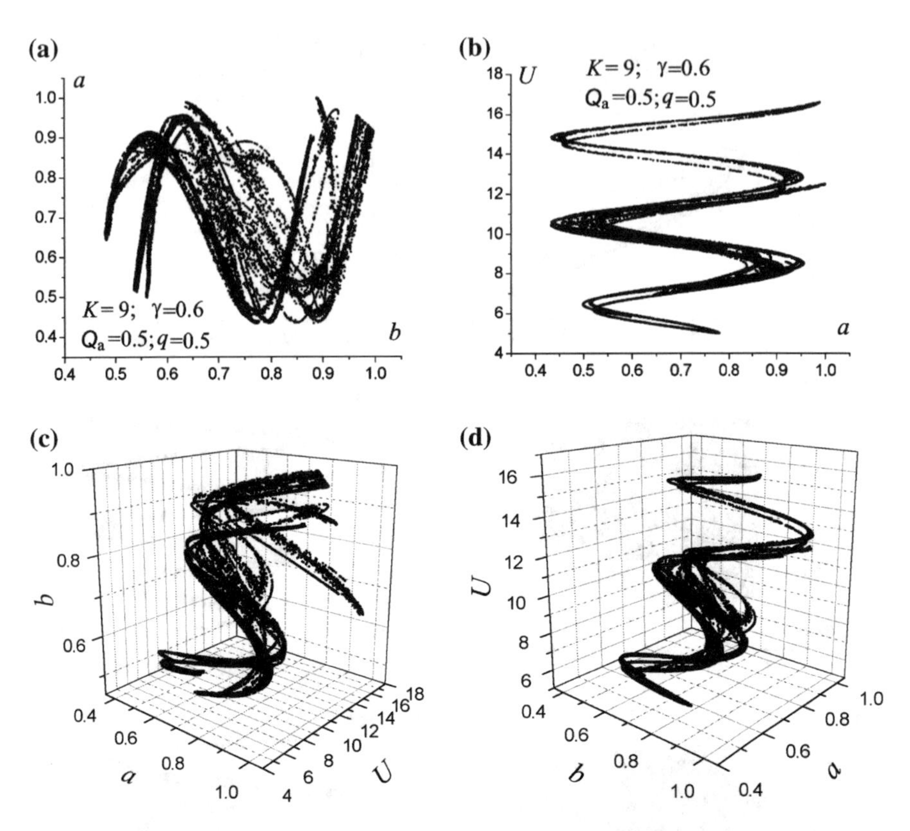

Fig. 2.40 2D-phase portraits (**a**, **b**) and 3D-phase portraits (**c**, **d**) of the attractor with fractal dimension $D_0 = 1.53023$ of the 4D-system

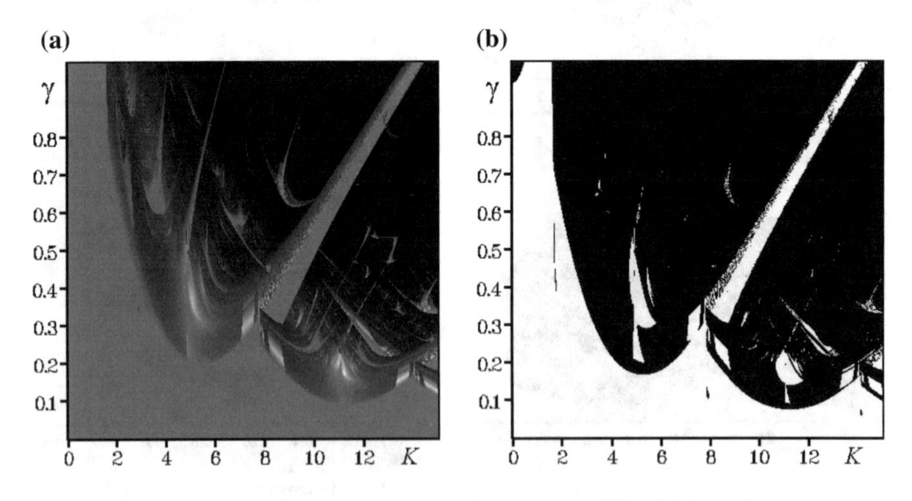

Fig. 2.41 Maps (in sense of charts) of Lyapunov's exponents $\Lambda(K, \gamma)$ (**a**) and the contrast map sign $[\Lambda(K, \gamma)]$ (**b**) for NRI model in the form of DM (2.36) for $q = 0$, $\Phi = 0$, $Q_a = 1$. The *white* color corresponds to the minimal value of Λ, the *black* color corresponds to maximal value

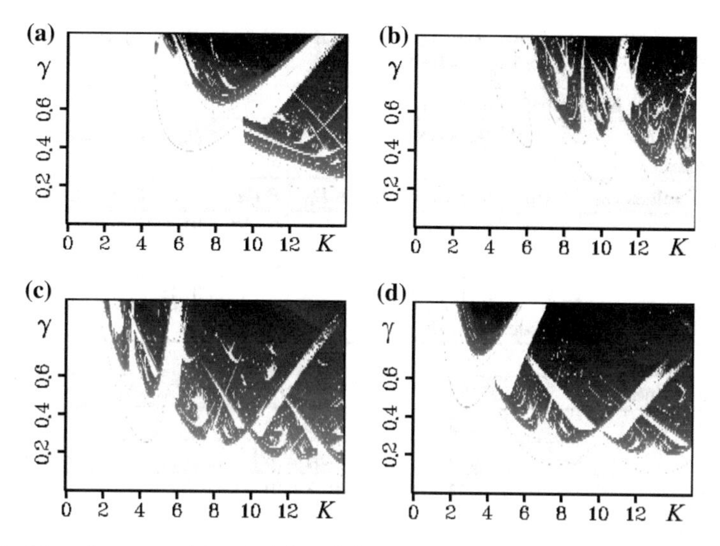

Fig. 2.42 Maps (in sense of charts) of fractal dimension of the attractor in the process model in NRI (2.36) at variation of bi-harmonic parameters: $q = 0.5$, $Q_a = 0.1$ (**a**); $q = 0.9$, $Q_a = 0.2$ (**b**); $q = 0.8$, $Q_a = 0.5$ (**c**); $q = 0.1$, $Q_a = 0.8$ (**d**)

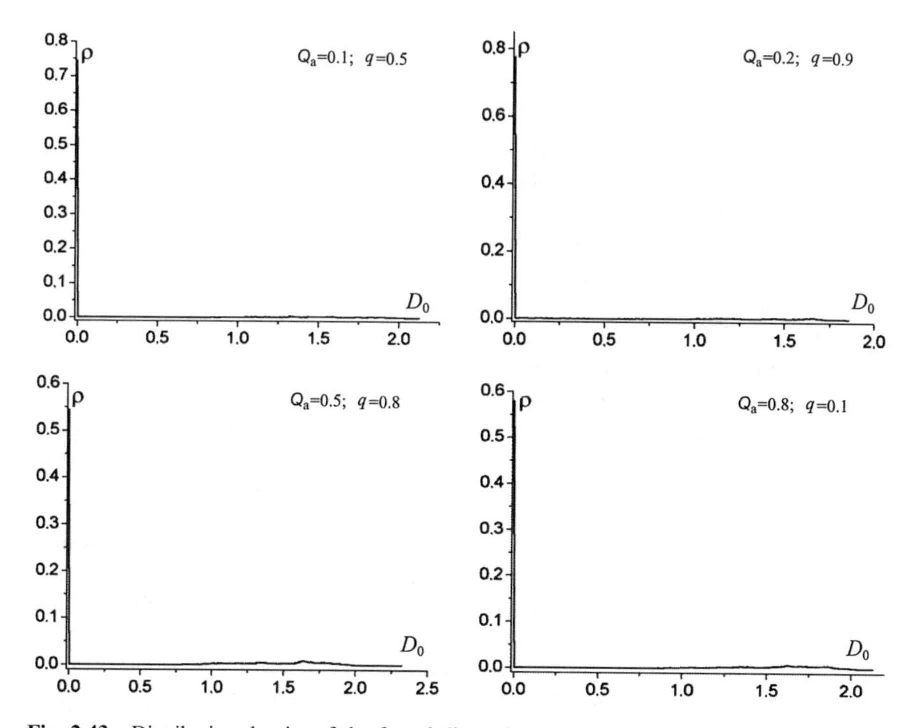

Fig. 2.43 Distribution density of the fractal dimension D_0 corresponding to the map $D_0(K, \gamma)$ corresponding to the appropriate model (2.36) at different values of bi-harmonic parameters: q and Q_a. $K \in [0; 15]$, $\gamma \in [0; 1]$

Table 2.5 A fraction P of map area of the fractal dimension $D_0(K,\gamma)$ corresponding to the interval of values D_0, $K \in [0; 15]$, $\gamma \in [0; 1]$

Characteristics of bi-chromatic spectrum					
$Q_a = 0.5$; $q = 0$ (monochrome)		$Q_a = 0.8$; $q = 0.1$		$Q_a = 0.5$; $q = 0.2$	
Interval of values D_0	P (%)	Interval of values D_0	P (%)	Interval of values D_0	P (%)
[0; 0.1]	54.93	[0; 0.1]	58.27	[0; 0.1]	66.51
[0.9; 1.53]	42.63	[0.9; 2.14]	39.94	[0.9; 2.49]	31.75
[0.1; 0.9]	2.45	[0.1; 0.9]	1.8	[0.1; 0.9]	1.74

Characteristics of bi-chromatic spectrum			
$Q_a = 0.1$; $q = 0.5$		$Q_a = 0.5$; $q = 0.5$	
Interval of values D_0	P (%)	Interval of values D_0	P (%)
[0; 0.1]	74.75	[0; 0.1]	59.53
[0.9; 2.13]	23.73	[0.9; 2.92]	38.38
[0.1; 0.9]	1.52	[0.1; 0.9]	2.11

Characteristics of bi-chromatic spectrum			
$Q_a = 0.5$; $q = 0.8$		$Q_a = 0.2$; $q = 0.9$	
Interval of values D_0	P (%)	Interval of values D_0	P (%)
[0; 0.1]	54.69	[0; 0.1]	77.89
[0.9; 2.33]	42.79	[0.9; 1.86]	19.96
[0.1; 0.9]	2.53	[0.1; 0.9]	2.53

of the attractor values D_0. Besides, it is also reasonable in this case to calculate the part P of the map area corresponding to this or that interval of D_0 values (Table 2.5). From the joint examination of Fig. 2.43 and Table 2.5, we can deduce how the percentages for $D_0 \in [0; 0.1]$, $D_0 \in [0.1; 0.9]$ and $D_0 \in [0.9; D_{0\max}]$ depend upon the set of values of the model parameters. As a whole, all these representations of the dynamic system behavior allow us to predict the most favorable ranges of parameters for the cryptosystem operation.

Let us proceed to the peculiarities of the DNRI model behavior.

Peculiarities of the process dynamics in the NRI point model with field rotation. Computer experiments show that the choice of DNRI parameters (the nonlinearity coefficient K, the transmission coefficient in FBL γ_i, delay time $t_{e,i}$ in FBL_1 and FBL_2, the field rotation angle Δ_i in each of the FBL) allows controlling the dynamics character. Thus, the control can be carried out by:

(1) the U mean values and the phase shift between U oscillations in (transposition) points of the transverse section of the laser beam;
(2) type and final state of the setting process whose possible form is the transient process, the transition being from the periodic mode to the quasi-periodic one (Fig. 2.44a); from periodic one to chaotic one (Fig. 2.44b); from static one to periodic one (Fig. 2.44c); from the periodic in-phase (at the two transposition points) mode to the periodic out-of-phase mode at the same points (Fig. 2.44d); from chaotic mode to periodic mode (Fig. 2.44e) [38].

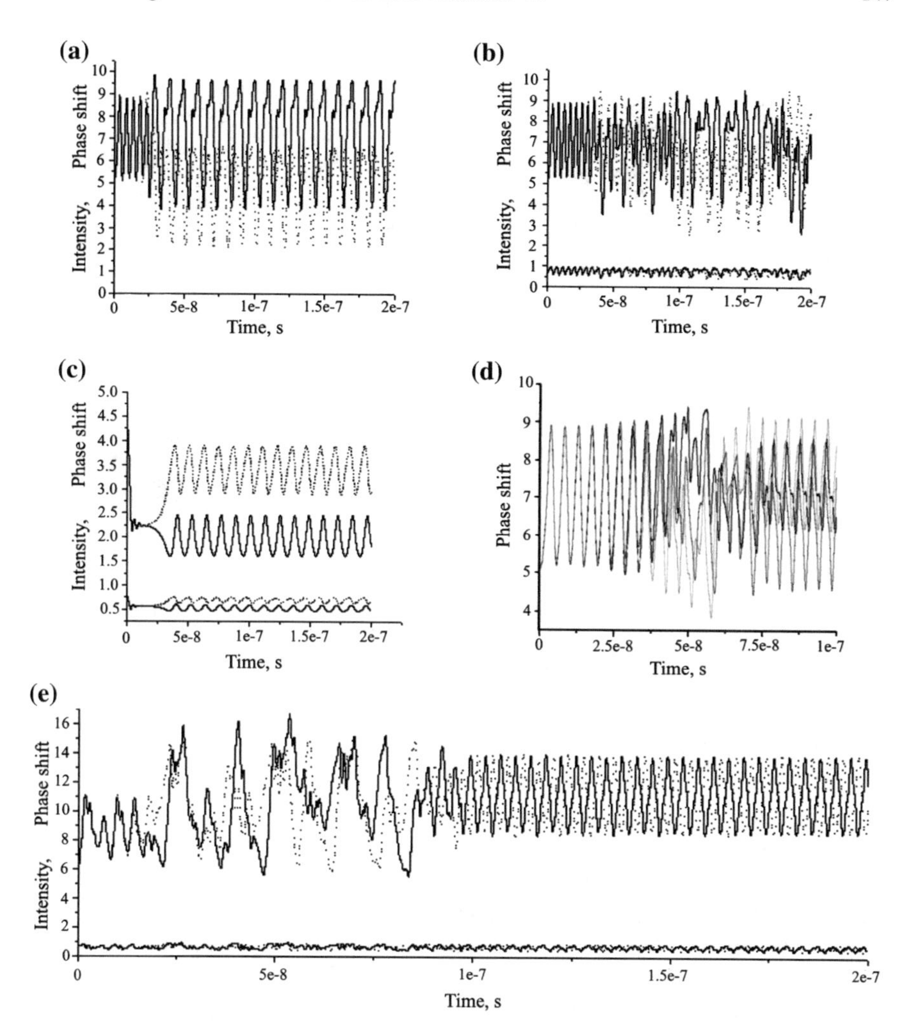

Fig. 2.44 Time realizations of the nonlinear phase shift in DNRI demonstrating the control (at the expense of the relation choice of parameter values) of character and final transient process for $\gamma_1 = (0.5\text{–}\gamma_2^2)^{0,5}$: $m = 2$, $\Delta_1 = \Delta_2 = 180°$, $t_{e1} = 1$, $t_{e2} = 2$ (*a–c*, *e*); $K = 5.5$, $\gamma_1 = 0.64$, $\gamma_2 = 0.3$ (**a**); $K = 5.5$, $\gamma_1 = 0.678$, $\gamma_2 = 0.2$ (**b**); $K = 3.5$, $\gamma_1 = 0.58$, $\gamma_2 = 0.4$ (**c**); $K = 10$, $\gamma_1 = 0.678$, $\gamma_2 = 0.2$ (**e**); $m = 3$, $\Delta_1 = 120°$, $\Delta_2 = 240°$, $K = 5.5$, $t_{e1} = 1$, $t_{e2} = 2$, $\gamma_1 = 0.66$, $\gamma_2 = 0{,}25$ (**d**)

From the structure of the phase portraits on the plane $A_{i+1}OA_i$, with varying parameters of the each of FBLs, we can see that for some model parameters the typical symmetric forms are possible (Figs. 2.45 and 2.46), along with such configurations, to which (quasi) periodic (Fig. 2.46) or aperiodic (Fig. 2.47) motions correspond. During the construction of the phase portraits, the transient process, which is capable of adding distortions to their structure, is eliminated.

For better understanding of how model parameters influence the character of the dynamic mode, we do the mapping of the fractal dimension $D_0(t_{e1}, t_{e2})$ of the

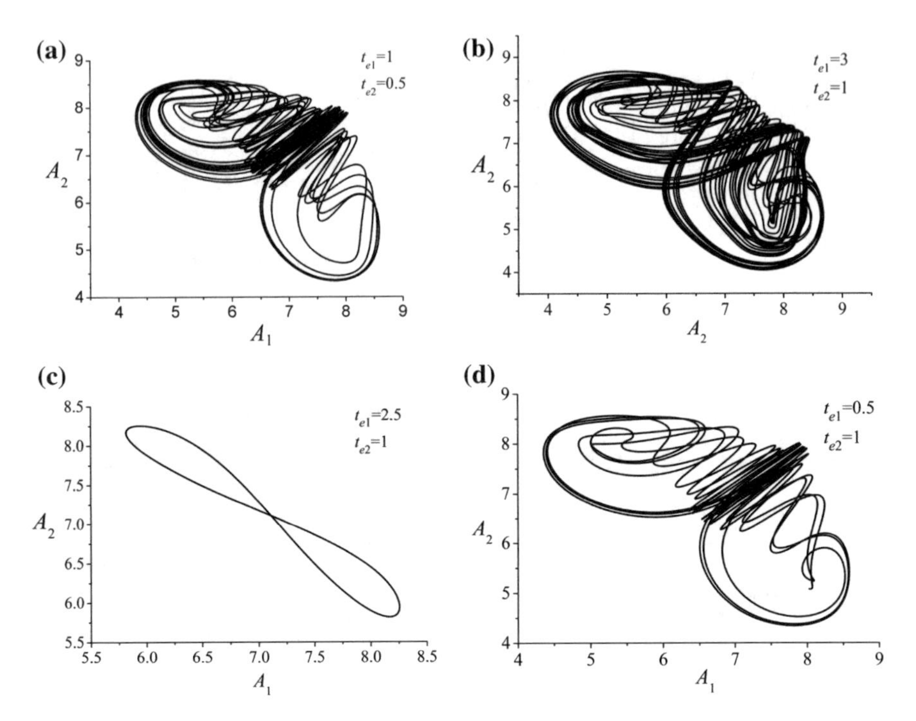

Fig. 2.45 Phase portraits of symmetric form; rotations $\Delta_1 = 180°$, $\Delta_2 = 180°$ (**a**, **b**, **d**); rotations $\Delta_1 = 180°$, $\Delta_2 = 0°$ (**c**)

attractor in the model (Fig. 2.48). As it is demonstrated above, the DNRI peculiarity is a diversity of geometric CTP structures. To find out how relations and angle values of field rotation Δ_1, Δ_2 in FBL_1 and FBL_2 affect the mapping structure of the fractal dimension $D_0(t_{e1}, t_{e2})$ and the fractional maps of the attractors $[|D_0 - round(D_0)|]$, it is necessary to calculate and construct its series. Here "round" means the rounding operation. Since these calculations are rather resource-capacious, it is expedient to decrease a number of calculated points. All maps (Figs. 2.49 and 2.50) are drawn under condition that FBL_1, FBL_2 in DNRI are isodynamic according to their influence on dynamics (i.e. double transmission coefficients γ_i are the same: $\gamma_1 = \gamma_2 = 0.25$).

Analysis of the map structure allows several generalizations. If the field rotation angle $\Delta \neq 0$ in both FBLs, the attractor fractal dimension does not decrease (as a rule, it is essentially increased) compared to the case when one of the angles $\Delta_i = 0$ [56]. The parameter areas where the attractor fractal dimension is high are themselves widening. Obviously, on the D_0 maps and on the fractional maps there are areas with the same "strangeness degrees" of attractors (the identical modes) that correspond to the various sets of time delay values t_{e1}, t_{e2} in FBL. These conclusions agree with the observations (Figs. 2.45, 2.46 and 2.47) of U behavior in

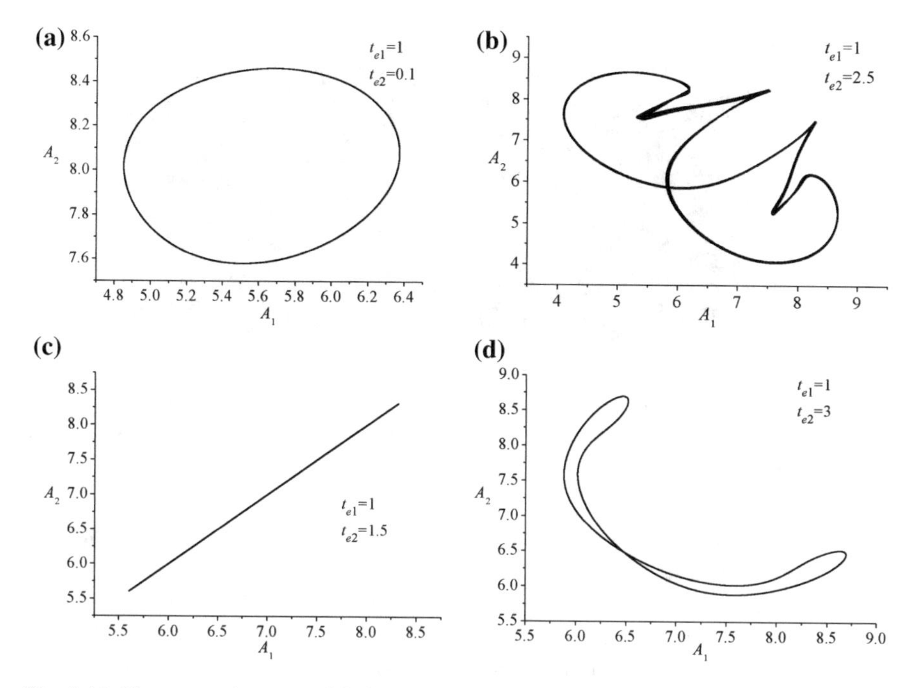

Fig. 2.46 Phase portraits responsible for periodic motions; rotations $\Delta_1 = 180°$, $\Delta_2 = 180°$ (**a**, **b**, **c**); rotations $\Delta_1 = 180°$, $\Delta_2 = 0°$ (**d**)

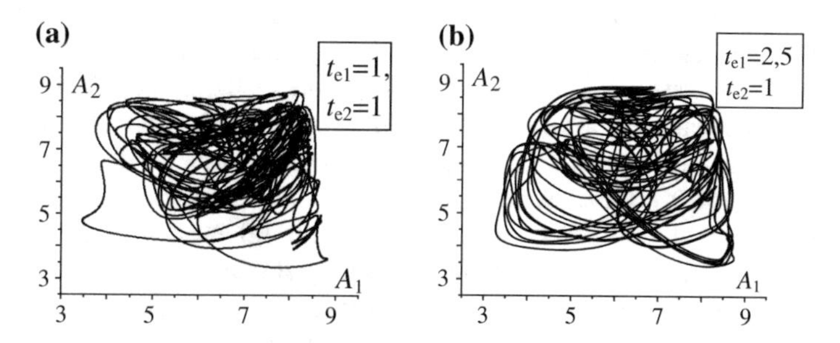

Fig. 2.47 Phase portraits responsible aperiodic motions; rotations $\Delta_1 = 120°$, $\Delta_2 = 120°$ for different values of t_{e1}, t_{e2}

various points of the beam transverse section: dynamics of U becomes complicated compared to the U behavior, when the one of the field rotation angles $\Delta_i = 0$. It is appropriate to remind that in accordance with publication [57], the great values of the fractal dimension $D_0(t_{e1}, t_{e2})$ are considered as a precondition for the high secretiveness of messages in the system of confidential communications.

Construction of the D_0 fractional map can serve as the reference point for the choice of DNRI parameters favorable for its functioning as an encoder applying the

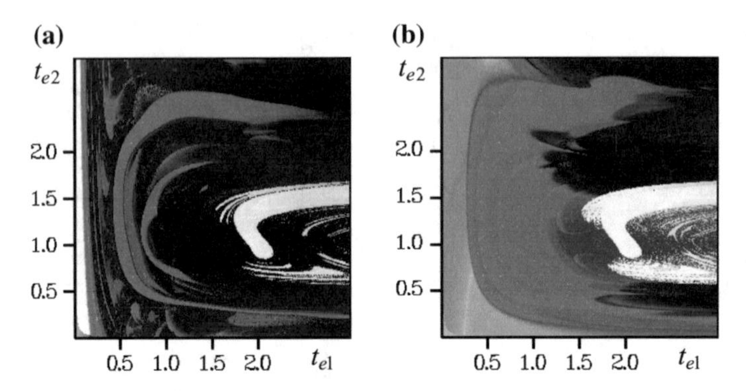

Fig. 2.48 Dependence of the fractal dimension $D_0(t_{e1}, t_{e2})$ upon the ratio of time delay in the feedback loops. The darker areas correspond to larger values of D_0. $m = 2$, $\Delta_1 = 180°$, $K = 5.5$, $\Delta_2 = 180°$(**a**), $\Delta_2 = 0°$(**b**)

deterministic chaos mode. We can control these areas sizes by choosing different combinations of DNRI parameters [34].

We have already considered the interferometers with two-dimensional feedback, often neglecting the molecule diffusion of the nonlinear medium, as well as the light diffraction, while transiting to the point models. Interferometers based on the optical fiber can be regarded as close to such models. However, the light dispersion and the field mode structure may have to be taken into account in this case.

Besides, having focused on such models, we can design radio electronic prototypes of the optical interferometers. With their aid—i.e., examining their properties, we can test the experimental modeling results, when construction of precision optical system is not expedient. Oscillators realizing the principle of these radio electronic analogs, are rather unusual in traditional radio electronics. The reason for this is the absence of the amplifier with the positive feedback. Oscillation generation, provided at the expense of interference amplification, occurs in the space of the signal parameters with some a priori given form—sinusoidal, for instance. Operation in this paradigm allows us to qualitatively confirm the modeling results and to receive evidence that chaos generation in these devices and their development for the diversification of the set of radio-frequency oscillators are possible [28, 29, 58].

For example, we will consider later in brief the simplest version of the fiber-optical interferometer and its description, while the radio-frequency analogs of interferometers will not be described to save time and space.

2.4.4 The Nonlinear Fiber-Optical Interferometer

The diagram of the nonlinear ring interferometer based on the optical fiber with nonlinear properties is shown in Fig. 2.51 [59, 60].

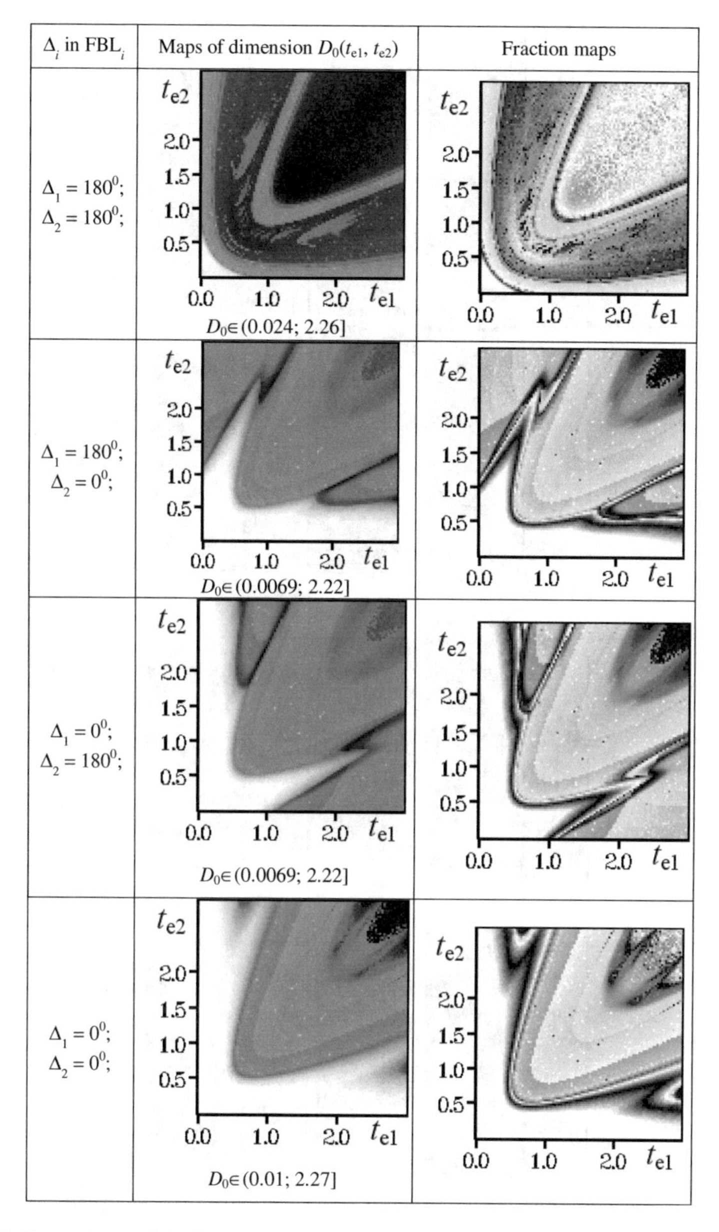

Fig. 2.49 Dependence of the fractal dimension maps structure $D_0(t_{e1}, t_{e2})$ and the fractional maps of attractors in the model (2.45) *versus* combination of angle values Δ_i of the light field rotation in the transverse plane of the laser beam in i-th FBL of DNRI, when the nonlinearity parameter $K = 5.5$, a point number in CTP $m = 2$, $\gamma_1 = \gamma_2 = 0.25$. On maps $D_0(t_{e1}, t_{e2})$ gradations of the *grey tins* correspond to increase of the value of $0 \leq D_0 \leq D_{0\max}$. On fractional maps the *light areas* correspond to integer dimension, and the *grey tins* symbolize the growth of D_0 deviation from the integer

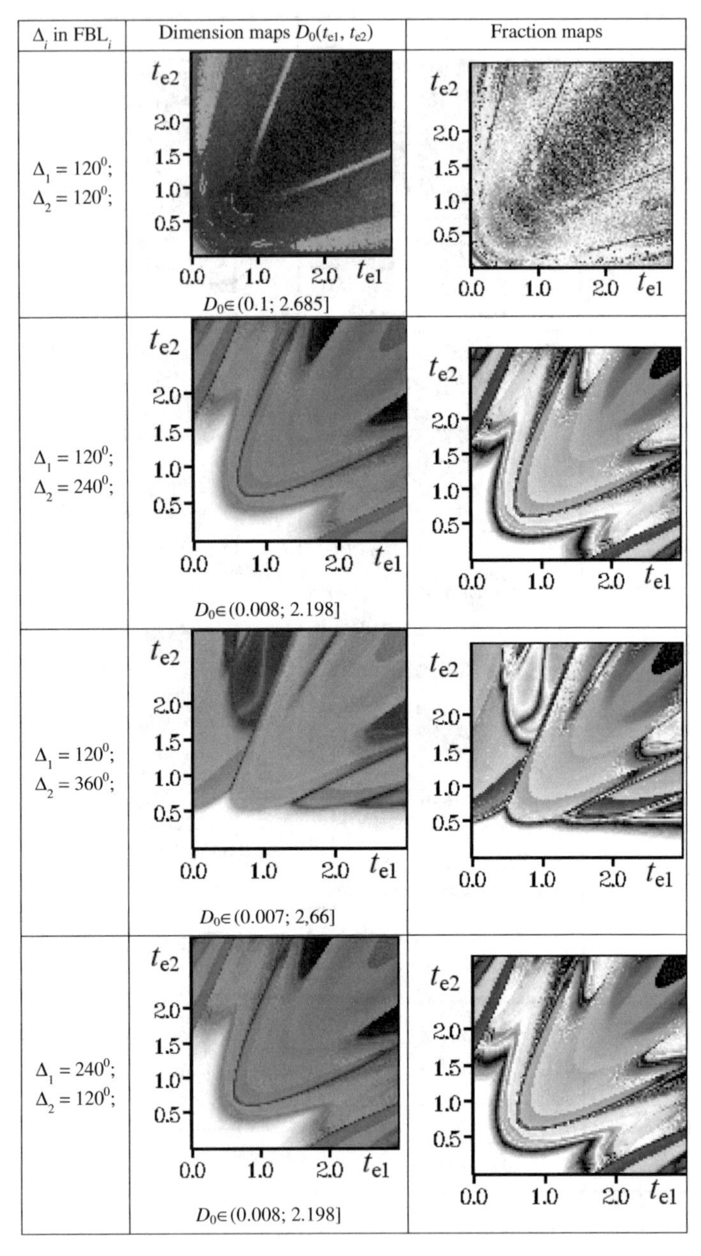

Fig. 2.50 Dependence of the fractal dimension maps structure $D_0(t_{e1}, t_{e2})$ and the fractional maps of attractors in the model (2.45) *versus* the combination of angle values Δ_i of the light field rotation in the transverse plane of the laser beam in *i*-th FBL on DNRI, when the nonlinearity parameter $K = 5.5$, a point number in CTP $m = 3$, $\gamma_1 = \gamma_2 = 0.25$. On maps $D_0(t_{e1}, t_{e2})$ the grey tin gradation corresponds to growth of the value of $0 \leq D_0 \leq D_{0\max}$. On fractional maps the light areas correspond to integer dimension, and the grey gradations symbolize the growth of D_0 deviation from the integer

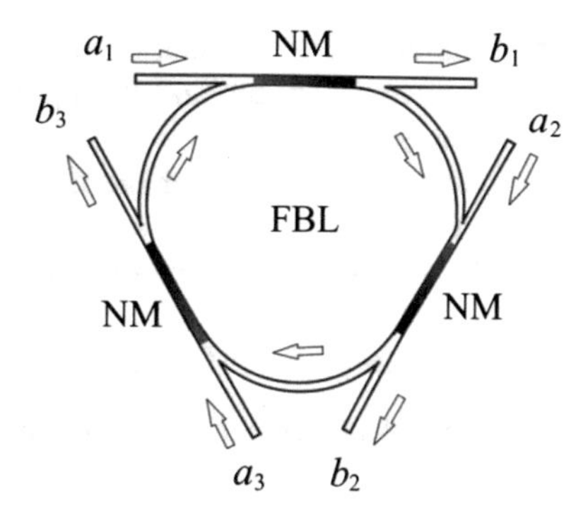

Fig. 2.51 The diagram of the nonlinear ring interferometer on the base of the optical fiber: a_i, b_i are inputs and outputs of three fibers, NM is the nonlinear medium, FBL is the feedback loop of the interferometer (marked by three arrows near a center)

The nonlinear fiber-optical interferometer represents several optical fibers connected in such a manner that a part of light field from the output of the one of fibers b_i would return to the input a_i of another fiber (Fig. 2.51). A number of such fibers may be large enough. The interferometer constructively consists of the two parts formed by the optical fiber: the nonlinear medium and feedback loop.

Optical fibers are made of quartz, which has very small nonlinear refraction index of the order $n_2 = 5 \times 10^{-13}$ cm^2/kW [44]. Therefore, in usual situations, this material is not suited for the application as the nonlinear element. However, the optical fiber may have a huge extension and in the wavelength region ($\lambda = 1.5$ μm), where the quartz glass has a low absorption factor, the relative powerful laser are applicable. The opportunities for obtaining deterministic chaos based on optical glass are presented in publication [61]. This makes us capable of realizing the suggested scheme. Of no little importance is the fact that glass doping can increase the nonlinear part of the refraction index. Thus, the glass doped by semiconductor CdHgTe has a parameter determining the nonlinear part of refraction index of $n_2 \approx 10^{-6}$ cm^2/kW. Under the nonlinear fiber we imply the optical one from the material with relaxation time of more than 10^{-12} s. There can be different polymer materials, or some kind of glass with the implanted impurities. The feedback loop is made of the fiber based on the quartz glass.

While designing the models for the dynamics of phase and amplitude of the optical field and for the refraction index of NM in nonlinear fiber-optical interferometer, we accept the following approximations:

- The fiber is single-mode and weakly directing, i.e. the difference between refraction indices of a core and an envelope is small;
- The fiber core and the envelope for the fiber with the stepping profile that forms the feedback loop are made of the materials with the identical dispersion, independent from the passing signal intensity;
- The nonlinear medium (a nonlinear element) represents, for instance, the optical fiber, in which the refraction index of the core and the envelope depends (linearly) on the intensity of the passing light signal (the Kerr nonlinearity);
- The chaotic signal spectrum (in the deterministic chaos mode) falls into the "transparent window" of the fiber;

- Damping in the feedback loop, the polarization dispersion, light reflection from the fiber interfaces and the light scattering are low, i.e. the model anticipates many single-directed light passes through the feedback loop;
- Losses at interfaces depend neither on the frequency, nor on the light amplitude in the fiber;
- The polarized molecule diffusion in the optical glass can be neglected;
- The relaxation time of the nonlinear medium is small ($\tau \approx 10^{-12}$ s) compared to the bypass time of the feedback loop (signal delay).

Then we can obtain the following mathematical model [59, 60]:

$$
\begin{aligned}
\tau\frac{du_{1,nel}(t)_i}{dt} =& u_{10,nel,i} - u_{1,nel}(t)_i + \rho_1(t)_{i-1}K_{0,core}(t)_i + \rho_1(t)_{i-1}\gamma K_{core}(t-t_k)_{i-1} \\
& + 2\rho_1(t)_{i-1}\sqrt{K_{0,core}(t)_i\gamma K_{core}(t-t_k)}_{i-1} \\
& \times \cos[\varphi_{nel}(t-t_k)_{i-1} + \varphi_{lin,i-1} + \varphi(t-t_k)_{i-1} - \varphi_{0,i}(t)],
\end{aligned}
$$

$$
\begin{aligned}
\tau\frac{du_{2,nel}(t)_i}{dt} + u_{2,nel}(t)_i =& u_{20,nel,i} + [1-\rho_1(t)_{i-1}]K_{0,fac}(t)_i \\
& + [1-\rho_1(t)_{i-1}]\gamma K_{fac}(t-t_k)_{i-1} \\
& + 2[1-\rho_1(t)_{i-1}]\sqrt{K_{0,fac}(t)_i\gamma K_{fac}(t-t_k)}_{i-1} \\
& \times \cos[\varphi_{nel}(t-t_k)_{i-1} + \varphi_{lin,i-1} + \varphi(t-t_k)_{i-1} - \varphi_{0,i}(t)],
\end{aligned}
$$

$$
\varphi_{nel}(t) = \left[u_{2,nel}(t) + \{u_{1,nel}(t) - u_{2,nel}\}\frac{d(V_{nel}B_{\text{lin}})}{dV_{lin}}\right],
$$

$$
\varphi_{lin} = u_{2,lin} + (u_{1,lin} - u_{2,lin})\frac{d(V_{lin}B_{lin})}{dV_{lin}}.
$$

$$
\phi(t)_i = \begin{cases} (\pi/2) - \arctan(A_{\text{Re}}/A_{\text{Im}}), & A_{\text{Im}} < 0 \\ (3\pi/2) - \arctan(A_{\text{Re}}/A_{\text{Im}}), & A_{\text{Im}} > 0 \\ 0, & A_{\text{Im}} = 0,\ A_{\text{Re}} \geq 0 \\ 3\pi/2, & A_{\text{Im}} = 0,\ A_{\text{Re}} < 0 \end{cases}
$$

$$
\begin{aligned}
A_{\text{Re}} =& A_0(t)_i\cos[\varphi_0(t)_i] + \gamma A(t-t_k)_{i-1} \\
& \times \cos[\varphi_{nel}(t-t_k)_{i-1} + \varphi_{lin,i-1} + \varphi(t-t_k)_{i-1}], \\
A_{\text{Im}} =& A_0(t)_i\sin[\varphi_0(t)_i] + \gamma A(t-t_k)_{i-1} \\
& \times \sin[\varphi_{nel}(t-t_k)_{i-1} + \varphi_{lin,i-1} + \varphi(t-t_k)_{i-1}].
\end{aligned} \tag{2.49}
$$

There i is the numeration of the optical fiber, t_k is the light field propagation time in the feedback loop, $\gamma = \theta \cdot \exp(\alpha l_f)$, θ is the coefficient of radiation losses at interfaces and due to its transition into modes of higher orders, α and l_f is the Bouguer damping exponent and the fiber length, $V = \frac{2\pi(n_1^2 - n_2^2)\cdot a}{\lambda}$ is the reduced spatial frequency, B is the phase parameter [62, 63], or reduced phase, φ is the phase of the sun of the input field and the field passed through the feedback loop, φ_0 is the field phase at the interferometer input, φ_{nel} and φ_{lin} are phase shifts in the nonlinear element and in the feedback loop, $A_{\rm Re}$ and $A_{\rm Im}$ are the real and imaginary parts of the complex amplitude of the optical field, $u_{1,nel}$, $u_{2,nel}$ are field phase variations during passing the core and the envelope of the nonlinear fiber, $I(t)$ is the intensity of the optical field, coefficients ρ_1 and $(1-\rho_1)$ determine which part of power propagates in the core and in the envelope, $K_{0,core}(t)$, $K_{0,fac}(t)$ and $K_{core}(t-t_k)$, $K_{fac}(t-t_k)$ are nonlinearity coefficients at the input and at the output of the interferometer; indices "core" and "fac" correspond to the core and the envelope of the fiber, and indices "lin" and "nel" correspond to the linear and nonlinear (because of the Kerr optical effect) parts of the refraction index.

Examples of the results of the system modeling (2.49), without taking into account the material dispersion, are shown in Fig. 2.52. In the model, the core diameter of the optical fiber varies so that the V parameter does not exceed 2.405. At core diameter $a = 2$ μm the optical field propagates mainly in the fiber core. At $a = 4.5$ μm when $V \approx 2.405$, the light portion propagating in the fiber envelope achieves 20 % in power. This power portion is enough to change the refraction index, whose value thus makes the general change in the conditions of the light propagation in the optical fiber possible. Therefore, dynamics in the system complicates judging by character of the time realization and the Fourier spectra (Fig. 2.52a, b).

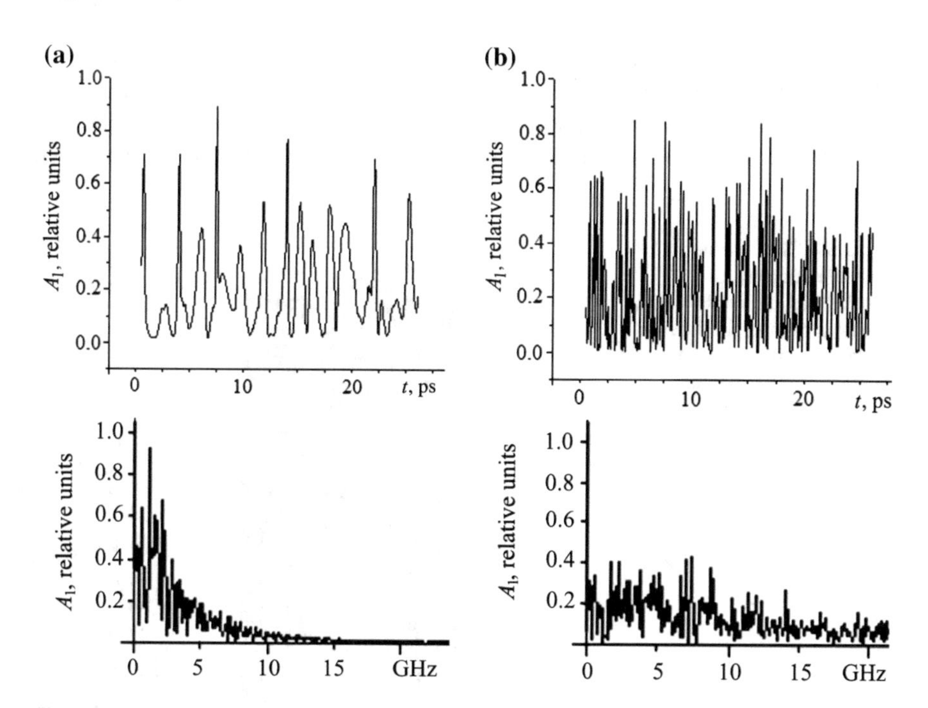

Fig. 2.52 The optical field amplitude $A_1(t)$ at interferometer output and its spectrum. The core diameter: *a*—2 μm *b*—4.5 μm

We now finish the discussion of the optical chaotic oscillators by addressing some exotic object, which may also be the source for chaotic processes. Although we will not use it later in constructing the models of the confidential communication, but it is, probably, still promising both for these purposes and for data processing.

2.4.5 *The Double-Circuit NRI and Structurally Connected NRIs: Prospects for Chaos Generating and Data Processing*

In our discussion of the nonlinear interferometers demonstrating the mono-, bi- and multi-stability, the nano-bio-photon object—the microtubule of the cytoskeleton (Fig. 2.53)—is unexpectedly but reasonably included. Each microtubule represents the albuminous polymer consisting of sub-units called tubulins. Tubulin dimers can exist in two different conformers (geometric configurations) and are capable of transfer from the one state to the other. In other words, tubulin molecules are *bistable,* and in this respect, they are similar to triggers in computers.

According to the premise of neurophysiologist Hameroff, the astrophysicist and mathematician Penrose, the microtubule is the hypothetic quantum "computing device" at the intracellular level [64–66].

The scientists came to such conclusion in 1996 discussing the possibility of relationships between the conscientious human behavior and the quantum phenomena in the brain cells. They introduced ideas prompting for the synthesis of quantum-mechanic and information approaches [65]. Experiments with isolated

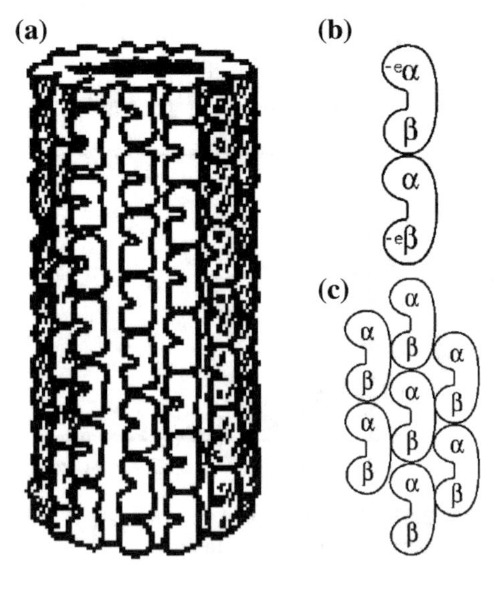

Fig. 2.53 The scheme of the molecule structure of cytoskeleton microtubule (**a**); tubulin molecules in different conformer with opposite dipole moments (**b**); the nearest vicinity of the tubulin molecule (**c**) [65, 67]

nervous cell confirm this approach [68]. They proved that even the isolated neuron being closed by the artificial feedback manifests complex and interesting behavior, and can learn and store results of obtained behavior tactics. Evidently, this fact contradicts wide-spread point of view referring to the role of neuron network in information processing.

Somewhat later, the interest to the properties of cytoskeleton microtubule leads to construction of its quantum-mechanic model by Slyadnikov. He considered the cytoskeleton microtubule as the object with the properties similar to those of ferroelectric and the spin glass. Thus, the transition of tubule molecule from the one conformer to the other is considered as the electronic barrier overcoming between two potential well [69–73].

Taking the above-mentioned into consideration in terms of abstract-system approach, the cytoskeleton microtubule presents itself as the *periodic sequence of bistable structure elements* (Fig. 2.53). Since the frequency values of the electron tunneling in the tubuline molecule are close to the optical range, the idea arises to start the search for the cytoskeleton microtubule analogs in optics. In our opinion, the double-circuit nonlinear ring interferometer can serve as the one of the possible structure analogs (Fig. 2.34) [38, p. 222; 52, 74–78].

Depending on the interferometer parameter combination as the dynamic system, the double-circuit NRI (Fig. 2.34) can be mono-, bi-, and multi-stable system, even when the field rotation in the feedback loop is absent (see [38, p. 208–210; 79]). In this case, the chain of the transposition points consists of only one transposition point (Fig. 2.54a), while the whole set of chains of the transposition points symbolizes a set of not-connected (between themselves) bi-stable elements. The latter we are to consider as the *functional analogs of tubulin molecules.*

The presence of the other field transformation in the second feedback loop causes the connection between these already formed (due to a presence of G_1 in the first circuit) elements (Fig. 2.54b). The similar connection can be caused by the polarized molecule diffusion in the nonlinear medium and/or by the light diffraction. It means that the double-circuit can be considered as the structural analog of the cytoskeleton microtubule.

It is easy to see that the purpose of the second FBL in the double-circuit NRI is the integration of the single-circuit "point" NRI into some integrity. We can achieve

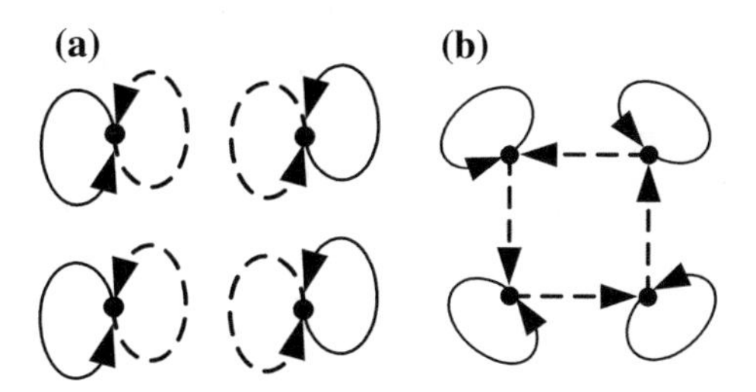

Fig. 2.54 CTP configuration corresponding to the angles of rotation Δ in the feedback loops of the double-circuit NRI: 0° and 0° (**a**); 0° and 90° (**b**). *Solid lines* indicate transitions between CTP points in the first FBL, *dashed lines* indicate in the second

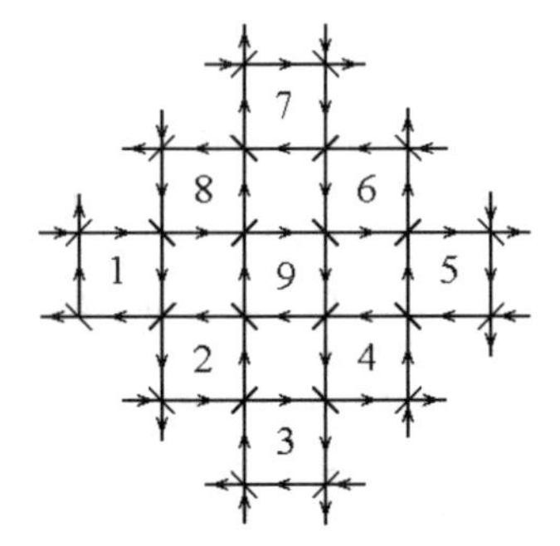

Fig. 2.55 The planar structure from NRI: *arrows* indicate the possible ways of input/output, and radiation propagation, numbers are the conditional NRI numbers [38]

such integration by connecting a series of NRI without addressing FBL for assistance, thus designing the structure reminding honeycombs or lattices, as shown in Fig. 2.55 [38]. Now the various 2D-surfaces: cylinder, torus, sphere etc. can be covered by interferometers (like polymers from tubulin molecule forming the microtubule. In this case one "elementary" NRI has not less than three neighbors (for the three-mirror interferometer) optically coupled with it by the one NRI.

We should improve this analog of bio-photon information technologies in accordance with the notions concerning the importance of self-organization in cognitive processes [80–82]. Obviously, the planar technology in Fig. 2.55 is easy to realize with the help of the integrated and fiber optics technologies, and, in prospect, with the support from the molecule technologies [83, p. 135–183, 261–293; 84, p. 279–281], keeping in mind nano-structure optics regulations [85].

Let us approximate the NRI composition in Fig. 2.55 to the cytoskeleton microtubule structure (Fig. 2.53). In the example in Fig. 2.55, each "internal" structural element (NRI) in the network (the planar structure) of NRI has four "neighbors"—NRIs—optically connected with it. Nevertheless, the structure of cytoskeleton microtubule is similar to the honeycomb (Fig. 2.53c). Therefore, it is logical to construct the network of NRIs, where each "internal" NRI would have six "neighbors" (Fig. 2.56a) [86].

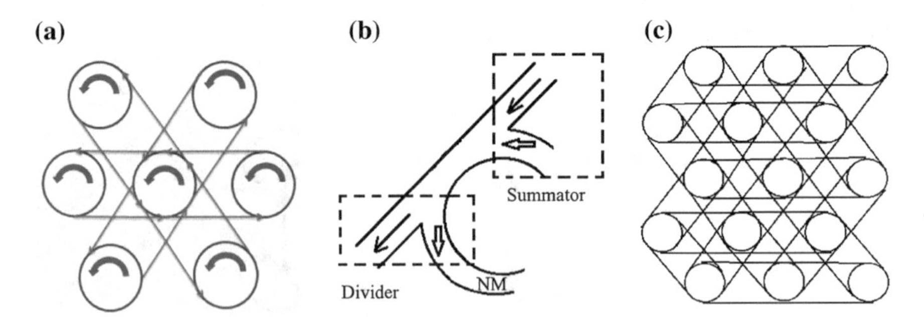

Fig. 2.56 The offered structural network component (an analog of Fig. 2.53, **c**) from seven coupled NRIs, where *arrows* indicate the propagation direction of the optical field; tangents to the circle correspond to optical fibers connecting NRIs are shown in (**a**). The construction of the one from six NRI links is shown in (**b**). The network from 5 lines and 6 columns ($i = 1\ldots5, j = 1\ldots6$) is shown in (**c**) [86]

Then, each of NRIs possessing the form of a circle (Fig. 2.56a) can be represented by six similar links formed by summator, divider and the nonlinear (Kerr) medium (Fig. 2.56b). For the internal NRI processes description, six differential equations are required with referring to the nonlinear phase shift of the optical field in the nonlinear medium of the one link.

In the simplest case, these equations can be derived from the same fundamental principles as the point models of NRI (2.32) or DNRI (2.45), designed earlier, if to neglect the waveguide effects in the fiber and the dispersion. For this, by analogy with the previous assumptions, we presuppose that the nonlinear medium in each link has a length l and the optical Kerr effect is demonstrated in it with nonlinear refraction index n_2. Moreover, we assume that the optical field is single-frequency but modulated. Let us designate the value of nonlinear phase shift in the k-th link of the interferometer of the i-th line, j-th column (Fig. 2.56c), as $U_{ijk}(t)$, and let $\varphi_{ijk}(t)$ and $a_{ijk}(t)$ be the designations for the phase and the light field amplitude normalized to $a_{\max}$ in the point in the middle between the adder and the coupler in this link ($a_{\max}$ is the maximal value of the light field amplitude at inputs of the network). Then, it appears that dynamics of these variables obeys to the following equations and relations

$$\begin{aligned}
\tau_n \frac{dU_{ijk}(t)}{dt} &= -U_{ijk}(t) + KR_{ijk}a_{ijk}^2(t),\\
a_{ijk}(t) &= \left(Ac^2 + As^2\right)^{0.5},\ \varphi_{ijk}(t) = \arg(Ac, As),\\
Ac &\equiv \frac{\gamma_{bi}}{2}\left(1 - R_{i'j'k'}\right)^{0.5} a_{i'j'k'}(t-\tau_{bi})\cos\left[\varphi_{i'j'k}(t-\tau_{bi}) - U_{ijk'}(t-\tau_{bi}) - \varphi_{0bi}\right]\\
&\quad + \frac{\gamma_{NM}}{2}\left(R_{ijk'}\right)^{0.5} a_{ijk'}(t-\tau_{bs})\cos\left[\varphi_{i'j'k}(t-\tau_{bs}) - U_{ijk'}(t-\tau_{bs}) - \varphi_{0bs}\right],\\
As &\equiv \frac{\gamma_{bi}}{2}\left(1 - R_{i'j'k}\right)^{0.5} a_{i'j'k}(t-\tau_{bi})\sin\left[\varphi_{i'j'k}(t-\tau_{bi}) - U_{ijk'}(t-\tau_{bi}) - \varphi_{0bi}\right]\\
&\quad + \frac{\gamma_{NM}}{2}\left(R_{ijk'}\right)^{0.5} a_{ijk'}(t-\tau_{bs})\sin\left[\varphi_{i'j'k}(t-\tau_{bs}) - U_{ijk'}(t-\tau_{bs}) - \varphi_{0bs}\right],\\
k' &= k-1 \text{ for } k>1, \text{ otherwise } k'=6,\\
i' &= i, j' = j+2 \text{ for } k=1,\ i' = i-1, j' = j+1 \text{ for } k=2,\\
&\quad i' = i-1, j' = j-1 \text{ for } k=3,\\
i' &= i, j' = j-2 \text{ for } k=4,\\
i' &= i+1, j' = j-1 \text{ for } k=5, j' = j+1, j' = j+1 \text{ for } k=6.
\end{aligned} \tag{2.50}$$

Here τ_n is relaxation time of NM; $K = l \cdot n_2 a_{\max}^2/2$ is the nonlinearity coefficient; R_{ijk} is the reflection coefficient of the "effective mirror" imitating the coupler operation in the k-th link of the interferometer of i-th line, j-th column; γ_{nm} and γ_{bi} are doubled loss coefficients in amplitude: in the nonlinear medium and at propagation between interferometers; τ_{bi} and τ_{bs} are light propagation times between interferometers and the time constant of the light propagation of DC component from the one link to the next in some particular interferometer ("intra-interferometer" and "intra-link" delay); $\varphi_{0,bi}, \varphi_{0,bs}$ are the phase shifts arising because of the presence of τ_{bi} and τ_{bs}.

Pilot simulations based on (2.50) demonstrate the periodic oscillations of the nonlinear phase shift $U_{ijk}(t)$, the field amplitude $a_{ijk}(t)$ and phase $\varphi_{ijk}(t)$. Rather high dimensionality of the phase space of the given dynamic system and the similarity between its nonlinearity type and the nonlinearity in NRI and DNRI allow expecting chaos generating in it [86].

Hereinafter, it is expedient to determine (by the means of the computer experiment and by the examination of the breadboard) the boundaries of analogy under investigation between the network from NRIs and the microtubule of the cytoskeleton, as well as a possibility of data processing with the help of the suggested networks made of NRIs. First of all, we must test its operability in the tasks of patterns recognition and clusterization [38].

Post factum, the reader may, probably, discover a parallel between our (Figs. 2.54, 2.55 and 2.56) structural analogs of cytoskeleton microtubule and the earlier synergetic models. Namely, those models describing the processes in closed and open chains (single-directing and double-directing) [87–91], two-dimension lattices [92–95] and other assemblies [96–100] of oscillators. This list of publications is, of course, incomplete, and comparison and discussion of their proximity to the phenomena in the microtubule of the cytoskeleton forms the separate problem to be solved. Here, we believe that the axiomatic approach to its modeling would be productive (see, for instance, [93, pp. 169–171; 101, pp. 256–258].

In the following chapters, the above-considered chaotic oscillators will allow designing the systems of nonlinear-dynamic cryptology based on them.

2.5 Conclusions

This chapter provides the theoretical examination of the radio-frequency oscillator of the deterministic chaos with the nonlinearity in the form of parabolas composition. The structure and mathematic model of the chaotic oscillator consisting of nonlinear element, amplifier, low-frequency filter and high-frequency filter of the first order, forming one closed ring, are described. The stability analysis of the equilibrium states is performed together with modeling of dynamic modes, as well as the bifurcation behavior in the chaotic oscillator. Results are presented in the graphical forms.

The experimental set up and the structural diagram of the breadboard deterministic chaos oscillator of the radio-frequency range is demonstrated. Experimental research results of the modes in the oscillator are examined. The dynamic modes are identified, together with the scenarios of transition to chaos in the oscillator.

Different variants of the nonlinear ring interferometers (with the Kerr nonlinearity) performing as the deterministic chaotic oscillators are considered theoretically. The appropriate mathematical models of the processes in these devices are constructed in the form of differential equations and the discrete maps. A presence of the second feedback loop, spatial transformation of the laser beam in its transverse plane, a presence of multiple beam passes through the feedback loop of the interferometer, all these aspects are accounted for during modeling. Stability, bifurcation behavior and the types of dynamic modes are also analyzed. The concept of the “spatial” deterministic chaos is introduced. Advantages of the double-circuit interferometer as a cipherer in the system of protected communication are predicted. Possibility for the construction of the radio-frequency analogs for the optical interferometers as the promising chaos sources is discussed.

The nonlinear interferometer based on optical fiber is suggested, its mathematical model is made, the possibility for the chaotic mode in it is shown.

In order to develop the themes referring to the optical chaotic oscillator, the Hameroff–Penrose hypothesis is involved suggesting the cytoskeleton microtubule as the intra-cellular computer. The functional microtubule analog in the form of the “fabric” composed of the structurally-coupled nonlinear ring interferometers is presented. Such “fabric” makes it possible to cover surfaces of various shapes. The pilot variant of the mathematical model is produced.

Obtained maps of fractal dimension (capacity) of attractors, maps of Lyapunov characteristic exponents and other representations of chaotic oscillator properties allow prediction of parameter value ranges, the ones that appear to be most favorable for the cryptosystem operation.

The next chapter is devoted to the theoretical analysis and experimental examination of the radio-physical systems of data transmission applying the considered chaotic oscillators.

References

1. Dmitriev AS, Kislov VYa, Starkov SO. Experimental research of formation and interaction of the strange attractors in the ring oscillator. J Tech Phys. 1985; 55(12):2417–2419.
2. Vladimirov SN. Dynamic instabilities of flows and maps. Glance of Radiophysicist. Tomsk: Tomsk University Publ.; 2008. p. 352 (in Russian).
3. Izmailov IV, Kokhanenko AP, Poizner BN. Romanov I.V. The deterministic chaotic oscillator of the radio frequency range with a delay line on the optical fiber. Russ Phys J. 2008; 51(9/2):178–179 (in Russian).
4. Romanov IV, Izmailov IV, Kohanenko AP, Poizner BN. The chaotic oscillator with the tri-modal nonlinearity as development of the logistic map construction principle. In: Proceedings of international conference information society: ideas, technologies, systems (May 2010, Taganrog city). Part 3. Taganrog: South Technical University Publ., 2010. p. 66–73 (in Russian).
5. Romanov IV, Izmailov IV. The chaotic oscillator with nonlinearity of N-type and the frequency doubler. In: Proceedings of IV Russian conference material sciences, technologies and ecology in 3rd millenium. Tomsk. Oct. 19–21, 2009. Tomsk: Institute of Atmosphere Optics of Siberian Branch of RAS Publ., 2009. p. 628–631 (in Russian).
6. Izmailov IV, Kokhanenko AP, Romanov IV, Shergin DA. Optical fiber in the deterministic chaotic oscillator of radio frequency range. In: Proceedings of VIII international conference applied optics–2008. Sankt-Petersburg, Oct. 20–24, 2008. Vol. 3: Computer technologies in optics. p. 77–78 (in Russian).
7. Romanov IV, Izmailov IV, Poizner BN. Stability of static states and dynamic s in the chaotic oscillator model containing a delay line. In: Proceedings of Russian conference of relevant problems of investigation of social and engineering systems. Part 3. Taganrog: South Technical University Publ., 2011. p. 45–52 (in Russian).

8. Romanov IV, Izmailov EIV, Kohanenko AP, Poizner BN. Nonlinear mixing of radio- and video-signals in the system of confidential communication using the dynamic chaos. Transactions of Tomsk Polytechnical University. 2011. Vol. 318, No 2. Mathematics and Mechanics. Physics. p. 53–58 (in Russian).
9. Romanov IV, Izmailov EIV, Kohanenko AP, Poizner BN. Modeling of signal/noise ratio dependence on the parameter offset of the communication system using the deterministic chaos. Russ Phys J. 2011;54(5):50–55.
10. Romanov IV. Generation and reception of the chaotic s of high-frequency range by the dynamic system with nonlinearity in the form of parabola composition. In: Proceeding of Tomsk State University of control systems and radio electronics. 2011. No 2 (24), part 1. p. 64–68 (in Russian).
11. Dmitriev AS, Panas AI. Dynamic: news for communication systems. Moscow. FizMatLit Publ., 2002. 252 p (in Russian).
12. Dmitriev AS, Kislov VYa. Stochastic oscillations in radiophysics and electronics. Moscow: Nauka Publ.; 1989. 280 p (in Russian).
13. Kuznetsov SP. Dynamic chaos (lecture course). Moscow: FizMatLit; 2001. 296 p (in Russian).
14. Dmitriev AS, Efremova EV, Nikishov AYu, Panas AI. Generation of microwave chaotic oscillations in the CMOS structure. Nonlinear Dyn. 2010;6(1):159–167 (in Russian).
15. Dmitriev AS, Efremova EV, Nikishov AYu. Generation of dynamic chaos of a microwave in oscillating structure on the base of SiGe. Lett J Tech Phys. 2009;35(23):40–46.
16. Nikishov AYu, Panas AI. Ultra-wide-band UHF chaotic oscillator of the ring structure on the amplifying micro-assemblies. In: Foreign Radio Electronics. Achievements of Modern Radio Electronics. 2008;1:54–62 (in Russian).
17. Chua LO, Ying R. Finding all solutions of piecewise-linear circuits. Int J Circuit Theory Appl;1982.10(3):201–229.
18. Chua LO, Hasler M, Neirynck I, Verburgh P. Dynamics of a piecewise-linear resonant circuit. IEEE Trans Circuits Syst. 1982;29(8):535–547.
19. Chua LO, Ayrom F. Designing nonlinear single circuits: a cookbook approach. Int J Circuit Theory Appl. 1985;13(3):235–268.
20. Graeme JG, Tobey GE, Huelsman LP. Operational amplifiers: design and applications. New York: McGraw-Hill Book Company; 1971. 473 p.
21. Anishchenko VS, Vadivasova TE, Astakhov VV. Nonlinear dynamics of chaotic and stochastic systems. Fundamentals and selected problems. Saratov: Saratov University Publ.; 1999. 368 p (in Russian).
22. Kholodniok M, Clich A, Kubichek M, Marek M. (eds) Analysis methods of nonlinear mathematical models. Moscow, MIR Publ.; 1991. 368 p (in Russian).
23. Izmailov IV, Ravodin VO. Influence of nonlinearity and delay in the ring interferometer upon bifurcations (calculation and modeling). Russ Phys J.1999;1:126 (in Russian).
24. Anishchenko VS, Astakhov VV, Vadivasova TE, Neiman AB, Strelkova GI, Shimanskiy-Gaer L. Nonlinear effects in chaotic and stochastic systems. Moscow-Izhevsk: Institute of computer researches Publ.; 2003. 544 p (in Russian).
25. Afraimovich VS, Shilnikov LP. Invariant two-dimension toruses, their destruction andstochastic properties. Methods of qualitative theory of differential equations. Gorky: Gorky State University Publ.; 1983. p. 3–26 (in Russian).
26. Afraimovich VS, Hsu SB. Lectures on chaotic dynamical systems. In: American Mathematical Society/International Press studies in advanced mathematics. Vol. 28. USA: AMS & IP; 2003. 361 p.
27. Neimark YuI, Landa PS. Stochastic and chaotic oscillations. Moscow: Nauka Publ.; 1987. 424 p (in Russian).
28. Izmailov IV, Poizner BN. The chaos on radio frequency device with quadratic phase modulator and interference amplification of quasi-harmonic signal: a model and the calculation experiment. Izvestiya VUZ. Appl Nonlinear Dyn. 2010;18(1):61–79 (in Russian).

29. Izmailov IV, Poizner BN. Axiomatic scheme of studying the dynamic s: from criteria of their diversity to self-change. Tomsk: STT Publ.; 2011. 574 p (in Russian).
30. Koronovskiy AA, Khramov AE. Continuous wavelet analysis and its applications. Moscow: FizMatLit Publ.; 2003. 176 p (in Russian).
31. Vladimirov SN, Negrul VV. On autoparametric route leading to chaos in dynamical systems. Int J Bifurcation Chaos. 2002;12(4):819–826.
32. Izmailov IV. Process modeling in the ring interferometer with noisy modulation of optical density of the Kerr medium. Russ Phys J. 1999;11:96 (in Russian).
33. Denisov PE, Izmailov IV, Poizner BN. The laser emission modulation on the base of the nonlinear ring: a model and the characteristic analysis. Atmos Oceanic Opt. 2006;19(2–3):238–243.
34. Vladimirov SN, Izmailov IV, Poizner BN. Nonlinear-dynamic: radiophysical and optical systems/Under edition of C.N. Vladimirov. Moscow: FizMatLit Publ.; 2009. 208 p (in Russian).
35. Izmailov IV. The process model in nonlinear ring interferometer taking into account the delay, losses, energy volume density transformation and many-passes of the non-monochrome field. Russ Phys J. 1998;4:112 (in Russian).
36. Izmailov IV, Magazinnikov AL, Poizner BN. Process modeling in the ring interferometer with, lag and diffusion at non-monochrome emission. Russ Phys J. 2000;2:29–35.
37. Izmailov IV, Poizner BN, Ravodin VO. Elements of nonlinear optics and synergy in the lecture course of opto-informatics: textbook. Tomsk: TML-Press Publ., Press; 2007. 92 p (in Russian).
38. Izmailov IV, Lyachin AV, Poizner BN. The deterministic chaos in models of the nonlinear ring interferometer. Tomsk: Tomsk State University Publ.; 2007. 258 p (in Russian).
39. Akhmanov, SA, Zhabotinskiy ME et al. (eds) Quantum electronics: small encyclopedia. Moscow: Soviet Encyclopedia Publ.; 1969. 492 p (in Russian).
40. Ditchburn RW. Light. NY: Dover Publ.; 2011. 690 p.
41. Godzhaev NM. Optics. Moscow: Vyshaya skola Publ.; 1977. 432 p (in Russian).
42. Landsberg GS. Optics. Moscow: Gostekhizdat Publ; 1947. 631 p.
43. Akmanov SA, Vorontsov MA. Instabilities and structures in coherent nonlinear optical systems. In: Nonlinear waves: dynamics and evolution: paper collection. Moscow: Nauka Publ.; 1989. p. 228–237 (in Russian).
44. Akhmanov SA, Vorontsov MA. New physical principles of optical information processing. Moscow: Nauka Publ.; 1990. p. 263–326 (in Russian).
45. Izmailov IV, Poizner BN, Ravodin VO. Shape formation in the ring interferometer: a forecast of static final type. In: Shekhonin AA (ed) Proceedings of international optical congress "optics-XXI century" (Oct. 16–20, 2000, Sankt-Petersburg). Conference "optics and education-2000" (Oct. 19–20, 2000, Sankt-Petersburg. Sankt-Petersburg: State University of Precision Mechanics and Optics Publ.; 2000. p. 67–68 (in Russian).
46. Arshinov AI, Mudarisov RR, Poizner BN. Shape formation in an interferometer with the: calculation experiment. Russ Phys J. 1994;6:102–104.
47. Arshinov AI, Mudarisov RR, Poizner BN. Mechanisms of the simplest optical structures formation in the nonlinear Fizo interferometer. Russ Phys J.1995;6:77–81.
48. Arshinov AI, Mudarisov RR, Poizner BN. A mechanism of optical structure formation in the nonlinear Fizeau's interferometer for mirror shift and variation of the beam sizes. Russ Phys J. 1997;7:67–72.
49. Izmailov IV. Optimization of interference interaction in the nonlinear interferometer by a choice of polarization parameters of quasi-monochrome light beams. Russ Phys J. 2000;7:101–103.
50. Izmailov IV, Lyachin AV, Poizner BN. Process description in the ring interferometer by: 's bifurcations and dimension. Bull Tomsk State Univ. Ser "Phys".2003;278:111–115 (in Russian).
51. Izmailov IV, Poizner BN. Property of an isodynamism as a principle of guaranteeing elimination of given system evolution. In: Fradkov AL, Churilov AN (eds) Proceedings of

international conference physics and control. Saint Petersburg, Russia. August 20–22, 2003. Saint Petersburg, 2003. Vol. 1 of 4: General problems and applications. p. 58–63 (in Russian).
52. Izmailov IV, Lyachin AV, Nazarov ME, Poizner BN, Shergin DA. Second circuit of two-dimensional feedback loop in ring interferometer as a way to create coupled oscillators system or couplings in a oscillator. In: Fradkov AL, Churilov AN (eds) Proceedings of IEEE. Cat. № 05EX1099C: Proceedings of 2nd international conference "physics and control" (August 24–26, 2005, S.-Petersburg). Saint Petersburg;2005. p. 841–846. http://www.ieee.org.
53. Izmailov IV, Poizner BN. Formalism and synthesis of the nonlinear-optical. Tomsk: Tomsk State University Publ.; 2001. 29 p (in Russian).
54. Izmailov IV, Lyachin AV, Poizner BN, Shergin DA. Discrete maps as a model of spatial deterministic chaos. Nonlinear Phenom Complex Syst.2006;9(1):32–42.
55. Izmailov IV, Lyachin AV, Poizner BN, Shergin DA. The spatial deterministic: a model and the phenomenon demonstration in the calculating experiment. Izvestiya VUZ. Appl Nonlinear Dyn. 2005;13(1–2):123–136 (in Russian).
56. Nazarov ME, Izmailov IV. Poizner BN. Attractor dimension estimation in the decipherer model on the base of the double-circuit ring interferometer. In: Proceeding of VII Russian conference problems of information safety of states, society and personality. Tomsk, Feb. 16–18, 2005. Tomsk, IOA Publ.; 2005. p. 34–36 (in Russian).
57. VanWriggeren GD, Roy R. Chaotic communication using time-delayed optical system. Int J Bifurcat Chaos. 1999;9(11):2129–56.
58. Izmailov IV, Poizner BN. Experiments with the chaotic source—the radio frequency device with a quadratic phase modulator and interference amplification of the quasi-harmonic signal. Izvestiya VUZ. Appl Nonlinear Dyn. 2010;18(2):39–50 (in Russian).
59. Izmailov IV, Poizner BN, Romanov IV. Nonlinear optical fiber interferometer: a model and simulation. In: Kulchin YN (ed) Fundamental Problems of Opto- and Microelectronics II (13–16 September 2004, Khabarovsk, Russia) Proceedings of SPIE Vol. 5851, p. 90–95 (2005).
60. Romanov IV, Izmailov IV, Poizner BN. A chaos and order in the model of the nonlinear optical-fiber interferometer: wavelet analysis and other emission methods. Atmos Oceanic Opt. 2007;20(7):631–634.
61. Nakatsuka H, Asaka S, Itoh H, Ikeda K, Matsuoka M. Observation of bifurcation to chaos in all—optical bistable system. Phys Rev Lett. 1983;2:109–112.
62. Unger H-G. Planar Optical Waveguides and Fibers. Oxford: Clarendon Press; 1977. 660 p.
63. Kazanie A, Flere G, Metr H. Optics and communications. Moscow: MIR Publ.; 1984. 256 p (in Russian).
64. Hameroff SR, Watt RC. Information in processing in microtubules. J Theor Biol. 1982;98:549–561.
65. Penrose R. Shadows of mind: a searching for the missing science of consciousness. N.-Y, Oxford: Oxford University Press.;1994. 656 p.
66. Hameroff S, Penrose R. Consciousness in the universe. A review of the 'Orch OR' theory. Phys Life Rev. 2014;11:39–78. (http://dx.doi.org/10.1016/j.plrev.2013.08.002).
67. Hameroff S. Quantum coherence in microtubules: a neural basis for emergent consciousness. J Conscious Stud. 1994;1:91–118.
68. Albrecht-Buehler G. Is cytoplasm intelligent too? Cell Muscle Motility. 1985;6:1–21.
69. Slyadnikov EE. Microscope model of informational bio-macro-molecule. Lett J Tech Phys. 2006;32(6):52–59.
70. Slyadnikov EE. Physical model and associative memory of the dipole system of cytoskeleton microtubule. J Tech Phys. 2007;77(7):77–86.
71. Slyadnikov EE. About interconnection of physical and informational characteristics in the vicinity of the ferroelectric transition point in the dipole system of the cytoskeleton microtubule. J Tech Phys. 2009; 79(7):1–12.

72. Slyadnikov EE. The microscopic model and the phase diagram of the dipole system of the cytoskeleton microtubule at finite temperatures. J Tech Phys. 2010;80(5):32–39.
73. Slyadnikov EE. Physical fundamentals, models of representation and image recognition in the neuron cytoskeleton microtubule. J Tech Phys. 2011;81(12):1–33.
74. Slyadnikov EE, Izmailov IV, Poizner BN, Sosnin EA. Microscopic model of conformational freedom degrees of cytoskeleton microtubule and its structural analog in optics. J Comput Technol. 2006;11(5):92–105 (in Russian).
75. Izmailov IV, Poizner BN, Slyadnikov EE, Sosnin EA. About calculation possibility in the protein nano-polymer from positions of the quantum mechanics and nonlinear optics. In: Proceedings of international conference information technologies in the modern world. Part 1—Taganrog, TGRU Publ.; 2006. p. 19–23 (in Russian).
76. Izmailov IV, Poizner BN, Slyadnikov EE, Sosnin EA. Quantum-synergy cyto-informatics as a possible direction in neuro-science. In: Scientific session of MIFI—2007. IX Russian conference neuro-informatics—2007 (Moscow, Jan 23–26, 2007): Proceedings collection in 3 volumes.—Part 2. Moscow: MIFI; 2007. p. 71–79 (in Russian).
77. Savelieva AV, Izmailov IV, Poizner BN. The double-circuit nonlinear ring interferometer and the microtubule of the cytoskeleton: search of the analog. Russ Phys J. 2008;51(9/2):206–207 (in Russian).
78. Izmailov IV, Poizner BN, Savelieva AV. Microtubules of the cytoskeleton as the development resource of bioengineering computing systems: to the problem statement. In: Proceedings of international conference innovations in society, engineering and culture (2008, Taganrog). Part 2.—Taganrog: SFU Publ.; 2008. p. 28–32 (in Russian).
79. Izmailov IV, Lyachin AV, Magazinnikov AL, Poizner BN, Shergin DA. Modeling of the laser beam transformation in the double-circuit nonlinear ring interferometer. Atmospheric and Oceanic Optics. 2007;20(3):275–282.
80. Maturana U, Varela F. The tree of knowledge. Boston and London: Shambhala New Science Library; 1992. 272 p.
81. Haken H. Principles of brain functioning: a synergetic approach to brain activity, behavior and cognition. Heidelberg: Springer; 1996. 332 p.
82. Arshinov VI, Budanov VG. Synergetic of complex perception. In: Archinov VI, Trofimova IN, Shendyapin VM (eds) Synergetic and psychology: texts: issue 3: cognitive processes. Moscow: Cognito-Center Publ.; 2004. p. 82–126 (in Russian).
83. Rakhman F (ed) Nanostructures in electronics and photonic. Moscow: Tekhnosfera Publ.; 2010. 344 p (in Russian).
84. Kiseliov GL. Quantum and optical electronics: textbook. Sankt-Petersburg: LAN Publ.; 2011. 320 p (in Russian).
85. Gaponenko SV, Rosanov NN, Ilchenko EL et al. Optics of nanostructures. In: Fiodorov AV. Sankt-Petersburg: Nedra Publ; 2005. 328 p (in Russian).
86. Kolesnikova II, Izmailov IV, Poizner BN, Slyadnikov EE. Mathematical model of a neuron cytoskeleton microtube functional analog. Russ Phys J. 2013;56(10/3):197–199.
87. Koronovskiy AA. Dynamics of single-dimension chain of the logical maps with single-direction threshold connection. Izvestiya VUZ. Appl Nonlinear Dyn. 1996;4(4–5):122–129 (in Russian).
88. Koronovskiy AA. Single-dimension map chain with single-direction threshold connection. Lett J Tech Phys. 1997;23(6):61–66.
89. Matrosov VV, Shmeliov AV. Nonlinear dynamics of the ring from three phase systems. Izvestiya VUZ. Appl Nonlinear Dyn. 2011;19(1):123–136 (in Russian).
90. Beloglazkina MV, Koronovskiy AA, Khramov AE. Numerical research of nonlinear non-stationary processes in the chain of coupled gyro-oscillators with the cross wave. Izvestiya VUZ. Appl Nonlinear Dyn. 2008;16(5):115–126 (in Russian).
91. Beloglazkina MV, Koronovskiy AA, Khramov AE. Nonlinear non-stationary processes in the chain of coupled hyro-oscillators with cross wave. J Tech Phys. 2009;79(6):13–20.
92. Koronovskiy AA. Dynamics of map lattice with the threshold connection. Tech Phys Lett. 1999;25(4):28–34.

93. Trubetskov DI, Mchedlova ES, Krasichkov LV. Introduction to self-organization theory of the open systems. Moscow: FizMatLit Publ.; 2002. 200 p (in Russian).
94. Pavlov EA, Osipov GV. Modeling of heart activity on the base of maps. Part II. An ensemble of coupled elements. Izvestiya VUZ. Appl. Nonlinear Dyn. 2011;19(3):116–126 (in Russian).
95. Bezuglova GS, Goncharov PP, Gurov YuV, Chechin GM. Discrete breathers in scalar dynamic ls on the plain square lattice. Izvestiya VUZ. Appl Nonlinear Dyn. 2011;19(3):89–103 (in Russian).
96. Nekorkin VI, Makarov VA, Kazantsev VB, Velarde MG. Spatial disorder and pattern in lattices of coupled bistable systems. Physica D. 1997;100:330–342.
97. Velarde MG, Nekorkin VI, Kazantsev VB, Ross J. The emergence of form by replication. Proc Nat Acad Sci USA. 1997;94:5024–5027.
98. Boccaletti St, Koronovskiy AA, Trubetskov DI, Khramov AE, Khramova AE. Stability of synchronous state of the arbitrary network of coupled elements. Izvestiya VUZ. Radiophys. 2006;49(10):917–924.
99. Koronovskiy AA, Khramov AE, Filatova AE. To a problem of synchronous behavior of coupled systems with discrete time. J Exp Theor Phys Lett. 2005;82(3):176–179.
100. Moskalenko OI, Ovchinnikov AA. Investigation of the noise influent on generalized chaotic synchronization in dissipative-coupled dynamics: stability of the synchronous mode with respect to external noise and possible practical applications. J Commun Technol Electron. 2010;55(4):436–449.
101. Karlov NV, Kirichenko NA. Oscillations, waves, structures. Moscow: FizMatLit Publ.; 2003. 496 p (in Russian).

Chapter 3
Radio Electronic System for Data Transmission on the Base of the Chaotic Oscillator with Nonlinearity in the form of Parabola Composition: Modeling and Experiment

Let us proceed to present of the original authors' results. This chapter is devoted to radio electronic system for data transmission. The chaotic oscillator with nonlinearity in the form of parabola composition is its base. At first, we study the communication system theoretically presenting its structural circuit, the mathematical model and computer modeling results. After that, we shall discuss a breadboard of the confidential communication system and about experiments with it. We estimate the system characteristics and factors, which restrict these characteristics.

3.1 Description of the Data Transmission System

It is naturally to begin consideration of the confidential communication systems from discussion of its structure and explanation of its processes.

3.1.1 The Structure of the Data Transmission System on the Base of the Chaotic Oscillator, Its Mathematical Model, and a Quality Criteria

The data transmission system using the chaotic oscillation as a carrier (Fig. 3.1) consists of a transmitter (the master system), a communication channel, a receiver (the slave system). It uses nonlinear mixing of the information signal m_1 for message hiding to the chaotic one in the transmitter *1*, and for extraction of the information signal m_2—the synchronous chaotic response in the receiver. In other words, it operates according above-discussed schemes in the Fig. 1.34 [1].

This data transmission system (Fig. 3.1), basically, has the chaotic oscillator (Fig. 2.1), which properties were discussed in Sects. 2.1–2.3. It plays the role of

I. Izmailov et al., *Cryptology Transmitted Message Protection*,
Signals and Communication Technology, DOI 10.1007/978-3-319-30125-9_3

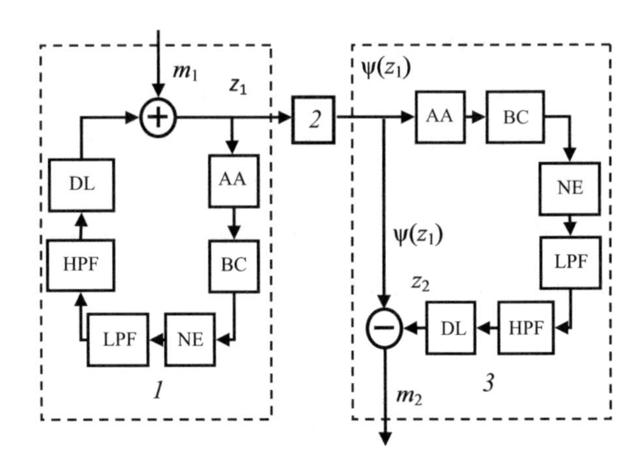

Fig. 3.1 The structural communication scheme on the base of he chaotic oscillator: *1* is a transmitter; *2* is a communication channel; *3* is a receiver. Taken designations: AA is an adjustable amplifier, BC is a bias circuits, NE is a nonlinear element, LPF is a low-pass filter of the first order, HPF is a high-pass filter of the first order, DL is delay line

both the transmitter and the master system. The z_1 sum of the information signal with the chaotic one at the summer output passes to the communication channel *2*. The following disturbances are superimposed on the z_1 signal during propagation through the communication channel: a noise, a filtering, the nonlinear distortions etc. The operator ψ performed the given circumstance account in our model. From the communication channel, the signal $\psi(z_1)$ passes to the receiver input *3*. The receiver (being of the non-ring system) consists of two arms. The one arm is "empty", and the second one (the slave system) consists of serially connected elements, which are identical on functioning, way of connecting and parameters to transmitter elements. The signal on the receiver output is formed by subtraction from the $\psi(z_1)$ signal by the "empty" arm of z_2, passing from the output of the slave system.

Mathematical model of the data transmission system is described by differential equations [1]:

$$\begin{aligned}
&\frac{dx_1(t)}{dt} = \frac{f(D_1 + K_1(z_1(t))) - x_1(t)}{T_1}, \quad \frac{dy_1(t)}{dt} = \frac{x_1(t) - y_1(t)}{T_2}, \\
&\frac{dx_2(t)}{dt} = \frac{f(D_2 + K_2\psi(z_1(t))) - x_2(t)}{T_3}, \quad \frac{dy_2(t)}{dt} = \frac{x_2(t) - y_2(t)}{T_4}, \\
&z_1(t) = x_1(t-\tau) - y_1(t-\tau) + m_1(t), \\
&z_2(t) = x_2(t-\tau) - y_2(t-\tau), \\
&m_2 = \psi(z_1(t)) - z_2(t).
\end{aligned} \tag{3.1}$$

In the model (3.1), the first two equations describe signals in the transmitter, and the last two—in the receiver. In them, T_1, T_2 are time constants of LPF and HPF in the transmitter, and T_3, T_4—in the receiver. Other variables with index "1" correspond to the transmitter, and "2"—to the receiver. The variable x_i is the voltage at LPF output, y_i is the voltage on the capacitor in first-order HPF, K_i is a gain of AA, D_i is the bias voltage of AA, $\tau = \tau_i$ is the delay, m_1 and z_1 are signals on input and output of the transmitter summator, $z_2(t)$ and $m_2(t)$ are signals on the delay line output in the receiver and on the receiver output.

Both external disturbing factors (nonlinear distortions, noises, filtering in the communication channel) and the inherent ones related to incomplete identity parameter values in the receiver and the transmitter (Fig. 3.1), for instance, parameters (K_2, D_2) and (K_1, D_1) influence on the communication quality in the systems using the chaotic oscillator. Let us introduce some quantitative characteristics of the communication quality.

In systems of type Fig. 3.1, there are the master and slave systems. If values of these relevant parameters are identical to each other, under condition $z_1(t) = \psi[z_1(t)]$ and for message absence $[m_1(t) = 0]$, the signal $z_2(t)$ on the delay line of the receiver accurately repeats the signal $z_1(t)$ at the transmitter output (Fig. 3.1). *Non-ideality of the synchronous chaotic response* leads to mistiming noise appearance at the receiver output. This noise is the main reason of received signal quality degradation [1, 2]. The following expression gives the *relative power of mistiming noise*

$$\eta = \frac{\langle \Delta z(t) \rangle^2}{\langle \psi(z_1(t)) \rangle^2} = \frac{P_{\Delta z}}{P_z}, \tag{3.2}$$

where the symbol "$\langle\rangle$" is a mean-square value in time; $\psi(z_1(t))$ is a signal at the receiver input; $\Delta z(t) = \psi(z_1(t)) - z_2(t)$ is a mistiming noise at receiver output; $P_{\Delta z}$ is the power of mistiming noise; P_z is the chaotic signal power at the slave system input.

Similarly, *the informational system power* P_m (at output and at input of the receiver) we measure [2] in the power units of chaotic signal P_z at the receiver input. In the other words, in units of the variable P_z, which would be at absence of the information signal ($P_m = 0$). Hence, we characterize the P_m level of *relative power of the information signal*

$$\mu = \frac{P_m}{P_z}. \tag{3.3}$$

As we know, the most important criterion of transmission quality of the signal is a *signal/noise ratio* (SNR) at the receiver output. Let assume that a presence of the information signal $[m_1(t) \neq 0]$ does not influence noticeably on the level of the mistiming noise ($P_{\Delta z}|_{m(t)\neq 0} \approx P_{\Delta z}$). Then, SNR approximately equals to the ratio of the information signal power at the receiver output P_m to the power $P_{\Delta z}$:

$$SNR = \frac{P_m}{P_{\Delta z}}. \tag{3.4}$$

Taking (3.2) and (3.3) into account, we obtain from (3.4) [2]:

$$SNR = \frac{P_m}{P_{\Delta z}} = \frac{P_m \cdot P_z}{P_z \cdot P_{\Delta z}} = \frac{\mu}{\eta}. \tag{3.5}$$

Let us introduce the *transfer coefficient of the chaotic carrier*:

$$\delta\alpha = \frac{\langle \Delta z(t) \rangle}{\langle \psi(z_1(t)) \rangle} \cdot 100\,\%.$$

It characterizes the reduction of the mean-square magnitude of the chaotic signal component $\psi[z_1(t)]$ after passing the receiver.

In the case of full coincidence of parameters of the transmitter and the receiver, evidently that $\eta = 0$, and SNR tends to infinity. Under real conditions, the relative white noise of mistiming is unavoidable, and hence, SNR $\neq \infty$. It follows from (3.5) that to increase the communication quality, it is necessary to increase the μ value and to reduce η. Unfortunately, the μ value is limited by the peculiarity of the signal $m_1(t)$ introduction (the method of nonlinear mixing): at large values of μ [i.e. $m_1(t)$] the chaotic mode failure is possible in the transmitter. This will lead to loss of communication confidentiality.

It is suitable to translate functions for other values of the relative power of the information signals, when reduction of μ by 10 times will move the SNR curves in the graphs by 1 dB downwards. Therefore, during further calculations of the communication quality we use the value $\mu = 1$. Obviously, that such a solution does not relates with peculiarities of the specific data transmission systems and the dynamic mode in it [1].

As we know, in the mathematical theory of communication (in the information theory) there is a concept of information and also its quantitative and qualitative features are substantiated together with properties, limited relations in the data transmission systems. If we have the digital communication channel, the one critical characteristic (a quality criterion of the signal transmission) can be served as the *carrying capacity* C [bit/s] of the channel. It can be calculated according to the Hartley formula

$$C = 2\delta f \cdot \log_2 M, \tag{3.6}$$

where δf is the spectral width of the channel [Hz], M a number of discretization levels (coding levels; there are two of them in the binary system). The Hartley formula does not show explicitly the noise influence upon the transmission speed (carrying capacity). But this influence can be described by the Hartley–Shannon formula:

$$C = \delta f \cdot \log_2(1 + SNR), \tag{3.7}$$

where SNR is a ratio of the full power of the useful signal to the full power noise *at receiver input* (an output of the communication channel) in the δf frequency band. It is important to note that the chaotic communication system plays a role of the communication channel with respect to end-user (useful signal). Therefore, to estimate its carrying capacity as SNR in (3.7), we must take SNR at the *receiver output of the chaotic communication system.* In other words, exactly those SNRs, which was introduced earlier in (3.4) will exist.

For multi-level and multi-phase coding methods, the Hartley–Shannon theorem connects the channel carrying capacity C with SNR. To be more exact, if the signal propagates with the average power P_m through the analog communication channel with the δf band, which is a subject of additive, normal distributed noise with the power $P_{\Delta z}$, then (3.7) gives the upper theoretical speed boundary of this data transmission.

Let us determine the dependence of the coding number quantity *versus* characteristics of the communication channel. For this, we equate expressions (3.6) and (3.7)

$$2\delta f \cdot \log_2 M = \delta f \cdot \log_2(1 + SNR). \tag{3.8}$$

Reducing δf and potentiating this equality (3.8), we obtain the relation

$$M = \sqrt{1 + SNR}. \tag{3.9}$$

From (3.9) it follows that under condition SNR $\gg$ 1 the number of coding levels M approximately equals a square root from SNR. We can estimate the optimal number of information coding levels corresponding to the maximal carrying capacity. In the Table 3.1, the estimation of the last presents the different conditions.

It follows from Table 3.1, that in the band δf = 6.5 MHz at SNR = 6 dB the theoretical limit of the carrying capacity C = 13 Mbit/s. At higher SNR, the transmission channel capacity grows as increase of coding level numbers.

The noise N assumes the uniform distribution of the spectral components within the limits of the channel carrying capacity. In conformity of a problem of the signal recovering from the chaotic carrier, the signal arising at the receiver output at absence of the information signal at the input of the transmitter, we can consider as the noise of such a type. It cannot have the uniform distribution of the spectral

Table 3.1 Estimation of the channel carrying capacity

	δf, MHz			δf, MHz			δf, MHz		
	4.5			5.5			6.5		
SNR, dB	6.0	10.0	20.0	6.0	10.0	20.0	6.0	10.0	20.0
M, pieces	2	3	10	2	3	10	2	3	10
C, Mbit/s	9.0	14.3	29.9	11.0	17.4	36.5	13.0	20.6	43.2

components, but nevertheless, in the future we shall use the formula (3.7) at estimation of the carrying capacity.

At conclusion, we introduce *characteristics of incomplete identity of parameters* of the transmitter and the receiver. We designate vectors of transmitter parameters $\mathbf{r}_1 = (K_1, D_1, T_1, T_2, \tau_1)$, the receiver $\mathbf{r}_2 = (K_2, D_2, T_3, T_4, \tau_2)$. Then, the *absolute* Δr_i and *relative* δr_i *lack of coincidence* (*offset*) of appropriate parameter are given by

$$\Delta r_i = r_{i1} - r_{i2}, \qquad \delta r_i = \frac{r_{i1} - r_{i2}}{r_{i1}} \cdot 100\,\% \tag{3.10}$$

Here, of course, not all parameters are indicated, but those, which easily checked or changes. Due to this, such parameters may serve as keys of the dynamic cryptosystem. For example, parameters of the transfer characteristic on the nonlinear elements are absent here. The temperature can influence on its values, which, at times, is not the only controllable factor.

3.1.2 Temperature Dependence of the Transfer Characteristics of the Nonlinear Element

External conditions influence on the parameters of the transmitter and the receiver when operation of the data transmission system. So one parts of the system may be in the premises, where the stable temperature is maintained. Other parts may be outside the premises, when temperature has a daily and season cycles. The electrical devices having the *p-n* junction are the most sensible to the temperature.

Running a few steps forward, we notice that at investigation of the breadboard of the data transmission system we found out the noise characteristic sensitivity of the recovered signal to the operation conditions (illumination and temperature). Alternating heating of the separate device units allowed determination of the unit, which temperature changing makes the extreme contribution into destroying the complete synchronization of the transmitter and the receiver. This unit is a voltage limiter on the Shottky barrier diode in the structure of the nonlinear element (Figs. 2.3, 2.19 and 3.1). Its temperature sensitivity causes noticeable variation of the sweep and the period of the transfer characteristic of the nonlinear element with variations of temperature T. Comparative analysis of the passport thermal stability characteristics of the system units justifies in favor of such conclusion. Therefore, it is expedient to anticipated measures for thermal stabilization of the voltage limiter.

Let us perform the theoretical analysis of the Shottky diode thermal sensitivity (more exactly, its saturation current) and its influence to the transfer characteristic of the nonlinear element, and, thus, to the communication quality. The volt-ampere characteristic of the semiconductor Shottky diode lays in the base of its transfer characteristic. This diode type is chosen from the practical considerations. The Shottky barrier is characterized by the small capacity (<1 pF), high operation speed, small opening voltage.

As we know, the volt-ampere characteristic of the diode with the Shottky barrier has clearly pronounced non-symmetrical view. In the region of direct biases, the current is exponentially grows with increase of applied voltage. On the contrary, in the region of reversed bias the current does not depend on the voltage. In both cases, the current is caused by the main carriers—electrons. Due to this reason, diodes are the high-speed devices since there are no recombination and diffusion processes in it. Non-symmetry of the volt-ampere characteristics of the Shottky diodes is typical for the barrier structures. Dependence of the current I versus the voltage U is caused in them by variation of the carriers number, which participate in the charge transfer processes, i.e. of the electron number, which transfer from the one part of the barrier structure to another [3].

$$I = I_s\left(\exp\left(\frac{U - IR_s}{kT}\right) - 1\right), \tag{3.11}$$

$$I_s = \frac{1}{4} q n_s \mathrm{v}_0,$$

where I_s is the saturation current, k is the Boltzmann constant, R_s is the total resistance of the semiconductor and the ohmic, n_s is the surface electron concentration in the semiconductor on the boundary with a metal, v is the heat speed of electrons.

Figure 3.2 shows the direct branches of the volt-ampere характеристик of the high-speed Shottky diode HSMS-2820, used in the experimental installation. They are calculated according to formulas from engineering specification [4]. Displacement of the volt-ampere characteristics with temperature growth is caused by saturation current I_s increase.

The equivalent circuit of the diode with the Shottky barrier can be presented in the form of Fig. 3.3. The total current through the diode is determined both the

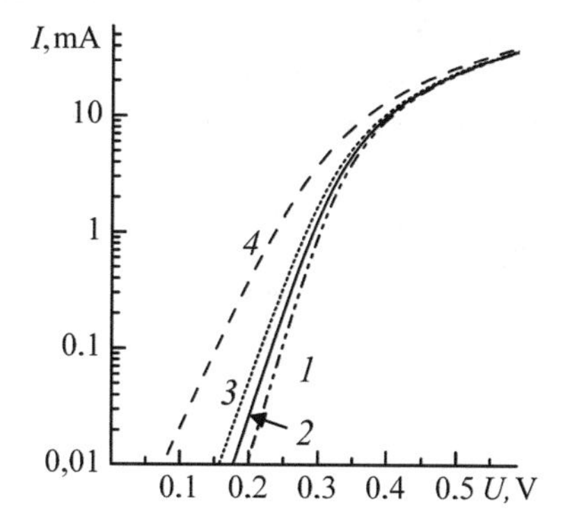

Fig. 3.2 The direct branches of the volt-ampere characteristics of the diode with the Shottky barrier at different temperature: *1* $T = -25$ °C; *2* 0 °C; *3* +25 °C; *4* +125 °C

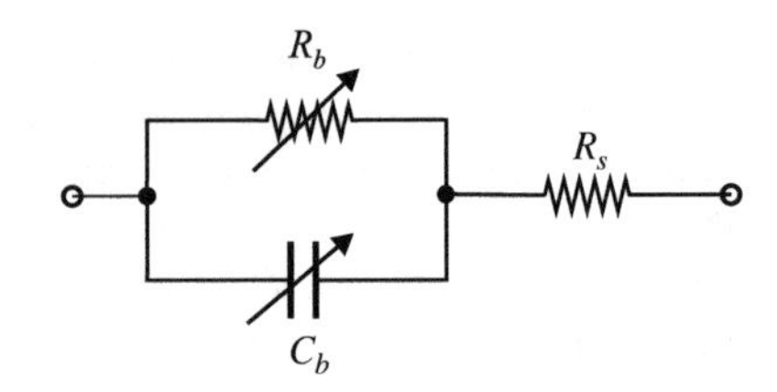

Fig. 3.3 Equivalent circuits of the Shottky diode: R_b and C_b are active resistance and capacitance of the barrier; R_s is the resistance of semiconductor thickness and the ohmic contact [3]

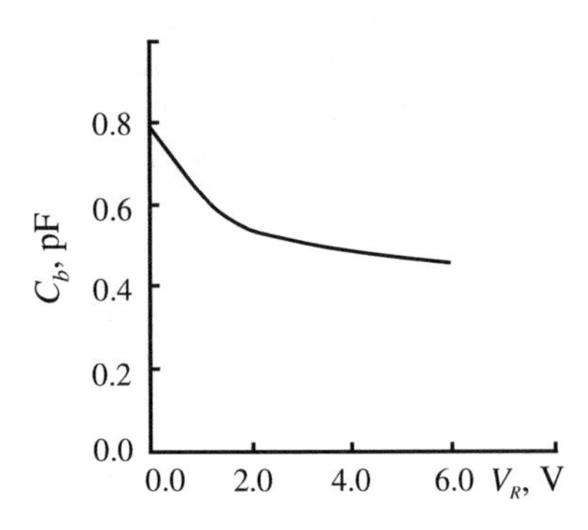

Fig. 3.4 Dependence of the capacitance C_b of the Shottky barrier versus the value of the negative bias voltage барьера V_R [4]

active and the reactive component [3]. The resistance R_b and the capacitor C_b (Fig. 3.4) of the barrier depend on the value and polarity of applied voltage of the diode. At reverse bias the value of R_b sharply increases and the alternate current flows mainly through the capacitor C_b.

The absolute diode capacitance C_b does not exceed 1 pF and it decreases with the growth of the reverse voltage V_R. From Fig. 3.4 it is clear that at variation V_R by 6 V, C_b changes approximately by 0.2 pF. For frequencies used in the experiments, we may neglect by the phase modulation at the expense of C_b variation, Modulation can be taken into consideration at frequencies higher that several hundred megahertz.

Let us examine how the temperature dependence of the volt-ampere characteristics affects to the transfer characteristics of the nonlinear element unit and the nonlinear element as a whole.

The unipolar case. Let us the dependence of the voltage on the diode u_D *versus* the power source voltage U in the voltage limiter (Fig. 3.5) is.

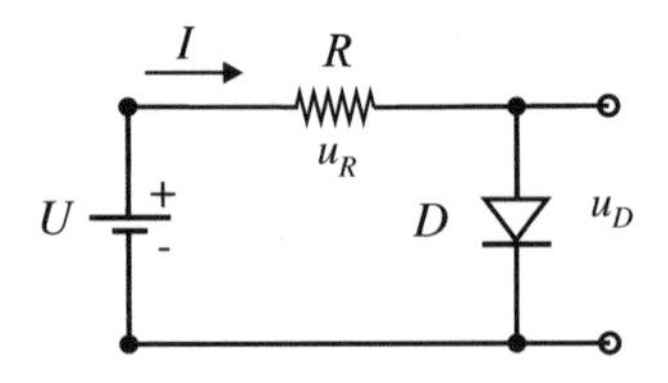

Fig. 3.5 The electric circuit of the unipolar limiter with the Shottky diode

Now we write expressions for voltages and currents, using (3.11):

$$U = u_R + u_D, \quad I = \frac{u_R}{R}, \; I = I_s\left[\exp\left(\frac{u_D - I \cdot R_s}{k \cdot T}\right) - 1\right].$$

Equalizing the current, we obtain:

$$\frac{u_R}{R} = I_s\left[\exp\left(\frac{u_D R - u_R R_s}{k \cdot T \cdot R}\right) - 1\right],$$

or, taking $u_R = U - u_D$ into account, we get the equation for determination of the voltage on the diode u_D for known voltage U:

$$\frac{U - u_D}{R \cdot I_s} - \exp\left(\frac{u_D R - (U - u_D)R_s}{k \cdot T \cdot R}\right) + 1 = 0. \tag{3.12}$$

This Eq. (3.12) connects the voltage U at the limiter input and the voltage u_D at its output.

The bipolar case. Let us consider the voltage limiter with connected additional back-diode (Fig. 3.6).

Assuming the diode reverse currents small compared to the direct one, and the diode parameters as identical, we can present the transfer characteristic of the double-sided limiter by the mutual solution of the two equation of the type (3.12):

$$\begin{aligned} &\frac{U - u_D}{R \cdot I_s} - \exp\left(\frac{U \cdot R - (U - u_D)R_s}{k \cdot T \cdot R}\right) + 1 = 0, \qquad U > 0, \\ &\frac{u_D - U}{R \cdot I_s} - \exp\left(\frac{-u_D \cdot R + (U - u_D)R_s}{k \cdot T \cdot R}\right) - 1 = 0, \qquad U < 0. \end{aligned} \tag{3.13}$$

Figures 3.7 and 3.8 shows the calculation results of the transfer characteristics of the nonlinear element and its units according to (3.13).

Calculation of transfer characteristics of the nonlinear element (Fig. 3.8) using (3.13) agrees with experimental data about strong influence of the temperature. As we see, growth of temperature leads to decrease of the maxima values as well as their displacement in the less voltages region.

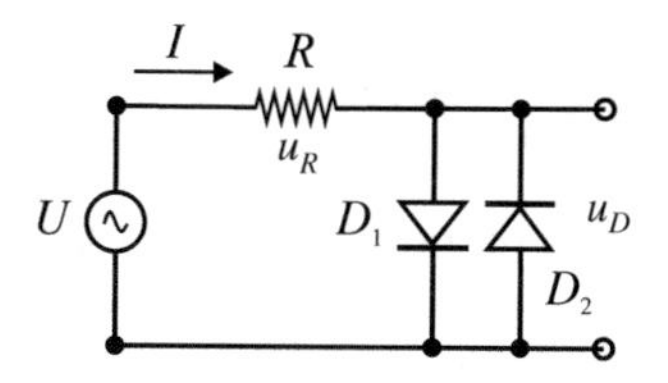

Fig. 3.6 The electric circuit of the voltage limiter with the Shottky diode

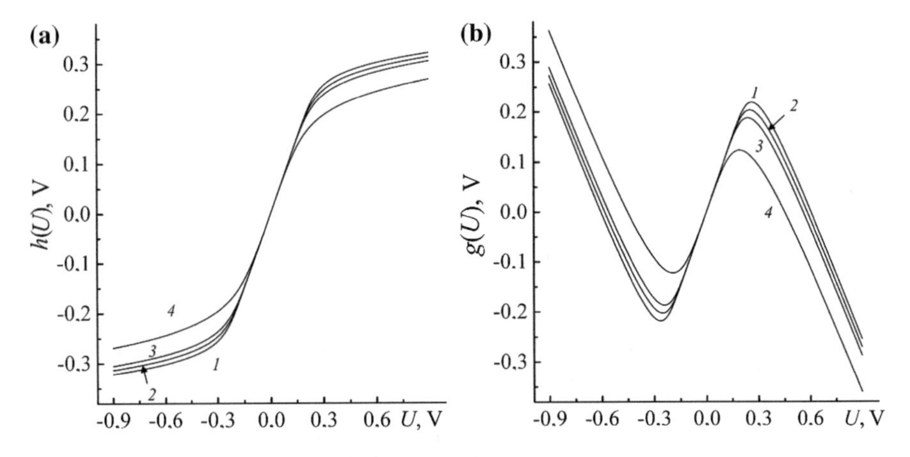

Fig. 3.7 Transfer characteristics of double-side diode limiter (**a**) and the discriminator on its base (**b**) for different temperatures: *1* $T = -25$ °C; *2* $T = 0$ °C; *3* $T = 25$ °C; *4* $T = 125$ °C

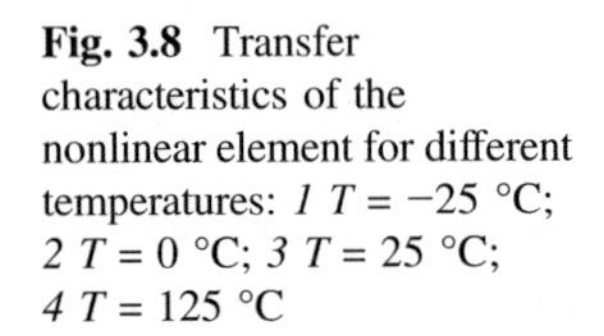

Fig. 3.8 Transfer characteristics of the nonlinear element for different temperatures: *1* $T = -25$ °C; *2* $T = 0$ °C; *3* $T = 25$ °C; *4* $T = 125$ °C

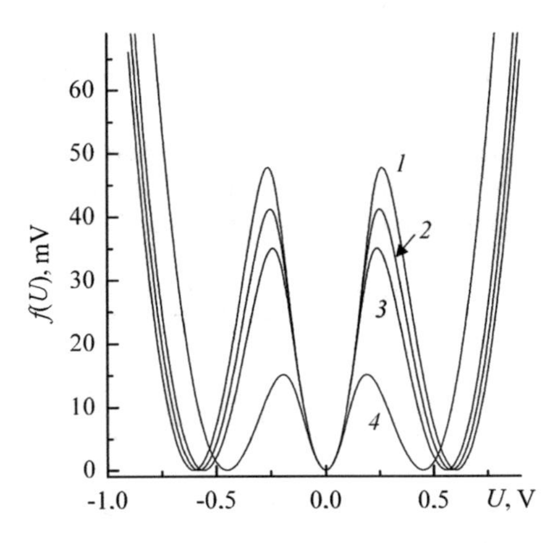

It is natural now to search the way to minimize the temperature influence to the transfer characteristics of the nonlinear element.

3.1.3 Temperature Compensation in the Voltage Limiter on the Shottky Diodes and a Choice of the Nonlinear Element Parameters

The unipolar case. To weaken the temperature influence, we offer modification of the diode limiter circuit in Fig. 3.9 [5–7].

Utilizing the first and second Kirchhoff laws, we may write equations for currents and voltages for the circuit parts (designations are clear from Fig. 3.9):

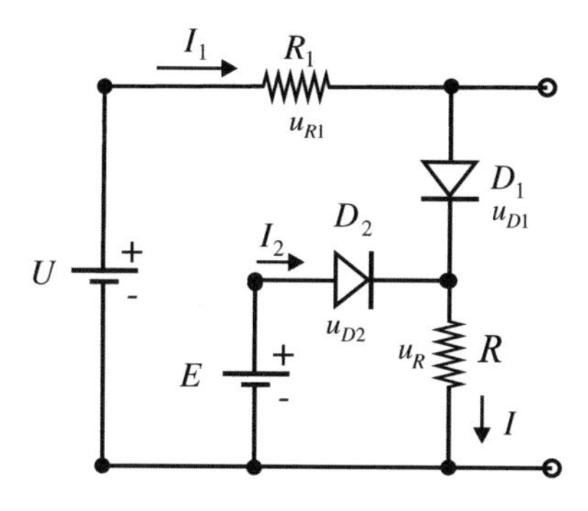

Fig. 3.9 Electric circuit of the thermal-stable unipolar limiter with the Shottky diode

$$\begin{aligned} &I_1 + I_2 - I = 0,\\ &u_{R1} + u_{D1} + u_R - u = 0,\\ &u_{D2} + u_R - E = 0, \end{aligned} \tag{3.14}$$

where $u_{R1} = R_1 \cdot I_1$, $u_R = R \cdot I$. From (3.14) we obtain the system of the equations

$$\begin{aligned} &(R_1 + R) \cdot I_1 + R \cdot I_2 + u_{D1} - U = 0,\\ &R_1 \cdot I_1 + R \cdot I_2 + u_{D2} - E = 0. \end{aligned} \tag{3.15}$$

Dependences for currents *versus* voltages on diodes D_1 and D_2 have a view

$$\begin{aligned} I_1 &= I_s\left[\left(\exp\left(\frac{u_{D1} - I_1 R_s}{kT}\right) - 1\right)\right],\\ I_2 &= I_s\left[\exp\left(\frac{u_{D2} - I_2 R_s}{kT}\right) - 1\right]. \end{aligned}$$

Taking (3.15) into account, we get the equation system for currents

$$\begin{aligned} &I_1 - I_s\left[\exp\left(\frac{U - (R_1 + R)I_1 - I_2 R - I_1 R_s}{kT}\right) - 1\right] = 0,\\ &I_2 - I_s\left[\exp\left(\frac{E - I_1 R_1 - I_2 R - I_2 R_s}{kT}\right) - 1\right] = 0. \end{aligned} \tag{3.16}$$

The equation system (3.16) is solved, for instance, by the methods of the simple iterations.

The bipolar case. By analogy with the unipolar case, we offer the circuit with the temperature compensation (Fig. 3.10).

We assume that the reverse current of diodes D_3 and D_4 much less than a direct one. We can write equations for currents and voltages in the case of the positive U voltage (designations are clear from Fig. 3.10):

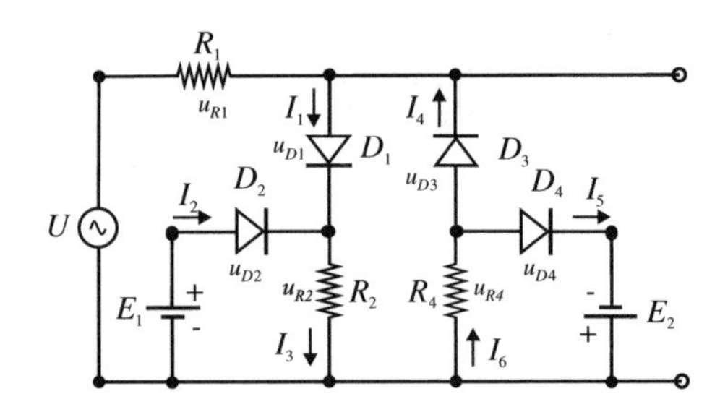

Fig. 3.10 Electric circuit of the thermal-stable bipolar limiter with the Shottky diode

$$
\begin{aligned}
&I_1 + I_2 - I_3 = 0,\\
&u_{R1} + u_{D1} + u_R - U = 0,\\
&u_{D2} + u_{R1} - E_1 = 0.
\end{aligned}
$$

For the negative U voltage, we have equations

$$
\begin{aligned}
&I_4 + I_5 - I_6 = 0,\\
&u_{R1} + u_{D3} + u_{R3} - U = 0,\\
&u_{D4} + u_{R2} - E_2 = 0,
\end{aligned}
$$

where $u_{R1} = R_1 \cdot I_1$, $u_{R2} = R_2 \cdot I_3$, $u_{R3} = R_3 = R_3 \cdot I_6$. Then we have the equation system

$$
\begin{aligned}
(R_1 + R_2) \cdot I_1 + R \cdot I_2 + u_{D1} - U = 0 \quad \text{for } U \geq 0,\\
R_1 \cdot I_1 + R_2 \cdot I_2 + u_{D2} - E_1 = 0 \quad \text{for } U \geq 0,\\
(R_1 + R_3) \cdot I_4 + R_3 \cdot I_5 + u_{D3} - U = 0 \quad \text{for } U < 0,\\
R_1 \cdot I_4 + R_3 \cdot I_5 + u_{D4} - E_2 = 0 \quad \text{for } U < 0.
\end{aligned}
\tag{3.17}
$$

Relations between currents and voltages for diodes D_1, D_2, D_3 and D_4 have a view

$$
\begin{aligned}
I_1 &= I_s \left[\exp\left(\frac{u_{D1} - I_1 R_s}{kT} \right) - 1 \right],\\
I_2 &= I_s \left[\exp\left(\frac{u_{D2} - I_2 R_s}{kT} \right) - 1 \right],\\
I_4 &= -I_s \left[\exp\left(-\left(\frac{u_{D1} - I_1 R_s}{kT} \right) \right) - 1 \right],\\
I_5 &= -I_s \left[\exp\left(-\left(\frac{u_{D2} - I_2 R_s}{kT} \right) \right) - 1 \right].
\end{aligned}
\tag{3.18}
$$

From (3.17) and (3.18) we have the following equation system for currents:

$$
\begin{aligned}
&I_1 - I_s\left[\exp\left(\frac{U-(R_1+R_2)I_1 - I_2R_2 - I_1R_s}{kT}\right) - 1\right] = 0,\ (U \geq 0),\\
&I_2 - I_s\left[\exp\left[\frac{E_1 - I_1R_1 - I_2R_2 - I_2R_s}{kT}\right] - 1\right] = 0,\ (U \geq 0),\\
&I_4 - I_s\left[\exp\left(-\left(\frac{U - I_4(R_1+R_3) - I_5R_3 - I_4R_s}{kT}\right)\right) - 1\right] = 0,\ (U < 0),\\
&I_5 - I_s\left[\exp\left(-\left(\frac{E_2 - I_4R_1 - I_5R_3 - I_5R_s}{kT}\right)\right) - 1\right] = 0,\ (U < 0).
\end{aligned}
\tag{3.19}
$$

The equation system (3.19), as (3.16), can be solved, for instance, by the method of the simple iterations.

Figures 3.11, 3.12, and 3.13 show (3.19) calculation results of the transfer characteristics on the nonlinear element units (Figs. 2.3 and 2.19). It is clear that

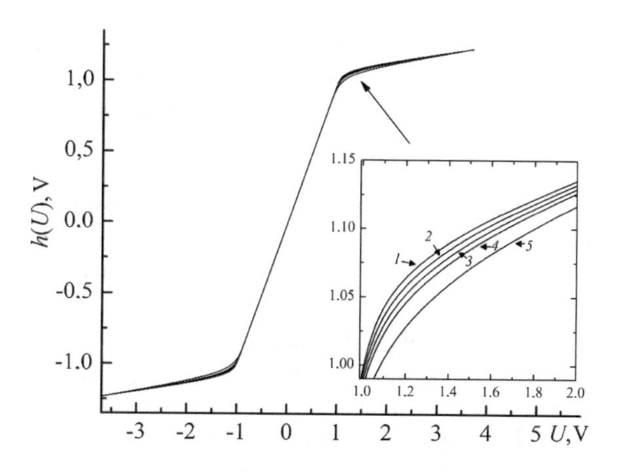

Fig. 3.11 The transfer characteristic of the thermal-stabilized voltage limiter for E_1 = 1.2 B, E_2 = −1.2 B, R_1 = 300 Ω, R_2 = R_3 = 65 Ω: *1* T = −25 °C; *2* T = 0 °C; *3* T = 25 °C; *4* T = 50 °C; *5* T = 125 °C

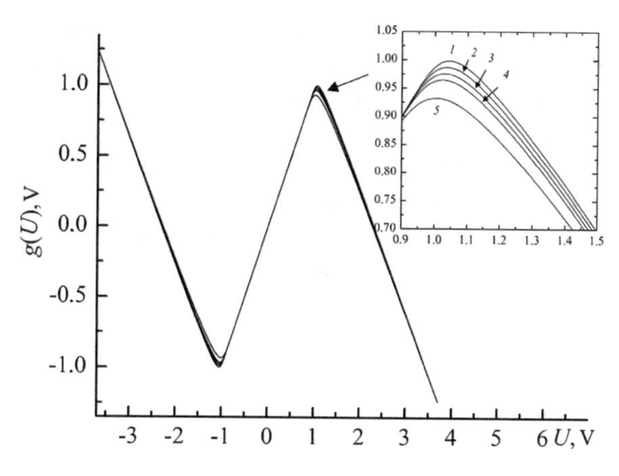

Fig. 3.12 The transfer characteristics of the thermal-stabilized discriminator (the *N*-element) for E_1 = 1.2 B, E_2 = −1.2 B, R_1 = 300 Ω, R_2 = R_3 = 65 Ω: *1* T = −25 °C; *2* T = 0 °C; *3* T = 25 °C; *4* T = 50 °C; *5* T = 125 °C

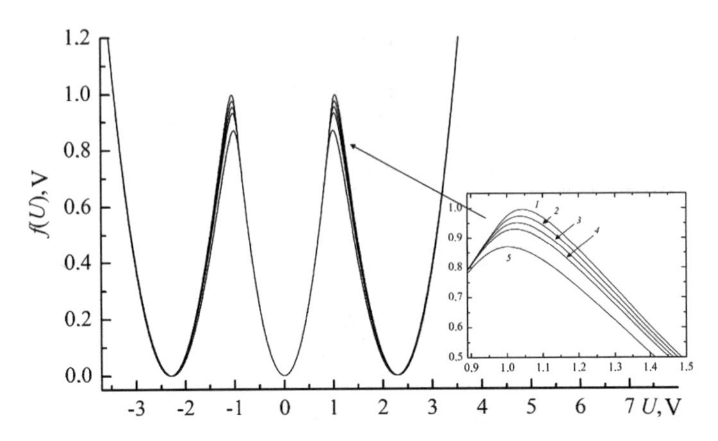

Fig. 3.13 The transfer characteristics of the thermal-stabilized nonlinear element for $E_1 = 1.2$ B, $E_2 = -1.2$ B, $R_1 = 300\ \Omega$, $R_2 = R_3 = 65\ \Omega$. A choice: *1* $T = -25$ °C; *2* $T = 0$ °C; *3* $T = 25$ °C; *4* $T = 50$ °C; *5* $T = 125$ °C

application of the additional diodes increased the nonlinear element stability to temperature variation compared to the circuit without compensation.

Another important property of nonlinear element implemented according the circuit in Figs. 3.9 and 3.10,—a possibility of its transfer characteristic control by voltages E_1, E_2.

A choice of the nonlinear element parameters. As we see from Fig. 3.14a, at R_2 variation, the coefficient of the temperature drift of the chosen point on the transfer characteristics of the nonlinear element $\Delta U/\Delta T$ varies from –80 to 240 $\mu V/^{\circ}C$, and for $R_2 = 65\ \Omega$, ΔU crosses a zero value. In connection with this, for further calculations we shall use exactly this R_2 value.

At variation of E_1, the coefficient $\Delta U/\Delta T$ varies within the limits of –40 to 180 $\mu V/^{\circ}C$ (Fig. 3.14b). Its value in the region of E_1 changing from 0 to 0.7 V is constant and relatively high (180 $\mu V/^{\circ}C$). This is explained by the fact that the diode D_2 (Fig. 3.10) at such a voltage E_1 is closed and does not participate in the mechanism of the temperature compensation.

At U variation, the coefficient $\Delta U/\Delta T$ achieves in the maximum ($U \approx 1.2$ V) of the value 260 $\mu V/^{\circ}C$ (Fig. 3.14c). The minimum of the coefficient $\Delta U/\Delta T = 0$ $\mu V/^{\circ}C$ takes place in the region of U from 0 to 0.9 V. Here the diode D_1 is closed, the ΔU drift remains constant close to zero and does not depend upon temperature (in approximation of R_1 ideality). At further U increase, the diode D_1 opens, and the current variation (then, $\Delta U/\Delta T$) is determined by D_1 characteristic.

Described results of Sect. 3.1 allow proceeding to the theoretical analysis of properties of the confident communication system.

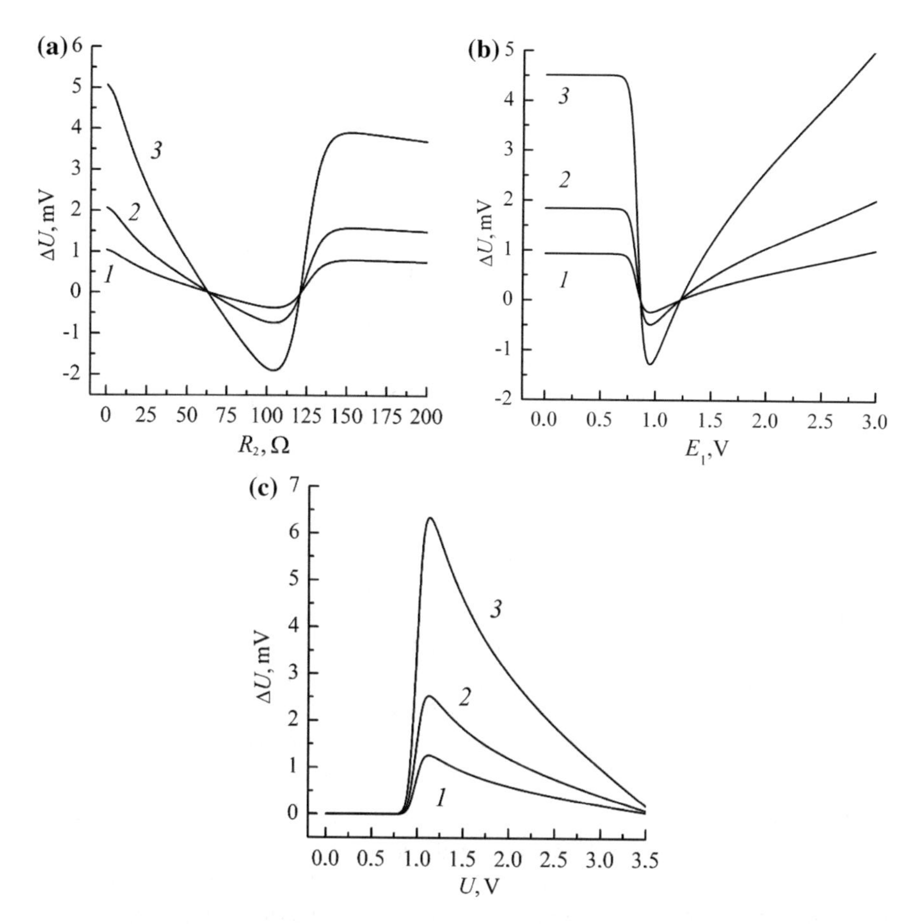

Fig. 3.14 Dependence of lack of coincidence on the limiter output (Fig. 3.10) in the receiver and in the transmitter ΔU upon the value: of the resistor R_2 for $U = 3.7$ V, $E_1 = 1.2$ V, $E_2 = -1.2$ V, $R_1 = 300$ Ω (**a**); of the bias voltage E_1 for the following circuit parameters $U = 3.7$ V, $R_1 = 300$ Ω, $R = 65$ Ω (**b**); of the input voltage value U—for $E_1 = 1.2$ B, $E_2 = -1.2$ B, $R_1 = 300$ Ω, $R = 65$ Ω (**c**). Curves *1, 2, 3* correspond to temperature lack of coincidence ΔT of the receiver and the transmitter: 5, 10, 25 °C

3.2 Numerical Modeling of the Data Transmission System Operation

In the real system of the chaotic communication the factors, which decrease the data transmission quality, act and manifest both in the receiver and the transmitter, and in the communication channel (see Fig. 3.15) [2, pp. 73–77]. We characterize in brief the most significant factors.

Lack of coincidence of the transfer characteristic form of the nonlinear element. The problem of precision oscillator development of the deterministic chaos relates

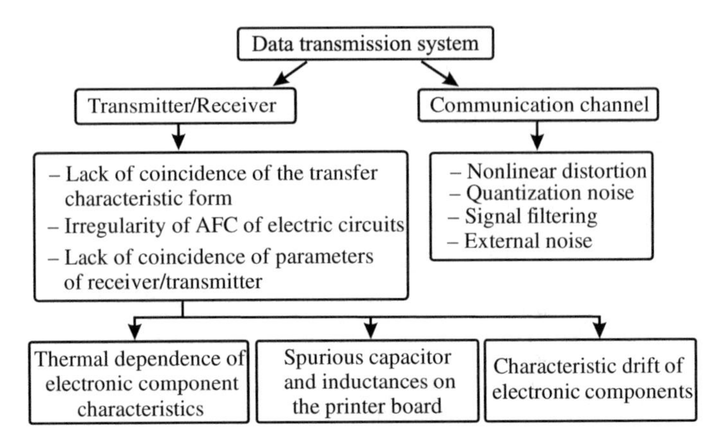

Fig. 3.15 Main factors influencing to communication quality in the data transmission system

mainly with complexity of transfer characteristic reproduction of the nonlinear element. The drift of parameters of resistors, capacitors, diodes (lack of coincidence of the transfer characteristics), op-amps, the sensitivity to the temperature of the semiconductor devices does not ensure identity of the transfer characteristics of the nonlinear elements in the transmitter and the receiver. For instance, in the broadband, which has been discussed in Sect. 3.3, the drift of voltage values of op-amps AD8056 is about 3 mV. For the volt-ampere characteristic of diodes-limiters used of HSMS-2820 with the Shottky barrier, the saturation current value I_s depends on temperature T according to the law: $I_s = I_0 \cdot \left(\frac{T}{298}\right)^{\frac{2}{n}} \cdot e^{-4060 \cdot \left(\frac{1}{T} - \frac{1}{298}\right)}$, where the initial current I_0 = 22 nA, n is a factor of the diode ideality. Stabilization of the temperature mode requires manufacture of all nonlinear elements of oscillators in the structure of the common semiconductor crystal for compensation the temperate drift of nonlinear element characteristics. In order to increase the chaotic oscillator, we may replace the op-amp AD8056 for more accurate amplifier with less value of input bias voltage, for instance, for OP37, the typical bias equals to 10 μV.

Irregularity of AFC of electric circuits. A presence of spurious capacitors in the amplifying stages and in other active components, as well as their mutual location leads to non-identity of these circuit characteristics. This adversely affects to repeatability devices and, hence, on the similarity of transmitter and receiver device parameters resulting in reduction SNR at receiver output.

Lack of coincidence of controlling parameters of the transmitter and the receiver. Vibrations, temperature difference and other conditions of transmitter and receiver operation, as well as the inaccurate parameter choice of their adjustable units lead to quality degradation of data transmission.

Noises. In electric circuits noises are always present. For instance, the thermal noise and the Johnson noise are unavoidable, since they relates to fundamental limitations of materials and the element base of the device. Nevertheless, noise measurements performed by us in the receiver and the transmitter showed that noise in the breadboard do not exceed ~ 1 mV. From this it follows that they are not the main factor of SNR reduction.

Quantization noise. It arises, when the chaotic signal is transmitted in the digital communication channel. In other words, if the signal in the communication channel is a subject to quantization over level. This noise has the same nature as the white noise. The magnitude of the quantization noise is inversely to the discretization level number. To decrease the noise influence, it is necessary to choose such a quantization step, which ensures the required transmission quality. As it will be shown, to perform SNR of 65 dB, it is sufficient to use 12-bit analog-digital converter (ADC) and digital-analog converter (DAC) for discretization of the chaotic signal (i.e. with the digit capacity of 12 bit).

Signal filtering. The spectrum of the chaotic signal at transmitter output has a continuous spectrum and uniformly decreasing envelope. Therefore, a part of the high-frequency spectral components in the communication channel is exposed to filtering. As it will be (theoretically and experimentally) shown, the filtering of the chaotic signal in the communication channel is the one of the main factors, which worsens the data transmission quality. Therefore, it is necessary to form the chaotic signal in the frequency band of the communication channel. For this, we may use filters of high order in the structure of the oscillator of deterministic chaos.

Signal nonlinear distortions in the communication channel. They can arise due to the following reasons: the wrong choice of the dynamic range of amplifying stages, a presence of amplitude restriction in the communication channel. These signal distortions, evidently, lead to distortion of transmitted chaotic signal.

Practical orientation of the data transmission system on the base of the chaotic oscillator requires investigations of characteristic stability for internal and external disturbances. We understand under internal disturbances the lack of coincidence of the following parameters of the receiver and the transmitter: an amplification coefficient K, the bias voltage D, time constants of LPF T_1 and HPF T_2, the delay time τ in the delay line, temperature T. External disturbances considering in the model (3.1) by the ψ function, are distortions and noise in the communication channel.

3.2.1 Lack of the Coincidence Influence of the Transmitter and Receiver Parameters on the Data Transmission Quality

It was demonstrated earlier a variety of operating modes in the chaotic oscillator. Now we have to proceed to study of the transmission system. In practice, it impossible to ensure the full identity of parameters values of the received and the transmitter. Therefore, the problem of finding out the permissible limits of mismatching of these values, which does not disturb its efficiency, is issue of the day.

Let us remind that a series of circumstances affects on the choice of system parameter values:

- The information inlet into the chaotic signal is performed at the expense of signal summation in the transmitter. In this case, the information signal participated in the chaotic signal formation, which allows controlling by the communication quality, adjusting the part of the signal power at the receiver input and the power of the chaotic signal in the transmitter (see (3.4) and (3.5)).
- Requirement of confidentiality applies restrictions on the spectral characteristics of the chaotic and information signals. The spectrum width of the chaotic signal should be more than the spectrum width of the information signal or equal to it. The control of the spectral band is performed by specification in the transmitter of the time constants T_1 and T_2 LPF and P3A as well as values of the delay time τ in the delay line.
- Vector of parameters $\mathbf{r}_1 = (D_1, K_1, T_1, T_2, \tau_1)$ in the transmitter и $\mathbf{r}_2 = (D_2, K_2, T_3, T_4, \tau_2)$ in the receiver unambiguously define the operation mode of data transmission system. Equality of each elements of the vectors ($\mathbf{r}_1 = \mathbf{r}_2$) leads to fulfillment of the passive synchronization condition, due to which the message transmission is possible. Mismatching of parameters ($\mathbf{r}_1 \neq \mathbf{r}_2$), characterizing by (3.10), partially or completely disturbs the mode of the synchronous chaotic response, which complicates (or impossible) its transmission. Therefore, values $\mathbf{r}_1$ are the analog of the cryptographic key. Its number the larger for less mismatching parameter values leads to impossibility to read the message.
- To choose parameter values (with the aim to implement the specific operation mode of the transmitter) it is expedient to use the mode maps of the chaotic oscillator. If we have no the specific mode of transmission operation, it is expedient to use the mode maps of the chaotic oscillators. In the case of their absence, we may use bifurcation diagrams (as some section of the first) as well as maps of the real parts of the Liapunov's characteristic exponent for the steady-states of the chaotic oscillator.

Let us begin to clarification the fact how a deviation of the one parameter value will reflect on characteristics of data transmission system, relying on the model (3.1) taking into account (3.19). Figure 3.16a shows that with changing of the δD offset, the bias voltage from 5 to 50 %, the relative power of mismatching noise η increases by two-orders (from 0.01 to 1) for amplification K values equaled 2 and 5. Growth of amplification ($K = 15$) displays the function η *versus* δD into the region of small η values. So, for $\delta D = 50$ %, the value $\eta = 0.05$. With variation offset δD from 1 to 50 %, SNR at transmitter output deceases by 30 dB (see Fig. 3.16b with typical dependences SNR (δD)). The growth of amplification $K = K_1 = K_2$ from 2 to 15 displaces the curve by 18 dB.

We note the asymmetric SNR dependence *versus* a sign of δD, more noticeable at small values of K. So, for $\delta D = -50$ % the slope SNR (δD) is more steeply. Probably, it relates by the fact that operation region on the transfer characteristic of

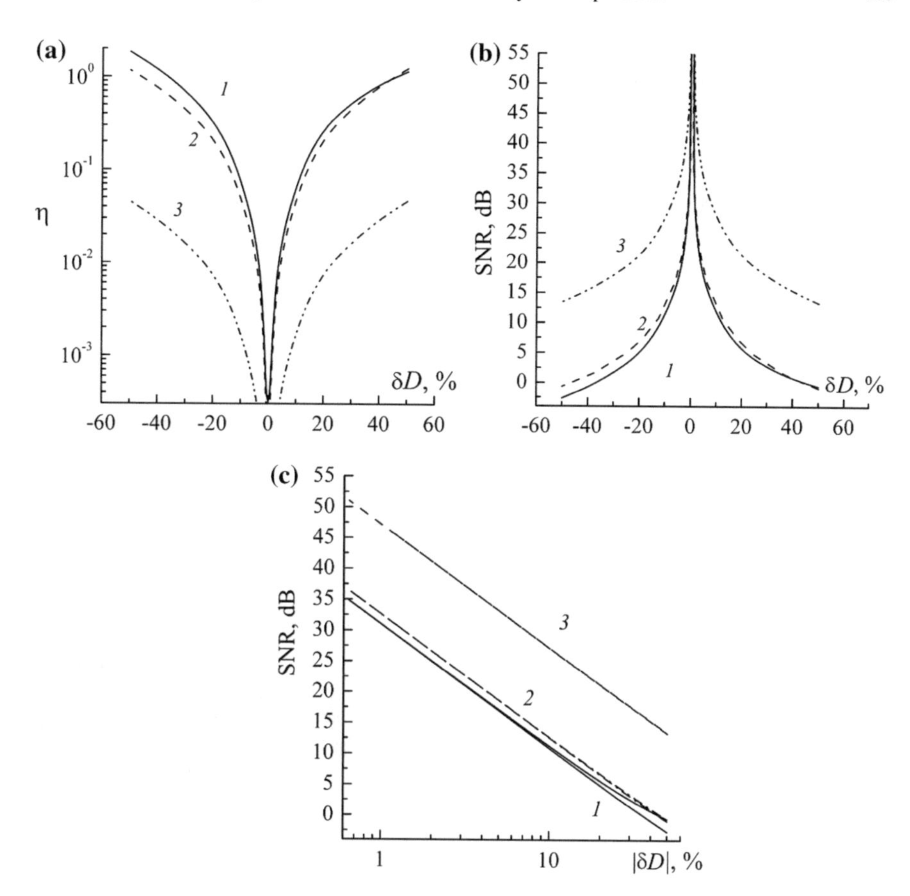

Fig. 3.16 Influence of the relative offset δD of the parameter D_2 on the relative power of mismatching noise η (**a**) and on SNR in logarithmic (**b**), double-logarithmic (**c**) scale for amplification. *1* $K = 2$; *2* $K = 5$; *3* $K = 15$. Here $D_1 = -0.8$ V, $T_1 = T_3 = 1$, $T_2 = T_4 = 100$, $\tau = 5$, $\mu = 1$, $R = 300\ \Omega$, $R_1 = 65\ \Omega$, $E_1 = 1.2$ V, $E_2 = -1.2$ V

the nonlinear element (in the form of three parabola composition and two quasi-parabola) has not always symmetric form and, accordingly, the equal slope.

At small K values the oscillation swing is small: they occupy only separate linear and nonlinear fragments of the operation region of the transfer characteristics. At that, displacement into one side (suppose, $\delta D > 0$) intensifies the "capture" of the linear characteristic fragment, and displacement into another side (suppose, $\delta D < 0$), *vice versa*, the "capture" its nonlinear fragment. In this fact we see a reason of asymmetry of the function SNR (δD).

At large values of K the oscillation's swing is large: they propagate to several linear and nonlinear fragments of the operation parts on the characteristics. Therefore, a sign of displacement δD of the operating point weaker influences on variation of the fraction of "equipped" linear (or nonlinear) characteristic fragments.

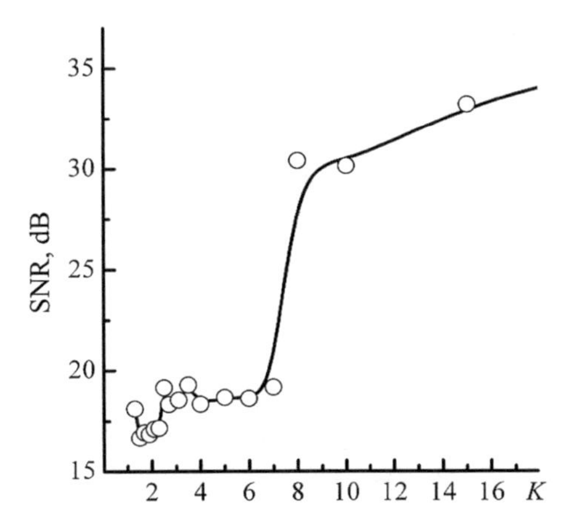

Fig. 3.17 The function of SNR *versus* the amplification coefficient $K = K_1 = K_2$ for $\delta D = 5\ \%$, $D_1 = -0.8$ V, $T_1 = T_3 = 1$, $T_2 = T_4 = 100$, $\tau = 5$, $\mu = 1$, $R = 300\ \ \Omega$, $R_1 = 65\ \ \Omega$, $E_1 = 1.2$ V, $E_2 = -1.2$ V

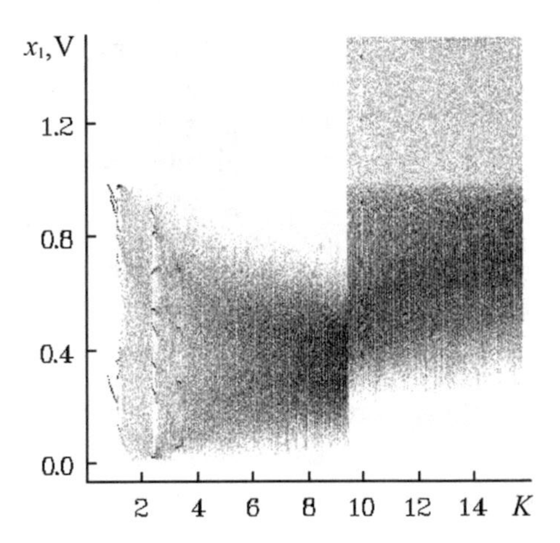

Fig. 3.18 Bifurcation diagrams in the model (3.1) on the plane $x_1 - K_1$, when $D_1 = -0.8$ V, $T_1 = 1$, $T_2 = 100$, $\tau = 5$, $R = 300\ \ \Omega$, $R_1 = 65\ \ \Omega$, $E_1 = 1.2$ V, $E_2 = -1.2$ V

The asymmetry SNR (δD) decreases (Fig. 3.16c). At noticeably larger K, the additional limitation is manifested at the ends of the transfer characteristics, and, therefore, the further restriction of SNR (δD) occurs.

Figure 3.17 shows SNR calculation results depending on amplification $K = K_1 = K_2$. We can distinguish two parts with low and high SNR values equaled to ≈18 dB (in the range of K values from 2 to 8) and not less than 30 dB (for $K > 8$). The stick-sleep nature of the K influence on SNR is connected with the dynamic mode replacement in the transmitter. Indeed, the bifurcation diagram (Fig. 3.18) justifies different dynamic modes in the investigating region of the parameter $K_1 \in [1; 25]$ values. The first mode is realized approximately in the region $K_1 \in (1.25; 8.5)$ and is characterized by the voltage variations at the LPF output in the region of its values $x_1 \in [0; 1]$. In addition, the stability regions (Fig. 3.18) are situated between boundaries $K_1 = [2; 4]$, which are characterized by the periodic mode. Reduction of SNR by 2–3 dB is typical for these regions

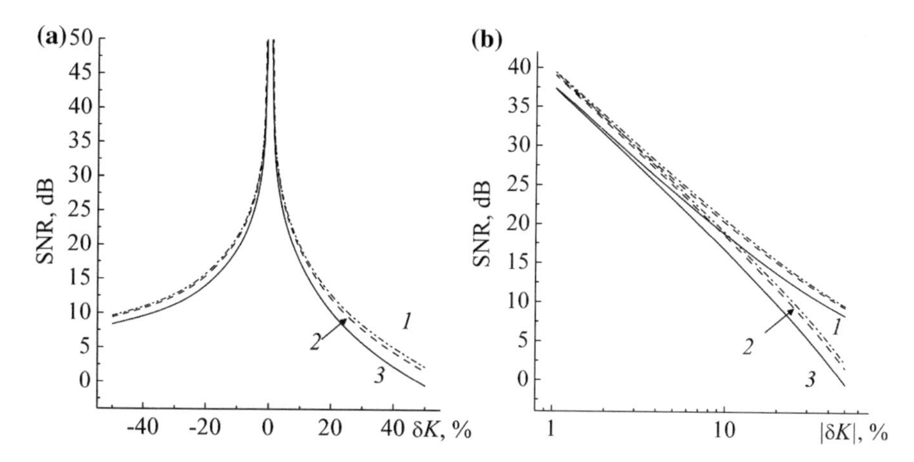

Fig. 3.19 The relative offset influence δK of parameter K_2 setting on SNR. Dependence is constructed in usual (**a**) and double logarithm (**b**) scale for different bias voltages *1* $D = -0.8$ V; *2* $D = -0.8$ V; *3* $D = -1.5$ V. Here $K_1 = 25.0, T_1 = T_3 = 1, T_2 = T_4 = 100, \tau = 5, \mu = 1, R = 300\ \Omega$, $R_1 = 65\ \Omega, E_1 = 1.2$ V, $E_2 = -1.2$ V

(Fig. 3.17). The second mode corresponds to the region of $K_1 \in [8.5; \infty)$ values and differs by more wide variations of $x_1 \in [0.2; 1.5]$. With K growth, the upper limitation is caused by a presence of offset part of 1.5 V on the transfer characteristic of the nonlinear element [1].

The asymmetry takes place for the function SNR *versus* δK. For $|\delta K| \geq 10$ it manifests in the various decrease slope SNR(δK) for negative and positive δK values (Fig. 3.19a). At changing of δK from 1 to 50 % SNR decreases from 40 to 0 dB, and for δK from -1 to -50 % it decreases from 40 to 10 dB. The similar dependence is typical for different bias voltages D (-0.5 V; -0.8 V; -1.5 V). At that, growth of D displaces SNR upward. Figure 3.19b shows asymmetry as a deviation from the linear function (in the double logarithm scale) for $\delta K < 0$. This is stipulated by signal reduction at the output of the slave system in the receiver. Besides, we can see that variation of δK in 10 times leads to SNR variation by 20 dB (or 100 times).

However, SNR *versus* D shown in Fig. 3.20 for $K_1 = 25$ justifies the fact that function displacement for different D is possible in downward also. Indeed, the curve for $K_1 = 25$ is notable by the fact that it is non-monotone. The maximal SNR value for it achieves 26 dB for $D \approx 0$. The minimal SNR value equals to 18 dB for $D = 3$ V. On the contrary, curves for $K_1 = 100$ and $K_1 = 200$ differ by the more monotone SNR behavior: it increases for the bias $D > -3$ V, achieving a maximum for $D = 2.2$ V [1].

Thus, we calculated functions of some parameters of the data transmission system (SNR, relative power of mismatching noise) *versus* lack of coincidence of the amplification factor and bias in the transmitter and the receiver. There is a correlation between SNR and a type of the dynamic mode. At that, it is important to

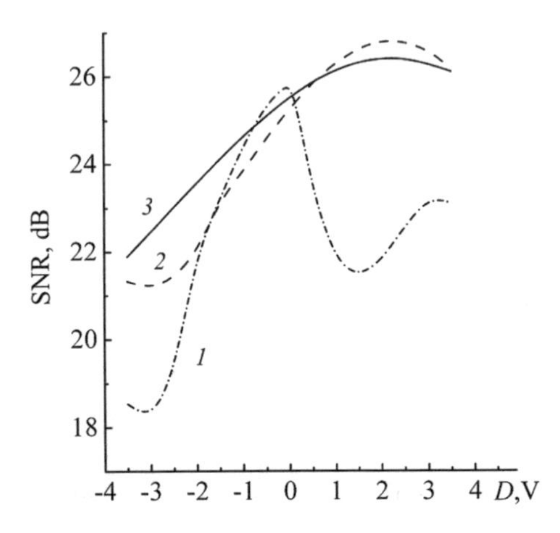

Fig. 3.20 A function SNR *versus* bias D for *1* $K_1 = 25$; *2* $K_1 = 100$ V; *3* $K_1 = 200$. Here $\delta K = 5$ %, $D_1 = -0.8$ V, $T_1 = T_3 = 1$, $T_2 = T_4 = 100$, $\tau = 5$, $\mu = 1$, $R = 300\ \Omega$, $R_1 = 65\ \Omega$, $E_1 = 1.2$ V, $E_2 = -1.2$ V

note that variation of SNR is inversed to the variation square of mismatching of the controlling parameters of the transmitter and the receiver. Tendencies found out during modeling, as it will be proved further, are demonstrated in the experiment as well. Mentioned regularities should be taken into consideration at development of the data transmission system.

3.2.2 Temperature Mismatching Influence of the Transmitter and the Receiver on Data Transmission Quality

To achieve the high quality of communication, the necessary accuracy of all structural system component implementation is required (Fig. 3.1). We considered above how a deviation of parameters D and K of the communication system affects on the data transmission quality, as well as we estimated, which mismatching of the output voltages of the transmitter and the received nonlinear elements are caused by inequality of its temperatures. However, it is yet unclear, how the last will reflect on the communication quality. Naturally, inequality of temperatures will lead non-identity of all structural components of the system. Therefore, we need to determine the most and the least critical system components in this aspect.

First of all, we should estimate the degree of characteristic offset due to temperature mismatching ΔT of the linear circuits of the receiver and the transmitter, which attracts appearance of the additional bias voltage δD: an adjustable amplifier (K) and the bias circuit (D).

So in the experimental breadboard, the microcircuit AD8056 was used, for which the typical of the temperature coefficient for output voltage bias does not exceed 6 $\mu V/^{\circ}C$. This magnitude is more than less by an order of values calculated above for relative error δD of the bias D setting. Really, for $D = -0.1$ V the value

δD at the expense of temperature mismatching does not exceed $6 \cdot 10^{-3}$ %/°C. But temperature mismatching, naturally, influences on δD stronger for less D, more ΔT and the number of microcircuits. For instance, the rough estimation for $D = 0.1$ V and $\Delta T = 50$ °C for 10 microcircuits gives: $\delta D = 10 \cdot 50$ °C $\cdot 6 \cdot 10^{-3}$ %/°C = 3 %. However, values of D, ΔT etc., chosen for our estimation, are non-typical and give some limited estimation.

Another possible reason of the mismatching noise is an error of the amplification coefficient K. Reducing this effect to the minimum in the radio electronic equipment is performed by arranging of necessary number of circuits of the deep negative feedback. Due to this, the thermal dependence of K is conditioned by the temperature sensitivity of only passive linear elements (resistors, capacitors). But thermal sensitivity of K at reasonable choice of the element base appeared to be insignificant. Due to the same reason, we can consider the thermal sensitivity of LPF, HPF and the delay line insignificant as well.

Therefore, in the future, we assume that the main mechanism of data transmitting system quality degradation is the temperature instability of the nonlinear element, if, of course, we do not apply in it the methods of temperature compensation, which includes them in the series of above-mentioned system components. Let us examine the manifestation force of this mechanism.

Mismatching noise power η with growth of temperature difference ΔT from 0 to 25 °C ($K = 5$) changes from $2 \cdot 10^{-5}$ to 10^{-3}, and for the values $K = 15$ and $K = 25$ from $7 \cdot 10^{-7}$ to $2 \cdot 10^{-5}$ (Fig. 3.21a), i.e. the power η depending on the mode can be changed by two orders. At that, the slopy $\eta(\Delta T)$ keeps in the wide mentioned interval of ΔT values. As we see from Fig. 3.21b, with variation of the receiver temperature with respect to the transmitter temperature by 20 °C, SNR deceases by 15 dB. The mode change of the dynamic system (owing to K variation)

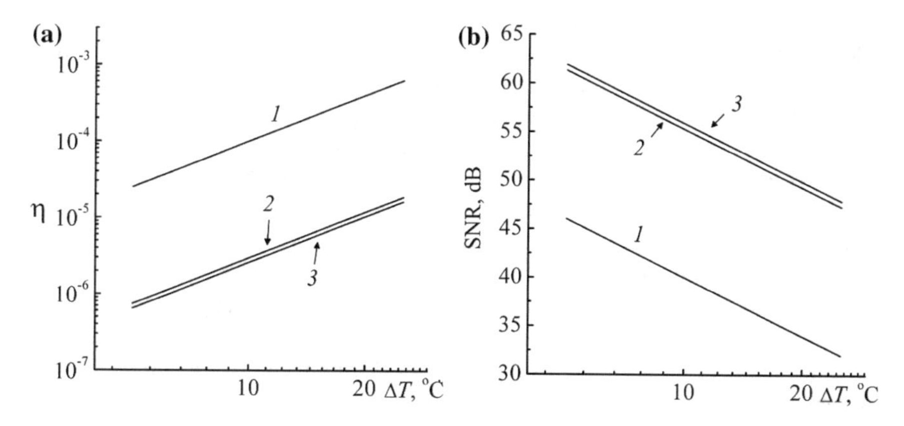

Fig. 3.21 Influence of temperature difference ΔT of the receiver and the transmitter on the relative power of mismatching noise η (**a**) and SNR (**b**). Here the amplification coefficient $K = 5$ (*1*), $K = 15$ (*2*), $K = 25$ (*3*); $D = -0.8$ V; $R = 300$ Ω, $R_1 = 65$ Ω, $E_1 = 1.2$ V, $E_2 = -1.2$ V, $T_1 = T_3 = 1$, $T_2 = T_4 = 100$, $\tau = 5$, $\mu = 1$

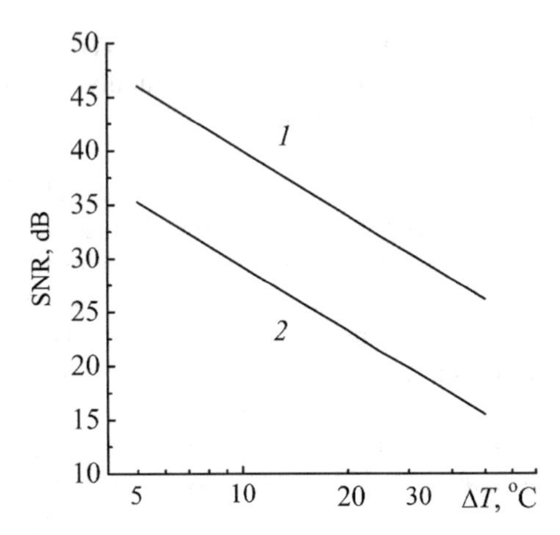

Fig. 3.22 Influence of the temperature difference ΔT of the receiver and the transmitter on SNR. The nonlinear element is thermal-compensated (*1*), the nonlinear elements is without compensation (*2*)

causes the parallel shift of the curve. So, difference between curves *1* ($K = 5$) and *2* ($K = 15$) is 16 dB.

From Fig. 3.22 we see that two-times variation of the temperature difference of the receiver and the transmitter decreases in 4 times (6 dB) the SNR.

For all subjects in Fig. 3.23 the regularity is typical: the growth of temperature mismatching ΔT decreases two times the SNR at the receiver output by 6 dB. For diagrams I Fig. 3.23a, where SNR is shown depending on the amplification coefficient *K*, two regions are typical: with low and high SNR values. As we mentioned earlier (concerning to Figs. 3.17 and 3.18), these regions corresponds to different dynamic modes in the chaotic oscillator.

For SNR function *versus* the *D* bias (Fig. 3.23b) at $K = 25$ the rather weak variation is typical. When approaching to the ends of the transfer characteristic of the nonlinear element ($D = -3.7$ V), the SNR value increases a little ($\sim$2 dB). Still more weak dependence is typical for SNR behavior with delay time growth τ in the delay line (Fig. 3.23c). The horizontal part is observed in the case of SNR dependence on the time constant T_2 of HPF (Fig. 3.23d). It should be taken into account that from the diagram in Fig. 3.23b the points are excluded, which corresponds to the static modes in the chaotic oscillator.

Let us examine the *D* bias influence on SNR for $\Delta T = 30$ °C and the medium values of *K*, when far from the whole nonlinear (VVV-type) part of the transfer characteristic of the nonlinear element participates. Owing to the last circumstances, the phase trajectories do not fill the maximal wide region of the phase space (see Fig. 2.14a, b). Curves *1* and *2* in Fig. 3.24a correspond to exactly this subject, and the curve 3, on the contrary, related to the case high values of *K* (see Fig. 2.14d, e). For calculation of these functions the point were eliminated (for example, $D = 0$), corresponding to static modes in the oscillator.

Curves *1* and *2* in Fig. 3.24a demonstrate the essential non-monotone dependence of SNR *versus* the *D* bias voltage. The local maxima are observed in the curve *1* in points $D \in [-1.1$ V; 0 V; 1.1 V], and the local minima in points

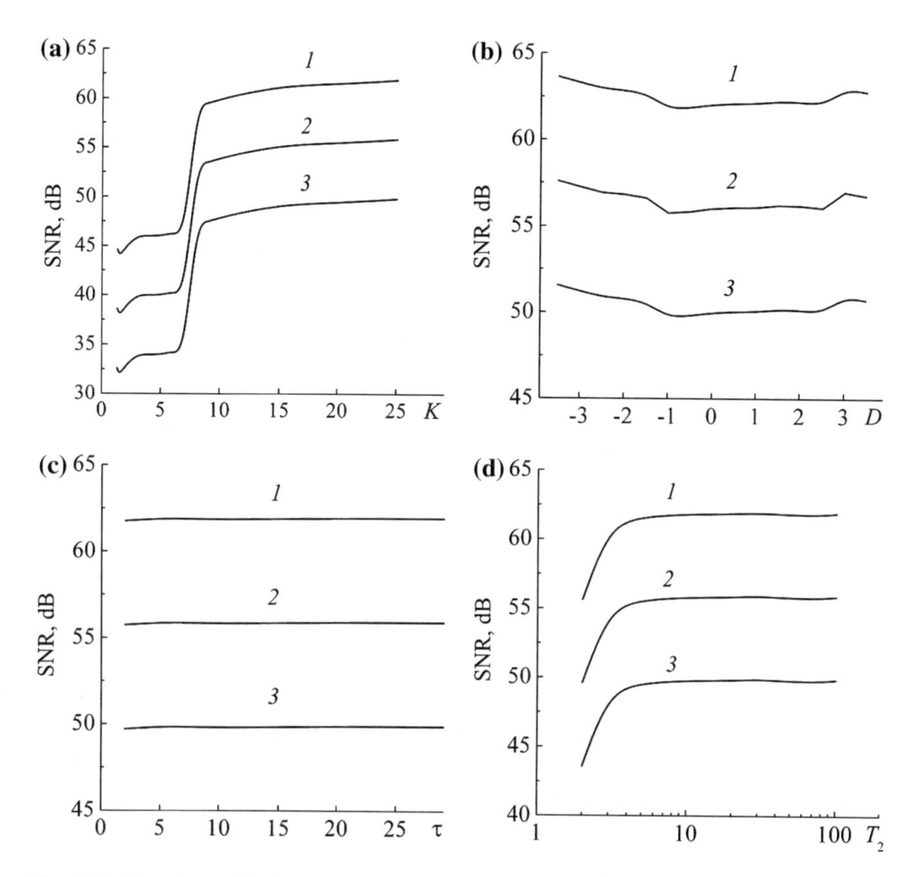

Fig. 3.23 Functions of SNR: *versus* the amplification coefficient K (**a**); *versus* D bias (**b**); *versus* delay time τ (**c**); *versus* time constant T_2 of HPF (**d**) at temperature mismatching: *1* $\Delta T = 5$ °C; *2* $\Delta T = 10$ °C; *3* $\Delta T = 20$ °C. Values of parameters, which are not an argument of this or that dependence, are: $K = 25$, $D = -0.8$ V, $\tau = 5$, $T_2 = 100$, $R = 300\ \Omega$, $R_1 = 65\ \Omega$, $E_1 = 1.2$ V, $E_2 = -1.2$ V, $T_1 = T_3 = 1$, $T_4 = 100$, $\mu = 1$

$D \in [-0.8$ V; 0.8 V]. Positions of the local extremes on the curve 1 coincide with minima and maxima on the transfer characteristic of the nonlinear element (Fig. 3.13, formula (3.19)). Growth of amplification coefficient K smoothes SNR (D)—see curves *2* ($K = 7$) and 3 ($K = 18$). Really, for the curve *3* the value of SNR practically does not change its value in all regions of the D bias variation (SNR $\approx 39 \pm 2$ dB, SNR $\in (38; 41)$ dB).

And how does amplification affects on SNR? Answering this question, let us rely in Fig. 3.24b. Commenting it, we should analyze jointly peculiarities of the receiver and the transmitter transfer characteristic construction and sizes of the phase space region filled by the phase trajectory at given dynamic mode. It is clear from the figure that with K growth, SNR at first sharply decreases and after that some

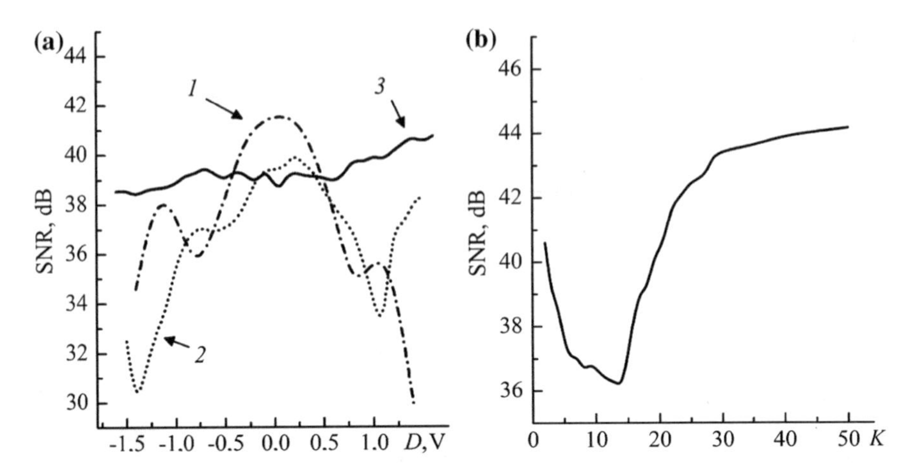

Fig. 3.24 Dependences of SNR: *versus* the bias voltage D for *1* $K = 4$; *2* $K = 7$; *3* $K = 18$ (**a**); *versus* amplification coefficient K for $D = -0.5$ V (**b**). For all diagrams $E_1 = 0.95$ V, $E_2 = -0.95$ V, $R = 300\ \Omega$, $R_1 = 15\ \Omega$, $T_1 = 7$, $T_2 = 1$, $\tau = 30$, $\Delta T = 30$ °C

saturation occurs (the portion from $K = 5$ to $K = 12$) with further increase and limitation.

At K growth (approximately from $K = 1$ to $K = 5$), sizes of the phase space region occupied by the attractor are increased. Accordingly, the operating region on the transfer characteristic of the nonlinear element increases gradually occupying not only the descending part, but the other parts of it, in which mismatching of transfer characteristics of the nonlinear elements (the transmitter and the receiver) is more. This leads to reduction of SNR (Fig. 3.24b).

The further increase of the amplification coefficient slows down the SNR reduction (a part from $K = 5$ to $K = 12$), which then replaces by SNR increase ($K > 12$). The thing is that the operating region begins cover the ascending parts of the transfer characteristic, where mentioned mismatching is less than in the maxima vicinity adjacent to the characteristic center. And then the operating region at first drops in the one horizontal part of characteristic, and then—in another part (i.e. due to the D bias presence non-simultaneously), where characteristics have no difference at all. Let us to remind: on the ascending parts the diodes in the limiter (Fig. 3.10) are open. Therefore, the transfer characteristic type is determined by almost exclusively by resistor divider and the multiplier, hence, it weakly depends upon temperature.

In other words, the further increase of K does not lead to infinitely growth of SNR. The thing is that due to limitation of the transfer characteristics, it becomes impossible to widen a region occupied by the attractor in the phase space—see part the saturation area in Fig. 3.24b. Naturally, in this case (at large enough K) the bias D value does not influence upon SNR.

Thus, presented function in Fig. 3.24 confirm assumption that mechanisms responsible for SNR variations relate to peculiarities of dynamic system modes stipulated a size of the phase space region filled by phase trajectories.

3.2.3 The Role of Noises, Filtering, Level-Discretization in the Communication Channel

Influence of the white noise in the communication channel on the transmission quality. In real operation conditions in the communication channel, noises of various natures are always present. In order to study the noise influence on the complete synchronization of the transmitter and the receiver, the numerical modeling of equation system (3.1) was performed in assumption of the white noise presence in the communication channel (Fig. 3.25). With noise power growth, we observe the linear (in the double logarithm scale) decrease of SNR. Its slope is such that noise power growth by 10 dB leads SNR decrease by 10 dB [8].

In communication system under consideration, as in any other, the multipathing can be presented as a consequence of electromagnetic wave reflections from various irregularities in the communication channel. Influence of such chaotic interference on the synchronous chaotic response in the communication system is discussed in [9].

Filtering influence in the communication channel. Assuming that the communication channel role is similar to action of the low-pass filter of the first order with time constant T_5, numerical modeling shows (Fig. 3.26): SNR takes values in the range from –14 to 58 dB at variation of T_5/T_1 from 0.01 to 5. And the variation of T_5/T_1 by 10 times changes SNR by ≈20 dB.

Quantization error influence in the communication channel of transmission quality. Let us analyze the case, when the chaotic signal is transmitted in the digital communication channels. The signal from the transmitter output passes to ADC,

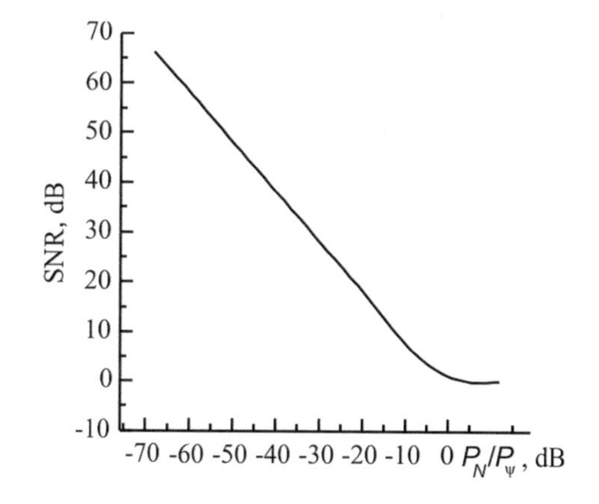

Fig. 3.25 Function of SNR *versus* the white noise power P_N in the communication channel normalized to the power P_ψ of the input receiver signal ψ for: $R = 300\ \Omega$, $R_1 = 15\ \Omega$ $E_1 = 0.667$ V, $E_2 = -0.667$ V, $K = 7$, $D = -0.4$ V, $T_1 = 1$, $T_2 = 122.2$, $\tau = 15.5$

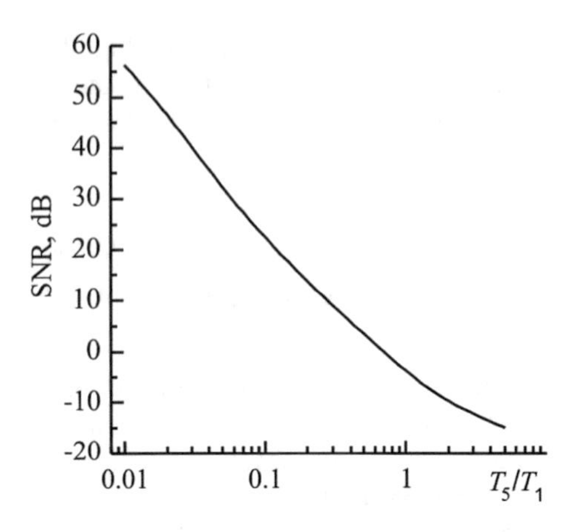

Fig. 3.26 Dependence of SNR *versus* a ratio of time constants T_5/T_1 of LPF in the communication channel and in the transmitter: $E_1 = 0.667$ V, $E_2 = -0.667$ V, $K = 7$, $D = -0.4$ V, $T_1 = 1$, $T_2 = 122.2$, $\tau = 15.5$

where the analog signal is converted in the digital (binary) code. After that, the binary data are transmitted through data transmission channel and pass to DAC, where they convert in the analog form and pass to the receiver input. The signal transmitted through the digital channel does not distort. But due to round-up or truncation of the digitized value of the sample (in ADC) the *quantization errors* arise. Due to this, quantization noises appear, which change the shape and the spectrum of the chaotic signal.

The quantization error value (and as a consequence, noises) depends on the quantization level number, the quantization step value in ADC and DAC. From Fig. 3.27 we may determine that minimal number of quantization levels, necessary to achieve SNR equaled approximately to 65 dB, is 4096. This corresponds to 12 bits' ADC. Besides, we may conclude that at growth of quantization level number by 10 times, SNR changes approximately by 20 dB [8]. Comparing functions in Figs. 3.25 and 3.27, we may notice the similar in shape bends pre-

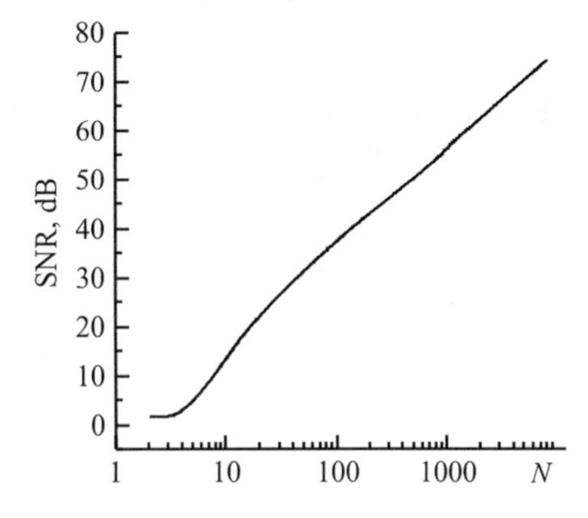

Fig. 3.27 Dependence of SNR at transmitter output *versus* the quantization level number in the data transmission channel: $E_1 = 0.667$ V, $E_2 = -0.667$ V, $K = 7$, $D = -0.4$ V, $T_1 = 1$, $T_2 = 122.2$, $\tau = 15.5$

senting on both figures near SNR values of 0 dB. This fact, probably, connects with that quantization noise in its nature is the random process (white noise), as in the case in Fig. 3.25.

3.2.4 *From Bias Voltage Manipulation in the Oscillator of the Deterministic Chaos to Transmission and Reception of Digital Signals*

Various ways of information insertion in the transmitter were described in the first chapter. Up to now, we studied the communication system with nonlinear mixing of the information signal in the transmitter to the chaotic one. Now we consider a system with parameter modulation or transmitter parameter manipulation for information entering (Fig. 3.28).

We remind that in the system in Fig. 3.28 the message enter is performed with the help of changing of one parameter in the vector $\mathbf{r}_1$ of the master system. If parameter vectors in the receiver and in the transmitter are equal ($\mathbf{r}_1 = \mathbf{r}_2$), we can think that the data transmission system is in the mode of *complete* synchronization, at that, the signal at receiver output $m_2 = 0$. At variation of transmitter parameter vector value, i.e. for $\mathbf{r}_1 \neq \mathbf{r}_2$, the operation mode of the chaotic system (transmitter) changes, which leads to destruction of the complete synchronization mode of the receiver and the transmitter. This is accompanied by appearance at the receiver output of the nonzero signal: $m_2 \neq 0$.

The transmitter and the receiver are built on the base of auto-oscillating system of the ring type. The time constants of LPF and HPF (T_1 and T_2), the delay time τ in

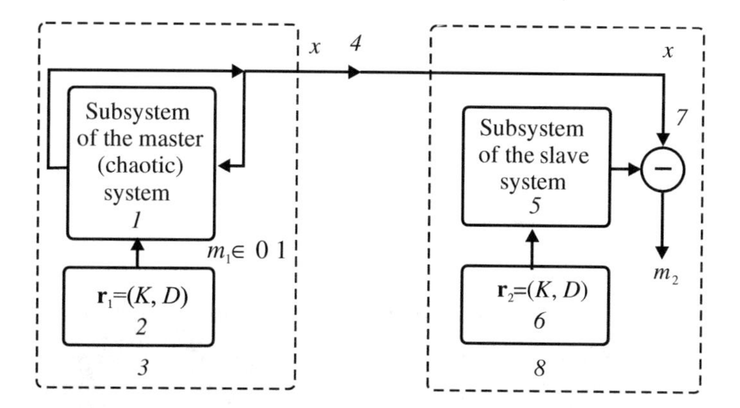

Fig. 3.28 The block-diagram of the data transmission system with digital modulation of chaotic system parameters (an analog of the communication system with chaotic modes switching in Fig. 1.35): *2* the control circuit of the transmitter parameters, *3* master system (transmitter), *4* the communication channel, *6* the control circuit of the receiver parameters, *7* subtractor, *8* slave system (receiver)

the feedback line serve as its time parameters. Their values determine the frequency and temporal properties of the chaotic system. Determination of the dynamic mode transient at stepping parameter changing $\mathbf{r}_1$ allows estimation of permissible speed of binary signal transmission.

The wavelet analysis is a convenient tool for complicate signal investigation. Owing to properties of basic expansion functions, the wavelet analysis allows investigations the signal features in the specific time intervals. In the chaotic signal, both slow and fast rhythms can be presented. On the wavelet-spectrogram the small time scales depict a character of fast-changing processes. On the contrary, the large scales correspond to the slow processes, which are characterized by the large typical time of changing. The wavelet coefficients, located at different scales, carry information about contribution of processes with appropriate typical time of variation into the investigated signal.

Therefore, the continuous wavelet conversion is used by us to estimate the transient time of the dynamic mode and diagnostics of time moments, in which the jumping variations of the bias voltage occur in the transmitter. At that, a choice of this or that real or complex mother wavelet is the independent problem. For simplicity, we are limited by wavelet-conversion with the basic function Simlet—a variety of the real mother wavelet. Nevertheless, the wavelet conversion with the complex basics (for instance, Morle-wavelet) [10] is the more fine tool for solution of the similar problem.

Figure 3.29a shows single stepping variation $\Delta D(t)$ of the bias D—the controlling parameter in the transmitter, and Fig. 3.29b shows the wavelet-spectrogram for $\Delta D(t)$. The time value is shown in the abscissa axis, which is normalized by the time constant T_1 of LPF. The wavelet-conversion is shown in the ordinate axis in the logarithm scale. Values of the wavelet-conversion coefficients normalized to the maximal value are represented by the grey color gradations: The white color corresponds to the maximal value, the black—to the minimal. From the

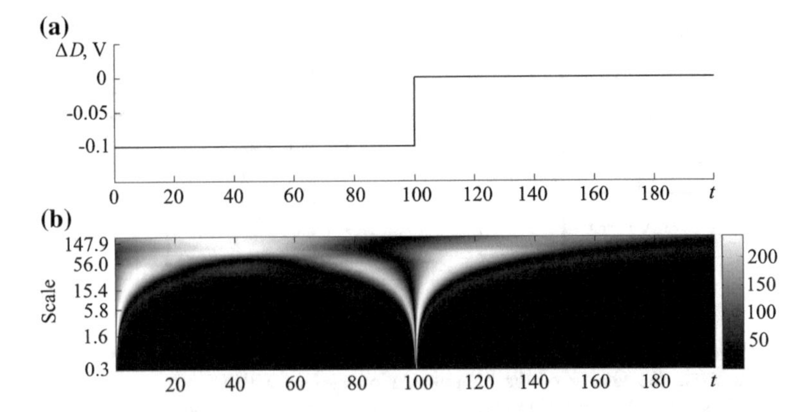

Fig. 3.29 The single stepping variations of the increment parameter of the bias voltage $\Delta D(t)$ of the transmitter (**a**); its wavelet-spectrum with the basic function Simlet with scaling order 4 (**b**)

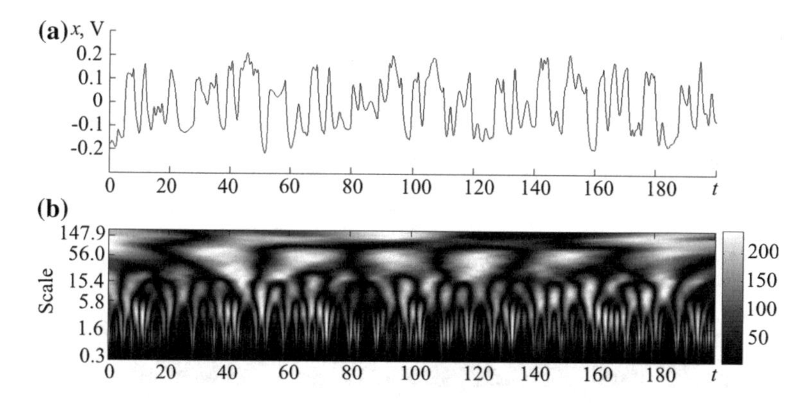

Fig. 3.30 The signal at transmitter output for single jumping variation of the parameter D (**a**); its wavelet-spectrum (**b**) with the basic function Simlet with the scaling order 4. Here $D = -0.4$ V, $K = 4.5$, $R = 300\ \Omega$, $R_1 = 15\ \Omega$, $E_1 = 0.95$ V, $E_2 = -0.95$ V, $T_1 = 7$, $T_2 = 1$, $\tau = 30$, $T = 23$ °C

wavelet-spectrogram, we see that in the region adjacent to the moment of fast (jumping) signal variation ($t = 100$ I Fig. 3.29a), the spectrum demonstrates a burst of wavelet-conversion coefficients which are responsible for small scales. In other words, the non-zero wavelet-coefficients arise in all range of scales (white color in the vicinity of the line $t = 100$ in vertical). On the contrary, coefficients for large scales decrease. As a result, the typical vertical white-black-white "icicle" is formed in the middle of the figure.

Figure 3.30a shows the signal at the transmitter output (Fig. 3.28) in the case of the single stepping variation of the bias voltage D, according to Fig. 3.29. On the wavelet-spectrogram (Fig. 3.30b) of the chaotic signal we can see non-uniformly scaled wavelet-coefficients. Having analyzed the spectrogram visually, we can extract three regions on the scale axis on the whole investigating time interval. The first region is located in the interval of the scale coefficients from 15 to 100. The second region is localized within the limits from 4–5 to approximately 30. The third region corresponds to just lesser values of its scale coefficients. Taking into account the delay existence $\tau = 30$, it is reasonable to assume that the first region corresponds to sub-harmonics, and two others—to harmonics of fundamental frequency of the chaotic oscillator, for which the phase excitation condition is satisfied.

On the wavelet-spectrogram (Fig. 3.30b), it is difficult to distinguish visually the existence of wavelet-coefficients related to jumping variations of the bias voltage D in the transmitter. This justifies a capability of this system (when using of the given way to message transmission) to withstand the attempts to uncover it using the wavelet-analysis with the basic function Simlet.

Figure 3.31a, b show the signal at the receiver output and its wavelet-spectrogram. The signal in the time scale can be divided into two parts. Before the moment $t = 130$, the signal represents the chaotic oscillations, after $t = 130$—the static mode (complete synchronization of the transmitter and the receiver). The moment of the last arising is displaced (with respect to D variation

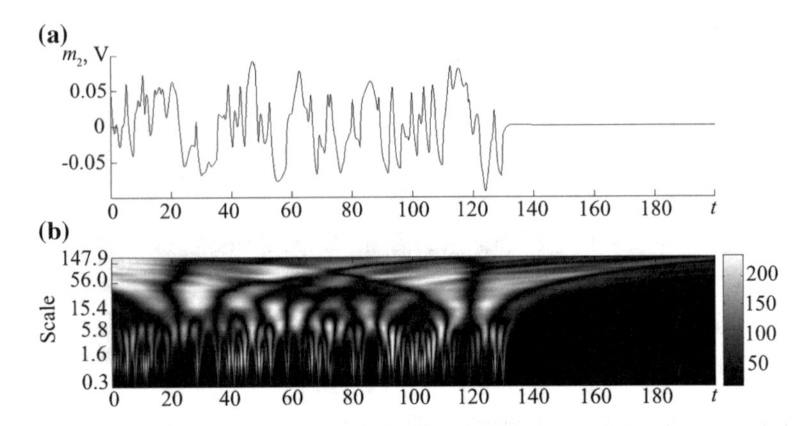

Fig. 3.31 The signal at the receiver output for single jumping variation of the *D* parameter (**a**); its wavelet-spectrum (**b**) with the basic function Simlet with the scaling order 4

moment) by the time value $\tau = 30$ of signal propagation in the delay line of the receiver.

Comparing Figs. 3.30b and 3.31b, we can notice differences in wavelet-coefficients distribution on the wavelet-spectrograms. If in Fig. 3.30b the coefficients in the scale range from 15 to 100 dominate, in Fig. 3.31b the more small-scale coefficients from 1 to 15 prevail. It means that in signal dynamics at receiver output the high-frequency oscillations predominate, than at *transmitter* output.

If we continue the numerical investigation of the data transmission system reaction, constructed according the circuit in Fig. 3.28, and apply to the transmitter input a signal in the form of a sequence of the negative rectangular pulses with the amplitude 0.1 V, which imitate the information signal (Fig. 3.32). As we see from Fig. 3.33, the information signal in the structure of the chaotic one of the transmitter output is invisible. In some time moment in the receiver, the complete chaotic synchronization mode is disturbed, which is accompanied by arising of the chaotic pulses at its output (Fig. 3.34). The last exist until the standard value of the bias voltage *D* will be recovered (and therefore, the complete chaotic synchronization), of course, with account of some delay. As a whole, modeling results confirm earlier made a conclusion related a possibility to ensure confidentiality of data transmission.

Taking into consideration all above-mentioned information, let us proceed the results of our breadboard of the communication system.

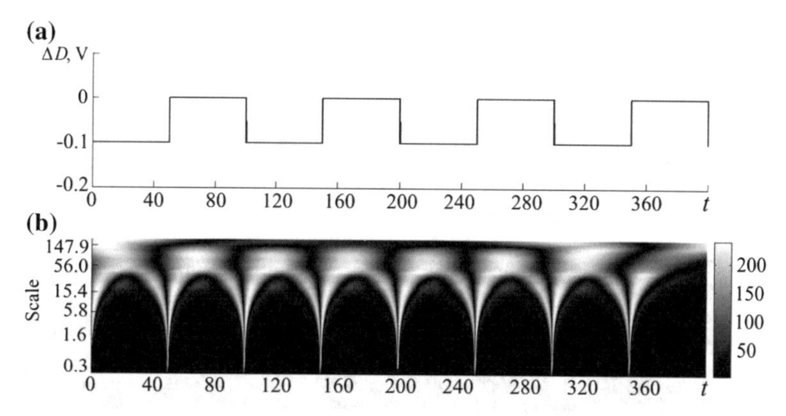

Fig. 3.32 Modulation of the increment $\Delta D(t)$ of the bias voltage in the transmitter by a sequence of negative rectangular pulses (**a**); its wavelet-spectrum (**b**) with the basic function Simlet with the scaling order 4

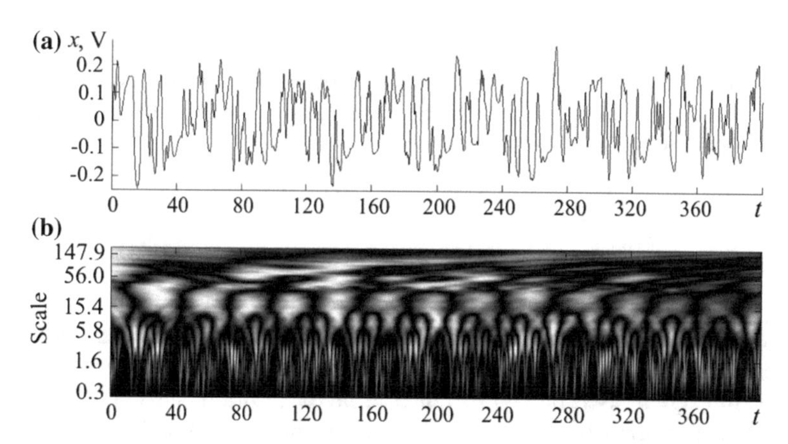

Fig. 3.33 The signal at transmitter output for modulation of the D parameter by a sequence of the negative rectangular pulses (**a**); its wavelet-spectrum (**b**) with the basic function Simlet with scaling order 4. Here $D = -0.4$ V, $K = 4.5$, $R = 300\ \Omega$, $R_1 = 15\ \Omega$, $E_1 = 0.95$ V, $E_2 = -0.95$ V, $T_1 = 7$, $T_2 = 1$, $\tau = 30$, $T = 23$ °C

3.3 Description and Characteristics of the Chaotic Communication System Breadboard, Experimental Reception-Transmission of Analog, Digital and Video Signals

For experimental investigation of data transmission system properties in the mode of complete chaotic synchronization and performing the experiments for transmission and reception of signals with the chaotic carrier, the laboratory system breadboard was developed. Three structural units form it: the transmitter (performs

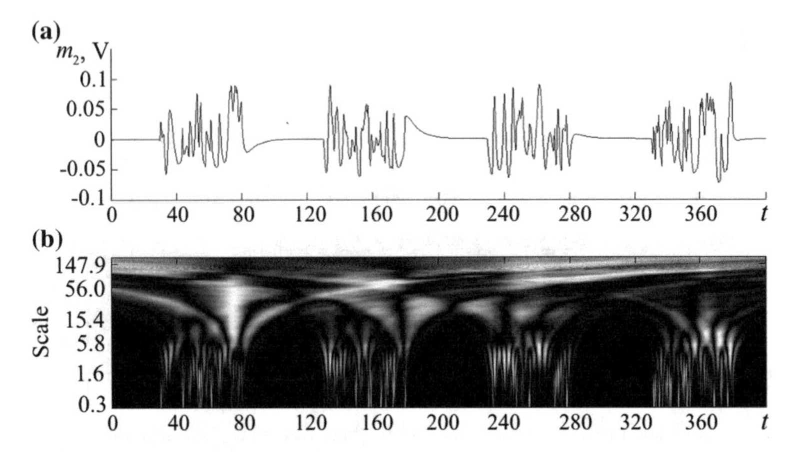

Fig. 3.34 The signal at the receiver output for modulation of D parameter by a sequence of negative rectangular pulses (**a**); its wavelet-spectrum (**b**) with the basic functions Simlet with scaling order 4

the nonlinear mixing of the information external signal to the generated chaotic one); the communication channel; the receiver (recover the information signal using the phenomenon of the chaotic response).

3.3.1 The Breadboard of the Data Transmission System

The transmitter. The structure of the transmitter repeats the structure and designations of the chaotic oscillator in Fig. 2.16, but it is added by the circuit of the information signal insertion and its outlet in mixture with he chaotic one. The last allowed the transmitter connection to the coaxial communication line, to fiber glass transmitter (Figs. 2.17 and 2.18), to the radio transmitter. In other words, we have a possibility in the laboratory breadboard for signal transmission through the cable channel, the fiber-optical channel and the radio channel.

The receiver. Figure 3.35 shows the structural circuit of the receiver. It consists constructively of two parts: the module *1* (Fig. 3.36) and the module *2*, which is a result of decomposition of the deterministic chaos oscillator in Fig. 2.16. The module *1* receives the mixture of the information signal with the chaotic one from the communication channel and from the module 2 output, amplifies them (sub-module *1.1*), performs the necessary delay of signal phase (sub-module *1.2*), recovers the information signal from its mixture with the chaotic one (sub-module *1.3*).

The cable communication channel. Here we use the segment of the coaxial cable RG-6 up to 100 m in length. The attenuation value in the cable depends on the frequency: for 5 MHz attenuation is 17 dB/km, for 100 MHz it is 64 dB/km,

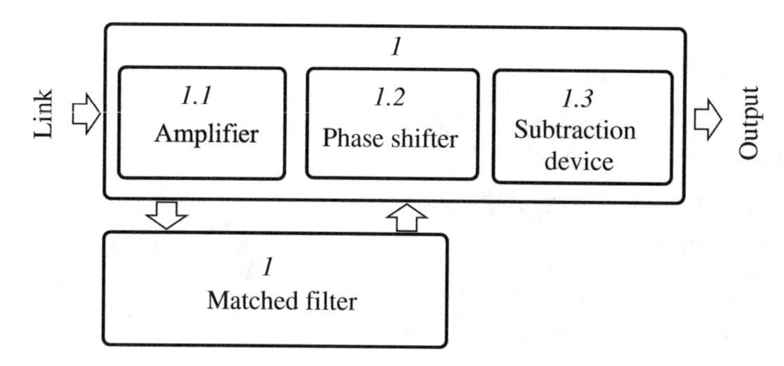

Fig. 3.35 Structural diagram of the of the data transmission

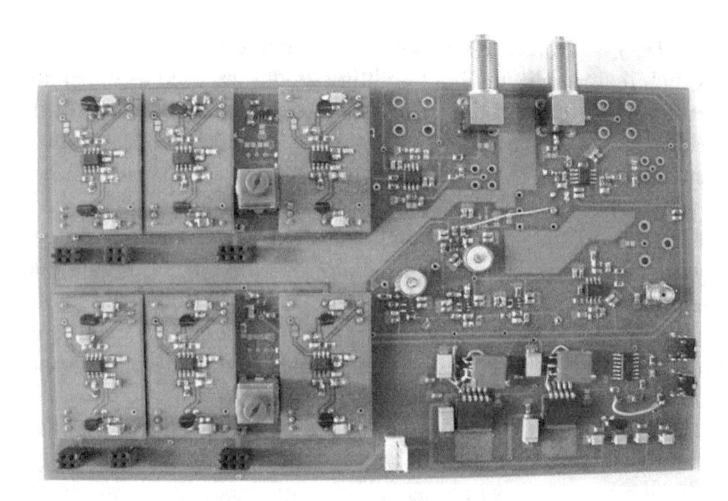

Fig. 3.36 The module *1* of the laboratory receiver

for 1 GHz–217 dB/km. Such characteristics of RG-6 allow transmission of the chaotic signal (in the band 10 MHz) with minimal distortions (not decreasing noticeably the quality of the recovered signal in the receiver) on the length not more than several hundred meters.

The fiber-optical communication channel. The attenuation value in this is significantly lower than in the coaxial cable. In the laboratory set up the multi-mode gradient optical fiber is used, which core diameter is 50, and 125 μm envelope. Attenuation on the wavelength $\lambda = 850$ nm is equal to 2.42 dB/km. We should note that the carrying capacity for $\lambda = 1.300$ nm exceeds the appropriate value for $\lambda = 850$ nm. This is explained by the fact that dispersion, which determines the carrying capacity, consists of inter-mode and chromatic components. If the inter-mode dispersion weakly depends on the wavelength, the chromatic dispersion is proportional to the radiation spectrum width. Coefficient of proportionality $D(\lambda)$ for given wavelengths in the vicinity of 1.300 nm is closed to zero, while on

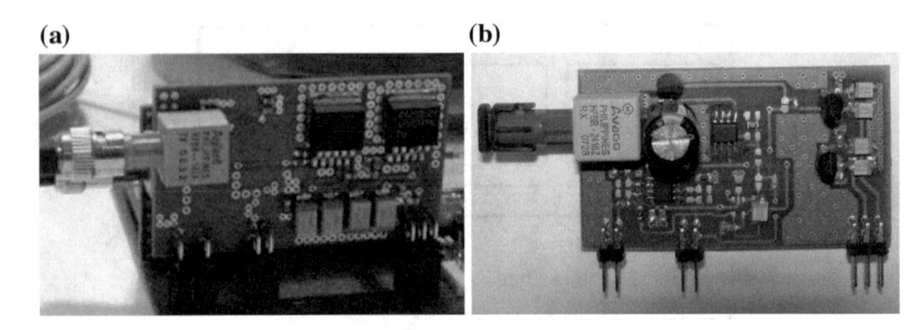

Fig. 3.37 The light-emitting diode HFBR-1414 on the transmitter plate (**a**); the fiber-optical receiver HFBR-2416 on the receiver plate (**b**)

$\lambda = 850$ nm $D(\lambda) \approx 100$ psec/(nm^2 km). The specific of the multi-mode fiber utilization is such that usually the light-emitting diodes are used as transmitters. They have widening of the spectral line of radiation (due to non-coherence of the source) approximately of $\Delta\lambda \sim 50$ nm, in contrast to laser diodes with widening $\Delta\lambda \approx 2$ nm and less. This leads to the fact that chromatic dispersion for $\lambda = 850$ nm begins to play the essential role together with inter-mode dispersion.

In construction of fiber-optical communication channel in the structure of the laboratory breadboard of the chaotic oscillator and the system of data transmission the light-emitting diodc HFBR-1414 is used (Fig. 3.37a) and the photo receiver HFBT-2426 (Fig. 3.37b), operating on $\lambda = 820$ nm. At that, in the frequency band of the transmitted signal 125 MHz the maximal length of the fiber-optical communication line is 2 km.

We can significantly decrease a chromatic dispersion using the laser diodes, which have essentially less spectral widening. But it is possible to realize this advantage of the laser diodes only using the single-mode fiber with the transparent windows $\lambda = 1.310$ nm and $\lambda = 1.550$ nm. In transparent windows the inter-mode dispersion is completely absent and we have only the chromatic dispersion.

The radio channel. The signal transmission with usage of radio waves is performed with utilization of the wireless transceiver BADA 2.4 GHz Wireless Long Distance A/V Sender Transmitter and Receiver (4 Channel)—see Fig. 3.38. The transceiver permits transmission of the analog audio and signals on the length up to 100 meters. The signal in the wireless communication line is transmitted with the help of frequency modulation in one of the four channels on the frequency 2.4 GHz (channels: 1—2.414 GHz, 2—2.432 GHz, 3—2.450 GHz, 4—2.468 GHz). From the output of the transmitter chaotic oscillator the signal with amplitude to 1 V is acted at the video input of the transmitter. After propagation over the ether, the radio signal passes in the receiver and from its video output it passes in the receiver input of data transmission system.

Probably, the professional engineer would like to know, which experimental approach of measurement of SNR and the mismatching noise magnitude in our system of data transmission. We shall touch this problem later.

Fig. 3.38 The receiver and the transmitter used in the bread

3.3.2 SNR Measurement in the Laboratory Experiment at Mismatching of the Transmitter and the Receiver Parameters

To measure the influence of parameter mismatching of the transmitter and the receiver on the relative mismatching noise level η (see (3.2)) and the SNR (according to (3.5) with account $\mu = 1$ and (3.2)) the experimental set up serves (Fig. 3.39), but without the unit 8.

The approach of measurement performing is the following. After switching-on of the power source and warming-up the equipment, the initial adjustment of the data transmission system is provided. For this, the dynamic mode of the chaotic oscillator is settled by the circuit 6 of transmitter parameters, and after this, the receiver parameters are settled with the help of the circuit 7. The value of the

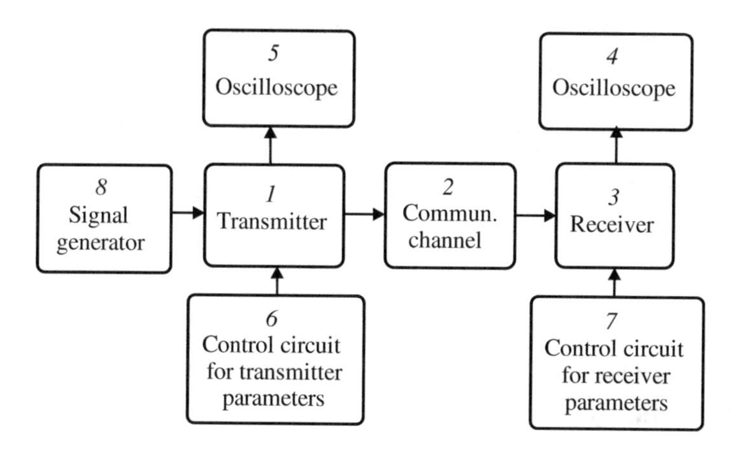

Fig. 3.39 The structural circuit of the set up for measurement of dependences: the mismatching noise and SNR *versus* parameters offset of the transmitter and the receivere (without the unit *8*); system characteristics versus the power ratios of the information and chaotic signals in the transmitter

square-mean power of the *noise mismatching signal* $P_{\Delta z}$ at the receiver output is checked by the digital oscilloscope *4*. After that, we perform the sequential adjustment of the receiver parameters (K, D) for minimization of the signal $P_{\Delta z}$ at its output. Then, in the operating journal, we write parameters of the transmitter and the receiver, the square-mean signal value of the signal at input P_z and output $P_{\Delta z}$ of the receiver. (If $\mu \neq 1$, we require to measure also the power of the information signal P_m at the receiver input).

After that, in order not to change the dynamic mode of the chaotic oscillator in the transmitter, its parameters are fixed and the offset is performed of receiver parameters only at the expense of the receiver parameter changing. A number of measurements and the frequency of point location on the curve "SNR—relative mismatching of parameter value" is chosen empirically in such a way to cover uniformly of dependence (usually, 10–12 points in the variation range of the relative mismatching of the controlling parameter 0–50 %).

In order to determine how data transmission system characteristics on the signal level at the transmitter input, the unit 8 must be connected to the measuring set up (Fig. 3.39). The measurement approach anticipates the warming-up and the preliminary adjustment of the system (see approach description above).

To measure the ratio μ of the power P_m of the information m_1 to the power P_z of the chaotic (z_1 for $m_1 = 0$) signal in the *transmitter*, at first, we register the square-mean of the information signal m_1 at the receiver output by the digital oscilloscope *5* (Fig. 3.39) without the chaotic signal in it (for this we break the feedback in the chaotic oscillator in the transmitter *1*). Then, with the help of *5* at the transmitter *1* output, we measure the square-mean value of the chaotic signal. For this, the unit 8 is switched-off, i.e. the information signal $m_1 = 0$, and the feedback of the chaotic oscillator in the transmitter is recovered. The square of a ratio of the square-mean values obtained is a ratio of μ of information and the chaotic signal powers in the transmitter (3.3).

To obtain the SNR value at the *receiver output* (3.4) with the help of the digital oscilloscope *4*, one sequentially measure the square-mean signal value at the *receiver output 3* at presence and at absence of the information signal m_1 at transmitter input *1*. These measurement repeats for the other power ratios of information TB chaotic signals in the transmitter.

Let us describe now experiments with the data protection system using the dynamic chaos.

3.4 Experimental Operation Studying of the Communication System with the Complete Chaotic Synchronization

Let us begin with discussion of experiments on transmission and reception of radio signals of different shapes [11].

3.4.1 *Transmission and Reception of Analog, Digital and Video Signals*

Experimental investigation of signal transmission regularities was performed with the help of the laboratory set up, which structure is shown in Fig. 3.39. The set up includes the transmitter on the base of the chaotic oscillator (Figs. 2.16, 2.17 and 2.18) and the receiver (Figs. 3.35 and 3.36). The transmitter operated in the mode of the nonlinear mixing. The signal was recovered in the receiver with the help the synchronous chaotic response.

Figures 3.40 and 3.41 show signals and their Fourier spectra for the case of the harmonic signal $m_1(t)$ with frequency $f = 1.31$ MHz at the transmitter input. The record of the signals was carried out in three units of the laboratory set up: at transmitter input (Figs. 3.40a and 3.41a), in the communication channel (Figs. 3.40b and 3.41b), at receiver output (Figs. 3.40c and 3.41c).

As we see from Fig. 3.40b, the useful signal spectrum $m_1(t)$ spectrum is “hidden” in the spectrum of chaotic type oscillations. In other words, it is hard distinguished on the background of the chaotic “interference” $[z_1(t) - m_1(t)]$, since the last spectrum is rich by components with amplitudes, which are larger than the components of the useful signal spectrum. The spectral component with the maximal amplitude has a frequency $f_0 = 1.16$ MHz. In the spectrum of recovered signal (Figs. 3.40 and 3.41c) the high-frequency distortions are observed [11].

In contrast to the harmonic signal, signals used for information transmission can be characterized by the finite frequency band width. Let us take as a testing signal

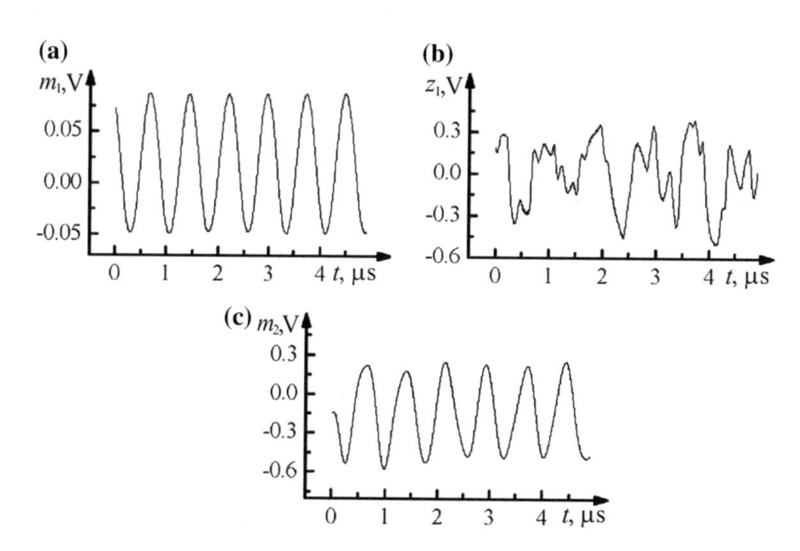

Fig. 3.40 Nonlinear mixing (**b**) and recovering (**c**) of the harmonic signal (**a**) in the communication system. Signals: **a** at the transmitter input $m_1(t)$, **b** in the communication channel $z_1(t)$, **c** at the received output $m_2(t)$

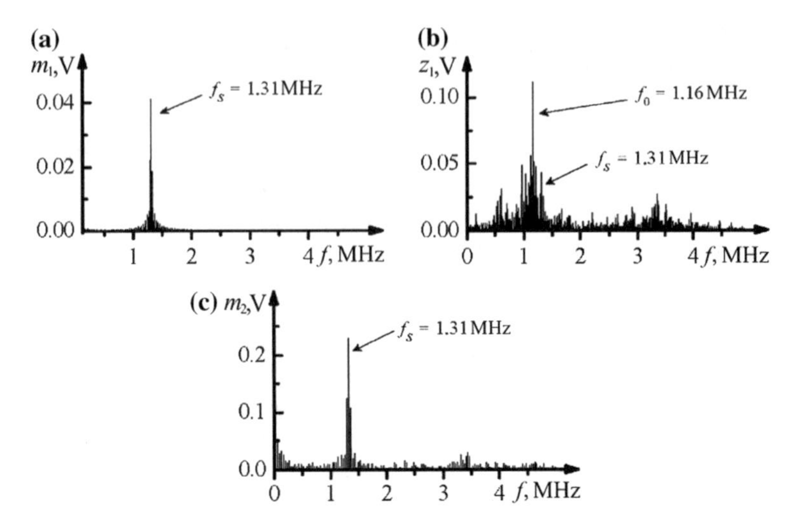

Fig. 3.41 The Fourier spectra of the signal at nonlinear mixing and recovering of the *harmonic signal* in the communication system: **a** before the transmitter, **b** in the communication system, **c** after the receiver. The form of signals is shown in Fig. 3.40. Here f_s is the frequency of the signal $m_1(t)$ at the transmitter input, f_0 is the frequency of the maximal in amplitude component of the chaotic oscillation $z_1(t)$ in the transmitter

the sequence of rectangular pulses. It is suitable for this aim because we know that it has a "lobe" spectrum, which is wider for shorter pulses.

In order to find out how the communication system transmits such signals, the periodic pulse sequence with duration of 1 μs and duty-off factor 30 was applied to the transmitter operating in the chaotic mode. Figures 3.42 and 3.43 are the same in the contents to Figs. 3.40 and 3.41. It is easy to see that the general conclusions were made during the analysis of Figs. 3.40 and 3.41, are true here as well.

Figure 3.44 shows results of transmission and reception of the video signal using the described chaotic communication. The initial signal with the frequency band up to 6 MHz (Fig. 3.44a) was created by the standard video camera. The computer with the standard device for video signal capture served as the signal receiver (a "visualizator"). Recovered image is well distinguished in Fig. 3.44c. High-frequency distortions observed are explained by a slight mismatching of the transmitter and the receiver parameters: dispersion in 5 % of parameters of passive discrete elements, a presence of non-zero output bias voltages of op-amps in the circuit of the nonlinear element. In addition, filtering and nonlinear distortions in the data channel transmission negatively manifests itself in the sense of the recovering image quality. Of course, a modernization and more fine adjustment of units are capable to significantly improve the transmission quality [11].

Since the key element of the communication system quality is SNR, we examine experimentally how it depends on different parameters of the communication system.

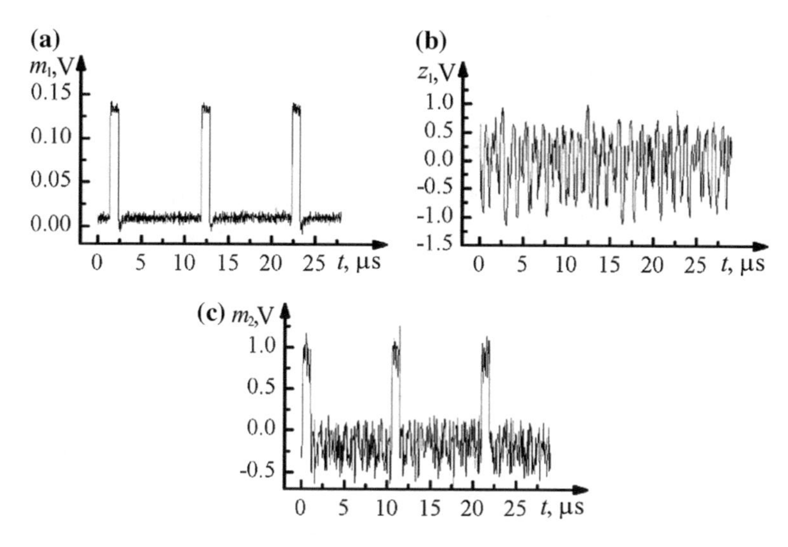

Fig. 3.42 Nonlinear mixing (**b**) and recovering (**c**) of the *rectangular pulse* sequence (**a**) in the communication system. Signals: **a** at the transmitter input $m_1(t)$, **b** in the communication channel $z_1(t)$, **c** at the receiver output $m_2(t)$

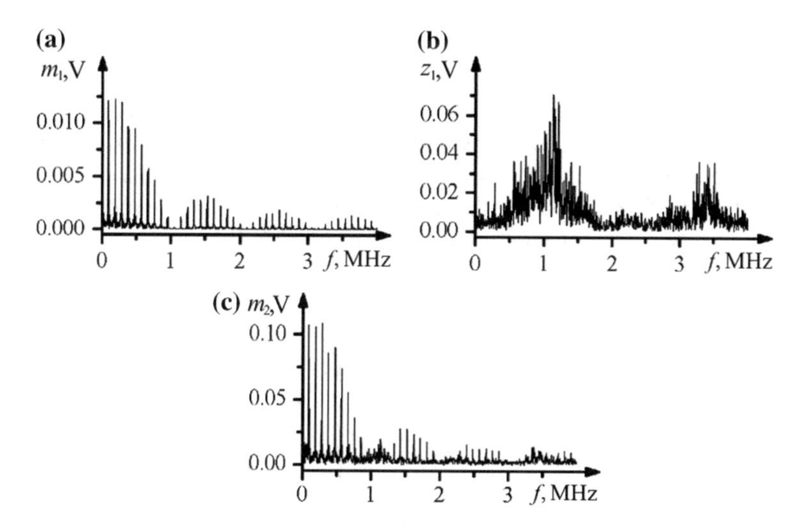

Fig. 3.43 The Fourier spectra of signals at nonlinear mixing and recovering of the *rectangular pulse sequence* in the communication system: **a** before the transmitter, **b** in the communication channel, **c** after the receiver. A view of signals is shown in Fig. 3.42

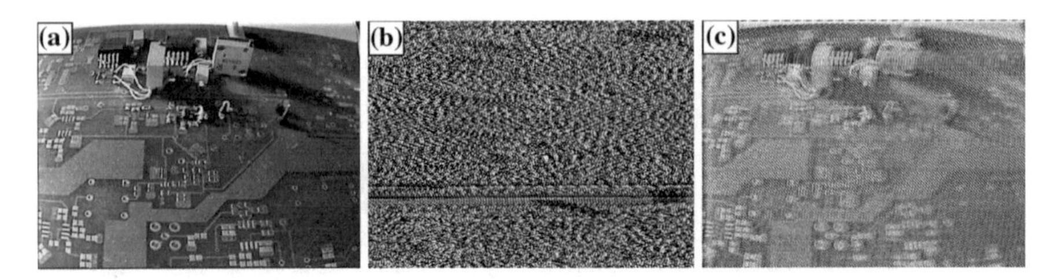

Fig. 3.44 The experience of image transmission (**a**) by a *video signal*, which is chaotized in the transmitter (**b**), and the recovering result of the pattern in the receiver (**c**)

3.4.2 Influence of Data Transmission System Parameters on SNR

Theoretical results obtained in Sect. 3.2, show that communication quality is mainly defined by mismatching of the transmitter and the receiver parameters as well as filtering in the communication channel. Here we give measuring data of the data transmission quality in the laboratory breadboard (Fig. 3.39) and comparison with mathematical modeling results. The harmonic signal with a frequency 600 kHz is used as an information signal.

To check the theoretical conclusions about a role of filtering in the communication channel discussed in the context of Fig. 3.26, the following experiment was fulfilled. The comparison result is presented in Fig. 3.45. Evidently, the quantitative correspondence of the experimental results (a sequence of the experimental points *1*) and the numerical modeling (curve *2*) within the limits of experiment error, takes place for $T_5/T_1 > 0.2$. At less degree of filtering (i.e. for $T_5/T_1 < 0.2$), the deviation of the experimental point sequence *2* from the theoretical dependence *1* can be explained by the fact that during the experiment, the offsets of the other parameters are unavoidably present, restricting the growth of SNR [8].

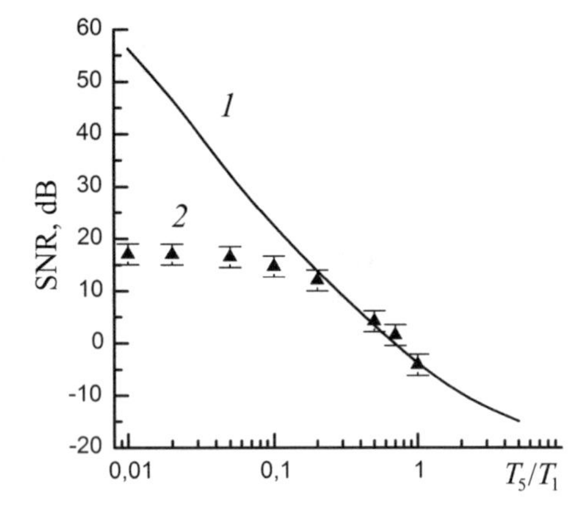

Fig. 3.45 Dependence of SNR *versus* a ratio of time constants T_5/T_1 of LPF in the communication channel and in the transmitter: *1* calculation, *2* experiment. Here $E_1 = 0.667$ V, $E_2 = -0.667$ V, $K = 7$, $D = -0.4$ V, $T_1 = 1$, $T_2 = 122.2$, $\tau = 15.5$

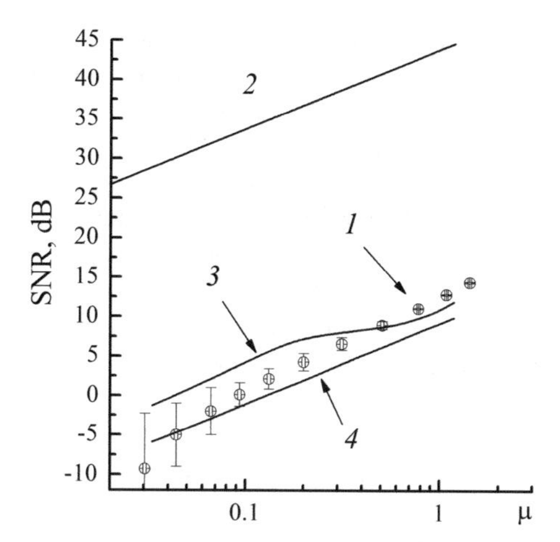

Fig. 3.46 Dependence of the normalized power of the information signal to the power of the chaotic signal in the transmitter: $K_1 = K_2 = 5.95$; $D_1 = D_2 = -400$ mV, $E_1 = 667$ mV, $E_2 = -667$ mV, $T_1 = T_1 = T_3 = 2.7 \cdot 10^{-8}$ s, $T_2 = T_4 = 3.3 \cdot 10^{-6}$ s, $\tau = 4.2 \cdot 10^{-7}$ s; *1* experiment; *2* calculation for the idealized case; *3* calculation for $\delta r_i = 10\ \%$ ($\delta \mathbf{r} = (\delta K, \delta D, \delta T_1, \delta T_2)$); *4* calculation for $\delta\tau = 1.3\ \%$

Figure 3.46 shows results of experimental measurements (a sequence of points *1*) and numerical modeling (curves *2–4*) of the function of SNR *versus* the normalized power of the information signal to the power of the chaotic signal in the transmitter, i.e. *versus* μ. For all curves, with growth μ value, the increase of SNR within the variation boundaries $\mu \in [0; 1]$ more than 20 dB is observed. In the experiment, for $\mu = 1$, the SNR is equal to 12 dB. The calculation shows that for $\mu = 1$ the theoretically achievable SNR value equals to SNR = 44 dB (curve *2*). In this (idealized) case, at SNR calculation, we took into consideration the mismatching of the transfer characteristics of nonlinear elements of the receiver and the transmitter in the laboratory breadboard. The curve *2* factually determines the maximal achievable limit at the absence of the other factors, which influence on the communication quality.

Mismatching of receiver and transmitter parameters, which negatively influences on SNR (see Sect. 3.2 and Fig. 3.46), stipulates for relatively low communication quality in the breadboard. The calculation shows that simultaneous mismatching of parameters of the receiver and the transmitter (besides τ) $\delta \mathbf{r} = (\Delta K, \Delta D, \Delta T_1, \Delta T_2)$ by 10 % (from values of the transmitter parameters) leads to the dependence shift by 30 dB (curve *3*). The permissible value of discrete element deviations of the laboratory breadboard was ±5 %, and this guaranteed that mismatching of receiver and transmitter parameters is $\delta r_i \leq 10\ \%$. This is agreed with the mutual location of the calculated curve 3 (for $\delta r_i = 10\ \%$) and experimental points (*1*) on Fig. 3.46. Really, basically, *1* lays below than curve 3 or very closed to *3*.

Modeling of mismatching δτ of the delay time (for the transmitter and the receiver) nothing but by 1.3 % (curve 4) justify about high sensitivity of SNR to inequality values of τ_1 and τ_2. The curve 4 is closed to the experimental dependence *1*. Evidently, the degradation of the communication quality should be determined by the total (not always by additive) contribution of mismatching of parameters of the transmitter and the receiver (δr_i, $\delta\tau$), especially, $\delta\tau$.

A choice of the specific value of μ depends on destination and a type of data transmission system with the help of a chaos. For instance, if we need to hide the information signal in the chaotic one, for our chaotic oscillator, we must restrict the value of μ by the value of 0.1–0.2. But the SNR becomes small in this case. In further experiments we assume $\mu = 0.25$. Such a value of the external signal does not lead to the chaotic mode loss in the transmitter and ensures confidentiality at relatively high SNR values at the receiver output.

Dependence of SNR *versus* an offset δD of bias D is presented in Fig. 3.47a for different values of the amplification coefficient K (a sequence of the experimental points *1*, *2* and *3*). All curves have maximum if D vales of the receiver and the transmitter coincide. With growth of offset δD (due to variation of bias D in the receiver), SNR at the receiver output decreases. At that, depending on the sign δD, the slope modulus has a various value: for negative values δD it is more than for the positive. According to our assumption, the asymmetry in the structure of the points *1*, *2* and *3* sequences in Fig. 3.47a, we can explain by different slope of the transfer characteristic of the nonlinear element in the vicinity of the point D [12].

Figure 3.47b, as Fig. 3.45, shows the quantitative agreement of experimental results and the numerical modeling within the limits the experimental error. But here, the boundary of this agreement on the abscissa axis δD serves the value 10 %. For lesser offset δD, the difference (as in Fig. 3.45) is explainable by the presence of the not-controlled offset of the other parameters.

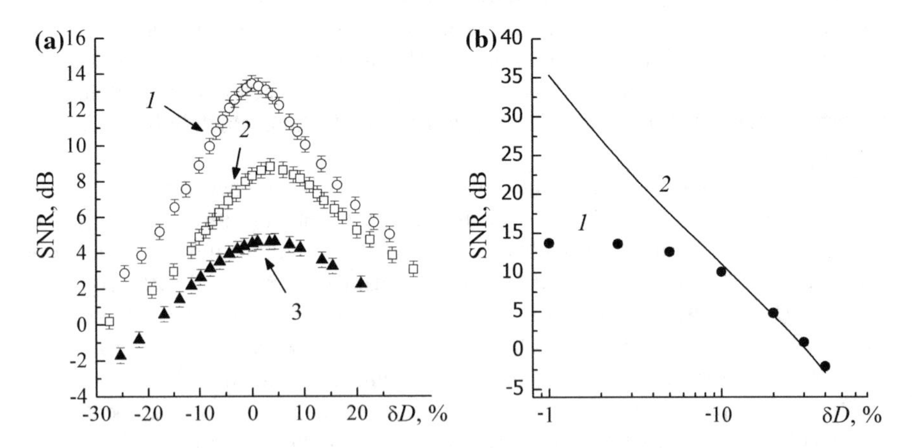

Fig. 3.47 Dependence of SNR *versus* the mismatching of bias voltages δD in the transmitter and the receiver: *1* $K = 4.55$; *2* $K = 4.95$; *3* $K = 6.0$ (**a**) and the logarithmic scale for $K = 4.45$: *1* experiment, *2* calculation (**b**). Here $D_1 = -400$ mV, $E_1 = 667$ mV, $E_2 = 667$ mV, $T_1 = T_3 = 2.7\cdot10^{-8}$ s, $T_2 = T_4 = 3.3\cdot10^{-6}$ s, $\tau = 4.2\cdot10^{-7}$ s, $\mu = 0.25$

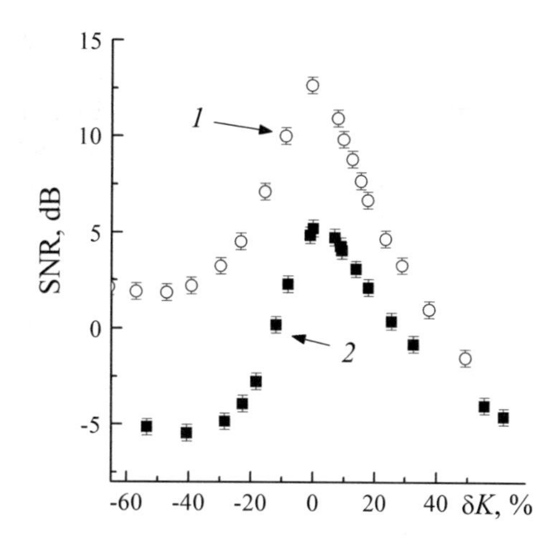

Fig. 3.48 Dependence of SNR *versus* mismatching δK of amplification coefficient in the transmitter and the receiver: *1* $K = 4.57$; *2* $K = 5.89$. Here $D_1 = D_2 = -400$ mV, $E_1 = 667$ mV, $E_2 = -667$ mV, $T_1 = T_3 = 2.7 \cdot 10^{-8}$ s, $T_2 = T_4 = 3.3 \cdot 10^{-6}$ s, $\tau = 4.2$ 10^{-7} s, $\mu = 0.25$

Figure 3.48 shows measurement results of SNR versus the δK values of amplification coefficient K for the receiver and the transmitter. Naturally, at $\delta K = 0$ curves take the maximal values. Values of $\delta K = 30$ % for $K = 4.57$ (curve *1*) and $\delta K = 20$ % for $K = 5.89$ (curve *2*) correspond SNR = 0 dB. For negative values of δK (approximately –40 %) the inflection presence is typical for the curve SNR(δK). Such an effect arises due to decrease of the contribution of the synchronous chaotic signal power into the signal formed at the receiver output [12].

Figure 3.15 shows results of measurements of SNR at output and input of the receiver for different amplification coefficient K. With K growth, the SNR at the receiver output (a sequence of points *1*) reduces. At that, the function *1* can be divided into two linear parts with different slope. The first one, steeper, corresponds to the value range of K from 5 to 7. The second part corresponds to the weak SNR variation and lays within the limits of K from 7 to 9.5. The function *2* presents a ratio of the useful signal to chaotic one in the data transmission channel (at the

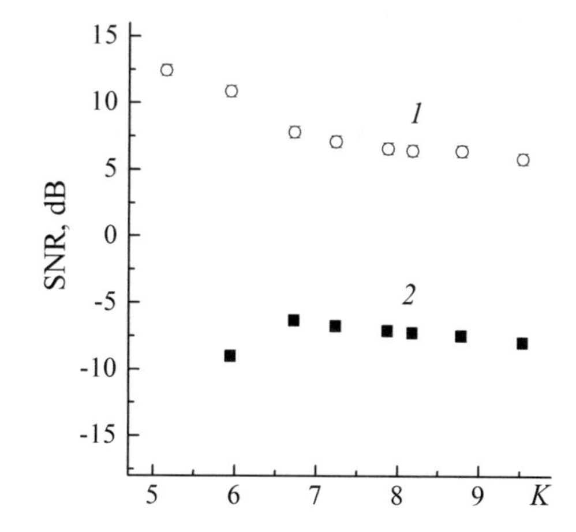

Fig. 3.49 Dependence of SNR at output (a sequence experimental points *1*) and at the receiver input (a sequence experimental points *2*) *versus* the amplification coefficient K for $D_1 = -400$ mV; $E_2 = 667$ mV; $T_1 = 2.7 \times 10^{-8}$ s; $T_2 = 3.3 \times 10^{-6}$ s; $\tau = 4.2 \times 10^{-7}$ s; $\mu = 0.25$

receiver input). It is shifted with respect to the function *1* into the region of negative SNR values. With K growth, the part of the useful signal at the receiver input, at first, increases from –10 to –5 dB. But, after achievement the value $K = 6.5$ the growth of SNR changes by its slowly decrease [12] (Fig. 3.49).

Restriction of amplification by the value $K = 9.5$ is dictated by the following circumstance. For values $K > 9.5$, the region of dynamics in the transmitter does not limited by quadratic parts of the nonlinear transfer characteristic (Fig. 2.4), and passes to its horizontal parts, the presence of which is caused by the cutoff in the op-amps (AD856). In this mode the chaotic dynamics from time to time is loss changing by the quasi-constant (quasi-static) state and vice versa. We get in a way a non-periodic switching.

3.5 Conclusions

Description of the structure and mathematical model of the data transmission system with nonlinear mixing of the information signal in the transmitter and with the synchronous chaotic response in the receiver is given. The summary of data transmission quality criteria is formed. The variants of electrical circuits of the nonlinear element parts with the transfer characteristic in the form of parabola composition are discussed. It is found out the temperature influence on the shape of the transfer characteristic of the nonlinear element. Methods of thermal compensation are offered. Numerical modeling of data transmission quality dependence the versus a series of factors (values of the controlling parameters and temperatures of the transmitter and the receiver, the white noise and the quantization noise as well as filtering in the communication channel etc.) is carried out. A series of wavelet-spectrograms is calculated.

The structural diagram of the communication system breadboard on the base of the deterministic chaos oscillator is given, experimental set up of the communication system is presented. The following principles are discussed: of measurement SNR (at mismatching parameters of the transmitter and the receiver); estimation of carrying capacity of the data transmission channel. Results of the experimental investigations of the data transmission system are presented. Influence of mismatching parameters of the transmitter and the receiver on the transmission quality was studied. The sessions of transmission of the analog and digital signals are described.

The following chapter is devoted to theoretical analysis of the optical systems for message transmissions with chaotic oscillators considered in Chap. 2 on the base of the single-circuit and double-circuit nonlinear interferometer.

References

1. Romanov IV, Izmailov IV, Kokhanenko AP, Poizner BN. Simulation of SNR dependence on the detuning of the communication system parameters using the deterministic chaos. Russ Phys J. 2011;54(5):50–55.
2. Dmitriev AS, Panas AI. Dynamic: news for communication systems. Moscow: FizMatLit Publication; 2002. 252 pp. (in Russian).
3. Gaman VI. Physics of semiconductor devices: text book. Tomsk: NTL Publication; 2000. 426 pp. (in Russian).
4. WebSite Avago technologies. http://www.avagotech.com/docs/AV02-1367EN.
5. Horowitz P, Hill W. The Art of Electronics. Cambridge University Press; 1989. 1128 pp.
6. Grabowski B. Aide-memoire Electronique. Paris: Dunod; 2001. 432 pp.
7. Brindley K, Carr J. Electronics engineer's pocket book. Reed Educational and Professional Publication Ltd.; 2000. 496 pp.
8. Romanov IV, Izmailov IV, Kokhanenko AP, Poizner BN Influence of noises and distortions in the communication channel on data transmission in the system with nonlinear nonlinear mixing of information and chaotic response. Russ Phys J 2012;55(8/3):207–208 (in Russian).
9. Romanov IV, Izmailov IV, Poizner BN, Grigoriev DV. Chaotic interference influence on the synchronous response in the communication system using the deterministic chaos. Russ Phys J. 2013; 56(8/2):318–320 (in Russian).
10. Koronovskiy AA, Khramov AE. Continuous wavelet analysis and its applications. Moscow: FizMatLit Publication; 2003. 176 pp. (in Russian).
11. Romanov IV, Kokhanenko AP, Poizner BN. Nonlinear mixing of radio and video signals in the system of confidential communication using the dynamic chaos (in Russian). In: Proceedings of Tomsk Polytechnical University. Series mathematics and mechanics. Physics, vol. 318(2), pp. 53–58. 2011 (in Russian).
12. Romanov IV, Izmailov IV, Kokhanenko AP, Poizner BN. The role of parameter mismatching of the transmitter and the receiver in the system of chaotic communication with nonlinearity in the form of parabola composition. Russ Phys J. 2012;55(8/3):211–212 (in Russian).

Chapter 4
Single- and Double-Circuit Nonlinear Ring Interferometer as a Cipherer in Optical Systems of Synchronous Chaotic Communications

Original theoretical results of this book authors systemized in this chapter, discover possibility of optical system development for confidential communication. The properties of the chaotic oscillator constructed on the base of the nonlinear ring interferometer, which is described in detail earlier, lay in its base. In this context, it will be clear a concept (introduced on its way) of the nonlinear-dynamic cryptology. Let us explain the operation of cipherer and decipherer (both in static and dynamic modes), offer appropriate models and suggest the mathematical means for formalization, analysis and synthesis of some class of systems. With the help of computer simulations, we find out the key properties of this system. We offer criteria of cryptoresistance of similar systems, and define the principles to improve it and demonstrate advantages of double-circuit interferometer (double circuit NRI (DNRI)).

However, before to describe of the confidential communication system on the base NRI, we make several notice of the general character and (for reader convenience): we give the simplified mathematical model of processes in NRI.

From the point of view of functional analysis, NRI realizes the image of totality of input spatial-temporal functions (signals) into the totality of output signals. Thus, devices operating in the deterministic chaos mode are able to transform the input (information) signal into the "complex" output signal, which makes it difficult to determine the input signal structure. Therefore, we can speak about the possibility of a secret information transmission by means of the processes in the nonlinear systems, or, in a wider scale, about the nonlinear-dynamic cryptography [1–4], capable of improving the quality of the confidential communication systems.

This idea presupposes the choice of the optical turbulence mode and the realization of the time-delay ring system forming a superposition of its dynamic variables. In this case the values of these variables correspond to the different spatial-temporal points $(\mathbf{r}_1, t_1)$ and $(\mathbf{r}_2, t_2)$, while the positions of these points are

I. Izmailov et al., *Cryptology Transmitted Message Protection*,
Signals and Communication Technology, DOI 10.1007/978-3-319-30125-9_4

determined by the (non) regular evolution of the system; and this evolution in turn depends on the external control and informational influences.

Hence, we now proceed to the analysis of the NRI. We assume that the monochromatic and plane-polarized light field with $B_{\text{in}} = 0$, $A_{\text{in}} = \text{const}$, $\psi = 0$ and $\Omega = 0$ acts at its input. Hence, in (2.21) and (2.22) $a(\mathbf{r},t) = b(\mathbf{r},t) = A(\mathbf{r},t)/2$, $\mathbf{E}(\mathbf{r},t) = \mathbf{e}_x A(\mathbf{r},t)\cos[\omega t + \varphi(\mathbf{r},t)]$. Let $n_2(\mathbf{r}) = n_2$. Then the model (2.24) of nonlinear phase shift evolution $U(\mathbf{r},t)$ in NRI in nonlinear medium (NM) and the Eq. (2.29) go over to:

$$\begin{aligned}
\tau_n(\mathbf{r})\frac{\partial U(\mathbf{r},t)}{\partial t} &= \frac{K}{1-R}\frac{A^2_{nm}(\mathbf{r},t)}{A^2_{\max(x,y,t)}} + D_e(\mathbf{r})\Delta U(\mathbf{r},t) - U(\mathbf{r},t);\\
A^2_{nm}(\mathbf{r},t) &= (1-R)A^2(\mathbf{r},t) + [\frac{\gamma(\mathbf{r}',t)}{\sigma/2}]^2 A^2_{nm}(\mathbf{r}',t-\tau)\\
&\quad + \frac{(1-R)^{1/2}\gamma(\mathbf{r}',t)}{\sigma}A(\mathbf{r},t)A_{nm}(\mathbf{r}',t-\tau)\\
&\quad \times \cos[\omega\tau + \varphi(\mathbf{r},t) - \varphi_{nm}(\mathbf{r}',t-\tau)];\\
\mathbf{E}_{nm}(\mathbf{r},t) &= (1-R)^{1/2}\mathbf{E}(\mathbf{r},t) + \frac{\gamma(\mathbf{r}',t)}{\sigma/2}\mathbf{E}_{nm}(\mathbf{r}',t-\tau)
\end{aligned} \tag{4.1}$$

where $K = (1-K)\cdot an \cdot n_2 lk \cdot A^2_{\max(x,y,t)}/2$ is the nonlinearity parameter; $A_{\max(x,y,t)}$ is the maximal value of the input field amplitude. Let us transform the form of the model (4.1) introducing designations $2a = A \equiv a_{in}$, $A_{nm} \equiv a_{nL}$ and normalizing the field amplitudes at the NRI input a_{in} and at the NM input a_{nL} with respect to the value $A_{\max(x,v,t)}$, to the following view:

$$\begin{aligned}
\tau_n(\mathbf{r})\frac{\partial U(\mathbf{r},t)}{\partial t} &= \frac{K}{1-R}A^2_{nL}(\mathbf{r},t) + D_e(\mathbf{r})\Delta U(\mathbf{r},t) - U(\mathbf{r},t);\\
A^2_{nL}(\mathbf{r},t) &= (1-R)A^2_{in}(\mathbf{r},t) + [\frac{\gamma(\mathbf{r}',t)}{\sigma/2}]^2 A^2_{nL}(\mathbf{r}',t-\tau)\\
&\quad + \frac{(1-R)^{1/2}\gamma(\mathbf{r}',t)}{\sigma}A_{in}(\mathbf{r},t)A_{nL}(\mathbf{r}',t-\tau)\\
&\quad \times \cos[\omega\tau + \varphi(\mathbf{r},t) - \varphi_{nL}(\mathbf{r}',t-\tau)];\\
\mathbf{E}_{nL}(\mathbf{r},t) &= (1-R)^{1/2}\mathbf{E}_{in}(\mathbf{r},t) + \frac{\gamma(\mathbf{r},t)}{\sigma/2}\mathbf{E}_{nL}(\mathbf{r}',t-\tau),
\end{aligned} \tag{4.2}$$

где φ_{in}, φ_{nL} and $A_{in} \equiv a_{in}/A_{\max(x,y,t)}$, $A_{nL} \equiv a_{nL}/A_{\max(x,y,t)}$ are phases and normalized field amplitudes at the NRI input and at the NM input, respectively.

4.1 Confident Communication System Based on NRI

According to the classification described in Chap. 1, the message enters the cipherer in the form of the phase-modulated (or amplitude-modulated) light beam in the offered communication system, i.e. according to the nonlinear adulteration principle. Thus, a decipherer should operate in the chaotic response mode.

It is clear that the specific character of the optical systems reveals itself through the 2D signal property, the spatial parameter distribution and the presence of spatial transformations of the optical field in the feedback loop. These specific properties come to light during the substantiation of the recovering possibility for the signal made chaotic by means of the NRI.

4.1.1 *Substantiation of the Recovering Possibility for the Signal Made Chaotic by Means of the NRI*

According to (4.2), the output NRI field E_{out} is a function of the input light field E_{in}, as well as the output field, but in another point ($\mathbf{r}'$) and in the previous time moment $(t - t_0 - t_n)$ already:

$$\begin{aligned} E_{out}(\mathbf{r}, t) &= C_n(\mathbf{r})(1 - R)E_{in}[r, t - t_n(\mathbf{r}, t)] \\ &\quad + \left[\frac{C_n(\mathbf{r})}{2\sigma C_n(\mathbf{r}')}\right]\gamma[\mathbf{r}', t - t_n(\mathbf{r}, t)] \\ &\quad \times E_{out}\{\mathbf{r}', t - t_0[\mathbf{r}', t - t_n(\mathbf{r}, t)] - t_n(\mathbf{r}, t)\}. \end{aligned} \tag{4.3}$$

Here $C_n(\mathbf{r})$ are the light field loss in the nonlinear medium, for instance, due to the Bouguer's law: $\exp(\alpha_B l/2)$.

Description of the field transformation in the decipherer requires that, while introduced, its parameters are compared with the NRI parameters. Let us assume that the nonlinear two-beam interferometer (Fig. 4.1a, b), which is structurally close to the Rozhdestvenskiy and Mach–Zehnder interferometer [5, pp. 173–174], is used as the decipherer. In this case, the following decipherer parameters can be included: R_d is the analog of R; $t_{u,d}(\mathbf{r}, t) = L_d \frac{n_d(\mathbf{r},t) - n_{0d}(\mathbf{r})}{c}$ is the analog of $t_{u,d}(\mathbf{r}, t)$; $t_{n,d}(\mathbf{r}, t)L_d/c$ is the total time for field propagation through the NRI; $t_{0d}(\mathbf{r}', t)$ is the time of field propagation arriving (through the delay line—the lower decipherer arm) to the time moment t to the point $\mathbf{r}$ of its output plane; $C_{n,d}(\mathbf{r})$ is an analog of $C_n(\mathbf{r})$; $\kappa_d(\mathbf{r}', t)$ is the loss during the propagation in the lower decipherer arm; $\gamma_d = \gamma_d(\mathbf{r}', t) = 2R_d\kappa_d(\mathbf{r}', t)$ is the twice loss during the transition subject to the mirror reflection.

We assume that the coordinate transformations in G elements of the cipherer and the decipherer are identical, therefore, we shall not distinguish $\mathbf{r}'$ from $\mathbf{r}'_d$ and σ from σ_d. Then, taking introduced designations into account, we state that the light

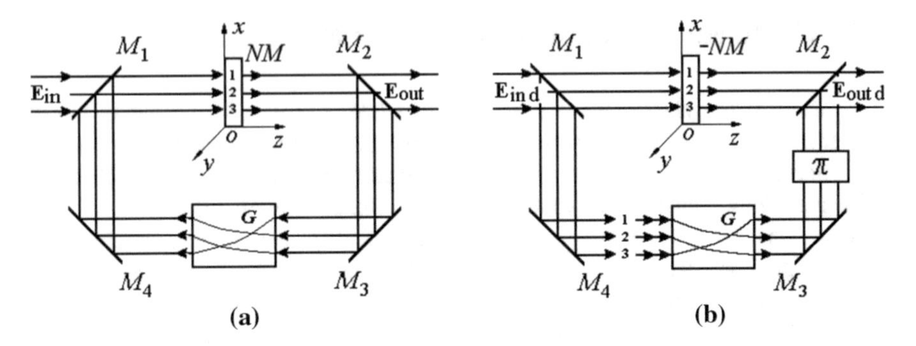

Fig. 4.1 The diagram of the nonlinear-dynamic cipherer (**a**) based on the nonlinear ring interferometer (NRI) and the decipherer to it (**b**). With the light field turned Δ = 120° (in the transverse *xOy* plane of the beam) by the *G* element, the trajectories of the three beams 1, 2, 3 are closed after three NRI bypasses

field $E_{out,d}$ at the decipherer output is a superposition of the fields $E_{in,d}$ at its input in different spatial-temporal points:

$$\begin{aligned} E_{out,d}(\mathbf{r},t) = C_{n,d}(\mathbf{r})(1-R_d)E_{in,d}\left[\mathbf{r}',t-t_{n,d}(\mathbf{r},t)\right] \\ +\gamma_d(\mathbf{r}',t)E_{in,d}\left[\mathbf{r}',t-t_{0,d}(\mathbf{r}',t)\right]/(2\sigma). \end{aligned} \tag{4.4}$$

In the case of the ideal transmission line (i.e. at the absence of noise, losses and optical irregularities), we can assume without generality limitation that

$$E_{in,d}(\mathbf{r}',t) = E_{out}(\mathbf{r}',t). \tag{4.5}$$

Using equalities (4.3)–(4.5), we obtain equation for the signal (supplied to the cipherer input, i.e. to the NRI) at the decipherer output, in the form:

$$\begin{aligned} & E_{out,d}(\mathbf{r}',t)\frac{E_{out,d}(\mathbf{r},t)}{C_n(\mathbf{r})C_{n,d}(\mathbf{r})(1-R)(1-R_d)} \\ & = E_{in}(\mathbf{r},t-t_{nn}) + \frac{1}{\sigma(1-R)}\Gamma E_{out}[\mathbf{r}',t-t_{nn}-t_0(\mathbf{r}',t-t_{nn})] \\ & + \Gamma_e E_{out}\left[\mathbf{r}',t-t_{0,d}(\mathbf{r}',t)\right], \end{aligned} \tag{4.6}$$

where $t_{nn} \equiv t_{nn}(\mathbf{r},t) \equiv t_{n,d}(\mathbf{r},t) + t_n\left[\mathbf{r},t-t_{n,d}(\mathbf{r},t)\right]$ is the propagation time in nonlinear medium of the cryptosystem; $\Gamma \equiv \Gamma(\mathbf{r}',t-t_{nn}) \equiv \gamma(\mathbf{r}',t-t_{nn})/[2C_n(\mathbf{r}')]$ is the loss parameter in the NRI feedback loop (FBL) with account of mirror reflections (similar to the decipherer parameter $\gamma_d(\mathbf{r}',t)/2$; $\Gamma_e \equiv \Gamma(\mathbf{r}',t) \equiv \frac{\gamma_d(\mathbf{r}',t)}{\left[2C_n(\mathbf{r})C_{n,d}(\mathbf{r})(1-R_d\right]}$ is parameter Γ_e for the decipherer equivalent to Γ. We would like to remind that the output field E_{out} of the NRI is considered as quasi-monochromatic:

$$E_{out}(\mathbf{r},t) = a(\mathbf{r},t)\cos[\omega t + \varphi(\mathbf{r},t)] \equiv A(\mathbf{r},t)A_{\max(x,y,z)}\cos[\omega t + \varphi(\mathbf{r},t)].$$

Let us assume the following equality for the condition of correct deciphering: $E_{out,d}(\mathbf{r},t) = E_{in}(\mathbf{r},t - t_{nn})/const$. From its comparison with (4.6) it follows that $const = C_n(\mathbf{r})C_{n,d}(\mathbf{r})(1-R)(1-R_d)$, and the deciphering error $\delta(\mathbf{r},t)$ supplied to the cipherer input is defined as:

$$\begin{aligned}
[\sigma(1-R)]\delta(\mathbf{r},t) &= \Gamma E_{out}[\mathbf{r}', t - t_{nn} - t_0(\mathbf{r}', t - t_{nn})] \\
&\quad + \Gamma_e E_{out}[\mathbf{r}', t - t_{0,d}(\mathbf{r}', t) \\
&= \Gamma a[\mathbf{r}', t - t_{nn} - t_0(\mathbf{r}', t - t_{nn})] \\
&\quad \times \cos\{\omega[t - t_{nn} - t_0(\mathbf{r}', t - t_{nn}] + \varphi[\mathbf{r}', t - t_{nn} - t_0(\mathbf{r}', t - t_{nn})]\} \\
&\quad + \Gamma_e a\left[\mathbf{r}', t - t_{0,d}(\mathbf{r}', t)\right]\cos\left\{\omega\left[t - t_{0,d}(\mathbf{r}', t)\right] + \varphi\left[\mathbf{r}', t - t_{0,d}(\mathbf{r}', t)\right]\right\}.
\end{aligned} \tag{4.7}$$

Deciphering correctness is to be ensured through the solution of the following equations:

$$\begin{aligned}
&\Gamma a[\mathbf{r}', t - t_{nn} - t_0(\mathbf{r}', t - t_{nn})] = \Gamma_e a\left[\mathbf{r}', t - t_{0,d}(\mathbf{r}', t)\right]; \\
&\cos\{\omega[t - t_{nn} - t_0(\mathbf{r}', t - t_{nn})] + \varphi[\mathbf{r}', t - t_{nn} - t_0(\mathbf{r}', t - t_{nn})]\} \\
&\quad = -\cos\{\omega[t - t_{nn} - t_0(\mathbf{r}', t)] + \varphi[\mathbf{r}', t - t_{nn} - t_0(\mathbf{r}', t)]\}.
\end{aligned}$$

Taking into account the t_{nn} form (see 4.6) and assuming that $\varphi[\mathbf{r}', t - t_{nn} - t_0(\mathbf{r}', t - t_{nn})] - \varphi[\mathbf{r}', t - t_{nn} - t_0(\mathbf{r}', t)] \approx 0$, we can prove that the following five conditions ensure the correct deciphering:

$$\omega\left|\delta t_{u,d}(\mathbf{r},t)\right| \equiv |\delta u_d(\mathbf{r},t)| \ll \pi, \tag{4.8}$$

$$\Gamma_e(\mathbf{r}',t)/\Gamma_{e0}(\mathbf{r}',t) = 1, \tag{4.9}$$

$$\delta t_{0,d}(\mathbf{r}',t) \approx 0, \tag{4.10}$$

$$a[\mathbf{r}', t - t_{nn} - t_0(\mathbf{r}', t - t_{nn})] \approx a\left[\mathbf{r}', t - t_{0,d}(\mathbf{r}', t)\right], \tag{4.11}$$

$$\varphi[\mathbf{r}', t - t_{nn} - t_0(\mathbf{r}', t - t_{nn})] \approx \varphi\left[\mathbf{r}', t - t_{0,d}(\mathbf{r}', t)\right], \tag{4.12}$$

where $\delta t_{u,d}(\mathbf{r},t) \equiv t_{u,d}(\mathbf{r},t) - t_{0,d0}(\mathbf{r},t)$ is the setting error of propagation time through nonlinear media compared to the case of complete matching, when $t_{u,d}(\mathbf{r},t) = t_{u,d0}(\mathbf{r},t) \equiv -t_u\left[\mathbf{r}, t - t_{n,d}(\mathbf{r},t)\right]$; $\delta u_d \equiv \delta u_d(\mathbf{r},t) - u_{d0}(\mathbf{r},t) \equiv \omega t_{u,d}(\mathbf{r},t) - \omega t_{d,u0}(\mathbf{r},t)$—the value of the misfit for the nonlinear phase incursion $u_d(\mathbf{r},t) \equiv \omega t_{u\,d}(\mathbf{r},t)$ in the nonlinear medium of the decipherer $u_{d,0}(\mathbf{r},t) \equiv \omega t_{u,d0}(\mathbf{r},t) \equiv -u(\mathbf{r}, t - t_{nd}(\mathbf{r},t))$; $\delta t_{0,d}(\mathbf{r}',t) \equiv t_{0,d}(\mathbf{r}',t) - t_{0,d0}(\mathbf{r}',t)$ is the setting error for $t_{0,d}(\mathbf{r}',t)$ with respect to the ideal value of $t_{0,d0}(\mathbf{r}',t) \equiv n_0(\mathbf{r})L/c + n_{0d}(\mathbf{r})L_d/c + t_0(\mathbf{r}', t - t_{nn}) - \pi/\omega$; $\Gamma_{e0} \equiv \Gamma_{e0}(\mathbf{r}',t) \equiv \Gamma(\mathbf{r}', t - t_{nn})$ is the optimal value of the loss parameter Γ_e of the decipherer.

We can prove [4, 6] that fulfillment of these conditions is practicable, if the values $a(\mathbf{r},t)$, $\varphi(\mathbf{r},t)$, $\omega t_{0,d}(\mathbf{r},t)$, $\omega t_0(\mathbf{r},t)$, $\kappa_d(\mathbf{r}',t)$, $\kappa(\mathbf{r}',t)$ change weakly during the time interval π/ω, which is close to 10^{-14} s for the visible range. For instance, on the assumption that during the light propagation through NM of the decipherer within a time interval $t_{n,d}(\mathbf{r},t)$ this interval itself does not vary noticeably, the inequality (4.8) is true, if $n_{2,d}(\mathbf{r})L_d = -n_2(\mathbf{r})\frac{L}{C_n^2(\mathbf{r})(1-R)(1-R_d)}$, i.e. $\delta K_d = 0$. This allows introduction of δK_d, which is the setting error of the nonlinearity parameter K_d with respect to the $K_{d,0}$ value matched with K:

$$\begin{aligned} \delta K_d &\equiv K_d - K_{d,0}, \\ K_d &= (1 - R_d)n_{2d}L_d k A^2_{\max(x,y,z)}, \\ K_{d,0} &\equiv -\frac{K}{[C_n(r)(1-R)]^2}. \end{aligned} \tag{4.13}$$

We have earlier assumed that the decipherer structure is a priori known. But the question remains, if it is possible to determine the required decipherer structure in terms of the optical scheme of the cipherer in a general case. Besides, it is not quite clear, how the 2D character of the optical field will influence the cryptosystem level of proficiency, from the point of view of a number (one, two, many) and a role (synchronizing, information) of communication channels. The following section is devoted to the answers to these questions.

4.1.2 *"Route-Operator Formalism" and Synthesis of the Cryptosystem Structural Scheme*

We have earlier presented the "route-operator formalism" [7]. Within the limits of the model (4.2), it allows substantiation of the NRI capability to perform a function of such a "chaotizer" of the light field, for which the synthesis of the "dechaotizer" is possible, recovering the spatial field structure at NRI input (terms are suggested in [4, 8, 9]).

Route description of the beam motion by the nonlinear ring interferometer. If to neglect the molecule diffusion in the NM (liquid crystal), then, from (4.2), we can obtain the "point" model of the processes $U(\mathbf{r},t)$ and $A_{nL}(\mathbf{r},t)$ in the transverse section of the laser beam in the *x*-*y* plane. It is convenient to use the concept of transposition point chain (CTP) (see Sect. 2.4.1), since the CTP structure represents the motion *route* through the NRI. At that, the number of NRI bypasses serves as a measure of the length of the route part passed. It is natural to treat CTP as graphs. We know that graphs can be defined by different ways: by adjacent and incident matrices, by lists, for example, of vertex pairs connected with edges (arcs), and by definition of the totality of adjacent vertices for each particular vertex [10, p. 162].

The specificity of the optical-physical processes in the NRI consists in the strict *sequence* formed by the events of laser beams propagation through NM, and the element G, where beam bifurcations (branching, fork) and their linear transformations and convergence into the joint beam are possible. In order to support this specificity through the formalism, we suggest the following language for the chains structure description:

(g_i) or (g) is a point (graph vertex) g_i or g, which is the "parent" (from Latin *generator*). In other words, the term input within the parentheses "(…)" is the "parent" for the next term following after the closing parenthesis ")".

(g) is i-th point following after the point g (i-th "descendant" of the parent g). $(g)0 \equiv g$ at that. But if the symbol "]" follows, then the value of i is interpreted otherwise.

$[(g)0]_m$, $[(g)]_m$ is a *fork* (a point of rout bifurcation) of capacity m: in the point g the one line branches out into m lines.

$[(g)d]_m\ i$ is i-th element (a point) of the d-th sub-sequence starting from the fork point $[(g)]_m$. In this case $[(g)d]_m 0 \equiv [(g)]_m \equiv g$.

$[(g)d]_m\forall$ is any d-th sub-sequence (a path, line) starting from the fork $[(g)]_m$ (all elements (points) of the d-th sub-sequence). Here $[(g)d]_m 0 \in [(g)d]_m\forall$.

$[(g)d]_m$ is the path segment (an edge or arc of the graph) connecting the point g with a point $[(g)d]_m 1$. In the future, for the sake of simplicity, we will not distinguish the concepts of the edge and arc.

$[(g)\forall]_m$ is any of the path segments emerging from the point g.

$\{(g)\}_m$ is a *convergence* (convergence point) of capacity m: in the point g m lines converge into the one.

$\{(g)\}_m i$ is the i-th element (point) of the sub-sequence starting from the convergence $\{(g)\}_m$. At that, $\{(g)\}_m 0 \equiv \{(g)\}_m \equiv g$.

fin_i is the final element of the chain, i.e. the point, at which the chain terminates, where the lower index is the identification number of the final point.

Evidently, any point g is the point of fork and convergence simultaneously, at least, with the capacity 1.

The absence of symbols between a symbol ")" and the following symbol of convergence "}" makes the presence of the parentheses "(…)" non-obligatory. In other words, the following identities take place: $\{((g))i\}_m \equiv \{(g)i\}_m$; $\{([(g)d]_n i)\}_m \equiv \{[(g)d]_n i\}_m$ etc.

In some cases, we can omit the opening brackets, but sometimes we should keep them. Let, for instance, there be a fork point with capacity 3 in the point $(g_1)4$. And let each from the "birthed" sub-sequences contain a various number of elements n_1, n_2, n_3, and they converge in the only point, after which the chain of points terminates on the fifth element, i.e. the fifth element is final. Then this situation can be expressed symbolically as: $\{[(g_2)1]_3 n_1; [(g_2)2]_3 n_2; [(g_2)3]_3 n_3\}_3 = fin$, where $(g_1)4 \equiv g_2$. If each of the sub-sequences contains the equal number of elements ($n_i = n$), this situation can be expressed more compactly: $\{[(g_2)\forall]_3 n\}_3 = fin$, or even as: $\{[(g_2)]_3 n\}_3 = fin$.

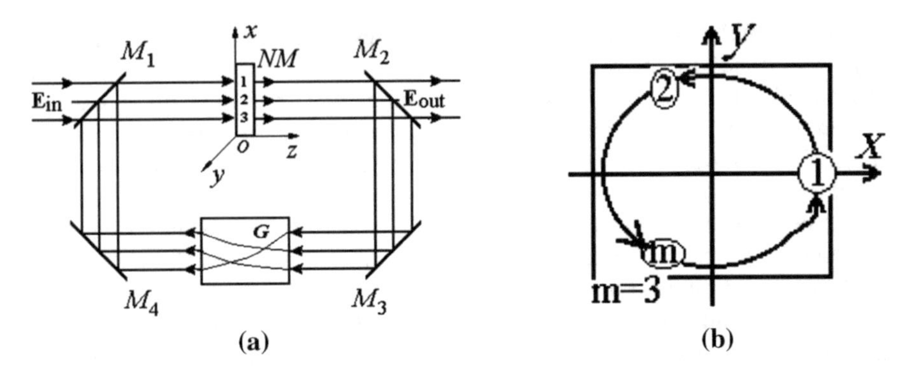

Fig. 4.2 NRI scheme. When light field rotation by the *G* element is Δ = 120° (in the transverse *xOy* plane of the beam), the trajectories of three beams *1*, *2*, *3* are closed after three NRI bypasses (**a**); the projection of closed trajectories of beams *1*, *2*, *3* on the *xOy* plane (**b**)

Any expression meaningful in the above discussed context and applying the formalism suggested, initializes some route of the laser beam in the NRI or even the totality of routes. Here this point is a route of zero length. This formalism allows us to estimate both the number of forks and convergences (or number of routes) and the length of the routes. In our opinion, the definite totality of routes can serve as a one more way of graph definition.

Interferometer as a graph structure system and the route-operator formalism. Let us describe the simplest CTP types, which can be implemented in the NRI. If the *G* element (Fig. 4.2) performs the mirror imaging relatively to a straight line lying in the transverse *x-y* plane of the laser beam and passing through its center *O*, all CTPs are described by expression $(g_k)2 = g_k$, where *k* is the CTP identifier and $g_k \neq g_l$ at $k \neq l$ (further the meaning of indices *k* and *l* remains former). For the points located on the line of mirror-image expressions $(g_k)1 = g_k$ are true.

With laser beam displacement distance Δx, the expression $(g_k)m_k = fin_k$, where $fin_k \neq g_k$ and m_k depend upon Δx and location of the point g_k, is true. For square aperture of the laser beam and the shift $\Delta x = a/m$ along the square side with a length *a*, all $m_k = m - 1$. It is clear that CTP is not closed and its configuration can be called as the *linear* one.

When laser beam rotation angle is $\Delta = 2\pi n/m$ in the *xOy* plane, the expression $(g_k)m = g_k$ is true, where *n*, *m* are mutually prime numbers. It is noteworthy that for the beam center g_c there is possible to have *m*=1 for any Δ. In this case, CTP closes (into a ring) and its configuration can be logically called as *ring*. Evidently, if $m = 2$ ($\Delta = 180°$), this expression reduces to the formula for the mirror image.

If the rotation angle $\Delta \neq 2\pi n/m$, the CTP contains the infinite number of points, and it is impossible to separate neither start nor end point of the chain: $(g_{k,i})1 = g_{k,i+1}$, where *i* is a number of the point in the chain, $i \in (-\infty; +\infty)$.

With laser beam compression (σ<1), expressions $\{(g_k)\infty\}_\infty = g_c$, $\{(g_c)1\}_\infty = g_c$ are true. Therefore, as for its (non)closeness, the CTP happens to be combined, and the image of the “converging” *star* with infinite beam numbers corresponds to its configuration.

In case of laser beam stretching (σ>1), expressions $[(g_c)k]_\infty = fin_k$, $(g_c)1 = g_c$ or expression $\{(fin_k)\infty\}_\infty = g_c, (g_c)1 = g_c$ are true. The last ones (in defiance of the relevant chronology of bypass by the NRI beams) describe the reverse route: from the final points fin_k of the chains located at the beam periphery to their common initial point g_c. CTP is also combined, but its configuration associates with the image of “diverging” *star* with the infinite beam number.

Let us consider that there are points of input g_{in} and output fin_{out} in NRI of the laser beam energy flow. For specificity, we shall hold that the energy output goes on in the first way: $[(g_i)1]_m$. Then, for the case of mirror imaging and if the k index is omitted, the expression $(g_k)2 = g_k$ may be supplemented in the following way: $\{(g_{in,i})1;\ [(g_j)2]_2 1\}_2 = g_i$ and $[(g_j)1]_2 1 = fin_{out,j}$ or $\left[\left(\{(g_{in,i})1;\ [(g_j)2]_2 1\}_2 = g_i\right)1\right]_2 1 = fin_{out,i}$, where i, j=1,2 or 2,1.

For the case of the beam shift, the expression $(g_k)m_k = fin_k$ is replaced by three: $[((g_{in,1})1 = g_1)1]_2 1 = fin_{out,1}$, $[(\{(g_{in,i})1;\ [(g_{i-1})2]_2 1\}_2 = g_i)1]_2 1 = fin_{out,i}$, $[(g_m)2]_2 1 = fin$, where $i \in [2; m]$. The third expression describes the case when CTP does not close, since the beam from the point g_m, having fallen into the NRI feedback loop through the route designated by symbol 2: $[(g_m)2]$, is absorbed in the point *fin* (on the diaphragm, for instance).

For the case of the beam rotation by the angle $\Delta = 2\pi n/m$, the expression $(g_k)m = g_k$ is replaced by the following: $[(\{(g_{in,i+1})1;\ [(g_i)2]_2 1\}_2 = g_{i+1})1]_2 1 = fin_{out,i+1}$, where $i \in [1; m]$, and if $i + 1 = m + 1$, such index value should be replaced by 1.

If $\Delta \neq 2\pi n/m$, the following is true: $[(\{(g_{in,i+1})1;\ [(g_i)2]_2 1\}_2 = g_{i+1})1]_2 1 = fin_{out,i+1}$, where $i \in (-\infty; +\infty)$.

In the case of the beam compression, instead of the expressions $\{(g_k)\infty\}_\infty = g_c$, $\{(g_c)1\}_\infty = g_c$, we get $[((g_{in1})1 = g_1)1]_2 1 = fin_{out,1}$—for the initial point g_1 of CTP located at the periphery;

$[(\{(g_{in,i})1;\ [(g_{i-1})2]_2 1\}_2 = g_i)1]_2 1 = fin_{out,i}$—for internal points g_i of CTP, where $i \in [2; m-1]$;

$\left[\left(\{(g_{in,m})1;\ [(g_{m-1})2]_2 1\}_\infty = g_m\right)1\right]_2 1 = fin_{out,m}$, $\{(g_m)1\}_\infty = g_m$—for the point g_m, being the limit g_c of the CTP sequence: for $m \to \infty$, $g_m \to g_c$.

In the case of the beam stretching, instead of the pair of expressions $[(g_c)k]_\infty = fin_k$, $(g_c)1 = g_c$, we get:

$\{(g_{in,c})1;\ [(g_c)2]_\infty 1\}_2 = g_c$, $[(g_c)1]_\infty 1 = fin_{out,c}$—for initial CTP point $g_1 = g_c$;

$$[(\{(g_{in,2})1;\ [(g_1)2]_\infty 1\}_2 = g_2)1]_2 1 = fin_{out,2};$$

$[(\{(g_{in,i})1;\ [(g_{i-1})2]_2 1\}_2 1 = g_i)1]_2 1 = fin_{out,i}$—for internal CTP points g_i, where $i \in [3; m]$;

$[(g_m)2]_2 1 = fin$—for the point g_m, being the limit of the CTP sequence for $m \to \infty$.

Instead of the expression for the "reverse" route: $\{(fin)\infty\}_\infty = g_c$, $(g_c)1 = g_c$, the following expressions are true:

$[(\{(fin_{out,m})1;\ (fin)1\}_2 = g_m)1]_2 1 = g_{in,m}$—for the point g_m, being the limit of the CTP for $m \to \infty$;

$[(\{(fin_{out,i})1;\ [(g_{i+1})2]_2 1\}_2 = g_i)1]_2 1 = g_{in,i}$—for internal CTP points g_i, $i \in [2; m]$;

$\left[\left(\{(fin_{out,c})1;\ [(g_2)2]_2\}_\infty 1 = g_c\right)1\right]_2 1 = g_{in,c}$—for the initial CTP point $g_1 = g_c$.

In the examples presented above, the route formulas completely specify the CTP structure in the NRI and the location of the NRI laser radiation input and output points. It is easy to see that for "non-singular" (internal) CTP points the following *route formula* is true:

$$[(\{(g_{in,i+1})1;\ [(g_i)2]_2 1\}_2 = g_{i+1})1]_2 1 = fin_{out,i+1}. \tag{4.14}$$

Evidently, we should include the description of the physical influences on the laser beam into the route formulas. For this purpose, it is necessary to introduce some *operators* (transmission coefficients, functionals etc.) describing the physical processes in the route elements, for the particular route points and path segments (vertices and edges of the graph). Thereby, we can interpret the transformations of the signal (the laser beam) in graph elements.

In conformity with the NRI (Fig. 4.2), the specified correspondence is implemented as follows. Transposition points g_i (convergence points) are localized in the nonlinear medium. In these points, the optical field summation occurs, and the resulting field has a delay, and the nonlinear phase incursion $U(\mathbf{r}, t)$ is its measure, developing under the influence of the resulting field intensity according to the differential equation in (4.2).

In the context of the suggested *route-operator formalism,* it means that in the appropriate graph vertices signals are summed, obtaining the phase shift $U_i(t)$. Due to the medium nonlinearity, the total signal changes the characteristics of the given vertex operator (transmission coefficient). It is easy to see that the transmission coefficient $(1 - R)^{1/2}$ can be brought to conformity with the edges $[(g_i)1]_2$ and $[(g_{in,i})1]_1$, while the transmission coefficient for the amplitude $\gamma/2$ and the delay $t_{e,i}(t)$ corresponds to the edge $[(g_i)2]_2$. The final beam point *fin*, which is located at the diaphragm, we would describe as an ideal absorber. All other route elements have the unitary transmission elements by default, because there is no interaction with the laser beam field attributed to them.

The operator corresponding to the input point $g_{in,i}$ of the laser beam, can be defined as a function of time and of the index *i* representing the spatial dependence. This function must describe the signal dynamics (of the amplitude $A_i(t)$) at the NRI input.

The output point $fin_{out,i}$ of the beam is naturally the interface location, i.e. where the NRI coupling with the successive devices. Knowing their characteristics, we could specify the operator corresponding to the point $fin_{out,i}$. At their absence, the point $fin_{out,i}$ should be described as an ideal absorber.

A basic general assumption that the significant (in the context of simulation and modeling) interactions in the systems can be reduced to the graph structure provides the wide range of applications for the route-operator formalism in the field of a mathematic models synthesis. Examples of the similar objects of research can be easily found out among optical, radio-electronic, communication (both technical and socio-cultural) systems.

Application of the route-operator formalism for creation of the decipherer model. The offered formalism is capable to serve as a means for a synthesis of the dynamic system, performing as a decipherer corresponding with the NRI as a ciphering device. It is logical to consider the expression for the route, connecting the NRI input and output signals with due regard for the operators implemented by the route elements, as the *equation relating the input signal* $A(\mathbf{r}, t)$. Let us present the experience of such a synthesis. It should be noted that now the non-ideal absorber by no means corresponds to the point $fin_{out,i+1}$, there is the decipherer input: $fin_{out,i+1} = g_{in,d,i+1}$ in its place.

From the structure of the route formula (4.14) and the given comparison of the operators with the elements of (4.14), it is easy to see that the signal passes via the edge $[(g_{i+1})1]_2$ with the transmission coefficient $(1-R)^{1/2}$ to the output of the cipherer $fin_{out,i+1}$. Hence, there must exist the decipherer element the radiation inlet point $\left(g_{in,d,i+1}\right)$, for instance, realizing the operator $(1-R)^{-1/2}$ reverse to the cipherer edge operator $[(g_{i+1})1]_2$.

In points g_i, cipherer signals arriving through edges $\left[\left(g_{in,i+1}\right)1\right]_1$ and $[(g_i)2]_2$ summarize, obtain the delay $U_i(t)$ and separate. It is important that, according to (4.14) and to the operators, the signals (at each divider outputs) identical to the signal at its input are the results of division. Thus, the division operation does not change the signal.

Accordingly, two decipherer points must exist: in the point $g_{d,i+1}$ the phase delay $U_{d,i+1} = -U_{i+1}(t)$ is realized. In the point $\left(g_{d,i+1}\right)1$ there occurs the substraction of the signal that is equal to the signal S_i, arriving through the cipherer edge $[(g_i)2]_2$ from the signal coming from the vertex $\left(g_{d,i+1}\right)$. Let the point $g_{d,i+1}$ be located on the edge $\left[\left(g_{in,d,i+1}\right)1\right]_m$, i.e. $\left[\left(g_{in,d,i+1}\right)1\right]_m 1 = g_{d,i+1}$, where the fork capacity m will be determined later.

We now take into account that to the edge $[(g_i)2]_2$ the signal arrives from the point (g_i), and it is equal to the signal arriving from the point $\left(g_{in,d,i}\right)$. Therefore, to form the signal S_i in the decipherer, it is sufficient to create the replica $\left[\left(g_{in,d,i}\right)2\right]_m$ of the edge $[(g_i)2]_2$. It will be a copy, but with that difference that the phase delay implemented by the edge $\left[\left(g_{in,d,i}\right)2\right]_m$, differs from $\omega t_{e,i}(t)$ by π. Thereby, subtraction operation is provided, now in the summator $\left(g_{d,i+1}\right)1$ already.

The cipherer edge $[(g_{in,i})1]_1$ corresponds to the transmission coefficient $(1-R)^{1/2}$, therefore, we need to have the final element in the decipherer, which would compensate for the radiation losses. Let it be the point $(g_{d,i+1})1$.

It is obvious that the necessary number of ways from the point $(g_{in,d,i})$ does not exceed two, i.e. we should consider $m = 2$. Taking this into account, we may construct a formula describing the signal transformations in the decipherer:

$$\{([(g_{in,d,i+1})1]_2 1 = g_{d,i+1})1;\ [(g_{in,d,i})2]_2 1\}_2 1 = fin_{out,d,i+1}, \tag{4.15}$$

where the point $(g_{in,d,i+1})$ has the transmission coefficient $(1-R)^{-1/2}$; the point $(g_{d,i+1})$ performs the phase delay $U_{d,i+1}(t) = -U_{i+1}(t)$, the edge $[(g_{in,d,i})2]_2$ has the transmission coefficient $\gamma/2$ and performs the phase delay $\omega t_{e,i}(t) + \pi$. The convergence point $(g_{d,i+1})1$ summarizes the input signals and transmits them with the amplification $(1-R)^{-1/2}$.

In order to provide the delay $U_{d,i+1}(t) = -U_{i+1}(t)$ in the decipherer, it is sufficient to perform the equation for the nonlinearity coefficients of the (de)cipherer $K_d = -K$ by the NM choice, for which $n_{2d} = -n_2$. But if we do not take the task of the NM choice, for which $K_d = -K$, it is enough to provide in the edge $[(g_{in,d,i})2]_2$ the fulfillment of the equalities: $\omega t_{e,d,i}(t) \approx \omega t_{e,i}(t) + \pi$ and $frac(\omega t_{e,d,i}(t)/(2\pi)) = -frac(\omega t_{e,i}(t)/(2\pi) + 0,5)$, where the symbol *frac* designates the fractional part of the number. Here the field at the decipherer output has the same amplitude, and its phase is shifted by π relative to the situation, when $K_d = -K$.

Comparing formulas (4.14) and (4.15), we see their essential differences: at the decipherer input there is the splitter (divider) of the laser beam present but not the summators as at the cipherer input. On the contrary, at the decipherer output there is a summator present instead of the divider; passing through decipherer edges $[(g_{in,d,i})\forall]_2$, all the beams leave it without returning back. In other words, the decipherer happens to be the nonlinear system *without feedback*, which is typical for the cipherer. Therefore, the differential equation of the model (4.2) is true for the decipherer, but the decipherer itself cannot generate a dynamic chaos. Hence, the decipherer operates in the mode of the *chaotic response*, or, according to terminology of [11], discussed in Chap. 1, in the passive synchronization mode. In the future we are limited to this case.

Synthesis of decipherer model (4.15) is provided based on the condition that the signal arriving from the NRI output is completely recovered up to the signal $A_{in,i}(t)$ applied at the NRI input: $A_{in,i}(t) = A_{out,d,i}(t)$. If to mitigate this requirement up to the condition $A_{out,d,i}(t) = const \cdot A_{in,i}(t)$, the contents of the operators realized by the route elements in the decipherer model (4.15) can be changed.

For instance, the point $(g_{in,d,i+1})$ has the transmission coefficient of 1; the point $(g_{d,i+1})$ performs the delay $-U_{i+1}(t)$. The convergence point $(g_{d,i+1})1$ summarizes signals and performs their transmission with the coefficient 1. Edges

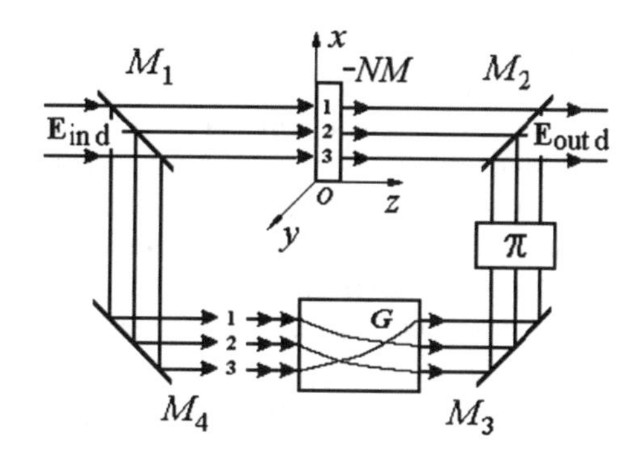

Fig. 4.3 The decipherer diagram. At light field rotation by the G element by $\Delta = 120°$ (in the transverse beam plane) trajectories of beams *1* and *3*, *2* and *1*, *3* and *2* after interferometer bypass are summed on the output mirror

$\left[\left(g_{in,d,i}\right)1\right]_2$ and $\left[\left(g_{i,d,i}\right)1\right]_1$ have the transmission coefficient $(1-R)^{1/2}$. In turn, in the edge $\left[\left(g_{in,d,i}\right)2\right]_2$ this requires the $(1-R)$ times amplitude attenuation. This edge has the transmission coefficient $\gamma(1-R)/2$ and provides the delay $\omega t_{e,i}(t)+\pi$. Edges $\left[\left(g_{i,i}\right)1\right]_2$ and $\left[\left(g_{in,d.i}\right)1\right]_2$ have the transmission coefficient $(1-R)^{1/2}$, therefore, it is required to correct the decipherer nonlinearity coefficient according to the rule: $K_d = -K/(1-R)^2$. For the decipherer with such parameters, the expression $A_{out,d,i} = (1-R)^2 A_{in,i}(t)$ is true. Its diagram is shown in Fig. 4.3, where the correcting variation of the amplitude is provided in the phase shifter π. As we can see, expression for K_d coincides with the similar one in (4.13) for $K_{d,0}$. This fact can be considered as a verification result.

Variants of nonlinear-dynamic cryptography devices: a classification aspect. The suggested formalism allows synthesizing of the decipherer diagram according to the common route-operator scheme of the cipherer, irrespective of its material-constructive implementation. Moreover, the CTP structure now, i.e. the structure of route formulas, in particular, (4.14) and, hence, (4.15), can serve as the classification sign for the arrangement of the typology of the known and for the prediction of the possible versions of cryptological methods and devices of nonlinear-dynamic cryptology [4, 12]. It is known that in cryptology such signs are: the number of keys (a presence of the open key) and the mathematical principles underlying the ciphering/deciphering (see Sect. 1.1 of Chap. 1).

It is evident that such characteristics of the CTP cipherer as the structure (closed, open, combined), the configuration (linear, the "diverging"/"converging" star type, ring, fractal-like etc.), and the number of points—communication channels (one, two, more than two, infinite number) can be expanded onto appropriate classified pairs of cipherer / decipherer.

It is necessary to take into consideration the mode of cipherer functioning (chaotic, static etc.). For example, the static mode is inevitable for the open CTP and for any constant signal at the cipherer input.

Possibilities of simultaneous transmission of different messages can be naturally considered as the classifying features: through the only communication channel of the given chain (simultaneous modulation of phase and amplitude); through different communication channels of the given CTP (through ways $g_{in,i} \rightarrow fin_{out,d,i}$); and through the different sets of communication channels from the various chains.

Along with this, the functions of the one communication channel (the one from the routes $fin_{out,i} \rightarrow g_{in,d,i}$) are the message transmission and synchronization and both performed simultaneously. Also, a fraction of the channels intended for the mentioned procedures can be also considered as classifying characteristic.

The signal $S_i(t)$, arriving at the point $g_{in,d,i}$ at time moment t, after various transformations in the decipherer, is the "masked information signal" $In_i(t) = F_{In,i}(S_i(t - \tau_{In,i}))$. But, it can serve as the "reference signal" $B_i(t) = F_{B,i}(S_i(t - \tau_{B,i}))$ as well. Besides, $S_i(t)$ is acting as the external influence on the different decipherer elements, i.e. performs the function of the synchronizing signal. Depending on this or that function, it is appropriate to call the signal $S_i(t)$ as the "information masked", "reference", or synchronizing signal.

Extraction of the proper information signal $I_i(t)$, applied to the cipherer input $g_{in,i}$ at previous time moments, happens as a result of the binary operation—subtraction: $I_i(t) = In_i(t) - B_j(t) = F_{In,i}(S_i(t - \tau_{In,i}) - F_{B,j}(S_j(t - \tau_{B,j})))$. It is noteworthy that the extraction procedure can be based on another binary ("+", "⊕" etc.) or, say, N-nary operation.

In the act of extraction of the signal $I_i(t)$, the signal $S_j(t - \tau_{In,i})$ is the information masked one; and the signal $S_j(t - \tau_{B,j})$ is the reference one; both signals may serve as synchronizing signal depending on the fact whether they really affect the synchronizing effect on the decipherer elements.

If CTP is closed and consists of the only point, then $i = j$. It corresponds to the single-channel (according to classification [11] given in Chap. 1) system of confidential communication, although, in the NRI case, the CTP number is not limited. For many cryptographic systems using the chaotic response at deciphering, the fulfillment of the equality $In_i(t) = S_i(t)$ is typical together with the location of the nonlinear element in the feedback loop of the decipherer (see, for example, Figs. 1.34, 1.40–1.42, 1.48).

Thereby, three functions of the S_i signal at $I_i(t)$ extraction *are separated in time*: the signal S_i arriving in the decipherer at time segment $(-\infty; t - \tau_{B,i})$ provides its synchronization (the first function) at first. After that, S_i plays a role of the reference signal $S_i(t - \tau_{B,i})$ (the second function). Then $S_i(t)$ transmits the element of the message $I_i(t)$ (the third function), extracted due to the presence of the reference signal in the preliminarily synchronized decipherer.

If the CTP is not closed and consists of two points, i.e. $i \neq j$ (such a situation is possible in NRI in case of light beam shift), $K = 0$ and the chaotic signal is applied at the first point, and the information signal—at the second point, then we deal with the double-channel (with separate channel of the (passive) synchronization) [11] system of confidential communication. In the case of NRI, a number of CTP is not limited as before.

In the act of the signal $I_2(t)$ extraction, the signal $S_2\left(t-\tau_{In,2}\right)$ is information masked one; and the signal $S_1\left(t-\tau_{B,1}\right)$ is the synchronization and the reference one.

Thereby, three functions of the signal S_i at the extraction of $I_2(t)$ *are separated in time and in space* (through channels 1 and 2): the signal S_1 arriving in the decipherer at the time segment $\left(-\infty; t-\tau_{B,1}\right)$, provides its synchronization at first (the first function). After that, S_1 plays a role of the reference signal $S_1\left(t-\tau_{B,1}\right)$ (the second function). Then, $S_2(t)$ transmits the message $I_2(t)$ (the third function, or carrier signal function), which is deciphered.

The separation of the functions of the S_i signal in time and/or in space affects the noise immunity of the transmission system. Let us assume that the additive noise impacts the communication channel. Then, with function S_i separated in space, the fatal influence on the deciphering result is manifested by the difference of averaged (for propagation path) values of the noise $\langle N_i(t)\rangle$ in channels (say, 1 and 2), where t is the time of signal S_i arrival at the *i-th* decipherer input. And for the separation in time, the variation of $\langle N_i(t)\rangle$ level is significant for the time $\Delta\tau=\left|\tau_{In,i}-\tau_{B,j}\right|$. It is clear, here we need to take into account the influence of transformation $F_{In,i}\left(S_i\left(t-\tau_{In,i}\right)\right)$ and $F_{B,j}\left(S_j\left(t-\tau_{B,j}\right)\right)$ performed over the signals S_i and S_j. But, we will limit ourselves by the cases of time moments $\tau_{In,i}$ and $\tau_{B,j}$ only.

Evidently, in the single-channel communication system, the variation of $\langle N_1(t+\Delta\tau)\rangle-\langle N_1(t)\rangle$ affects only the noise immunity. The value of $\Delta\tau$, generally speaking, is defined by the time delay in the feedback loop of the cipherer.

In the double-channel communication system with the special synchronization channel, the value of $\left\langle N_2\left(t-\tau_{In,2}\right)\right\rangle-\left\langle N_1\left(t-\tau_{B,1}\right)\right\rangle$ affects the noise immunity, i.e. both factors. However, separation in time of S_i functions can be eliminated by means of delay lines in both channels at the decipherer input and the cipherer output. When an equality $\tau_{In,2}=\tau_{B,1}$ is achieved, only the separation in space will matter, i.e. the value of $\langle N_2(t)\rangle-\langle N_1(t)\rangle$.

Thus, if conditions $\langle N_1(t+\Delta\tau)\rangle-\langle N_1(t)\rangle \ll \langle N_2(t)-N_1(t)\rangle$ or $\langle N_1(t+\Delta\tau)\rangle-\langle N_1(t)\rangle \gg \langle N_2(t)-N_1(t)\rangle$ are satisfied, the single- or double-channel system of confidential communication will have an advantage in respect of noise immunity. By the way, experimental results [11] represented in the form of Fig. 3.9a, b in [4], indicate the advantage of double-channel system, perhaps, because the second condition is implemented.

It is obvious that in the communication system based on the NRI, the various relationships between the number of channels (the number of points in CTP) specified for the message transmission and for the synchronization are possible. This causes the differences from the common systems. Let, for instance, the laser beam in NRI turn by the angle $\Delta=2\pi n/m$ or undergo its mirror imaging, i.e. the closed CTP is realized. If in the cipherer or decipherer, the NM is located directly before the G element along the beams pass (Fig. 4.4, where correcting variation of the field amplitude occurs in the π phase shifter), and $K_d=K/(1-R)$, the following situation is possible.

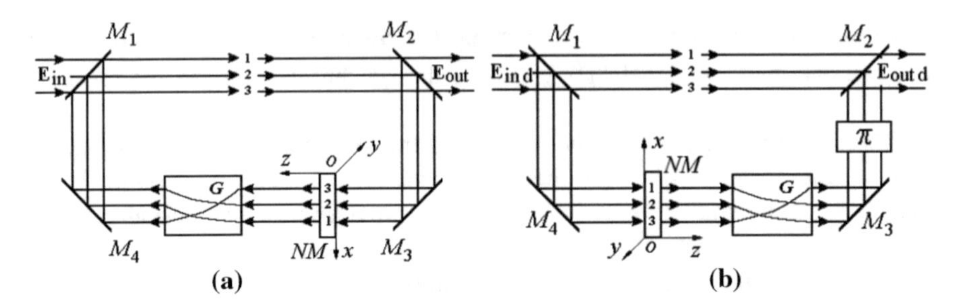

Fig. 4.4 Structure of the cipherer (**a**) and decipherer (**b**), when NM is located into the feedback loop of the cipherer. Beam trajectories correspond to the light field rotation by $\Delta = 120°$

The communication channel $fin_{out,i} \to g_{in,d,i}$ is used for the transmission of the signal S_i, which is a synchronizing one and a reference one relating to the masked information signal S_{i+1} transmitted through the channel $fin_{out,i+1} \to g_{in,d,i+1}$. The signal S_{i+1} is in turn used as the synchronizing and reference in respect of the masked information signal S_{i+2} transmitted through the channel $fin_{out,i+2} \to g_{in,d,i+2}$, etc. In other words, each signal S_i (the channel $fin_{out,i} \to g_{in,d,i}$) carries information and simultaneously serves as the synchronizing and reference signal for S_{i+2}. Thereby, the share of auxiliary (exclusively synchronizing) channels is equal to zero.

If the laser beam in NRI has a shift in the transverse xOy plane, i.e. open CTP is implemented, it is reasonable that the S_1 signal (the channel $fin_{out,1} \to g_{in,d,i}$) serves as the synchronizing and the reference one, but does not transmit information. The other signals and channels, as in the previous case, may play all three roles. In this topic, the share of the auxiliary channels is determined as a ratio $1/m$, where m is the number of points in the CTP. In double-channel systems with the special channel for (passive) synchronization according to the terminology [11] introduced in Chap.1, this share equals to 1/2.

Thus, answering to the questions stated at the end of Sect. 4.1.1, we have demonstrated the heuristic potential of the route-operator formalism in terms of the signal transmission study in case of the nonlinear ring systems of optical range [4].

4.1.3 Simulation of Secret Transmission of Images: Modes of Deterministic Spatial-Temporal and Spatial Chaos

For the principal testing of the results in Sects. 4.1.1 and 4.1.2, the calculating experiments are carried out based on the model (4.2). Below follow examples of the images ciphering/deciphering simulation [4, 13, 14] (another variant of the images being ciphered/deciphered is given in [15]).

The case of the closed CTP with four points ($\Delta = 90°$, $\tau_n = 10^{-9}$ s, $R = 0.5$, $t_e = \tau_n$, $\gamma = 0.5$) is shown in Fig. 4.5. Here the deepness of the spatial modulation

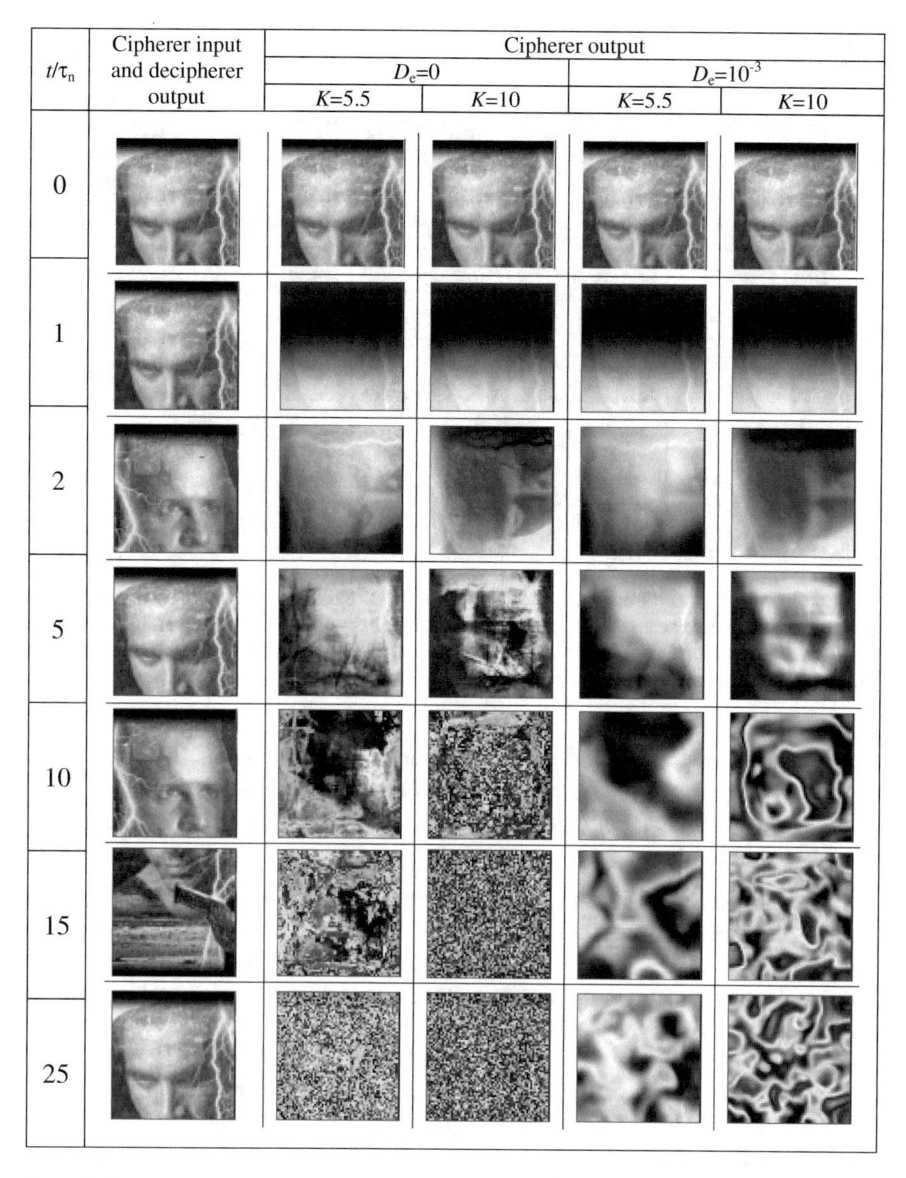

Fig. 4.5 Frames of image simulation processes (ciphering/deciphering) in the dynamic chaos mode for different nonlinearity coefficients K and the normalized diffusion coefficient D_e in NRI

of the laser beam amplitude equals to 0.048, which is about 10 times greater than the one for the model [6, 16]. Figure 4.5 demonstrates the possibility of the ciphering/deciphering of 2D image, represented by the frame sequence at the cipherer (i.e. NRI) operating in the mode of the deterministic spatial-temporal (in other words, –dynamic) chaos. From the visual images analysis in Fig. 4.5 we can conclude:

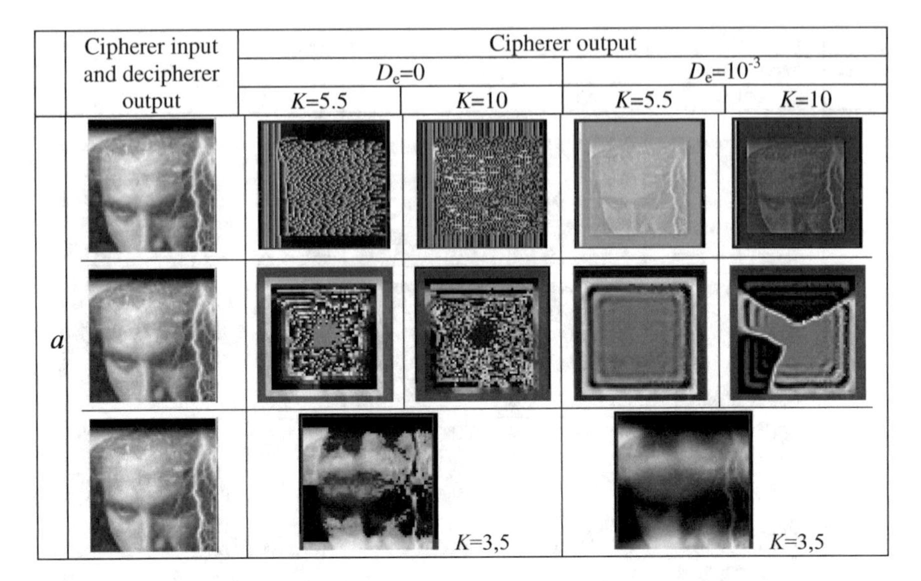

Fig. 4.6 Imitation of the image ciphering/deciphering in the static NRI mode for the different values of nonlinearity coefficient K and the normalized diffusion coefficient D_e. The laser beam in the NRI feedback loop is subject to: shift along x-axis equaled to 1/80 (a); compression $\sigma = 0.9$ (b); mirror imaging with respect to x-axis

- with time growth t/τ_n, a degree of message secrecy increases;
- the "warming-up" time of the cipherer is no less than $5\tau_n$;
- the secrecy degree and its growing speed depend on the NRI parameters combinations.

The communication secrecy degree increase with the growth of the NRI nonlinearity coefficient K can be estimated by the spectral analysis methods using Figs. 2.36 and 2.37. These figures demonstrate the temporal realizations, phase portraits, Fourier spectra of the wave amplitude A_i at the output, where i is a number of the transposition point in CTP, $D_e = 0$.

Figure 4.6 demonstrates the possibility of the ciphering/deciphering of the 2D image during the cipherer operation in the static mode. Here the message transmission secrecy depends upon the NRI parameters combination. As mentioned in Sect. 2.4.3, the static mode, while assuming the absence of the time-variations, is nevertheless capable of providing the structure chaotization (2D or 3D) in space. The visual analysis of Fig. 4.6 testifies to the principal opportunity of the spatial deterministic chaos existence [4, 12].

It is clear from physical considerations that the cipherer functioning can be characterized by the warming-up time τ_h (in the mode of the spatial-temporal chaos) and by the transient time τ_r (in the static mode), and in case of the decipherer such characteristic is the synchronization transient time τ_s. Let us review the procedure for the evaluation of the effectiveness of these devices in the above mentioned modes and in a situation when we need to transmit only the one image or the set of images.

If it is required to transmit only the one image, then, in the static mode, the time necessary for the ciphering is determined by the transient time τ_r of the processes in the cipherer (NRI), which particularly depends on the CTP length. In this case, the receiving side needs the stable (established) image at the cipherer output (a cryptogram) for the correct deciphering. Therefore, if the decipherer structure includes a device for the cryptogram storing, the message duration τ_m is determined by the operation speed of this device. If such a device is absent, then $\tau_m = \tau_s$. The time necessary for the deciphering is defined by the synchronization transient time τ_s.

In the mode of spatial-temporal chaos, there can be two variants. If the cipherer and decipherer are preliminarily synchronized, i.e. the appropriate initial conditions are established for both of them, the time necessary for the ciphering and for the deciphering and τ_m also are determined by the warming-up time τ_h. Otherwise, the deciphering time and τ_m are determined by τ_s time, and the ciphering time equals the highest of values τ_h, τ_s.

If the transmission of the set of the images is required, then, in the static mode, the transmission process for each message will not differ from the transmission of the single image. In the spatial-temporal chaos mode, the transmission of the first message will also not differ from the transmission of the single message. Nevertheless, the further images can be transmitted as fast as it is allowed by the characteristics of the devices for the image changing and registration (we do not consider here the possible decrease or increase of crypto-resistance at the similar fast transmission).

It is quite reasonable to consider that the following inequality fulfills: $\tau_s < \tau_h < \tau_r$. Then it turns out that the static mode is *more preferable*, if the carrying capacity of the communication channel is really a limiting factor by its cost, or if our task is to save and store an encrypted information.

Evidently, the experience of the simulation of the optical device in the nonlinear-dynamic cryptography leads to a problem of multi-parameter optimization of such device, along with its possible counterparts and variants.

Here arises the question concerning the evaluation of the quality (crypto-resistance) of the cryptosystem in Fig. 4.1. The evaluation does rely on the two criteria basically: (1) the probability of the direct selection of the deciphering key by the enemy; (2) the possibility of the key determination by means of the analysis of the intercepted cryptograms (ciphering texts), i.e. signals at the NRI output (Fig. 4.1a) [17]. Moreover, it is natural that the richness of the output signal spectrum, which exceeds the spectral width of the information signal, is also a criterion. Estimation of the quantity of the essential components in the NRI output signal spectrum is performed in [18].

The second criterion presupposes the involvement of any inverse solution methods. Below we will discuss the possibility of the determination of the delay time in NRI and the CTP structure (with the help of the correlation analysis), as the most important components of the key, which defines chaotic synchronization of transmitter and receiver. Evaluation procedure conforming to the first criterion is reduced to the determination of the power of the set of the keys. We mean the number of sets of the cryptosystem parameters that provide the *difference* between two signals at the system output in case of the *identity* of the two signals at the system input. Under the system we should understand both the cipherer and the decipherer. The conception of the identity and the difference of the two signals should be considered in the context of the transmitted information recognition task. To quality estimation based on the first criterion, we need to understand what parameters and in what way affect deciphering error.

4.1.4 Deciphering Error δ(r, t) as a Wave Process and Its Normalizing Amplitude A_δ as a Function of Setting Errors of the Decipherer. Evaluation of A_δ

We transform (4.7) for the case of quasi-monochromatic field. Let us assume that for the t_{nn} time of the light propagation through NM in NRI and the deciphering time t_0 of the field propagation through the NRI feedback loop does not change noticeably, i.e. $t_0(\mathbf{r}', t - t_{nn}) \approx t_0(\mathbf{r}', t)$. Then, granted what we have in (4.11), (4.12), we get [4, 19], according to which deciphering error $\delta(\mathbf{r}', t)$ behaves itself as a wave process with the amplitude $a_\delta \equiv a_\delta(\mathbf{r}, t) \equiv A_\delta(\mathbf{r}, t)A_{\max(x,y,z)}$, the circular frequency ω and the phase $\varphi_\delta \equiv \varphi_\delta(\mathbf{r}, t)$:

$$\begin{aligned} \delta(\mathbf{r},t) &= a_\delta(\mathbf{r}, t)\cos[\omega t + \varphi_\delta(\mathbf{r}, t)], \\ a_\delta &= \frac{1}{\sigma(1-R)}\left(Ac^2 + As^2\right)^{1/2} \\ &= \frac{1}{\sigma(1-R)}\left[a_1^2 + a_2^2 + 2a_1a_2\cos(\varphi_1 - \varphi_2)\right]^{1/2}, \\ \varphi_\delta &= \arg(Ac, As), \end{aligned} \tag{4.16}$$

where

$$\begin{aligned} a_1 &= \Gamma a[\mathbf{r}', t - t_0(\mathbf{r}', t)], \\ \varphi_1 &\equiv \varphi\left[\mathbf{r}', t - t_{0,d}(\mathbf{r}', t)\right] - \omega[t_{nn} + t_0(\mathbf{r}', t)], \\ a_2 &\equiv \Gamma_e a\left[\mathbf{r}', t - t_{0,d}(\mathbf{r}', t)\right], \\ \varphi_2 &\equiv \varphi\left[\mathbf{r}', t - t_{0,d}(\mathbf{r}', t)\right] - \omega t_{0,d}(\mathbf{r}', t), \\ As &\equiv a_1\sin(\varphi_1) + a_2\sin(\varphi_2), \\ Ac &\equiv a_1\cos(\varphi_1) + a_2\cos(\varphi_2). \end{aligned} \tag{4.17}$$

It is clear that the decipherer output signal written in the form of (4.6), when fed to the NRI input can be rewritten as (4.16). For this, having designated the amplitude and phase at the decipherer output fed to the NRI input as $a_{d,e} \equiv a_{d,e}(\mathbf{r}, t) \equiv A_{d,e}(\mathbf{r}', t)A_{\max(x,y,z)}$, $\varphi_{d,e} \equiv \varphi_{d,e}(\mathbf{r}, t)$, we obtain expressions:

$$\begin{aligned} a_{d,e} &= \{a_{in}^2(\mathbf{r}, t - t_{nn}) + a_\delta^2(\mathbf{r}, t) + 2a_{in}(\mathbf{r}, t - t_{nn})a_\delta(\mathbf{r}, t) \\ &\quad \times \cos[\varphi_{in}(\mathbf{r}, t - t_{nn}) - \omega t_{nn} - \varphi_\delta(\mathbf{r}, t)]\}^{1/2}, \\ \varphi_{d,e} &= \arg(Ac_d, As_d), \end{aligned} \tag{4.18}$$

where $As_d \equiv a_{in}(\mathbf{r}, t - t_{nn}) \sin[\varphi_{in}(\mathbf{r}, t - t_{nn}) = \omega t_{nn}] + a_\delta(\mathbf{r}, t) \cos[\varphi_\delta(\mathbf{r}, t)]$, $Ac_d \equiv a_{in}(\mathbf{r}, t - t_{nn}) \cos[\varphi_{in}(\mathbf{r}, t - t_{nn}) = \omega t_{nn}] + a_\delta(\mathbf{r}, t) \cos[\varphi_\delta(\mathbf{r}, t)]$.

Let us introduce the relative deciphering error

$$\delta\alpha(\mathbf{r}, t) \equiv \frac{a_{d,e}(\mathbf{r}, t) - a_{in}(\mathbf{r}, t - t_{nn})}{a_{in} a_{d,e}(\mathbf{r}, t)} \equiv \left[\frac{A_{d,e}(\mathbf{r}, t) - A_{in}(\mathbf{r}, t - t_{nn})}{A_{in}(\mathbf{r}, t - t_{nn})}\right].$$

Substitution of (4.18) for $a_{d,e}(\mathbf{r}, t)$ into $\delta\alpha(\mathbf{r}, t)$ produces:

$$\begin{aligned} \delta\alpha(\mathbf{r}, t) &= -1 + \left\{1 + [\frac{A_\delta(\mathbf{r}, t)}{A_{in}(\mathbf{r}, t - t_{nn})}]^2 + 2\frac{A_\delta(\mathbf{r}, t)}{A_{in}(\mathbf{r}, t - t_{nn})} \right. \\ &\quad \times \cos[\varphi_{in}(\mathbf{r}, t - t_{nn}) - \omega t_{nn} - \varphi_\delta(\mathbf{r}, t)]\}^{1/2}. \end{aligned} \tag{4.19}$$

As we see, $\delta\alpha(\mathbf{r}, t)$ is the function of variables $A_\delta(\mathbf{r}, t)$ and $\varphi_\delta(\mathbf{r}, t)$, which, in their turn, according to (4.16) and (4.17), depend upon the chaotically evolving field amplitude $a(\mathbf{r}, t)$ and phase $\varphi(\mathbf{r}, t)$ at the NRI output. We take into account that according to (4.16), $A_\delta(\mathbf{r}, t) \geq 0$. Then from (4.19) we get:

$$\begin{aligned} -A_\delta(\mathbf{r}, t) \leq \delta\alpha(\mathbf{r}, t) \leq A_\delta(\mathbf{r}, t) \quad & \text{for } 1 \geq A_\delta(\mathbf{r}, t), \\ A_\delta(\mathbf{r}, t) - 2 \leq \delta\alpha(\mathbf{r}, t) \leq A_\delta(\mathbf{r}, t) \quad & \text{for } 1 < A_\delta(\mathbf{r}, t). \end{aligned} \tag{4.20}$$

For cryptosystem synthesis, the first variant is more preferable, because the estimation of the $\delta\alpha(\mathbf{r}, t)$ value is symmetric relative to zero.

Now we are to find out the influence of the main physical factor on the deciphering error amplitude A_δ. For this, we establish phase difference $\varphi_1 - \varphi_2$ in (4.16) under assumption that $t_0(\mathbf{r}', t - t_{nn}) \approx t_0(\mathbf{r}', t)$ and accounting for the designations in (4.8)–(4.10), (4.6):

$$\varphi_1 - \varphi_2 = \omega\delta t_0(\mathbf{r}', t) - \delta u_d(\mathbf{r}, t) - \pi = \delta\varphi(\mathbf{r}, t) - \pi,$$

where $\delta\varphi(\mathbf{r}, t) \equiv \omega\delta t_{0,d}(\mathbf{r}', t) - \delta u_d(\mathbf{r}, t) - \pi$. Then, given (4.17), we obtain from (4.16)

$$\begin{aligned} a_\delta &= \frac{1}{\sigma(1 - R)} a\left[\mathbf{r}', t - t_{0,d}(\mathbf{r}', t)\right] \\ &\quad \times \Gamma_{e0}\left\{1 + \left[\frac{\Gamma_e}{\Gamma_{e0}}\right]^2 - 2\frac{\Gamma_e}{\Gamma_{e0}} \cos\left[\omega\delta t_{0,d}(\mathbf{r}', t) - \delta u_d(\mathbf{r}, t)\right]\right\}^{1/2}. \end{aligned} \tag{4.21}$$

In case of the absence of the molecule diffusion in NM, we can show that discrepancy δu_d (4.8) of the nonlinear phase shift u_d in the nonlinear medium of the decipherer is expressed as

$$\delta u_d \equiv \delta u_d(\mathbf{r},t) = \delta u_d(\mathbf{r},t_s)\exp\left[-\frac{(t-t_s)}{\tau_{n,d}}\right] + \frac{\delta K_d C_d(\mathbf{r},t)}{\tau_{n,d}},$$
$$C_d(\mathbf{r},t) = \int_{t_s}^{t} A^2(\mathbf{r},t')\exp\left[-\frac{t-t'}{\tau_{n,d}}\right]dt'. \qquad (4.22)$$

From (4.21) and using (4.22), we can obtain the total (with contribution of the following factors: Γ_e/Γ_{e0}, $\delta t_{0,d}(\mathbf{r}',t)$, $\delta u_d(\mathbf{r},t_s)$, δK_d in mind) formula for the deciphering error amplitude A_δ normalized to $A_{\max(x,y,z)}$:

$$A_\delta = A_e\frac{\gamma(\mathbf{r}',t-t_{nn})}{2\sigma}\times\left\{1+[\frac{\Gamma_e}{\Gamma_{e0}}]^2-\right.$$
$$\left.2\frac{\Gamma_e}{\Gamma_{e0}}\cos\left[\omega\delta t_{0,d}(\mathbf{r}',t)-\delta u_d(\mathbf{r},t_s)\exp[-\frac{t-t_s}{\tau_{n,d}}]+\frac{\delta K_d C_d(\mathbf{r},t)}{\tau_{n,d}}\right]\right\}^{1/2}, \qquad (4.23)$$

where $A_e \equiv A_e\left[\mathbf{r}',t-t_{0,d}(\mathbf{r}',t)\right] \equiv A\left[\mathbf{r}',t-t_{0,d}\right]/[(1-R)C_n(\mathbf{r}')]$ is the amplitude in the transmission channel fed to the cipherer input.

In (4.23) the factors $\delta t_{0,d}(\mathbf{r}',t)$, $\delta u_d(\mathbf{r},t_s)$, δK_d are summarized in the argument of the cos function. This circumstance causes the similarity in their influence upon A_δ, while the inequality in these factors actual impact on it is determined by the multipliers difference. The multiplier $\exp\frac{-(t-t_s)}{\tau_{n,d}}$, with the initial setting error value of the nonlinear phase shift $\delta u_d(\mathbf{r},t_s)$, will in time smoothly weaken the role of $\delta u_d(\mathbf{r},t_s)$. Therefore, the value $\delta u_d(\mathbf{r},t_s)$ can serve as a key only for $(t-t_s)<\tau_{n,d}$. But, in the initial period, its role happens to be rather significant. The factor $\delta t_{0,d}(\mathbf{r}',t)$ acts permanently and its level is completely controlled by the experimenter. The factor δK_d is also controlled. But, its contribution depends upon the multiplier $C_d(\mathbf{r},t)$, which varies in a rather complicated manner,—either in time, according to (4.22) (in the dynamic chaos mode), or in space (in the static chaos mode).

Let us estimate an interval of possible values of the normalized amplitude A of the decipherer input field, when $\gamma(\mathbf{r},t)\leq\gamma_{max}<2\sigma$ and $A_{in}\left[\mathbf{r}',t-t_{0,d}(\mathbf{r}',t)-t_n\left(\mathbf{r}',t-t_{0,d}(\mathbf{r}',t)\right)\right]\geq A_{in,min}(\mathbf{r}')$. Taking into account that $A_{in}\leq 1$, we get the following inequalities from the cipherer structure in Fig. 4.1a

$$A_{in,min}(\mathbf{r}') = -\frac{\gamma_{max}}{2\sigma-\gamma_{max}} \leq \frac{A\left[\mathbf{r}',t_{0,d}(\mathbf{r}',t)\right]}{\left[(1-R)C_n(\mathbf{r}')\right]} \leq \frac{2\sigma}{2\sigma-\gamma_{max}}. \qquad (4.24)$$

Entering designation

$$\delta_\Sigma \equiv \left\{ \left(1+\left[\frac{\Gamma_e}{\Gamma_{e0}}\right]^2 - 2\frac{\Gamma_e}{\Gamma_{e0}} \right.\right.$$
$$\left.\left. \times \cos\left[\omega\delta t_{0,d}(\mathbf{r}',t) - \delta u_d(\mathbf{r},t_s)\exp\left[-\frac{t-t_s}{\tau_{n,d}}\right] + \frac{\delta K_d C_d(\mathbf{r},t)}{\tau_{n,d}}\right]\right\}^{1/2} \tag{4.25}$$

we get from (4.6), (4.21), (4.23), (4.24) the inequality for A_δ,

$$\delta_\Sigma \gamma(\mathbf{r}',t-t_{nn})\frac{A_{in,min}(\mathbf{r}')-\gamma_{max}}{2\sigma-\gamma_{max}}\frac{1}{2\sigma} \le A_\delta \le \frac{\delta_\Sigma\gamma(\mathbf{r}',t-t_{nn})}{2\sigma-\gamma_{max}}. \tag{4.26}$$

Setting in (4.26) $\gamma(\mathbf{r}',t-t_{nn}) = const = \gamma_{max}$, $A_{in,min}(\mathbf{r}') = 1$, $\sigma = 1$, we can provide another form for the inequality

$$\frac{\delta_\Sigma\gamma(1-\gamma)}{2-\gamma} \le A_\delta \le \frac{\delta_\Sigma\gamma}{2-\gamma}. \tag{4.27}$$

The set of inequalities (4.24), (4.26), (4.27) obtained allows, at the given values of δ_Σ, the evaluation of the range of possible values of the deciphering error amplitude A_δ normalized to $A_{max(x,y,z)}$ and fed to the NRI input.

4.1.5 Statistical Characteristics of the Relative Deciphering Error Amplitude $\delta\alpha(\mathbf{r}, t)$: Simulation Data and Theoretical Estimations

Preliminary estimation of the decipherer parameters setting error influence upon the statistical characteristics of the deciphering relative error $\delta\alpha(\mathbf{r},t)$ is undertaken in [4, 19, 20]. It is based on the direct simulation (numerical modeling) of the ciphering/deciphering processes using (4.2). As the statistical characteristics we use the following: the averaged in space $\langle\rangle_\mathbf{r}$ and in time $\langle\rangle_t$ value of the deciphering relative error $\delta\alpha \equiv \langle\langle\delta\alpha(\mathbf{r},t)\rangle_\mathbf{r}\rangle_t$; the averaged in time square root for the averaged (in space) deciphering square error $d \equiv \left\langle\left\{\left\langle[\delta\alpha(\mathbf{r},t)]^2\right\rangle_\mathbf{r}\right\}^{1/2}\right\rangle_t$; the averaged in time mean deviation in space (the mean-square error) of deciphering error $s \equiv \left\langle\left\{\left\langle\{\delta\alpha(\mathbf{r},t) - \langle\delta\alpha(\mathbf{r},t)\rangle_\mathbf{r}\}^2\right\rangle_\mathbf{r}\right\}^{1/2}\right\rangle_t$. For the case of averaging in discrete representative set of m spatial points $\mathbf{r}_i$, the deviation is $s \equiv \left\langle\left\{\left[\frac{m}{m-1}\right]\left\langle\left[\delta\alpha(\mathbf{r}_i,t) - \langle\delta\alpha(\mathbf{r}_i,t)\rangle_m\right]^2\right\rangle_m\right\}^{1/2}\right\rangle_t$. The relative error $\Delta K_d \equiv \frac{\delta K_d}{K_{d0}} \equiv$

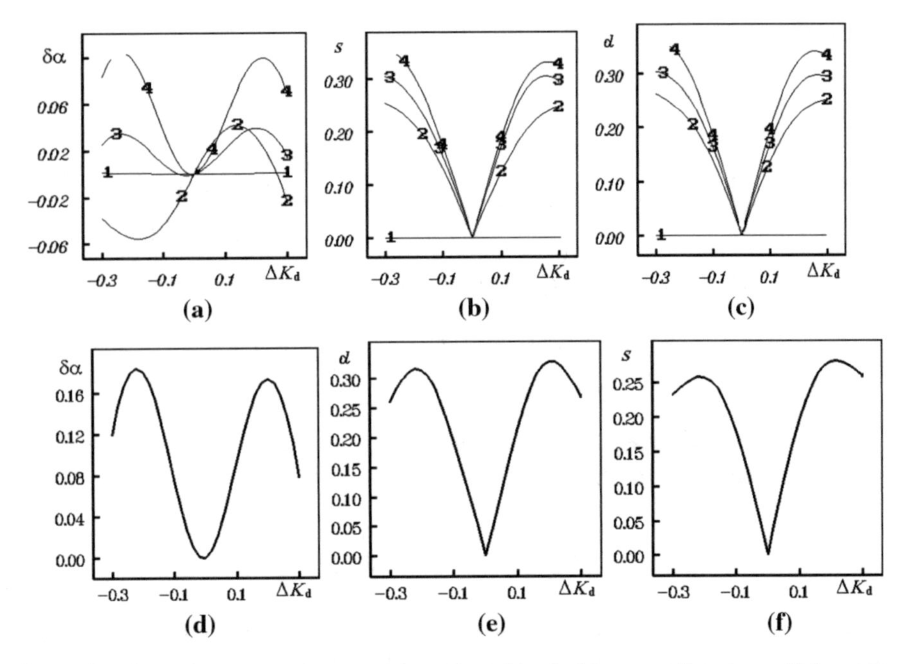

Fig. 4.7 Influence of the relative error of the setting of the decipherer nonlinearity coefficient ΔK_d on the averaged (in space and in time) value of the deciphering relative error $\delta\alpha$ (**a**, **d**); the square root from the deciphering averaged square deciphering error d (**b**, **e**); the mean deviation of deciphering error s (**c**, **f**), $K = 10$, $D_e = 0$, $\gamma = 0.5$, $R = 0.5$. In dynamic operation mode of the cryptosystem, the field rotation in the xOy plane of $\Delta = 90°$ (**a**, **b**, **c**). In the static mode (**d**, **e**, **f**) the light beam shift along the x-axis is 1/80 part of its diameter. Time-averaging intervals t/τ_n equaled to [0; 1], [0; 10], [0; 25], [0; 300] correspond to curves *1–4* in figures **a**, **b**, **c**

$\frac{K_d - K_{d0}}{K_{d0}}$ (see (4.13)) is used as a measure of the setting error of the decipherer nonlinearity parameter K_d.

The following regulations follow from the form of the plots in Fig. 4.7. During the cryptosystem operation, the values of d, s and the secrecy coupled with them grow in time (Fig. 4.7a–c). If we take their values as the maximal achievable values of d, s for $t/\tau_n = 300$, then it turns out that the cryptosystem warming-up time is close to $t/\tau_n = 10$ (compare with $t/\tau_n > 5$ Sect. 4.1.3). The fulfillment (during cryptosystem operation) of the condition $|\delta\alpha| \ll d$ is prerequisite for the high degree of secrecy. It means that the *modulus* of the $\delta\alpha(\mathbf{r}, t)$ value is rather high (commensurable to d), and the averaged value $\delta\alpha$ of the variable $\delta\alpha(\mathbf{r}, t)$ is low. Hence, the influence of the setting error ΔK_d is not reduced to either increase or decrease of intensity of the whole image.

By means of the plots similar to those in Fig. 4.7, we can estimate the number of cryptosystem keys [20]. Under assumption that some parameter (for instance, K_d) can be changed within the limits of $\pm c\%$, the algorithm for the cryptosystem key number N_K estimation in terms of this parameter is: (a) we specify some threshold value d_{th} for the parameter d; (b) considering d as a function of normalized

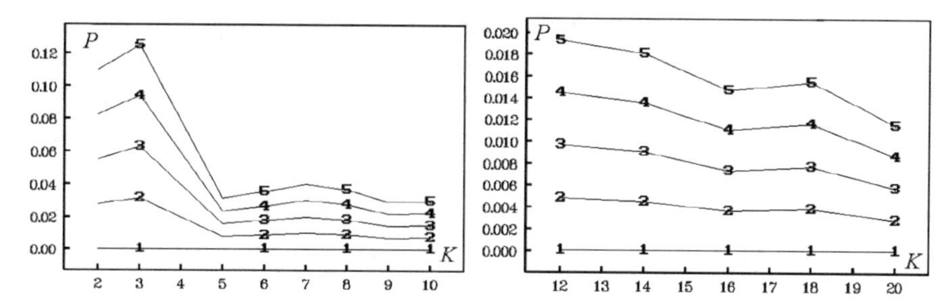

Fig. 4.8 The probability P of the direct selection of the decipherer nonlinearity parameter K_{d0} (a key part) depending on the nonlinearity coefficient K in NRI for different threshold values of the deciphering error, d_{th} = 10^{-5} (*1*), d_{th} = 0.005 (*2*), d_{th} = 0.01 (*3*), d_{th} = 0.015 (*4*), d_{th} = 0.02 (*5*). The other NRI parameters are $\Delta = 360°$, $D_e = 0$, $\gamma = 0.5$, $\sigma = 1$, $R = 0.5$. The coefficient K varies from 2 to 10 (*left*) and from 12 to 20 (*right*)

mismatch ΔK_d of this parameter, we find out a set of values $\Delta K_{d,th,i} = \Delta K_d(d = d_{th})$; (c) we choose a pair of values $\Delta K_{d,th,1}$ and $\Delta K_{d,th,2}$, which are close to one, such as $\Delta K_{d,th,2} < 1$, and $\Delta K_{d,th,1} \Delta K_{d,th,2} > 1$; (d) we calculate $N_K = 0.02c/(\Delta K_{d,th,2} - \Delta K_{d,th,1})$ and the probability $P = 1/N_K$ of the direct selection of K_d parameter. As a result of algorithm application, we obtain Fig. 4.8. It demonstrates the probability P decrease with K growth. That is, the key number increases. So, if d_{th} = 0.02, then $N_K(K = 2) \approx 9$, $N_K(K = 12) \approx 51$, $N_K(K = 20) \approx 82$. If d_{th} = 0.01, then $N_K(K = 2) \approx 18$, $N_K(K = 12) \approx 100$, $N_K(K = 20) \approx 163$.

We should note that the keys number can significantly differ for the different parameters. Thus, the element G has a larger resource in this respect; the types of field spatial transformations and their combinations must also be considered as a key. Evidently, in the case of DNRI, a number of such keys increases due to the presence of the two G elements in the feedback loops. Besides, it would perhaps be relevant to refer to the self-organized criticality mode that was actively investigated in 2000s in the context of the risk control [21].

Let us estimate the boundaries of the possible variation of the value $\delta\alpha(\mathbf{r}, t)$ and the scale of d in the specific case corresponding to Fig. 4.7: $\sigma = 1$, $\gamma(\mathbf{r}', t - t_{nn}) = const = 0.5$, $A_{in,min}(\mathbf{r}') = 1$ and $\Gamma_e = \Gamma_{e0}$. Then $\gamma_{max} = 0.5$, $\sigma = 1 > \gamma_{max}/2 = 0.25$ and $\sigma = 1 > \gamma_{max} \frac{1 + \frac{1}{A_{n,min}(\mathbf{r}')}}{2} = 0.5$. And because, according to (4.25), $0 \le \delta_\Sigma \le 2$, we can estimate the deciphering error amplitude fed to the cipherer input and normalized to $A_{max(x,y,z)}$ in the form of $0 \le A_\delta(\mathbf{r}, t) \le 0.66(6)$, as it follows from (4.27). Then $1 \ge A_\delta(\mathbf{r}, t)$ and from (4.20) we get: $|\delta\alpha(\mathbf{r}, t)| \le 0.66(6)$. At last, formally majorizing the value of $d \equiv \left\langle \left\{ \left\langle [\delta\alpha(\mathbf{r}, t)]^2 \right\rangle_{\mathbf{r}} \right\}^{1/2} \right\rangle_t$ as $0.66(6)/2 = 0,33(3)$, we obtain the value closed to maxima of d in Fig. 4.7a, b, f: $d = 0.35$ or $d = 0.32$.

Estimating the maximum point position in plots $d(\Delta K_d)$ (Fig. 4.7b, e) and the influence of the setting error of nonlinearity parameter ΔK_d on the deciphering error, we now refer to (4.23). Since the detuning δK_d is included into the expression (4.23) in the structure of the product $\frac{\delta K_d C_d(\mathbf{r},t)}{\tau_{n,d}}$, it is the last that we should estimate.

We arrange this using (4.24) for $\sigma > \gamma_{max}/2$ and $\sigma > \gamma_{max}\frac{1+\left[\frac{1}{A_{in,min}(\mathbf{r}')}\right]}{2}$ and assuming, as earlier, that $\gamma_{max} = \gamma = 0.5$, $\sigma = 1$, $A_{in,min}(\mathbf{r}') = 1$. Besides, let $R = 0.5$, $C_n(\mathbf{r}') = 1$. Hence, according to (4.13), $K_{d0} = -\frac{K}{[C_n(\mathbf{r})(1-R)]^2} = -4K$. Then for time moments $(t - t_s) > \tau_{n,d}$, when the system does not "remember" the value of the initial detuning of the nonlinear phase shift, taking into account that $\delta K_d = K_{d0}\Delta K_d$, we obtain

$$0.44(4)|K\Delta K_d| \leq \frac{|\delta K_d| C_d(\mathbf{r},t)}{\tau_{n,d}} \leq 1.77(7)|K\Delta K_d|. \tag{4.28}$$

For $\delta t_{0,d}(\mathbf{r}',t) = 0$ due to the presence of $\Delta K_d \neq 0$, the deciphering error amplitude $A_\delta \neq 0$ (see 4.23). Taking into consideration (4.23), we show that as the inequality of *necessary condition* for the A_δ reaching its own maximum-possible value (at fixed Γ_e/Γ_{e0}, $\gamma(\mathbf{r}', t - t_{nn})$, σ), specified by equation $A_\delta = A_e \frac{\gamma(\mathbf{r}',t-t_{nn})}{2\sigma}\left\{1 + \frac{\Gamma_e}{\Gamma_{e0}}\right\}$, we get:

$$\pi \leq 1.77(7)|K\Delta K_d| \text{ or } |\Delta K_d| \geq \frac{\pi}{1.77(7)|K|}. \tag{4.29}$$

According to the situation in Fig. 4.7, we choose $K = 10$. Then (4.28) and (4.29) assume the form of $4.44(4)\Delta K_d \leq \frac{\delta K_d C_d(\mathbf{r},t)}{\tau_{n,d}} \leq 17.7(7)\Delta K_d$, $|\Delta K_d| \geq 0.177$. The last does not contradict the fact that in Figs. 4.7b, e the value of d increases, while $|\Delta K_d| \leq 0.177$, and achieves the maximum in the region $|\Delta K_d| \geq 0.177$. Apparently, the maximum point position in plots $d(\Delta K_d)$ can be estimated after equating the arithmetic mean of variable $|\delta K_d| C_d(\mathbf{r},t)/\tau_{n,d}$ in (4.28) to π: $(0.44(4) + 1.77(7))|K\Delta K_d| = 1.11(1)|K\Delta K_d| = \pi$. From it $|\Delta K_d| = \frac{\pi}{1.11(1)|K|}$. For $K = 10$ we obtain the require estimation: $|\Delta K_d| = \frac{\pi}{11.1(1)} \approx 0.28$, which corresponds to the modeling data in Figs. 4.7b, e: $|\Delta K_d|$ equals to 0.24 or 0.20. The closeness of the estimates obtained analytically and during modeling does not only explain the modeling data but gives us a reason to consider the analytical investigation fulfilled as a correct one, i.e. produces the verification.

Now we will discuss a possibility of the delay time determining for the NRI performed with the help of the correlation analysis.

4.1.6 *Imitation of "Cracking" of the Delay Time in NRI*

A central moment function of the second order is the correlation function, which is widely applied for the estimation of statistical connections between values of a random process at different time moments [22–24]. If the system is linear, then, knowing the cross-correlation function connecting signals at input and output of a system, we can determine the delay time according to the cross-correlation function peak position at the shift value in the correlator equal to delay time [23, 24].

Trying to disclose ("to crack") the delay time in NRI, i.e. having investigated the crypto-resistance according to the second criterion, we now refer to expression (4.3) introducing in it: $t_0(\mathbf{r}', t - t_n(\mathbf{r}, t)) = const = t_0$, $\gamma(\mathbf{r}', t - t_n(\mathbf{r}, t)) = const = \gamma$, $C_n(\mathbf{r}') = const = C_n$, $\sigma = 1$, $n_0(\mathbf{r}) = const = n_0$. In the case of optical field rotation in the transverse beam xy-plane by the angle $\Delta = 120°$ (i.e. $\mathbf{r}''' = \mathbf{r}$), we may show that

$$E_{out}(\mathbf{r}, t) = C_n(1 - R)\left\{E_{in}(\mathbf{r}, t_1) + \left(\frac{\gamma}{2}\right)^3 E_{in}(\mathbf{r}, t_1 - 3t_e - t_2)\right\} + C(\mathbf{r}, t),$$

where $t_1 = t - \frac{u(\mathbf{r},t)}{\omega}$; $t_2 = \frac{u(\mathbf{r}',t-t_e)}{\omega} + \frac{u(\mathbf{r}'',t-2t_e)}{\omega} + \frac{u(\mathbf{r},t-3t_e)}{\omega}$;

$$\begin{aligned} C(\mathbf{r}, t) = {} & C_n(1 - R)\{\tfrac{\gamma}{2} u(\mathbf{r}', t - t_e) E_{in}\left[\mathbf{r}', t_1 - t_e - \frac{u(\mathbf{r}', t - t_e)}{\omega}\right] \\ & + \left(\frac{\gamma}{2}\right)^2 \left\{E_{in}\left(\mathbf{r}'', t_1 - 2t_e - \frac{u(\mathbf{r}', t - t_e)}{\omega}\right) - \frac{u(\mathbf{r}'', t - 2t_e))}{\omega}\right\} \\ & + |o\left[\frac{\gamma}{2}\right]^4| \le |E_{in,max}| \frac{\left[\frac{\gamma}{2}\right]^4}{1 - \gamma/2}. \end{aligned}$$

Hence, the contribution of the input signal $E_{in}(\mathbf{r}_1, t_{\mathbf{r}1} < t)$ into the output signal $E_{out}(\mathbf{r}, t)$ decreases, according to the geometric series law with common ratio $\gamma/2$ during growth of $t - t_{\mathbf{r}1}$ (a number of signal $E_{in}(\mathbf{r}_1, t_{\mathbf{r}1})$ bypasses in NRI). Here the component of the NRI output signal $E_{out}(\mathbf{r}, t)$, caused by the presence of $E_{in}(\mathbf{r}_1, t_{\mathbf{r}1})$, arises quasi-periodically with a period equal to about $3t_e$. That is, the field $E_{in}(\mathbf{r}_1, t_{\mathbf{r}1})$ affects $E_{out}(\mathbf{r}, t)$ quasi-periodically. This can be explained by the fact that the light beam trajectories 1 or 2 or 3 close after three bypasses of the NRI: $\mathbf{r}''' = \mathbf{r}$ (see Fig. 4.1a).

Deviation from the strict periodicity is caused by the inconstancy of the nonlinear phase shift $u(\mathbf{r}, t)$. But because the value $u(\mathbf{r}, t)/\omega$ is small, it does not give the significant contribution to delay time τ. This means the possibility of "cracking" of the delay time by means of calculating the field correlation coefficient at NRI output. Since in practice it is much easier to measure the light field amplitude than to determine a phase (direct measurements of which are impossible), instead of the optical field analysis we examine its amplitude only.

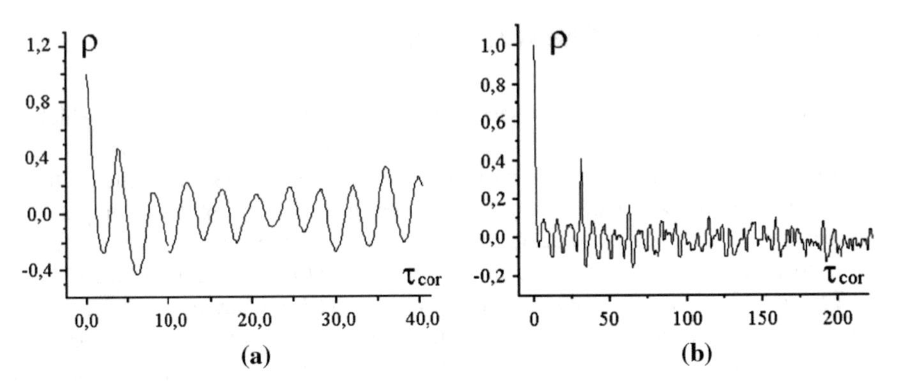

Fig. 4.9 Auto-correlation coefficient $\rho(\tau_{cor})$ of the NRI output field amplitude for: $t_e/\tau_n = 1$ (**a**), $t_e/\tau_n = 1$ (**b**). The case of field rotation in FBL NRI in the transverse *xOy* plane by an angle $\Delta = 120°$, $K = 5.5$, $D_e = 0$, $\gamma = 0.5$, $R = 0.5$

Application of the auto-correlation analysis. Thus, for "cracking' we must calculate the normalized correlation time $\tau_{cor,p} = \frac{t_{cor,p}}{\tau_n}$ corresponding to the main maximum ρ_p of the correlation coefficient ρ of the field *amplitude* at NRI output. For the case $\Delta = 120°$, $K = 5.5$, $D_e = 0$, $\gamma = 0.5$, $R = 0.5$ examples of these functions are presented in Fig. 4.9, and the "cracking" results—in the Table 4.1. We see that with growth of the delay time, the cracking error decreases from 30 % at $t_e/\tau_n = 1$ to 3.7 % at $t_e/\tau_n = 10$ [4, 6].

Influence of nonlinearity, delay time and the rotation angle of the optical field in FBL on the accuracy, with which the delay time is determined by the correlation analysis, is presented in Table 4.2. The delay time t_e/τ_n determination error value of the auto-correlation function varies from ±1 to ±48 %, depending on the model

Table 4.1 Results of the imitation of the equivalent delay time $t_e$cracking in the NRI

Normalized correlation time $\tau_{c\ p}$, corresponding to the main maximum ρ_p	The main maximum $\rho_p = \rho(\tau_{c\ p})$	Normalized equivalent delay time t_e/τ_n	Cracking of normalized equivalent delay time $\tau_{c\ p}/3$	Cracking error $[(\tau_{c\ p}\tau_n/3)-t_e]/t_e$, %
3.899	0.4657	1	1.299	29.9
7.099	0.2489	2	2.366	18.3
10.098	0.4778	3	3.366	12.3
13.099	0.4173	4	4.366	9.15
16.099	0.2964	5	5.366	7.32
18.999	0.3728	6	6.333	5.55
21.999	0.4036	7	7.333	4.76
25.099	0.3374	8	8.366	4.57
27.999	0.4079	9	9.333	3.7
31.099	0.4130	10	10.366	3.66

Table 4.2 Estimation of delay time t_e in FBL of the single-circuit NRI by the auto-correlation analysis for different Δ and K

Δ	K	t_e/τ_n	τ_{cor1}/m	$\lvert t_e/\tau_n - \tau_{cor1}/m \rvert$
Δ = 0°	K = 13.5	1	1.41	0.41
	K = 14.5		1.44	0.44
	K = 15.5		1.23	0.23
Δ = 120°	K = 9.5		1.48	0.48
	K = 11.5		1.2	0.2
	K = 14.5		–	–
Δ = 180°	K = 12.5		1.09	0.09
	K = 13.5		1.41	0.41
	K = 15.5		1.34	0.34
Δ = 0°	K = 5.5	5	5.33	0.33
	K = 9.5		5.39	0.39
	K = 15.5		–	–
Δ = 120°	K = 5.5		5.36	0.36
	K = 9.5		5.35	0.35
	K = 14.5		–	–
Δ = 180°	K = 5.5		5.09	0.09
	K = 9.5		5.36	0.36
	K = 15.5		–	–
Δ = 0°	K = 5.5	10	10.29	0.29
	K = 9.5		10.49	0.49
	K = 14.5		–	–
Δ = 120°	K = 5.5		10.33	0.33
	K = 9.5		10.46	0.46
	K = 14.5		–	–
Δ = 180°	K = 5.5		10.09	0.09
	K = 9.5		10.29	0.29
	K = 14.5		10.19	0.19

parameters. Nevertheless, in some cases, when $K \geq 14.5$ (see Table 4.2), we cannot estimate the delay time t_e in NRI FBL even for the known field rotation angle Δ in FBL provided by the G element. The reason for this is an absence of the second peak (after the main one), for which the estimate of the inter-peak interval is possible (Fig. 4.10a, b). Hence, at the given K, the system is crypto-steady to cracking by the auto-correlation analysis method.

On the contrary, at the lesser values of the nonlinearity parameter (Fig. 4.10a), it is possible to estimate $t_e/\tau_n = \tau_{cor,1}/m$, and, hence, in this mode the crypto-system is vulnerable. To verify this assumption, we have carried out the simulation of the ciphering/deciphering process for the 2D image in the case corresponding to Fig. 4.10a, when the delay time in the cipherer is $t_e = 10\tau_n$, and in the decipherer $-t_{ed} = 10.266\tau_n$ (Fig. 4.11). It is clear that the similar value of the cracking error t_e (2.66 %) leads to the impossibility of the transmitted message deciphering [4, 25].

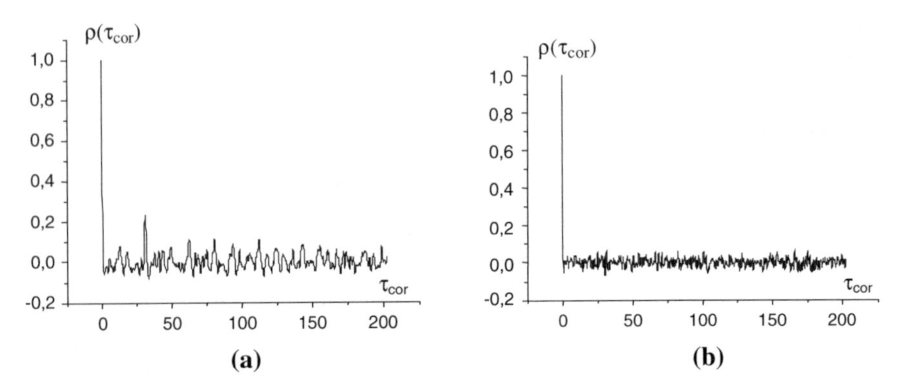

Fig. 4.10 Auto-correlation coefficient $\rho(\tau_{cor})$ of the NRI output field amplitude for: $K = 6.5$ (**a**), $K = 14.5$ (**b**). The case, when $t_e/\tau_n = 10$, $\Delta = 120°$, $D_e = 0$, $\gamma = 0.5$, $R = 0.5$

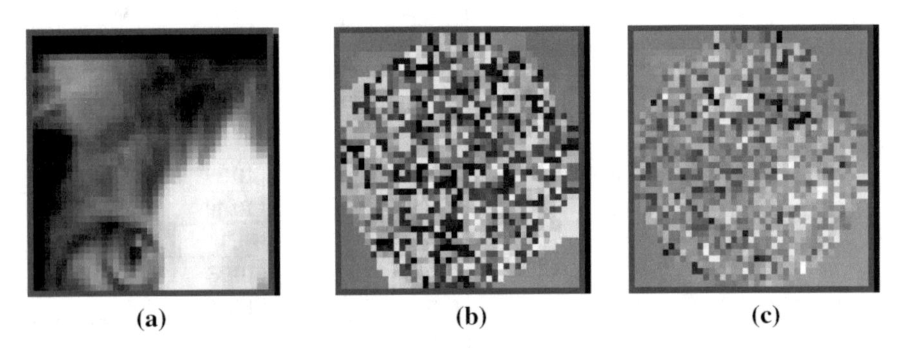

Fig. 4.11 Imitation of ciphering/deciphering processes at time setting error $t_{e\ d}$ presence $(t_{e\ d}-t_e)/t_e = 0.0266$ for the case $t_e = 10\tau_n$, $K = 6.5$; $\Delta = 120°$: **a**, **b**, **c** are 2D images at the cipherer input, at decipherer input, at decipherer output, relatively

Thus, the delay time estimate in NRI FBL by the auto-correlation analysis method can be performed for the following parameters $K \in [5.5, \ldots, 15.5]$, $m = [1, 2, \ldots, 6]$ at $\gamma = 0.5$. The lower bound of the nonlinearity coefficient K is produced by the arising dynamic chaos, and the upper bound—by the impossibility of auto-correlation analysis application for t_e estimation. The upper bound m is also the result of the impossibility of auto-correlation analysis application because after six passes the losses influence in NRI is such already that the second peak after the main one disappears at the auto-correlogram, and the situation becomes similar to the one presented in Fig. 4.10a, b.

Until now, we have considered that CTP are closed (the field rotation takes place) and the angle Δ is a priori known. In practice, the transverse beam section is a continuum of points with different intensity and with a varied profile of the phase shift. These points belong to different CTP. The CTP type is determined by the field transformation in FBL (Fig. 2.32). Therefore, the potential cracker will face a problem of CTP type (transformation) determination and identification of the definite CTP among the variety of others. Thereby, the following question is logical: how is it possible? It turns out, that the cross-correlation analysis is effective here.

Application of the cross-correlation analysis. Let us explain the principle of cross-correlation analysis examining the case when the rotation angle in the feedback loop is $\Delta = 120^{\circ}$. A part of the beam reflected in the i-th transposition point from the output mirror, transfers to i + 1-th point during the time equal to the time τ needed for the beam to bypass the FBL, and to i + 2-th point during the time 2τ, and returns to i-th point after the three bypasses of the FBL. More strictly, in case $i = 1$, this law is expressed by the following notation:

$$E_{out}(\mathbf{r}_2, t) \sim C_n(1-R)\frac{\gamma}{2}E_{out}(\mathbf{r}_1, t - t_e - u(\mathbf{r}_2, t)/\omega),$$

$$E_{out}(\mathbf{r}_3, t) \sim C_n(1-R)\left(\frac{\gamma}{2}\right)^2 E_{out}\left(\mathbf{r}_1, t - 2t_e - \frac{u(\mathbf{r}_2, t)}{\omega} - \frac{u(\mathbf{r}_3, t)}{\omega}\right),$$

$$E_{out}(\mathbf{r}_1, t) \sim C_n(1-R)\left(\frac{\gamma}{2}\right)^3$$
$$\times E_{out}\left(\mathbf{r}_1, t - t_e - \frac{u(\mathbf{r}_2, t - 2t_e)}{\omega} - \frac{u(\mathbf{r}_3, t - t_3)}{\omega} - \frac{u(\mathbf{r}_1, t)}{\omega}\right),$$

where $E_{out}(\mathbf{r}_2, t)$, $E_{out}(\mathbf{r}_3, t)$, $E_{out}(\mathbf{r}_1, t)$ are the field values at NRI output in the 1st, 2nd, and 3rd transposition points.

In other words, the field components at the NRI output are displaced in these points (delayed in time) relative to each other by the values equal to or multiple to t_e, if we can neglect (due to their smallness with respect to t_e) the values $\frac{u(\mathbf{r}_i, t)}{\omega}$, where i = [1, 2,…, m]. It means that the cross-correlation function should have a peak at shift τ_{cor} equal to or multiple to t_e. This conclusion confirms the results of the calculation of the field amplitude cross-correlation in the different CTPs in Fig. 4.12. As before, here and further, instead of the statistical analysis of the field at the NRI output, we examine its amplitude only.

From the interpretation of the cross correlograms shapes obtained (Fig. 4.12), it does follow that the main maximum on them corresponds, with some admissible error, to the time, during which the field propagates from the one analyzed point to the other: for the points *1* and *2*, *1* and *3*, *3* and *1*–τ_{cor1} = 5.32, τ_{cor1} = 10.68, τ_{cor1} = 5.32. We would like to emphasize (in contrast to auto-correlation case) that cross-correlation analysis allows estimation of the delay time t_e for large K—see Fig. 4.12 and Table 4.2.

Let us assume that there is such a mode in the NRI, when for $\forall i$ there exists at least one (the main one) assuredly identifiable peak in the cross correlograms corresponding to the points i and $i+1$. Moreover, in this mode correlation is absent, if the points belong to the different CTP, and if for the points belonging to the same CTP there are no "false" (directly not related to the field delay in FBL) maxima in the correlogram. (The above mentioned conditions are usually satisfied, when in the NRI model the dynamic chaos is realized). Then, cross-correlation analysis (i.e. the

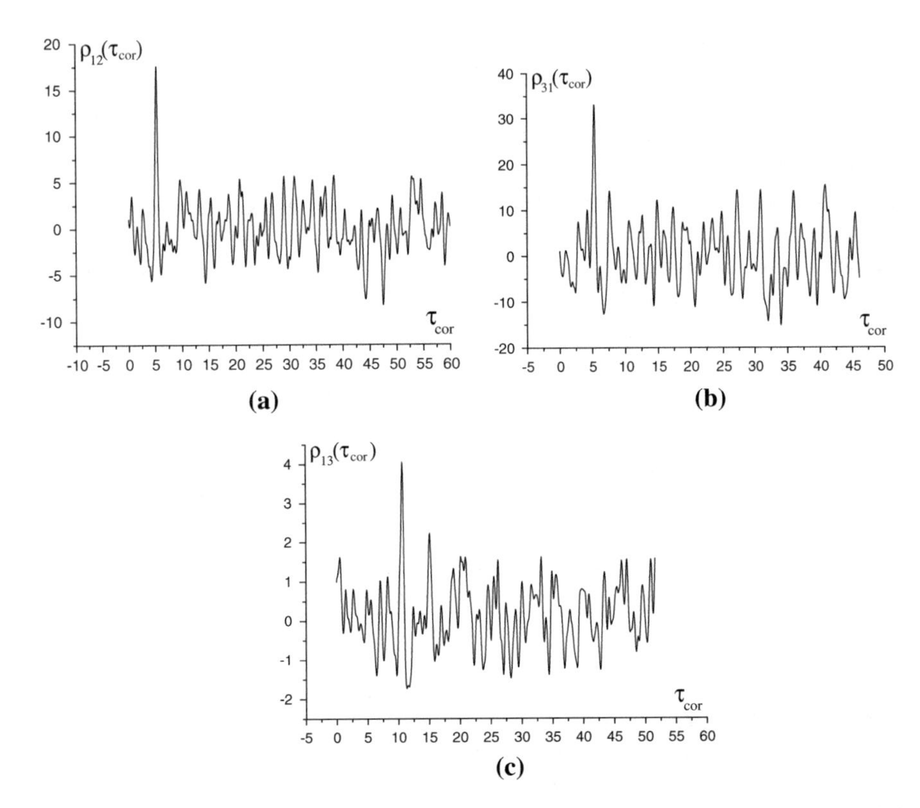

Fig. 4.12 Cross correlograms of the field amplitude in points *1* and *2* (**a**), *1* and *3* (**b**), *3* and *1* (**c**) for the case Δ = 120°, K = 15.5, γ = 0/5, $t_e = 5\tau_n$

analysis of the maxima "spectrum" for all the possible auto- and cross-correlation functions) allows us to determine, if the points belong to the same CTP, as well as the sequence and the duration of the beam transition from the one point to the other.

For this purpose, it is expedient to use the following operational procedure concerning the set of the correlograms. (1) Depending on the minimal correlation time $\tau_{cor,min}$, the delay time in NRI is determined. (2) Depending on the presence and the proximity of the main peak of correlogram (for the certain points i and j) to the value of $\tau_{cor,min}$, we can assess the situation: whether these points belong to the same CTP and whether the points numbers form the ratio $i = k + j$, where k is the integer. Here, the absence of the main peak does not give a reason to conclude that these points *do not belong* to the same CTP.

Naturally, if we know the class of the spatial field transformations in FBL, then a strategy of the CTP structure and delay time cracking can be optimized. So, if it is known that the field rotation by the angle Δ takes place, then, with an adequate choice of the analyzed points and in the case of the lower orders of transposition—$m =$ 2, 3 ($\Delta =$ 180°, 120°), it is sufficient to calculate the field cross-correlation coefficient in two points only, that is $\rho_{12}(\tau_{cor})$ in Fig. 4.13, for instance.

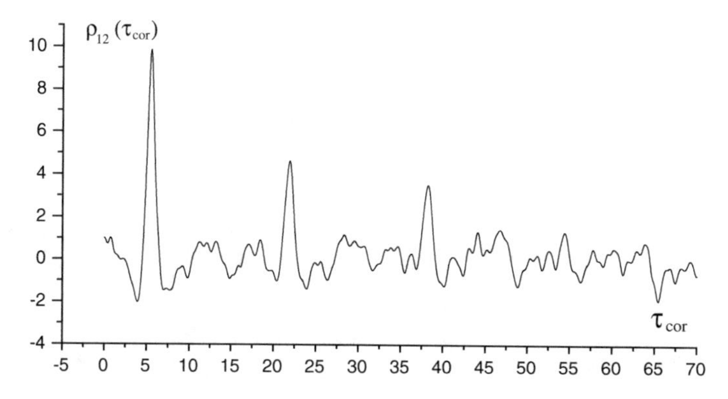

Fig. 4.13 The cross-correlation picture for the case $\Delta = 120°$, $K = 9.5$, $\gamma = 0.5$, $t_e = 5\tau_n$

The cross correlogram in Fig. 4.13 presents the three clearly distinguishable peaks. The first peak (maximal) corresponds to the time shift $\tau_{cor,1} = 5.46$, which, with regard to error, is the estimation of the delay time in the NRI FBL t_e. The second peak corresponds to $\tau_{cor,2} = 21.88 \approx \frac{4t_e}{\tau_n}$; the third peak (minimal) corresponds to the time shift $\tau_{cor,3} = 38.19 \approx \frac{7t_e}{\tau_n}$. The second peak is present due to the fact that the field component outcoming from the point *1* after an initial visit to the point *2* and having made the three bypasses of the NRI FBL ($m = (3)$ appears in the point *2* again. Similarly, we can explain the position of the third maximum on the correlation picture. It is clear that $m \approx \frac{(\tau_{cor,2} - \tau_{cor,1})}{\tau_{cor,1}} = 3.007$.

However, the possibilities of such an optimization are limited: with the growth of the nonlinearity coefficient K (starting from $K > 9.5$), or with the loss coefficient decrease γ, the main peak (at $\tau_{cor,1}$) is saved only. We see that the accuracy t_e of the estimation methods of cross- and auto-correlation analysis is practically the same. For the case studied, $\tau_{cor,1} = 5.46$ against $\tau_{cor,1} = 5.453$, which is in fact 8 % accuracy. The degree of influence of the delay time setting error $t_{e,d}/\tau_n$ in the decipherer upon the deciphering process is reflected by Table 4.3. The results are obtained for $\Delta = 120$, $K = 9.5$, $\gamma = 0.5$, $t_e = 5\tau_n$. This simulation experiment once more confirms that for the value t_e (the part of the key) the correlation analysis is capable of providing the approximate estimation only.

For comparison, we produce the data [4, 12, 26] referring to the NRI with the saturable absorber with the maximal amplitude of the entering message of 0.5 %. The deciphering quality decreases to 50 %, if the interferometer length detuning exceeded 1 %.

It is reasonable to ask the question about the properties of the double-circuit NRI in terms of the development of the cryptosystem.

Table 4.3 Influence of the delay time setting error t_{ed}/t_n in the decipherer on the deciphering process (delay in the cipherer $t_e/t_n = 5$)

Duration of system operation t/τ_n	Signal at decipherer output at delay time setting error values $t_{e\,d}/\tau_n$ in decipherer, %					Decipherer output signal
	0	1	2	4	8	
0	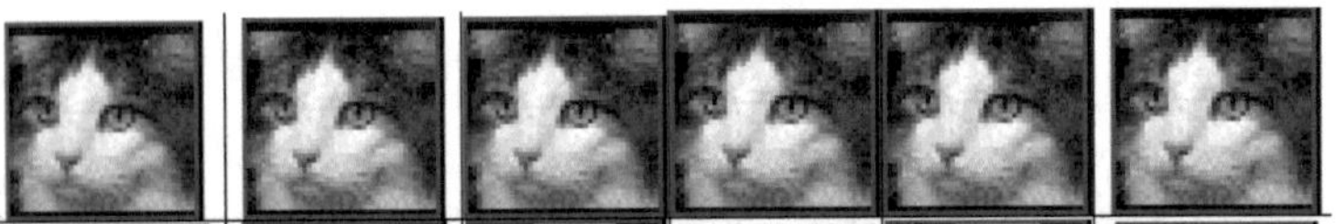					
5	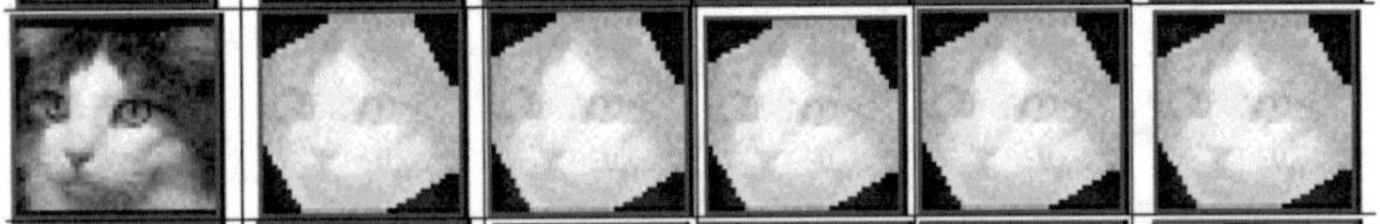					
10	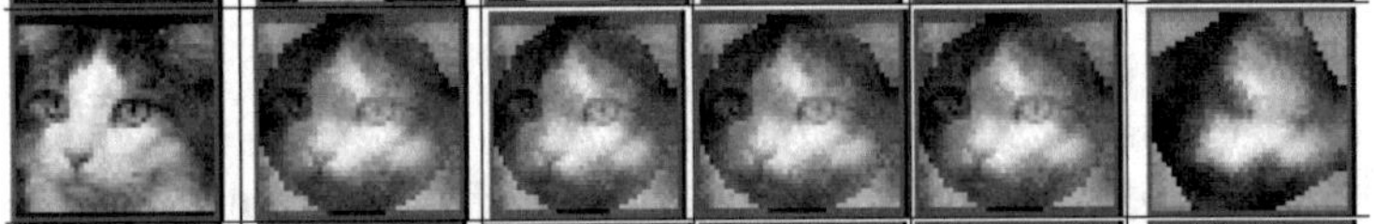					
15	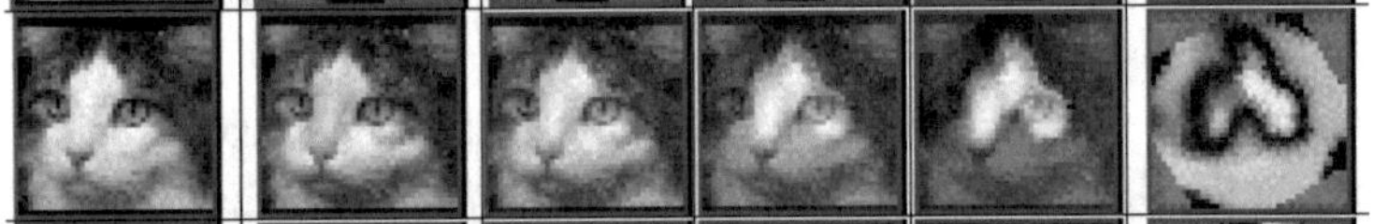					
20	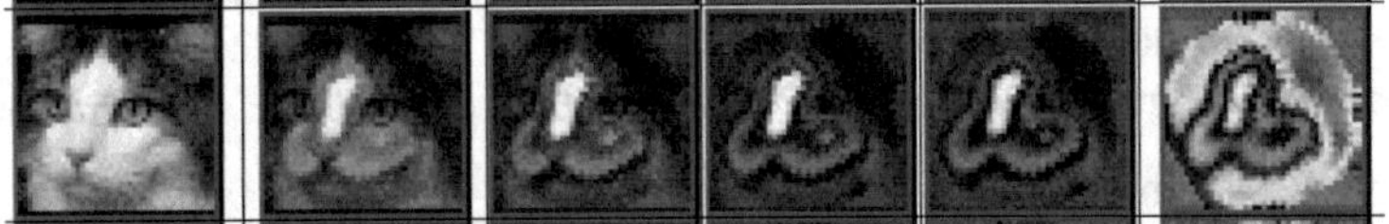					
25	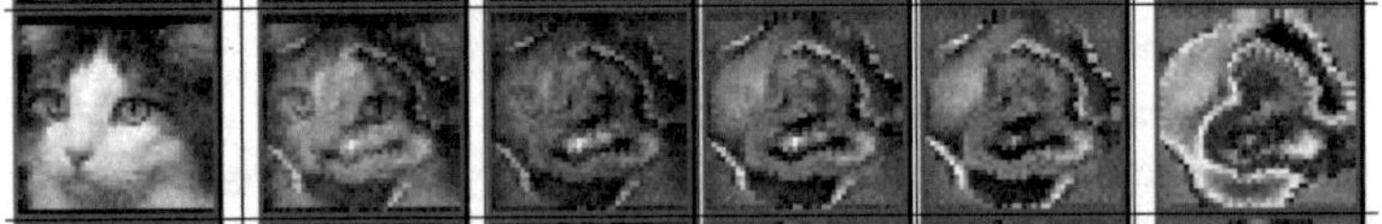					
50	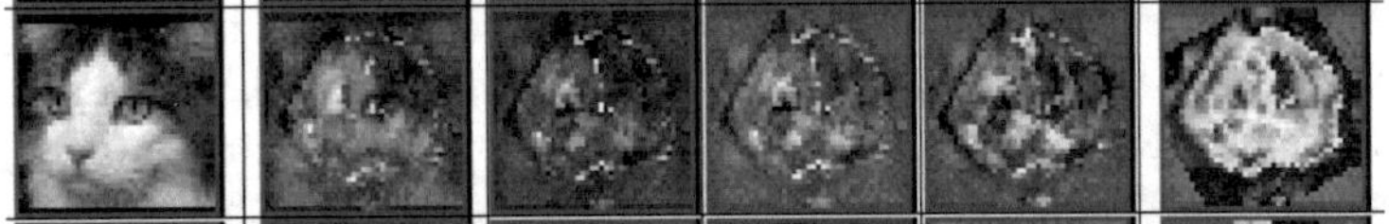					
75	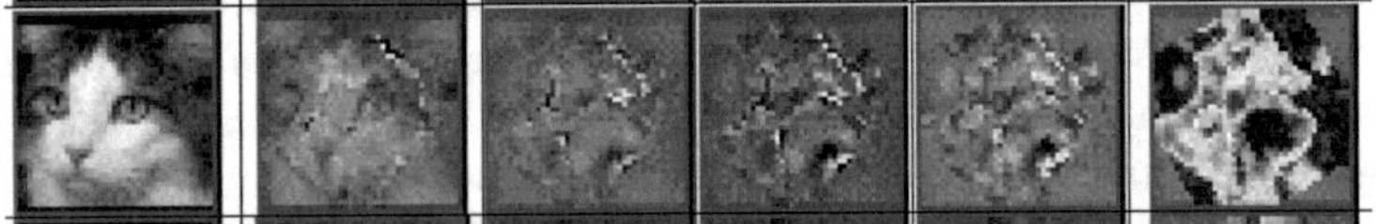					
100	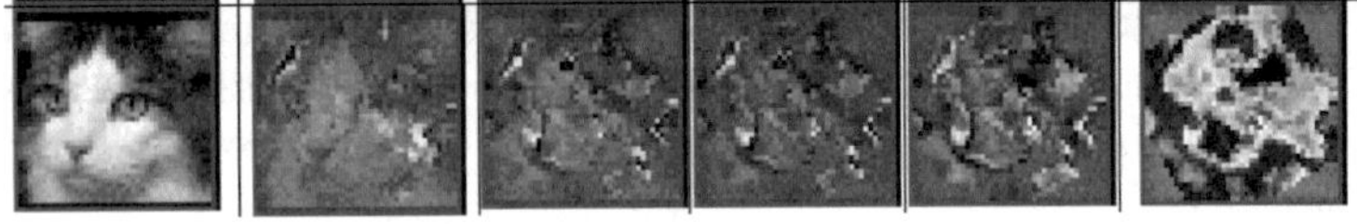					

4.2 Imitation of the DNRI Parameters Cracking Based on the Correlation Analysis: Discussion of Advantages

Similarly to the NRI, its double-circuit modification (DNRI) is capable of operating as a cipherer in the cryptosystem structure. For instance, using the route-operator formalism (described in Sect. 4.1.2), we can prove that the decipherer using the chaotic response mode should be created as based on the scheme depicted in Fig. 4.14. Evidently, a presence of two FBL in DNRI with different delay times $t_{e,i}$ and two different field transformers G_i complicates the problem of cipherer parameter cracking by the correlation analysis. We shall discuss this below.

4.2.1 *The Case of Field Transformation in FBL (Time Delay Estimation)*

Since the simplest case is the case of the field transformation absence in NRI FBL, i.e. CTP consists of the one point, for the delay time estimation in FBL we can only refer to the auto-correlation analysis method. Calculated auto-correlograms (Fig. 4.15) demonstrate a presence of the one clearly expressed maximum. However, its position $t_{cor,1} \approx 4$ does correspond neither to the value $t_{e,1} = \tau_n$, nor to the value $t_{e2} = 5\tau_n$. It is easy to notice that $t_{cor,1} \approx \frac{t_{e2} - t_{e1}}{\tau_n} = 4$. This non-trivial regulation is also reproduced for the other values of the model parameters, for instance, when, in contrast to Fig. 4.15, $K = 50$ or $K = 100$, as well as if $t_{e,1} = \tau_n$, $t_{e,2} = 2\tau_n$, or $t_{e,1} = 4\tau_n$ and $t_{e,2} = 5\tau_n$ for $K = 100$.

Thus, instead of the time moments t_{e2} and t_{e1} cracking, the auto-correlation analysis gives estimation of their *difference* $\Delta t_e = t_{e2} - t_{e1.}$ This circumstance significantly increases the crypto-resistance of the modified NRI system against the similar types of attacks, since for this difference Δt_e any amount of different

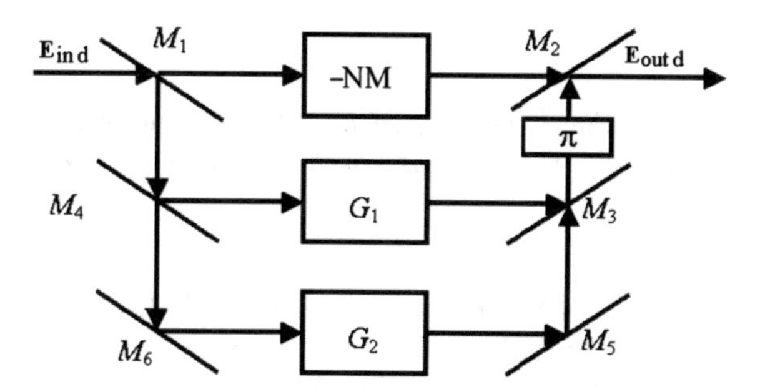

Fig. 4.14 Scheme of the decipherer for the DNRI: M_i are mirrors, NM is nonlinear medium, G_i are linear elements performing the spatial transformation of the optical field in the transverse plane of the light beam, π is a device of phase shift by π and compensating attenuation of the field amplitude

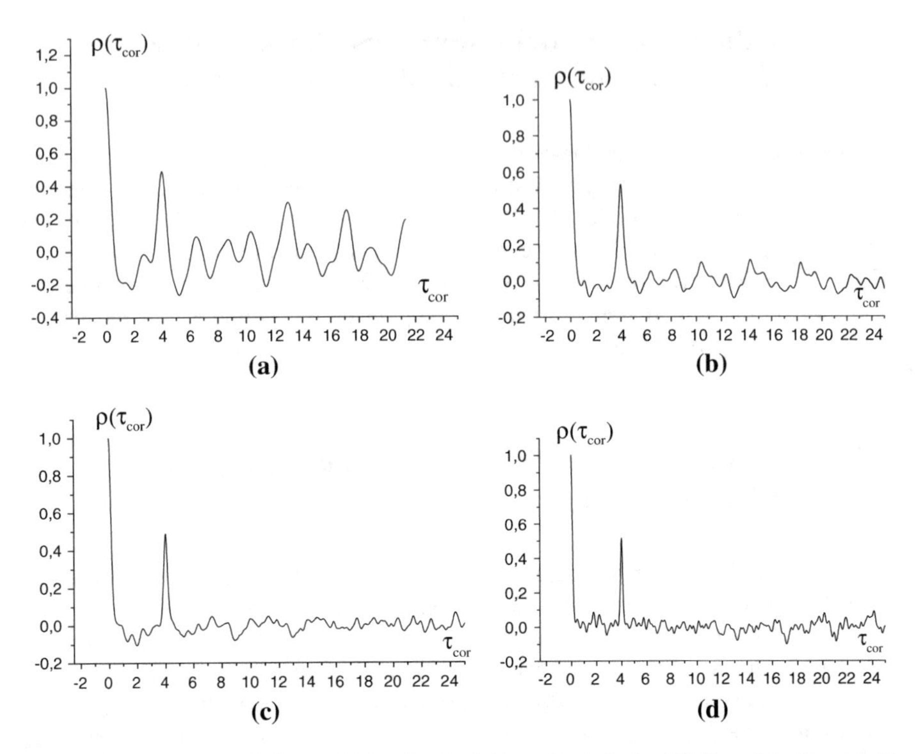

Fig. 4.15 The auto-correlation coefficient for the field amplitude in the DNRI model without field transformation in FBL and with delay time in FBL $t_{e1}/\tau_n = 1$ and $t_{e2}/\tau_n = 5$, loss coefficients $\gamma_1 = \gamma_2 = \sqrt{0.125}$ and various nonlinearity coefficients: $K = 10$ (**a**), $K = 15$ (**b**), $K = 20$ (**c**); $K = 30$ (**d**)

combinations of t_{e2} and t_{e1} exist. As opposed to the single-circuit NRI, the auto-correlation analysis is not sensible to the nonlinearity coefficient K. In this regard, DNRI loses to the single-circuit NRI as a base for the cipherer.

4.2.2 Cases with the Field Rotation in the One Feedback Loop with the Same and Various Field Rotations in FBL

Let the field rotation be absent in the one NRI FBL, and is 180° in the other one, i.e. CTP has a form 212 or 221 in Table 2.3.—see Sect. 2.4.2. Then, plots of cross-correlation functions ρ_{12} and ρ_{21} (Fig. 4.16) convince us that the delay time difference can be found out, as earlier: $\tau_{cor} = \frac{t_{cor}}{\tau_n} = 4.01$. In this case a queue of the certain CTP points selection in the transverse plane of the beam is not significant.

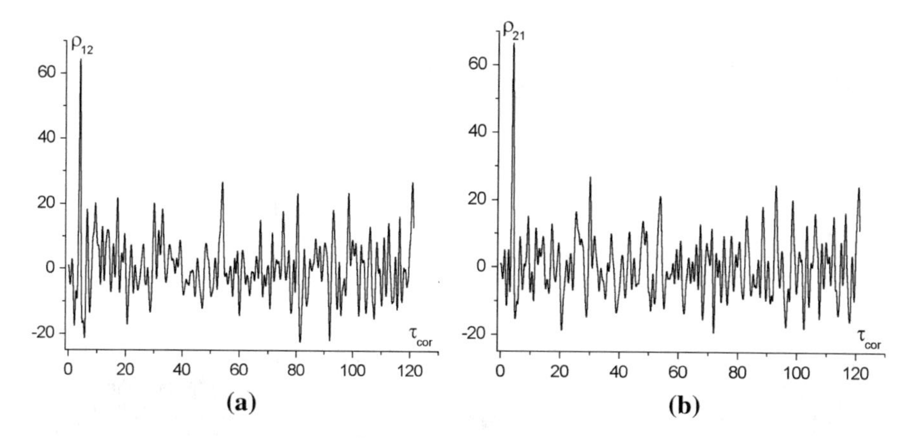

Fig. 4.16 The cross-correlation pictures of the field amplitude in the points *1* and *2* (**a**), *2* and *1* (**b**) for the DNRI model at $\Delta_1 = 0°$, $\Delta_2 = 180°$, $t_{e1}/\tau_n = 1$ and $t_{e2}/\tau_n = 5$; $\Delta t_e/\tau^n = 4$, $\gamma_1 = \gamma_2 = \sqrt{0.125}$, K = 10

Figure 4.17 shows the cross-correlation functions ρ_{ij} for the case, when in the one loop the field is rotated by $\Delta_2 = 120°$, and in the other one the rotation is absent ($\Delta_1 = \ 0°$), i.e. CTP has a form 331 or 313 in the Table 2.4. Obviously, the time delay difference Δt_e can be successfully determined only in the case of amplitude correlation in the points directly following each other in the order of their numeration, i.e. points *1* and *2*, *2* and *3*, *3* and *1*. If to change the direction of the points selection (i.e. *1* and *3*, *3* and *2*, *2* and *1*), it will be difficult to reveal Δt_e with the help of the cross-correlation analysis.

Simulation modeling shows that in the cases, when *the difference* of delay times can be determined from the cross-correlation function, the error does not exceed $\pm 0.06\ \tau_n$ (± 1.25 %)—see Table 4.4. We would like to remind that for the single-circuit NRI the delay time determination error t_e/τ_n from the auto-correlation function varies (depending on the model parameters) from ± 1 to ± 44 % (see Table 4.2 in Sect. 4.1.6).

When the field rotation in both loops is fulfilled by *the same* angles ($\Delta_1 = \Delta_2 = 180°$; $120°$ или $90°$), we can make the following conclusions based on the plots of the auto- and cross-correlation pictures calculated for $K = 10$, $\gamma_1 = \gamma_2 = \sqrt{0.125}$, $t_{e1}/\tau_n = 1$ and $t_{e2}/\tau_n = 5$, where $\tau_n = 10^{-9}$s. Firstly, the **auto**-correlation analysis of the light field amplitude dynamics in the points of the transverse plane allows determining of the delay time difference Δt_e in the feedback loops with the accuracy no less than $\pm 0.04\ \tau_n$. Secondly, the larger part of **cross**-correlation function plots has a form of the regular peak sequence. Thirdly, the distance between peaks on the **cross**-correlation plots varies in the interval from 3.31 τ_n до 5.51 τ_n depending on the conditions. In the most cases this distance has the value Δt_e with accuracy $\pm 0.2\ \tau_n$.

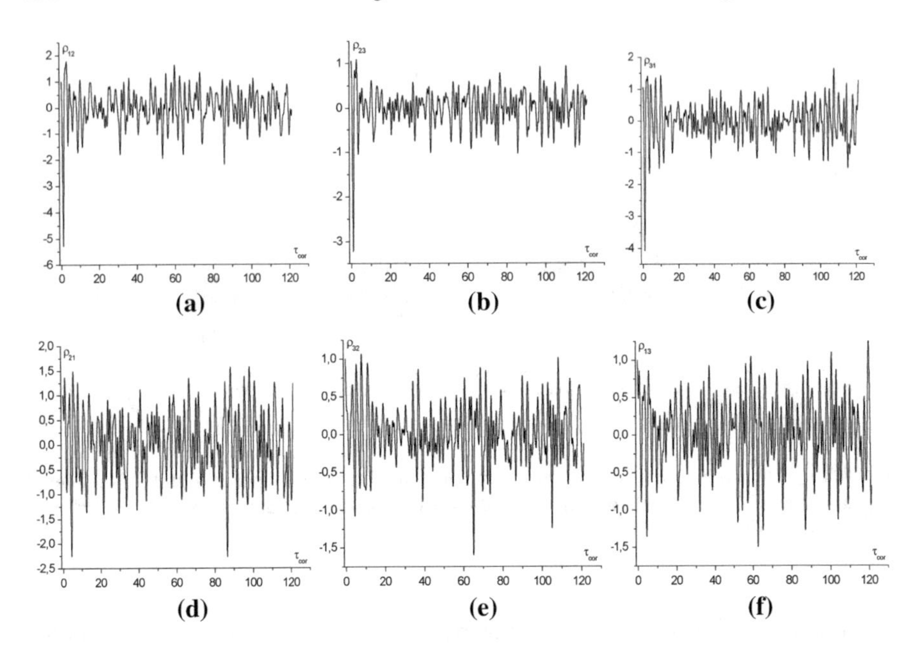

Fig. 4.17 The cross-correlation pictures of field amplitude in the points *1* and *2* (**a**), *2* and *3* (**b**), *3* and *1* (**c**), *2* and *1* (**d**), *3* and *2* (**e**), *1* and *3* (**f**) for the DNRI model with $\Delta_1 = 0°$, $\Delta_2 = 120°$ and the delay time: $t_{e1}/\tau_n = 1$ and $t_{e2}/\tau_n = 2$, $\Delta_{te}/\tau_n = 1$; $\gamma_1 = \gamma_2 = \sqrt{0.125}$, $K = 10$

Let it be the angle rotation in both feedback loops by angles, which differ from 0° and are *not equal* to each other (for instance, $\Delta_1 = 180°$ and $\Delta_2 = 120°$, i.e. CTP has a view of Fig. 2.35, or $\Delta_1 = 180°$ and $\Delta_2 = 90°$). Then, from the structure of auto- and cross-correlograms plots calculated for $K = 10$, $t_{e1}/\tau_n = 1$ and $t_{e2}/\tau_n = 5$, $\Delta t_e/\tau_n = 4$; $\gamma_1 = \gamma_2 = \sqrt{0.125}$, we can make the following conclusion. The cross-correlation analysis allows "revealing" $\Delta t_e = |t_{e2} - t_{e1}|$ with accuracy $\pm 0.05\ \tau_n$, only in the case if we analyze the amplitude in CTPs following directly one after another in the order of its numeration. Auto-correlation analysis is not effective here.

Thus, computer simulation of the DNRI parameter "cracking" (within the limits of point model) with the help of correlation analysis of the field amplitude leads to the following conclusions. Determining of the delay times in FBL t_{e1} and t_{e2} is impossible. But calculations of auto-correlation coefficients ρ_{ii} or cross-correlation ρ_{ij} $(i \neq j)$ allows us to reveal the value of $\Delta t_e = |t_{e2} - t_{e1}|$, if the cross-correlation amplitude analysis is provided in points i, j following in the CTP directly one after another (in the order of their numeration) $(j = i + 1)$.

These regularities theoretically allow us to determine the CTP structure, i.e. what points belong to the certain CTP (to perform fragmentation of the transverse beam section points to the certain CTP), and what is the order of their numeration, but at the cost of the calculating resources. Let, for example, the light field distribution in the transverse beam plane be represented by a matrix of $m \times N$ elements, where m

Table 4.4 Estimation of the delay time difference Δt_e in the NRI FBL by the method of the cross-correlation analysis for the different Δ_1 and Δ_2, K, Δt_e

Δ_1 and Δ_2	K	ρ_{ij}	$\Delta t_e / \tau_n$	$\tau_{cor} = t_{cor} / \tau_n$	$\lvert\Delta t_e - t_{cor}\rvert / \tau_n$
$\Delta_1 = 0°$ $\Delta_2 = 180°$	95	ρ_{12}	4	4.01	0.01
		ρ_{21}		–	–
	10	ρ_{12}	1	1.03	0.03
		ρ_{21}		0.99	0.01
		ρ_{12}	4	3.95	0.05
		ρ_{21}		3.96	0.04
		ρ_{12}	8	7.99	0.01
		ρ_{12}	12	11.99	0.01
	15.5	ρ_{12}	4	4.00	0.00
		ρ_{21}		4.00	0.00
$\Delta_1 = 0°$ $\Delta_2 = 120°$	9.5	ρ_{12}	4	4.00	0.00
		ρ_{23}		4.03	0.03
		ρ_{31}		4.01	0.01
		ρ_{13}		–	–
		ρ_{32}		–	–
		ρ_{21}		–	–
	10	ρ_{12}	1	0.98	0.02
		ρ_{23}		0.96	0.04
		ρ_{31}		0.99	0.01
		ρ_{13}		–	–
		ρ_{32}		–	–
		ρ_{21}		–	–
		ρ_{12}	4	3.94	0.06
		ρ_{23}		3.99	0.01
		ρ_{31}		3.97	0.03
		ρ_{13}		–	–
		ρ_{32}		–	–
		ρ_{21}		–	–
	15.5	ρ_{13}	4	–	–
		ρ_{31}		4.00	0.00
		ρ_{21}		–	–

is a number of points in each out of N isomorphic CTPs. Then, in the worst (for the cracking) combination of circumstances, the revealing of these parameters requires $mN(mN-1)$ acts of the cross-correlation coefficient calculation: both ρ_{ij} and ρ_{ji}. But, even in the best combination of circumstances, the calculation of $(m-1)$ functions $\rho_{i,i+1}(\tau_{cor})$ is necessary to crack the structure of the one particular CTP, and the same is valid for all the others.

In contrast to the single-circuit interferometer model, the correlation analysis does not permit cracking values of t_{e1} and t_{e2} in FBL. We can conclude that the double-circuit NRI is more stable-resistent compared to the single-circuit one from the point of view of the possibility to reveal its parameters by the correlation analysis [4, 27, 28]. We would like to add that when parameters of the cipherer and the decipherer do not coincide, the visual comparison of the signal shapes at the decipherer output proves that in case of the DNRI the result of the deciphering is less sensitive to detuning.

4.3 Conclusions

The main results of the analytical research and the numerical experiment can be reduced to the following.

It is shown that we may speak about secretive information transmission with the help of processes in the nonlinear system as about nonlinear-dynamic cryptology. We may prove a possibility to construct on the NRI base of the communication system, where a message is introduced into encipherer according to the principle of nonlinear mixing (in the form of modulated in phase and (or) amplitude of light beam). A decipherer operates in the mode of the chaotic response. Appropriated circuits are presented. The description of field conversion in cipherer and decipherer is given. The expression for dechiphering error reduced to the cipherer input, which is caused by the difference on conjugate parameters (de)cipherer is suggested. Mathematical conditions of deciphering correctness are offered and criteria of its practical realization.

Interferometers of the communication systems are interpreted as objects having the graph structure. For their structural designing analysis the route-operator formalism is suggested. Its heuristic potential is demonstrated on the example of studying of signal transmitting in mentioned nonlinear systems of the optical range.

Imitation of secretive image transmission in the modes of deterministic spatial-temporal and the spatial chaos is carried out. Interconnection of times of "warming-up", establishment, synchronization locking, and time, which is necessary for (de)ciphering are found out. Two possible criteria for estimation of system crypto-resistance are described. Deciphering error is interpreted as a wave process and its amplitude as the error function of setting the decipherer parameters. Deciphering error estimation is fulfilled.

Imitation of "cracking" of delay time value in NRI is carried out. For this, we use the auto- and cross-correlation analysis. The influence of the setting error for time delay in the decipherer on deciphering process in NRI is studied. The optical circuit and the decipherer model for double-circuit NRI are presented. The essentially increased resistance of the double-circuit NRI to the "cracking" of the time delay value in it is shown.

The next chapter switches the material description from predominantly cryptographic context (to be more exactly, from the context of nonlinear-dynamic cryptology) into the context of physical and steganographic methods of information protection. However, we shall speak not about all kinds of methods of such a type, but only about the approach offered by us. It is based on the specific properties of laser beams, containing the optical vortices, mentioned in the section. Here we should not that this method is limited by only optical systems. You see, for electromagnetic waves of the arbitrary range (microwave, terahertz etc.) it is possible the beam formation with the screw dislocation of the wave front.

References

1. Izmailov IV, Shulepov MA. Simulation of signal enciphering by means of nonlinear ring and decoding In: Proceedings of first Asia-Pacific conference "Fundamental problems of opto- and microelectronics". Vladivostok. 11–15 September 2000. pp. 131–133.
2. Izmailov IV, Poizner BN, Shulepov MA. Simulation experience of the optical device for nonlinear-dynamic In: Proceedings of international optical congress "Optics XXI" (16–20 of October, 2000, Sankt-Peterburg). The conference "Fundamental Problems of Optics" (17–19 October 2000, Sankt-Peterburg). 2000. pp. 30–31 (in Russian).
3. Izmailov IV, Poizner BN, Shulepov MA. Modeling a processes in synergetic cryptosystem. In: Mathematics, computer, education: proceeding of VIII international conference (Pushchino. 31 of January–5 of February, 2001)/Under edition of G.Yu. Reznichenko. Moscow: Progress-Tradition Publ.; 2001. no 8. pp. 157 (in Russian).
4. Vladimirov SN, Izmailov IV, Poizner BN. *Nonlinear-dynamic: radio physical and optical systems*/Under edition of C.N. Vladimirov. Moscow: FizMatLit Publ.; 2009. 208 p. (in Russian).
5. *Physical Encyclopedia*/Under edition of A.M. Prokhorov, vol. 2. Moscow: Soviet Encyclopedia Publ.; 1990. 703 p. (in Russian).
6. Avdeev SM, Izmailov IV, Poizner BN, Shulepova AA. Analysis of nonlinear optical cryptosystem stability: number of keys bounding and cracking of delay time. In: Proceedings of second Asia-Pacific conference "fundamental problems of opto- and microelectronics" (September 30–October 4, 2002, Vladivostok); Far Eastern State Technical University. Vol. Label: APCOM'2002 (82 247 680 bytes); SN: B52F-E938; text files (PAPERS\2-10.pdf (9 p., 122 001 bytes); DOC\2-10.doc (9 p., 301 568 bytes)). Vladivostok: Far Eastern State Technical University, 2002. 1 CD-ROM/ 12 cm. System Requirements: IBM PC; monitor/Windows 9x.
7. Izmailov IV, Poizner BN. Description of the open systems, where interactions have the structure of a graph. In: Proceedings of international conference "structure arranging in the open systems" (Kazakhstan, Almaaty. 24–27 September 2001). Almaty; 2001. pp. 13. (in Russian).
8. Izmailov IV, Poizner BN, Shulepov MA. Control of dechaotization of the oscillation-wave process formed in the nonlinear with feedback. In: Proceeding of IV international scientific conference "Mathematical models of nonlinear excitations, transition, dynamics, control in condensed systems and other media. (Moscow. 27 of June–1 of July 2000). Moscow: Stankin Publ., 2000. pp. 50 (in Russian).
9. Izmailov IV, Poizner BN. Optical generator of the deterministic spatial and dechaotizator to it. In: Abstracts of the 6th intern. school on Chaotic oscillations and pattern formation (Saratov. 2–7 October 2001). Saratov: Publ. "College"; 2001. pp. 74–75 (in Russian).

10. Mathematical encyclopedic dictionary/Under edition of Yu.V. Prokhorov. Moscow: Soviet Encyclopedia Publ.; 1988. 847 p. (in Russian).
11. Vladimirov SN, Negrul VV. Communication systems with passive chaotic synchronization. In: Proceeding of V international conference "Actual problems of electronic instrumentation APEI-2000" (Novosibirsk. 26–29 of September, 2000), vol. 7. Novosibirsk; 2000. pp. 39–41. (in Russian).
12. Izmailov IV, Poizner BN. Implementation variants of nonlinear-optical device of secretive information transmission. In: Optics of atmosphere and ocean, vol. 14(11); 2001. pp. 1074–1086. (in Russian).
13. Izmailov IV, Shulepov MA. Simulation of signal enciphering by means of nonlinear ring and decoding. In: Yuri NK, Oleg BV, editors. Optoelectronic information systems and processing (11–15 September 2000, Vladivostok), Proceedings of SPIE vol. 4513, pp. 46–51 (2001).
14. Izmailov IV. Model of confidence communication device using electro-optical element in ring interferometer. In: Proceedings of 1-st Siberian student IEEE/LEOS conference-competition (20 December, 2001, Novosibirsk); Novosibirsk State Technical University. vol. Label: 020118_2145 (3 p., 3 145 728 bytes); SN: 20CB-A23A; text files (LEOS19.doc (365 056 bytes)). Novosibirsk: Novosibirsk State Technical University, 2001. 1 CD-ROM / 12 cm. System requirement: IBM PC; monitor/Windows 9x.
15. Izmailov IV, Poizner BN. A simple principle of enciphering of images being carried by laser signal. In: Mayer GV, Soldotov AN, editors. The Proceedings of the 5-th Russian-Chinese symposium on laser physics and technologies (Tomsk. 23–28 October, 2000). Tomsk: Publishing Tomsk State University; 2000. pp. 62–63.
16. Garcia-Ojalvo J, Roy R. Spatiotemporal communication with synchronized optical Chaos. http://xxx.lanl.gov/abs/nlin.CD/0011012. 2000 6 Nov. 4 p.
17. Avdeev SM, Izmailov IV, Shulepova AA. Crypto-stability analysis of nonlinear-optical device for information protection. In: Proceedings of III seminar of young researchers "Modern problems of physics and technologies (Tomsk. 30 of January–1 of February, 2002). Tomsk University Publ.; 2002. pp. 116–119 (in Russian).
18. Izmailov IV, Shulepov MA. Spectral analysis of nonlinear dynamics in the ring interferometer model. In: Proceedings of II seminar of young researchers "Modern Problems of Physics and Technologies (Tomsk. 5–7 of February, 2001). Tomsk University Publ.; 2002. pp. 173–176 (in Russian).
19. Izmailov IV, Poizner BN. Theoretical analysis of parameters mismatch influence on deciphering error in nonlinear optical cryptosystem. In: Technical digest of the international conference on lasers, applications, and technologies "LAT'2002". Moscow. June 22–27; 2002. Moscow; 2002. pp. 206.
20. Izmailov IV. Nonlinear-optical devices for information protection. Opt J 2002; 69(7):62–67.
21. Risk control: Risk. Sustainable development. Moscow: Nauka Publ.; 2000. 431 p. (in Russian).
22. Gribanov YI, Mal'kov VL. Errors and parameters of the digital spectral-correlation analysis. Moscow: Radio i Sviaz Publ.; 1984. 160 p. (in Russian).
23. Bendat JS, Piersol AG. Engineering applications of correlation and spectra analysis. Newyork: Wiley; 1993. 472 p.
24. Bendat JS, Piersol AG. Random data: analysis and measurement procedures. Newyork: Wiley; 2010. 640 p.
25. Avdeev SM, Izmailov IV, Poizner BN, Shulepova AA. Analysis of nonlinear optical stability: number of of delay time. In: Yuri NK, Oleg BV, editors. Fundamental problems of optoelectronics and microelectronics. (September 30–October 4, 2002, Vladivostok, Russia). Proceedings of SPIE vol. 5129; 2003. pp. 153–161.
26. Garcia-Ojalvo J, Roy R. Parallel communication with optical spatiotemporal chaos. IEEE Trans Circuits Syst Fundament Theory Appl 2001; 48(12):1491–1497.

27. Shergin DA, Izmailov IV, Lyachin AV, Poizner BN, Romanov IV. Simulation of transformation in two-circuit nonlinear ring interferometer with account of many passes. In: Klimkin A, Kozlova E, editors. The 7-th international conference “Atomic and molecular pulsed lasers” (September 12–16 2005, Tomsk): Conf. Proc. Tomsk: Inst. of Atmospheric Optics SB RAS; 2005. pp. 66–67.
28. Izmailov IV, Lyachin AV, Poizner BN, Shergin DA. Second circuit of 2-D feedback loop in ring interferometer as a way to high quality of cryptosystem. In: Technical abstracts of international congress on optics and optoelectronics (28 August–2 September 2005, Warsaw, Poland, conference “systems of optical security”). Warsaw; 2005. pp. 124.

Chapter 5
Optical Vortices in Ring and Non-ring Interferometers and a Model of the Digital Communication System

It is time to discuss the communication system idea offered by the authors, which uses the properties of the screw dislocations of the wave front of electromagnetic field (vortical or singular-optical communication system). It has characters of steganographic and physical stableness. For this, we shall theoretically show how the *nonlinear ring interferometer* (NRI) can distinguish the order (topological charge) of the optical vortex. After that, we simplify the interferometer till the Rogdestvenskiy one. Let us explain the operation principle of such a detector of the vortex charge and the communication system on its base. We perform calculations of the data transmission such a system. We show how the photo receiver noises, phase and amplitude distortion of the beam during its propagation in the optical non-uniform medium, including the turbulent medium will influence on system operation. Since above-mentioned problems—separately, in combinations and completed are enough difficult, we shall be limited for their solution by analytical and numerical methods.

5.1 The Idea of the Singular-Optical Communication System

General presentation about properties of the optical vortex was done in the Sect. 1.2. In this section, we lead the reader to the conception that the optical vortex can be us as data carrier in the digital communication system called as singular-optical communication or vortical. Here we try to answer the following questions: how to construct the appropriate communication system, from which devices it should be included. At that, we should take into consideration that the laser beam expects to propagate in the communication channel with distortions, for instance, through an atmosphere. This introduces error in the data transmitted.

I. Izmailov et al., *Cryptology Transmitted Message Protection*,
Signals and Communication Technology, DOI 10.1007/978-3-319-30125-9_5

The degree of distortions and the probability of error must be examined in each particular case. These errors depend not only on the properties of the medium, but also on the principles of communication system performance and imperfections of its practical implementation. It is natural that for communication systems with optical vortices used to encode information, it is necessary to investigate the influence of the medium on the propagation of vortex laser beams. As an example, we can refer to [1], where the influence of random aberrations caused by the atmospheric turbulence on the optical communication system performance was analyzed.

Thus, the operating principle and the corresponding structural scheme of the adaptive optical communication system were suggested in [2] (Fig. 5.1). The system includes a transmitter formed by the generator of the optical vortex and the wave front pre-corrector, the adaptive receiver consisting of the corrector and the wave front sensor $S(\mathbf{r},t)$, a detector of the topological charge $V_d(\mathbf{r},t)$ and the optical vortex location $\mathbf{r}_d(t)$ as well as the comparator. The system comprised a transmitter formed by generator of optical vortices and preliminary wave front corrector, an adaptive receiver formed by a corrector and a sensor of the wave front $S(\mathbf{r},t)$, a detector of the topological charge $V_d(\mathbf{r},t)$ and position $\mathbf{r}_d(t)$ of the optical vortex, and a comparator.

The offered system novelty is caused by a presence not only the wave front sensor $S(\mathbf{r},t)$, but the *detector of vortex parameters* $V_d(\mathbf{r},t)$, $\mathbf{r}_d(t)$. At that, it is assumed that the detector cannot recover the phase, but it is able to diagnostic urgently and with high probability factor, whether the vortex in the transmitted was or not. Due to this, we can know the undistorted beam phase $S_0(\mathbf{r},t)$ and amplitude distribution (of course, the set of the etalon information signals is a priori known, and hence, the bit value ("0" or "1") which is transmitted.

It seems expedient to construct such detector based on devices registering (measuring) or reconstructing:

(1) distribution of the phase gradient (using the Shack–Hartmann [3–10] or pyramidal sensor [11–13]);
(2) interference patterns (for further analysis of the topology of interference bands [14, 15], in particular, resorting to branching interferometry [16–21] including non-monochromatic one [22, 23]);
(3) intensity distribution of a *non-monochromatic* singular beam (for further analysis of initial and normalized chromoscopic patterns [24–26] as well as their more complicated transformations [27]);
(4) phase distribution (from measurements with the Shack–Hartmann [8–10, 28–34], pyramidal, and interferometric sensors [21, 35–37]);
(5) phase distribution (from a comparison of the intensity distributions (in two planes) obtained by the methods based on the radiative transfer equations);
(6) the vortex charge (by means of holographic separation of vortex beams based on the diffraction maxima [38]).

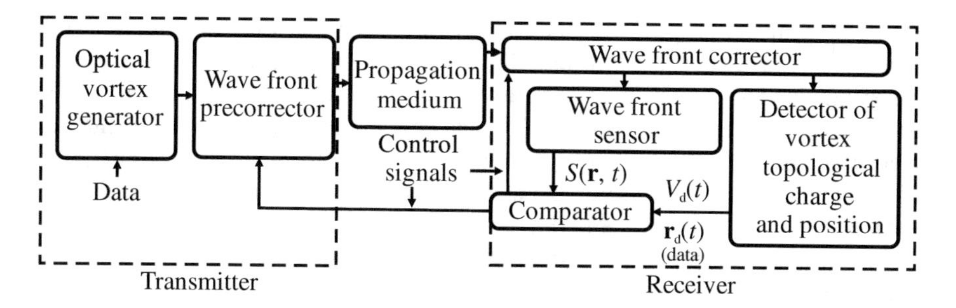

Fig. 5.1 Structural diagram of the adaptive optical communication system

However, the speed of response of devices in items 1–5 does not exceed several kilohertz (one of the reasons of this limitation is the application of a CCD camera), which does not meet the requirements to modern communication systems. In addition, in some of the above-mentioned cases, the established laws and effects:

(1) have not yet been implemented in algorithms of determining the topological vortex charges and their positions,
(2) have not yet been culminated in the metrological principles for the development of special measuring (microprocessor or computer) technology,
(3) have not allowed the vortex charge to be determined and closely localized vortices with identical charges to be resolved.

The most suitable here is the method based on holographic separation of vortex beams. Nevertheless, we shall suggest below another approach, which uses the ring or non-ring interferometer. (In addition to the point 4, we must note that the phase can be measured using the holography methods [39].)

According to our idea [2], in the communication system (Fig. 5.1) the vortex detector should be *less inertial* and, possibly, *less complicate* device than the wave front sensor. Thus, the last, for instance, the Shack–Hartman sensor, is necessary only for phase measurement $S(\mathbf{r}, t)$ of the received (distorted) beam. Comparison result of $S(\mathbf{r}, t)$ and $S_0(\mathbf{r}, t)$ is interpreted by the comparator, and it generates the control signals for the corrector and the pre-corrector. Owing to this, the distorting influence of propagation medium, and, for its turn, the vortex recognition by the detector is simplified.

A large number of publications (see, for example, [40–42]) are devoted to correction methods of the wave front in the adaptive optical systems, and, in particular, to investigation of the Shack–Hartman sensor characteristics. A part of these methods uses the algorithm for recovering of the phase distribution of the light beam in its transverse plane according to measurement data of the transverse phase gradient as the fundamental base. The last ones, as it is known, can be obtained with the help of the Shack–Hartman sensor. In this context, the Freed algorithm is the one of the best [29, 43]. Nevertheless, we tried to make the Freed algorithm modernization in several aspects: (a) increase of the phase recovering accuracy, including at its very high gradients, (b) natural introduction into the algorithm of the unwrapping potential, or the vortical phase Freed algorithm components, (c) union of the and the algorithm of the vortex search in the feedback mode [10, 32–34, 44–48].

So, how to construct this high-speed and more simple detector for the topological vortex charge? It seems to us, the detector requiring the only beam measurement intensity (or its power), which, hence can operate with the speed of several gigahertz and can be constructed on the base of the NRI or the interferometer of Rogdestvenskiy. We start discussion from the operation principle of more complicate NRI.

5.2 Nonlinear Ring Interferometer as an Option Detector for the Screw Dislocation Order

How does the optical vortex influence on the processes in the NRI? General presentations about properties of the NRI are given in Sect. 2.4. Let us try to answer the question: what will be if on the input of the nonlinear interferometer's the light field (1.1) arrives with a screw dislocation of the wave front, having the order (the optical vortex charge V_d)? Let us assume that the singular point, where the beam intensity is equal to zero, coincides with the optical axis of the interferometer.

When searching the answer, we can rely on the evolution model of the nonlinear phase shift $U(\mathbf{r},t)$ and the optical fields (2.31), assuming in it:

$$A_{\text{in}}(\mathbf{r},t) = C\,r^{V_{\text{d}}},\ \varphi_{\text{in}}(\mathbf{r},t) = V_d\vartheta,\ B_{\text{in}}(\mathbf{r},t) = 0,\ \psi_{\text{in}}(\mathbf{r},t) = 0,\ \Omega = 0. \quad (5.1)$$

Really, now the field on the input of the NRI (2.21) (or (2.22)) will be reduced to (1.1).

Nevertheless, for simplification of analytic calculations and modeling, we transform the model (2.31). For this simplification, we may use both approximation of the only one light pass through the feedback of the NRI (i.e. $a_{\text{NL}}(\mathbf{r}',t-\tau) = (1-R)^{0,5}a(\mathbf{r}',t-\tau)$, $b_{\text{NL}}(\mathbf{r}',t-\tau) = (1-R)^{0,5}b(\mathbf{r}',t-\tau)$, $\varphi_{\text{NL}}(\mathbf{r}',t-\tau) = \varphi(\mathbf{r}',t-\tau)$, $\psi_{\text{NL}}(\mathbf{r}',t-\tau) = \psi(\mathbf{r}',t-\tau)$), and approximation of large losses (i.e. $[\gamma(\mathbf{r}'{,}t)/\sigma/2]^2 \approx 0$, $\varphi_{\text{NL}}(\mathbf{r}',t-\tau) = \varphi(\mathbf{r}',t-\tau)$, $\psi_{\text{NL}}(\mathbf{r}',t-\tau) = \psi(\mathbf{r}',t-\tau)$). Then, from the model (2.31), we obtain the following system of equations [49, 50, pp. 32–44]:

$$\begin{aligned}
&\tau_{\rm n}(\mathbf{r})\mathrm{d}U(\mathbf{r},t)/\mathrm{d}t = D_{\rm e}(\mathbf{r})\Delta U(\mathbf{r},t) - U(\mathbf{r},t) + f(\mathbf{r},t),\\
f(\mathbf{r},t) &= n_2(\mathbf{r})Lk\, a_{oe}\left\langle \mathbf{E}^2_{\rm NL}(\mathbf{r},t)\right\rangle_{\rm T} = a_{oe}\, n_2(\mathbf{r})Lk\,[a^2_{\rm NL}(\mathbf{r},t) + b^2_{\rm NL}(\mathbf{r},t)]\\
&= Kab(\mathbf{r},t,\mathbf{r}) + pKab(\mathbf{r}',t-\tau,\mathbf{r}) + [\gamma(\mathbf{r}',t)]\\
&\quad\times\{Ka(\mathbf{r},t,\mathbf{r}',t-\tau)\cos[(1+q)\omega\tau + \varphi(\mathbf{r},t) - \varphi(\mathbf{r}',t-\tau) + \psi(\mathbf{r},t) - \psi(\mathbf{r}',t-\tau)\\
&\quad + Kb(\mathbf{r},t,\mathbf{r}',t-\tau)\cos[(1-q)\omega\tau + \varphi(\mathbf{r},t) - \varphi(\mathbf{r}',t-\tau) - \psi(\mathbf{r},t) + \psi(\mathbf{r}',t-\tau)]\}.
\end{aligned} \tag{5.2}$$

Here $p = 0$ for approximation of the large losses, and $p = [\gamma(\mathbf{r}',t)/\sigma/2]^2$ for approximation of the single pass; "mixed" (Kab) и and "partial" (Ka, Kb) nonlinear parameters:

$$\begin{aligned}
Kab(\mathbf{r},t,\mathbf{r}_n) &\equiv (1-R)\, a_{oe}\, n_2(\mathbf{r}_n)Lk\,[a^2(\mathbf{r},t) + b^2(\mathbf{r},t)],\\
Ka(\mathbf{r},t,\mathbf{r}',t-\tau) &\equiv (1-R)\, a_{oe}\, n_2(\mathbf{r})Lk\, a(\mathbf{r},t)a(\mathbf{r}',t-\tau)],\\
Kb(\mathbf{r},t,\mathbf{r}',t-\tau) &\equiv (1-R)\, a_{oe}\, n_2(\mathbf{r})Lk\, b(\mathbf{r},t)b(\mathbf{r}',t-\tau)],
\end{aligned}$$

$a_{oe} = 1$ (circuit in Fig. 2.31b); $a_{oe} = 2$, $\Omega = 0$, $\psi = \text{const}$ (circuit in Fig. 2.31a).

Within the limits of mentioning assumptions, we can prove the following statement [50–52]. Let the light field in the feedback loop of the NRI rotates on an angle $\Delta = 2\pi M/m$ in the transverse plane of the beam xOy, where M and m are mutually distinct integer numbers. Then the *map periodicity* of the set of optical fields with different values $V_{\rm d}$ into the set of the optical structures on NRI with the period m. That is, one NRI sorts the vortices into m groups according to the rule [50, 51].

This rule for the specific case of the field rotation by the angle то $\Delta = 120°$ ($m = 3, M = 1$, Fig. 5.2) can be formalized in the form

$$N_s = V_{\rm d} \bmod 3,$$

where N_s is the serial number of the group (structure), mod is the operation of calculating modulus of the number. In other words, for $\Delta = 120°$ the NRI distinguishes vortices, for example, with $V_{\rm d1} = 1$ and with $V_{\rm d2} = 2$, as well with $V_{\rm d1} = 0$ and with $V_{\rm d2} = 1$. These simulations in Fig. 5.3 demonstrate clearly this conclusion.

In the general case of rotation to an angle $\Delta = 2\pi M/m$, the following equation is true [53]

$$N_s = \{m + [(V_{\rm d} + L) \bmod m]\} \bmod m, \tag{5.3}$$

$$\varphi_{\rm fb1} = \varphi_{\rm fb10} + 2\pi(LM)/m, \tag{5.4}$$

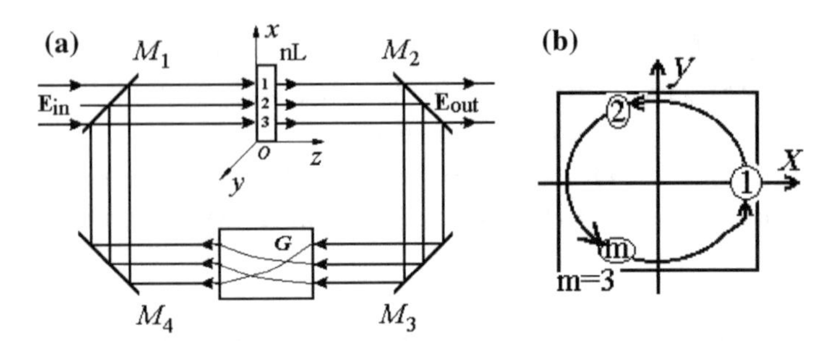

Fig. 5.2 Diagram of the NRI: M_i are mirrors, nL is the nonlinear medium (with the Kerr effect). At rotation by the element G the light by $\Delta = 2\pi M/m = 120°$ (in the transverse plane xOy of the beam) the trajectory of three beams *1*, *2*, *3* are closed after three passes of the NRI (**a**). Projection of closed trajectories of the beams *1*, *2*, *3* onto the plane xOy (**b**) gives the presentation about a construction of the chain (from three) of transposition points

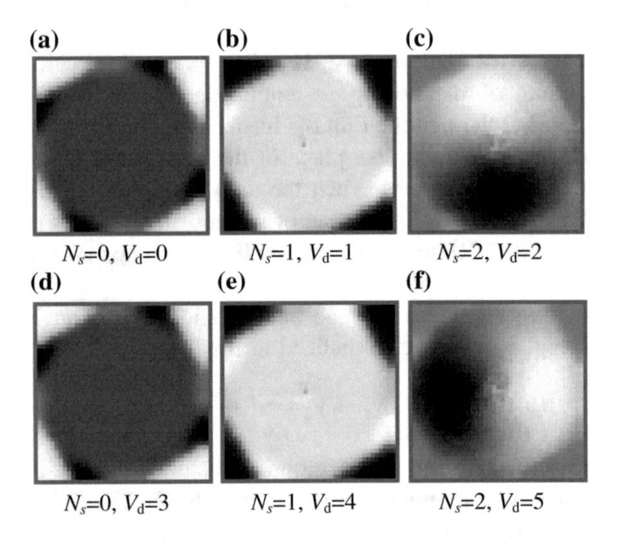

Fig. 5.3 Structures of the nonlinear phase shift $U(\mathrm{r}, t)$ in the plane xOy in the process model (5.2) in the NRI for the diffusion coefficient $D_e = 10^{-6}$, delay $t_e/\tau_n = \Delta z/(c\tau_n) \approx 0.1$ (but for $\varphi_{\mathrm{fb10}} = 0$ and $L = 0$), the nonlinearity $K = 3.5$, field rotation $\Delta = 120°$ (i.e. $M = 1$, $m = 3$), $\gamma = 0.5$, $R = 0.5$. The structure is static at $N_s = 0$, $N_s = 1$ and rotates for $N_s = 2$, the value of $\langle U(\mathrm{r}, t)\rangle_{\mathrm{r}} = 2.467;\ 4.675;\ 3.591$ for $N_s = 0;\ 1;\ 2$

where L, M and m are integer numbers (M and m—are mutually prime number), φ_{fb1}—is the phase shift in the feedback loop, φ_{fb10}—is the value of φ_{fb1} for $L = 0$ or $M = 0$.

Such a fact that in (5.3) the values V_d, L appear in the form of a sum $V_d + L$ is a consequence of influence equivalence of the value of V_d order of the screw dislocation and the value of some phase delay φ_{fb1} of the optical field in the nonlinear ring interferometer (this equivalence can be strictly proved within the limits of the approximated model (5.2)). If to take into consideration: (a) this equivalence, (b) periodicity of the phase delay influence $\varphi_{\text{fb}1}$ on the result of the field interference on the input mirror of the NRI (at quasi-monochromatic light field), the periodicity of the influence of the value V_{d} upon N_s becomes clear.

Strictly speaking, this property is true *not for all* optical structure in the transverse beam section at the NRI output. This is true for this area, where we can neglect by the dependence of the vortex field amplitude on the order of the screw dislocation V_{d}. In other words, the presence of this dependence $C\, r^{V_{\text{d}}}$ in (5.1) does not provoke to bifurcation appearance of the equation solution, which forms the model (5.2).

Relying on the property (5.3) and (5.4), we can offer a principle of obtaining of large enough V_{d}. Let a recognition of the vortex of higher order $V_{\max}$ is required, but the development of the NRI with $m = V_{\max}$ is a challenge (or makes no sense for some reasons). Then it is necessary to construct N NRI devices with $m_1, m_2, \ldots, m_N$. Moreover, all numbers m_i of these NRI devices must be pairwise coprime, and their product is $\prod \equiv m_1 \cdot m_2 \cdot \ldots \cdot m_N \geq V_{\max}$. Then the radiation intensity distribution (with a certain average beam intensity) corresponding to the number $N_{s\,i}$ will be formed at the *i*th NRI output. The combination N of $N_{s\,i}$ numbers will unambiguously determine the $V_{\text{d}} \in [1, \prod]$ value. By which algorithm we can do this? Consideration of this problem is out of the limits of our work. We remind only about a possibility of expansion of the integer number V_{d} (the vortex order) over mutually prime numbers m_i.

We assume implicitly above that the essential different distributions of the beam intensity at the output of the given nonlinear ring interferometer or its different mean values (over the beam section) corresponds to the different numbers $N_{s\,i}$ of the one interferometer ($i = \text{const}$). At that, in the last case the nonlinear ring interferometer can be used of the *high-speed detector of the value* $V_{\text{d}}(\mathbf{r}, t)$. Really, you know, we need not to perform an analysis of the complete structure for determination a presence and the order of the optical vortex, but it is sufficient to analyze of the averaged over the transverse section $\langle U(\mathbf{r}, t)\rangle_{\mathbf{r}} \equiv \frac{1}{S_b} \iint_{(S_b)} U(\mathbf{r}, t) d\mathbf{r}$ the nonlinear phase shift $U(\mathbf{r}, t)$ or he *averaged beam intensity*, and the value $U(\mathbf{r}, t)$ (see Fig. 5.3 [50–52]) is proportional. Here the symbol "S_b" is the area of the transverse section with square S_b, occupied by the beam.

Simulation results of the processes in the NRI (Fig. 5.2) on the base of the model (5.2), taking into account (5.1), justify the fact that at necessary choice of nonlinear ring interferometer, the different distributions of the beam at the NRI output correspond to the different N_s or its mean values (which can be estimated according to the distribution $U(\mathbf{r}, t)$) [50–57]. Let us discuss some of these results obtained in approximation of large losses and demonstrating the vortex influence to formation of optical structures in the transverse plane xOy of the beam in the interferometer and to dynamic mode in the NRI.

Evidently, computing experiments in Fig. 5.3 justifies that varying of the V_{d} order of the screw dislocation is able to change:

- the type of the mode (for instance; dynamic to static: compare Fig. 5.3c, f with Fig. 5.3a, b, d, e);
- the mean value (over the beam section) of nonlinear phase shift $\langle U(\mathbf{r},t)\rangle_{\mathbf{r}}$, proportional to the mean value of the beam intensity (for instance, compare the value $\langle U(\mathbf{r},t)\rangle_{\mathbf{r}}$ for Fig. 5.3a, d and for Fig. 5.3b, e).

Mentioned regulations relate to double-dimension model. It is suitable to explain using the "point" model, i.e. the model, which follows from (5.2), if we neglect by molecule diffusion of the nonlinear medium ($D_e = 0$). Let us remind that exactly in this manner we obtained the model (2.32) from the model (2.26), (2.27), (2.30) and (2.31).

Let us construct the bifurcation diagram of static states of the nonlinear phase shift U in the nonlinear medium for the case corresponding to Fig. 5.3, but for $D_e = 0$. At that, we take into consideration the above-mentioned equivalence of the influence of the V_d and φ_{fb1}, and therefore, we choose here as the bifurcation parameter not discrete variable V_d, but the continuous phase delay φ_{fb1} (Fig. 5.4). Values of phase φ_{fb1}, equal to $0(2\pi/3)$, $1(2\pi/3)$, $2(2\pi/3)$, where N is the integer, correspond to the phase with $V_d = 0 + 3N$, $V_d = 1 + 3N$, $V_d = 2 + 3N$. These values are shown in Fig. 5.4 by dotted vertical lines. In addition to these diagrams, we construct diagrams with bifurcation parameters K for three values of V_d (Fig. 5.5).

From the diagram structure in Fig. 5.4a and diagrams in Fig. 5.5 we see that for $V_d = 0 + 3N$ or for $V_d = 1 + 3N$ (for example, when $N = 0$, i.e. dislocation is present) the static state is the stable singular point. Hence, oscillations are impossible. On the contrary, for $V_d = 2 + 3N$ the static states is unstable singular point. It loses stability (even at small delay $v = 0,06$). It means that in the real optical device subjected by the noise action the static final is unachievable. Therefore, for $V_d = 2 + 3N$ the oscillation motion is inevitable.

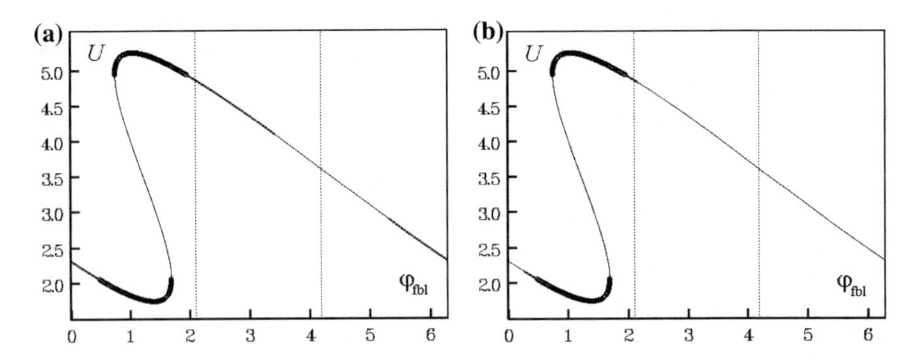

Fig. 5.4 Bifurcation diagrams of the static states in the NRI model (5.2). The diagram is constructed on the plane: phase delay φ_{fb1} in NRI—the nonlinear phase shift U for $\Delta = 120°$, $K = 3.5$, $\gamma = 0.5$, $R = 0.5$. Normalzed delay time $\nu \equiv t_e/\tau_n \approx 0.1$ (**a**), $\nu \approx 1$ (**b**). Solutions are shown by thick lines if they are stable for any v, by lines of the medium thickness if solution are stable for given v, by thin lines if they are unstable at given v

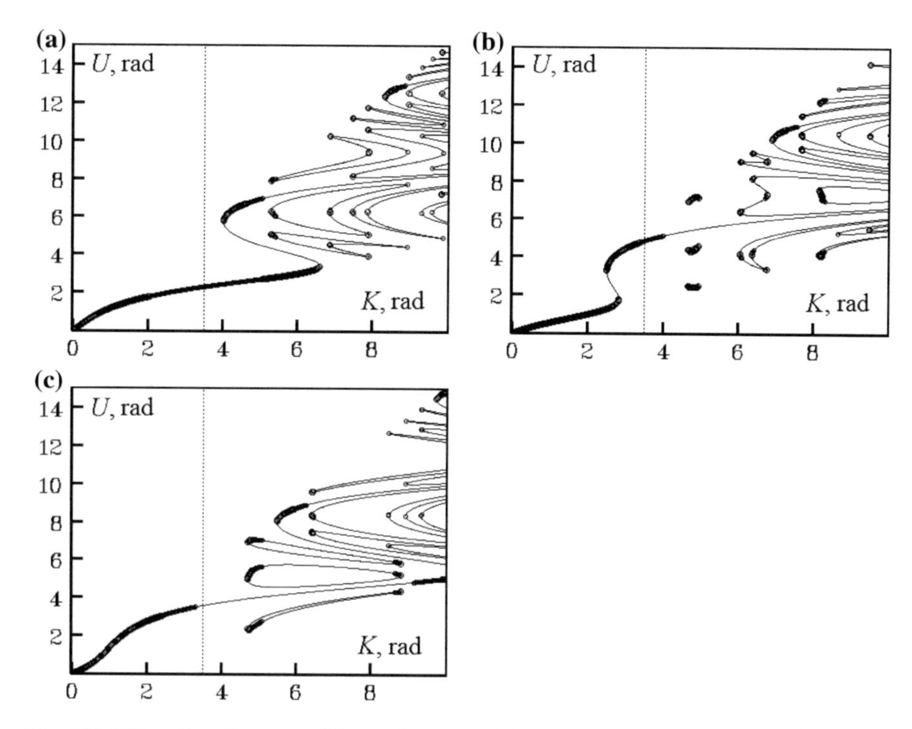

Fig. 5.5 Bifurcation diagrams of the static states of the nonlinear phase shift in the NRI model (5.2) for $m = 3\,\Delta = 2\pi/3$, $\varphi_{fb10} = 2\pi N$, $\nu \equiv t_e/\tau_n \approx 0,1$ and the field on input with screw dislocation period $V_d = 0$ (**a**); $V_d = 1$ (**b**); $V_d = 2$ (**c**). *Thin lines* correspond to unstable states, medium in thickness—to stable at given value of t_e/τ_n, thick—to always stable

Phase portraits of the structure-formation process in NRI created for these two cases are shown in Fig. 5.6. As we see from Fig. 5.6a, for $V_d = 0 + 3N$ the singular point of stable focus type takes place. The similar situation takes place for $V_d = 1 + 3N$. But for $V_d = 2 + 3N$, this is also a focus but unstable (Fig. 5.6b). Establishing (final) dynamic mode is characterized by the limit cycle presented in Fig. 5.6b.

From this it is clear why in Fig. 5.3a, d the optical structure is fixed, but in Fig. 5.3c, f it rotated. Difference in brightness of structures $U(\mathbf{r}, t)$ in Fig. 5.3 can be explained with the help of bifurcation diagrams because the brightness is proportional to U on diagrams.

With delay time growth ($v = 1$), the singular point of the focus type corresponding to vortices with the dislocation order $V_d = 0 + 3N$ loses a stability, and stability (or instability) of other singular points does not change (Fig. 5.4b). That is why, for $V_d = 0 + 3N$ the limit cycle appears, i.e. the Andronov–Hopf bifurcation happens (Fig. 5.7a). With variation V_d to $V_d = 2 + 3N$ or with growth of the delay time v to $v = 1$, the view of the limit cycles is transformed (Fig. 5.7b). This justifies about anharmonicity (about growth of energetic part of higher harmonics) compared to the case of the vortex with $V_d = 0 + 3N$ (when $v = 1$) or to the case

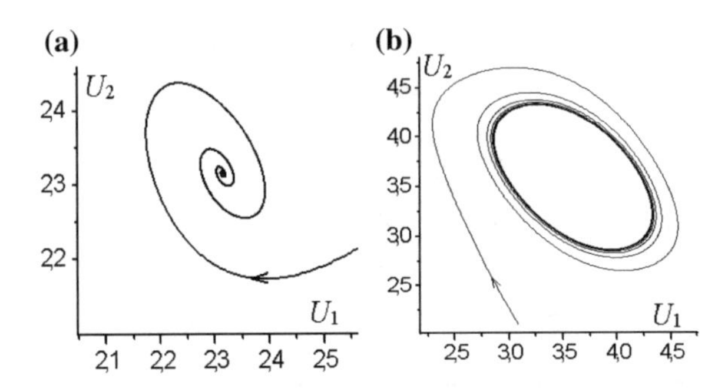

Fig. 5.6 Phase portrait of the structure-formation process in NRI constructed of the plane U_1, U_2. Normalized delay time $v = 0.1$, $\Delta = 120^\circ$, $K = 3.5$, $\gamma = 0.5$, the screw dislocation order is $V_\mathrm{d} = 0 + 3N$ (**a**), $V_\mathrm{d} = 2 + 3N$ (**b**)

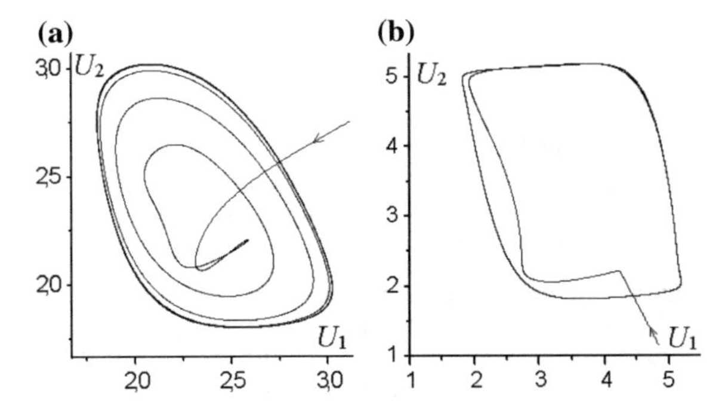

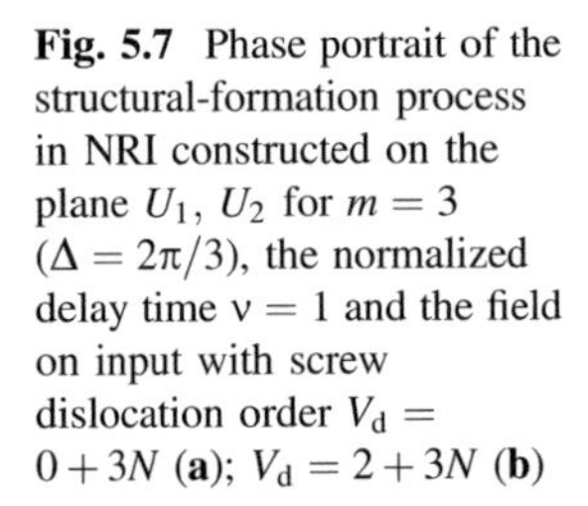

Fig. 5.7 Phase portrait of the structural-formation process in NRI constructed on the plane U_1, U_2 for $m = 3$ ($\Delta = 2\pi/3$), the normalized delay time $\nu = 1$ and the field on input with screw dislocation order $V_\mathrm{d} = 0 + 3N$ (**a**); $V_\mathrm{d} = 2 + 3N$ (**b**)

$v = 0.1$ (when $V_\mathrm{d} = 2 + 3N$) [53]. Apparently, for the larger nonlinearity coefficients K the variation of the dislocation order V_d is able to arise of the doubling-period bifurcation or even transition to the dynamic chaos.

How does the noise influents on a possibility of the vortex recognition with the help of NRI? Let us explain the reasons of a reliability of V_d recognition with the help of NRI. It turn out that a recognition mechanism agrees with the neuro-network paradigm. Really, due to action of the element G (Fig. 5.2), the set of points of the transverse section divides in independent (at neglecting of the field diffraction and molecule diffusion of the nonlinear medium) subsets from m points —the chain of transposition points (CTP). That is, CTP may be treated as some space area, where physical processes happens quasi-independently (for $D_\mathrm{e} = 0$) from processes in the other similar areas. Then CTP is similar to artificial neuron responsible for processing of the spatial fragment of 2D-signal. Such a processing is performed in parallel by all CTP (neurons), and for $D_\mathrm{e} = 0$ they do it independently from each other. "Generation of problem solution" based on measurement of the beam intensity is similar to processing results averaging of all signal fragments by all CTP set.

Described CTP activity can be interpreted as an analog of input neuron signal multiplication by synaptic weights and the nonlinear signal transformation in the neuron. And spatial averaging of the beam intensity—as summation of output neuron signals. An analog of the procedure of "NRI learning" considered by us as adjustment of system parameters to solution of chosen problem, is logical to state the parameter specification: K, Δ, φ_{fb10} etc. Similar algorithm of the device operation realized at the physical level provides the limit high operation speed, and fulfillment at the same physical level averaging—*noise immunity*. We note that mentioned neuro-network theme is agreed with the Renascence of interest to it expressed, for instance, by the project FACETS (Fast Analog Computing with Emergent Transient States (technique of fast analog calculations with smoothly changing transients [58].

At pilot checking of the thesis about noise-immunity of the vortex recognition in the NRI model, we imitate the field phase deviation of the input signals from the "etalon" field with screw dislocation (1.1). As a result, the Figs. 5.8, 5.9, 5.10 are constructed [52].

Let these phase are regular (within CTP) deviations (non-stochastic) and the field rotation angle $\Delta = 120°$. Then the average value of the nonlinear phase shift as a function of the field phase in 2nd and 3rd transposition points, which is counts out with regard to the phase in 1st point, is presented in Fig. 5.8. In interception points of the dotted lines in Fig. 5.8, lying on straight line $\varphi_3 = 2\varphi_2$, values of φ_3, φ_2, correspond to the cases of vortical field with screw dislocation, which is increased as far as moving off from the point (0; 0). A presence of relatively lengthy areas with almost constant value of $\langle U(\mathbf{r}) \rangle_{\mathbf{r}}$ near mentioned interceptions of the dotted lines we should interpret as the *reason of satisfactory recognition* of the order V_d *in the presence of phase distortions* in the input beam.

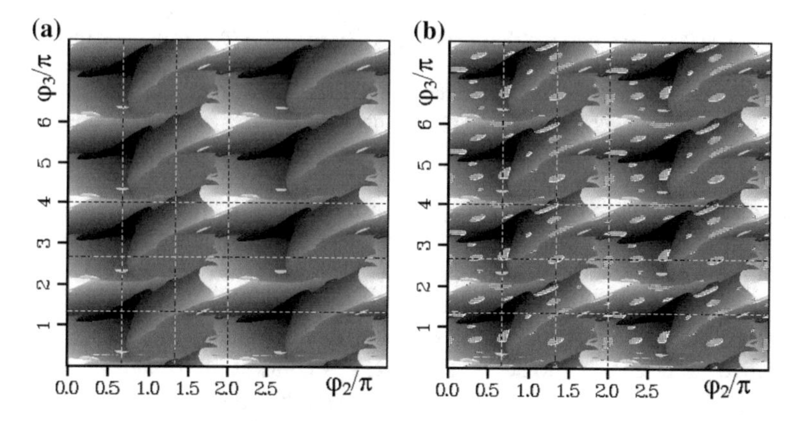

Fig. 5.8 Dependence of average (over transposition points) value of the nonlinear phase shift $U(\mathrm{r})$ in static states *versus* the phase value of the input field in 2nd and 3rd transposition points: $K = 3.5$; $\Delta = 120°$ ($m = 3$); $t_e/\tau_n \approx 0.1$; $\omega t_e = 2\pi N$. Area in Figure (**b**), which differs from areas in Figure (**a**), correspondent to the dynamic mode, i.e. at these values of φ_2 and φ_3 there are not the stable static states $U(\mathrm{r})$

$A_{nS} = 0{,}4\pi$			$A_{nS} = \pi$		
V_d=0	V_d=1	V_d=2	V_d=0	V_d=1	V_d=2

Fig. 5.9 Spatial distribution of the input signal phase, distorted by the phase noise with amplitude A_{nS}

V_d	Exposition duration (T_t) of randomly non-uniform transparency and amplitude (A_{nS}) of the phase noise on the input signal			
	$T_t = 0.1\tau_n$		$T_t = 3\tau_n$	
	$A_{nS} = 0.4\pi$	$A_{nS} = \pi$	$A_{nS} = 0.4\pi$	$A_{nS} = \pi$
0	 $\langle U\rangle_{\mathbf{r}\,\mathrm{Min}} = 2.749$ $\langle U\rangle_{\mathbf{rt}} = 2.757$ $\langle U\rangle_{\mathbf{r}\,\mathrm{Max}} = 2.763$	 $\langle U\rangle_{\mathbf{r}\,\mathrm{Min}} = 3.488$ $\langle U\rangle_{\mathbf{rt}} = 3.500$ $\langle U\rangle_{\mathbf{r}\,\mathrm{Max}} = 3.509$	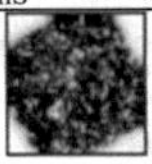 $\langle U\rangle_{\mathbf{r}\,\mathrm{Min}} = 2.778$ $\langle U\rangle_{\mathbf{rt}} = 2.796$ $\langle U\rangle_{\mathbf{r}\,\mathrm{Max}} = 2.816$	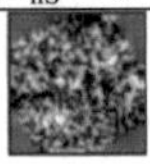 $\langle U\rangle_{\mathbf{r}\,\mathrm{Min}} = 3.486$ $\langle U\rangle_{\mathbf{rt}} = 3.503$ $\langle U\rangle_{\mathbf{r}\,\mathrm{Max}} = 3.520$
1	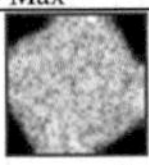 $\langle U\rangle_{\mathbf{r}\,\mathrm{Min}} = 4.303$ $\langle U\rangle_{\mathbf{rt}} = 4.311$ $\langle U\rangle_{\mathbf{r}\,\mathrm{Max}} = 4.318$	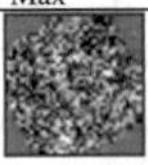 $\langle U\rangle_{\mathbf{r}\,\mathrm{Min}} = 3.488$ $\langle U\rangle_{\mathbf{rt}} = 3.501$ $\langle U\rangle_{\mathbf{r}\,\mathrm{Max}} = 3.514$	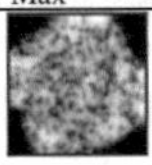 $\langle U\rangle_{\mathbf{r}\,\mathrm{Min}} = 2.778$ $\langle U\rangle_{\mathbf{rt}} = 2.796$ $\langle U\rangle_{\mathbf{r}\,\mathrm{Max}} = 2.816$	 $\langle U\rangle_{\mathbf{r}\,\mathrm{Min}} = 3.493$ $\langle U\rangle_{\mathbf{rt}} = 3.511$ $\langle U\rangle_{\mathbf{r}\,\mathrm{Max}} = 3.543$
2	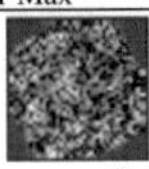 $\langle U\rangle_{\mathbf{r}\,\mathrm{Min}} = 3.561$ $\langle U\rangle_{\mathbf{rt}} = 3.572$ $\langle U\rangle_{\mathbf{r}\,\mathrm{Max}} = 3.584$	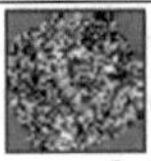 $\langle U\rangle_{\mathbf{r}\,\mathrm{Min}} = 3.488$ $\langle U\rangle_{\mathbf{rt}} = 3.497$ $\langle U\rangle_{\mathbf{r}\,\mathrm{Max}} = 3.512$	 $\langle U\rangle_{\mathbf{r}\,\mathrm{Min}} = 3.562$ $\langle U\rangle_{\mathbf{rt}} = 3.569$ $\langle U\rangle_{\mathbf{r}\,\mathrm{Max}} = 3.579$	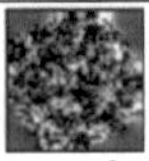 $\langle U\rangle_{\mathbf{r}\,\mathrm{Min}} = 3.487$ $\langle U\rangle_{\mathbf{rt}} = 3.523$ $\langle U\rangle_{\mathbf{r}\,\mathrm{Max}} = 3.548$

Fig. 5.10 Optical structures $U(\mathrm{r}, t)$ obtained as a solution of (5.2) and justifying about a possibility or impossibility (for $A_{nS} = \pi$) of recognition of the screw dislocation order

A presence of these relatively lengthy areas turns by a presence of the transient areas with more sharp variation of $\langle U(\mathbf{r})\rangle_{\mathbf{r}}$. This says about the fact that NRI has *trigger* properties: it amplifies the contrast of its reaction on some external influences. And vice versa: NRI is capable the decrease this contrast for other influence. Therefore, NRI fulfill functions of decision device and in this sense, the verb "recognize" is true. So, if to see in Fig. 5.4, this contrast of U values are large for $V_d = 0 + 3N$ and $V_d = 1 + 3N$. On the contrary, differences U for $V_d = 2 + 3N$ and, for example, $V_d = 0 + 3N$ is approximately two times less.

Speaking more exact, each CTP (similarity of the artificial neuron) has the trigger properties. And NRI as a whole (similarity neuro-network) together with measuring system at the interferometer output inherent this property. But, besides this inheritance, due to *averaging* of solutions of separate riggers, as we already mentioned, NRI acquires *additional immunity* against influence of input beam distortions on the vortex recognition.

Now we assume that this phase distortion of the input field is caused by *stochastic property* of the optical density (for example, accidently non-uniform transparency), through which the field propagates, which arrives on NRI input. In the similar context, we usually speak about the phase (Fig. 5.9). Imitation modeling shows (Fig. 5.10), that, for instance, for $T_t = 0.1\tau_n$ and $A_{nS} = 0.4\pi$, if $V_d = 0$, then $\langle U \rangle_{\mathbf{r}} \in [2.749;\ 2.763]$, $\langle U \rangle_{\mathbf{r\,t}} = 2.757$; if $V_d = 1$, then $\langle U \rangle_{\mathbf{r}} \in [4.303;\ 4.318]$, $\langle U \rangle_{\mathbf{r\,t}} = 4.311$; if $V_d = 2$, then $\langle U \rangle_{\mathbf{r}} \in [3.561;\ 3.584]$, $\langle U \rangle_{\mathbf{r\,t}} = 3.572$. Mentioned values are close to appropriate values of $\langle U \rangle_{\mathbf{r}}$: 2.467; 4.675; 3.591 (Fig. 5.3), when there is no a phase noise ($A_n = 0$). This fact tells about a possibility to recognize on order of the screw dislocation with the help of NRI. If $T_t = 0.1\tau_n$ and $A_n = \pi$, then such closeness of values значений $\langle U \rangle_{\mathbf{r}}$ does not observe, moreover, values $\langle U \rangle_{\mathbf{r}}$ for different V_d are practically undistinguishable (Fig. 5.10). This tells about impossibility of order recognition of the screw dislocation for so much phase noise [52].

Let us repeat that screw dislocation order can be recognized by analyzing the average nonlinear phase shift $\langle U(\mathbf{r}) \rangle_{\mathbf{r}}$ (or averaged beam intensity), when the level of additive "white" noise is small enough ($A_{nS} = 0.4\pi$), but this is impossible if it closed to the value $A_{nS} = \pi$ [52]. But the last is not surprise because the condition $A_{nS} = \pi$ means that the noise completely disturbs the coherence of the light beam on NRI input.

So, using NRI, we can:

- *identify* the order of screw dislocation of the optical vortex according to structure view $U(\mathbf{r}, t)$, $A^2_{in.NL}(\mathbf{r}, t)$ in the transverse beam section and for values $\langle U \rangle_{\mathbf{r\,t}}$, $\langle A^2 \rangle_{\mathbf{r\,t}}$. This method of identification presents the interest to improve experimental approaches of the singular optics [18];
- selecting the value L, satisfying the condition (5.3) at given N_s and V_d, *compensate* the vortex influence on the process and result of structural-formation in the interferometer. The similar compensation is relevant in the context of development of correctors of the wave front [41, pp. 70–71], which is fulfilled in the atmospheric adaptive optics;
- owing to the expression (5.3) create the *arithmetic-logic device* realizing the addition operation modulo m.

After this, we shall suggest the some simpler detector of the screw dislocation order based on application of the *linear* interferometer.

The next section is written at active participation of Valerii P. Aksenov and Feodor Yu. Kanev—researchers from V.E. Zuev Institute of Atmospheric Optics, Russian Academy of Sciences, Siberian Branch. They are coauthors of papers [59–62], which materials are used below. Using this case, authors express our thanks to V.P. Aksenov and F.Yu. Kanev.

5.3 Rozhdestvenskiy's Interferometer as a Vortex Detector

5.3.1 *A Principle and Description of Vortex Detection with the Help of Rozhdestvenskiy's Interferometer at Noise Presence*

The foregoing suggests [2, 59–61] that the modified Rozhdestvenskiy's interferometer with a unit for light beam rotation about the longitudinal axis through the angle $\Delta = 2\pi M/m$ (element G) and a unit for constant phase shift φ connected into one of its arms can be used as a vortex charge detector (Fig. 5.11). We remind that Rozhdestvenskiy's intereferometer (Fig. 5.11a) is similar to Mach–Zehnder interferometer. Offered modification (Fig. 5.11b) of mentioned interferometers simplifies alignment of the optical lengths of interferometer arms. Let us open the operation principle of the device in Fig. 5.11 as a detector of vortex charge relying on publications [59–62].

The field $\mathbf{E}_{\text{in}}(\mathbf{r},\ t)$ incident on the interferometer has two components. Propagating through different arms, these components undergo different attenuation values and different diffraction transformations and acquire different phase shifts. The fields are superimposed at the interferometer output. Let the reflection coefficients of the mirrors and the energy losses in elements G and φ be such that values of the optical field attenuation in both arms will be identical. Let the optical lengths of these arms differ slightly, that is, by several orders of magnitude less than the diffraction length of the incident field. Due to this, the phase shifts of the field in the interferometer arms will differ only by $\varphi \equiv \omega\delta t$ (Fig. 5.11a or b), where ω is the circular light field frequency ($T = 2\pi/\omega$), and δt is the time delay of the field due to a larger optical length of the interferometer arm with elements G and φ compared to the second arm.

Let us designate by $\mathbf{E}(\mathbf{r},\ t)$ the field into which the incident field $\mathbf{E}_{\text{in}}(\mathbf{r},\ t)$ is transformed after passage of the distance equal to the optical length of the interferometer arm without elements G and φ, being attenuated in one of the interferometer arms. That is, $\mathbf{E}(\mathbf{r},\ t)$ is the output field of the empty Rozhdestvenskiy's interferometer with equal arms whose optical lengths are equal to the initial ones. In other

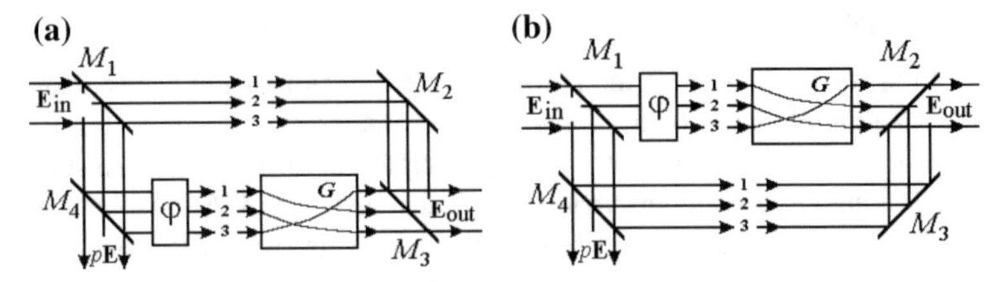

Fig. 5.11 Rozhdestvenskiy's interferometer (**a**) and its modification (**b**) as a detector of vortices. G is the rotation device of the light beam by the angle $\Delta = 2\pi M/m$ and φ is the phase shift. The trajectories of rays *1*, *2*, and *3* are shown for the angle $\Delta = 120°$

words, $\mathbf{E}(\mathbf{r},\ t)$ is the incident field of the interferometer $\mathbf{E}_{\rm in}(\mathbf{r},\ t)$ reduced to its output disregarding its splitting into two components and their subsequent interference.

Without loss of generality, we can consider below the field $\mathbf{E}(\mathbf{r},\ t)$ instead of $\mathbf{E}_{\rm in}(\mathbf{r},\ t)$. However, to generalize conclusions, characteristics, and properties of the detector to actual optical systems, the lengths of the light ray propagation paths must be considered with allowance for the optical length of each interferometer arm. Based on the foregoing, the field $\mathbf{E}(\mathbf{r},\ t)$ is also said to be the incident field. Information on the $\mathbf{E}(\mathbf{r},\ t)$ characteristics is retrieved from its component $p\mathbf{E}(\mathbf{r},\ t)$ at the exit from mirror M_4 (Fig. 5.11a or b). Then the output field of the interferometer (the interference field) $\mathbf{E}_{\rm out}(\mathbf{r},\ t) \equiv \mathbf{E}_{\Sigma}(\mathbf{r},\ t)$ is described by the expression

$$\mathbf{E}_{\Sigma}(\mathbf{r},\ t) = (0.5^{1/2}[\mathbf{E}(\mathbf{r},\ t) + \mathbf{E}(\mathbf{r}',\ t - \delta t)]).$$

Here $\mathbf{r}' \equiv (x',\ y')$ specifies the point in the laser beam cross section at the input of element G from which the ray comes to the point $\mathbf{r} \equiv (x,\ y)$ at the exit of element G (for a plane incident wave). Let we have

$$\mathbf{E}(\mathbf{r},\ t) = \mathbf{e}_x A(\mathbf{r}) \cos(\omega t + S(\mathbf{r})), \tag{5.5}$$

where $A(\mathbf{r})$ and $S(\mathbf{r})$ are the light field amplitude and phase, and $\mathbf{e}_x$ is the unit vector of the Ox axis (the polarization vector). Then the field at the exit from the interferometer obeys the expression

$$\begin{aligned}\mathbf{E}_{\Sigma}(\mathbf{r},\ t) &= \mathbf{e}_x(0.5^{1/2}\{A(\mathbf{r})\cos[\omega t + S(\mathbf{r})] + A(\mathbf{r}')\cos[\omega t + S(\mathbf{r}') - \varphi]\}) \\ &\equiv A_{\Sigma}(\mathbf{r})\cos[\omega t + S_{\Sigma}(\mathbf{r})].\end{aligned}$$

Since the electromagnetic wave intensity in vacuum is given by the expression $(\varepsilon_0/\mu_0)^{0.5}\langle\mathbf{E}^2(\mathbf{r},t)\rangle_{\rm T}$, where ε_0 and μ_0 are the electric and magnetic constants in the SI units of measurement, the intensity distribution at the detector input, $I(\mathbf{r})$, and output, $I_{\Sigma}(\mathbf{r})$, can be written (disregarding the multiplier $0.5(\varepsilon_0/\mu_0)^{1/2}$) as follows:

$$\begin{aligned}I(\mathbf{r}) &= 2\langle\mathbf{E}^2(\mathbf{r},t)\rangle_{\rm T} = A^2(\mathbf{r}), \\ I_{\Sigma}(\mathbf{r}) &= 2\langle\mathbf{E}_{\Sigma}^2(\mathbf{r})\rangle_{\rm T} \\ &= 0.5I(\mathbf{r}) + 0.5I(\mathbf{r}') + [I(\mathbf{r})I(\mathbf{r}')]^{0.5}\cos[S(\mathbf{r}) - S(\mathbf{r}') + \varphi] = A_{\Sigma}^2(\mathbf{r}),\end{aligned}$$

where $\langle\mathbf{E}^2(\mathbf{r},\ t)\rangle_{\rm T} \equiv \frac{1}{T}\int_t^{t+T}\mathbf{E}^2(\mathbf{r},\ t')dt'$ and the multiplier $0.5(\varepsilon_0/\mu_0)^{1/2}$ is omitted hereinafter for compact representation. The field intensity averaged over the beam cross section (that is, the beam intensity) at the input of the detector, $I \equiv \langle I(\mathbf{r})\rangle_{\mathbf{r}}$, and at the exit from the detector, $I_{\Sigma} \equiv \langle I_{\Sigma}(\mathbf{r})\rangle_{\mathbf{r}}$, are

$$I \equiv \langle A^2(\mathbf{r}) \rangle_{\mathbf{r}},\ I_\Sigma \equiv I + \left\langle [I(\mathbf{r})I(\mathbf{r}')]^{0.5} \cos[S(\mathbf{r}) - S(\mathbf{r}') + \varphi] \right\rangle_{\mathbf{r}} = \langle A_\Sigma^2(\mathbf{r}) \rangle_{\mathbf{r}},$$

where $\langle I(\mathbf{r},\ t) \rangle_{\mathbf{r}} \equiv \frac{1}{S_b} \iint_{(S_b)} I(\mathbf{r},\ t) d\mathbf{r}$, (S_b) is the cross sectional area S_b, occupied by the beam.

Let us consider the relative output beam intensity of the interference field

$$I_{\rm r} \equiv I_\Sigma / I = 1 + I^{-1} \left\langle [I(\mathbf{r})I(\mathbf{r}')]^{1/2} \cos[S(\mathbf{r}) - S(\mathbf{r}') + \varphi] \right\rangle_{\mathbf{r}}. \tag{5.6}$$

It is obvious that $I_{\rm r}$ is also equal to the ratio of light beam powers at the exit from and input to the interferometer. Hence, $I_{\rm r}$ can be easily determined in the field experiment.

Let us consider an ideal vortex field $\mathbf{E}(\mathbf{r},\ t)$ at the detector input comprising only one screw wave-front dislocation (at the origin of coordinates in the beam cross section). In this case, the beam amplitude, $A(\mathbf{r})$, and phase, $S(\mathbf{r})$, entering into Eq. (5.5) assume the form

$$A(\mathbf{r}) = A(r),\ S(\mathbf{r}) = S_r(r) + V_{\rm d}\vartheta(\mathbf{r}), \tag{5.7}$$

where $\mathbf{r} \equiv (x,\ y)$ is the radius-vector of the point in the beam cross section, $r \equiv |\mathbf{r}|$ is the distance from the screw dislocation (SD), $V_{\rm d}$ is the SD order, and $\vartheta(\mathbf{r}) \equiv \arg(\mathbf{r})$ is the azimuth angle. In particular, $A(\mathbf{r})$ and $S(\mathbf{r})$ can have the form

$$A(\mathbf{r}) = C \exp(-r^2/r_0^2)(r/\rho_{\rm V})^{|V_{\rm d}|},\ S(\mathbf{r}) = V_{\rm d}\vartheta(\mathbf{r}), \tag{5.8}$$

where r_0 is the Gaussian beam radius for $V_{\rm d} = 0$ and $\rho_{\rm V}$ is the phenomenologically introduced (unlike (1.1)) parameter that regulates the slope of the radial amplitude dependence. Since $\mathbf{r}$ and $\mathbf{r}'$ are related by the transformation of rotation through the angle $\Delta = 2\pi M/m$, then $r = r'$ and $\vartheta(\mathbf{r}) = \vartheta(\mathbf{r}') + M\Delta$, and after substitution of (5.7) into (5.6), we obtain

$$I_{\rm r}(V_{\rm d}) \equiv I_\Sigma/I = 1 + \cos[2\pi M V_{\rm d}/m + \varphi]. \tag{5.9}$$

Let us indicate consequences following from (5.9).

1. When the charge $V_{\rm d}$ changes, $I_{\rm r}$ takes a discrete number of values, and $I_{\rm r}(V_{\rm d}) = I_{\rm r}(V_{\rm d} + im)$, where i is an integer.
2. The influence of the vortex of any arbitrary order on the $I_{\rm r}$ value can be compensated by changing the φ value. By analogy with (5.4), it is convenient to express φ in the form $\varphi = \varphi_0 + 2\pi(LM)/m$, where L, M, and m are integers and φ_0 is the φ value for $L = 0$.
3. The $I_{\rm r}$ value is independent of the concrete form of $A(r)$ and $S_r(r)$ in (5.7). Moreover, (5.9) is valid in a more general case of rotational symmetry of the mth order set by element G when for any arbitrary $\mathbf{r}$, $A(\mathbf{r}) = A(\mathbf{r}')$ and

$S(\mathbf{r}) - S(\mathbf{r}') = \text{const} + 2\pi i$. That is, when $A(\mathbf{r})$ and $S(\mathbf{r})$ coincide in sectors of the cross sectional plane bounded by different rays with angles $\vartheta = \Delta i$.

Consequences 1 and 2 are strictly described by (5.3) and (5.4). Here the former principle (5.3) of structure numbering remains valid, but now it can also be interpreted as a rule for numbering the I_{r} values (N_s is the serial number of the structure to which the $I_{\mathrm{r}.Ns}$ value corresponds).

It is obvious that if a remote source produces field with symmetry discussed in item 3, this symmetry is not broken when the field propagates in vacuum. Hence, it is unimportant at which distance from the sensor the source of field (5.7) is located.

Let us prove a more general statement about the role of the path length, namely, that I_{r} is independent of the distance between the field source and the detector located in vacuum or in any linear, isotropic, and absorbing medium uniform in the transverse cross section.

Indeed, the diffraction field in this medium is described by a linear operator L_D. The field transformation in the detector (beam splitting L_{Spl} into three beams, rotation L_G of one of the beams in element G, and summation L_Σ of two beams) is also linear (except the operation of intensity measurement). Let us permute these operators. It is obvious that due to linearity of the operator L_D, the result of diffraction of the sum of fields will be equal to the sum of the diffracted fields. This means that light splitting mirrors M_1 and M_4 of the interferometer can be shifted directly to the field source, that is, $L_{Spl}L_D\mathbf{E} \equiv L_D\mathbf{E}_1 + L_D\mathbf{E}_2 + L_D\mathbf{E}_3 = L_D(\mathbf{E}_1 + \mathbf{E}_2 + \mathbf{E}_3) \equiv L_D L_{Spl}\mathbf{E}$ (the operators L_{Spl} and L_D are commutative). Analogously, the order of application of operators L_D and L_G can be changed: $L_G L_D \mathbf{E} = L_D L_G \mathbf{E}$, since by virtue of the isotropic space, the result of field diffraction and its subsequent rotation is indistinguishable from the result of field rotation and subsequent diffraction (of course, for the same distances). Physically, this means that element G can be displaced directly to the field source. Taking again the advantage of L_D linearity from which it follows that the result of diffraction of two fields with their subsequent summation is equal to diffraction of the sum of initial fields, we conclude that L_D and L_Σ are commutative: $L_\Sigma(L_D\mathbf{E}_1, L_D\mathbf{E}_2) = L_D\mathbf{E}_1 + L_D\mathbf{E}_2 = L_D(\mathbf{E}_1 + \mathbf{E}_2) = L_D L_\Sigma(\mathbf{E}_1,\ \mathbf{E}_2)$. Hence, mirrors M_2 and M_4 can be displaced directly to the filed source.

As a result of these procedures, we placed the interferometer directly behind the field source and left the intensity meters (for $\mathbf{E}_\Sigma$ and $p\mathbf{E}$) at the receiver. In this case, two light beams: $\mathbf{E}_\Sigma$ at the exit from the interferometer and $p\mathbf{E}$—the portion of the field deflected by M_4—are transmitted from the source to the receiver. It is obvious that the ratio of powers (or intensities) of these beams is independent of the lengths (the same) of the paths they passed (in the same medium). Hence, $I_{\mathrm{r}} \equiv I_\Sigma / I$ is independent of the path length.

Now we consider that noise is always present in actual systems and study a non-ideal case. Let the phase (additive white) and amplitude (multiplicative white) noise components with amplitudes A_{nS} and A_{nA} are superimposed on the field described by (5.7):

$$A(\mathbf{r}) = A(r)[1 + A_{nA}\xi_A(\mathbf{r})],\ S(\mathbf{r}) = S_r(r) + V_d\vartheta(\mathbf{r}) + A_{nS}\xi_S(\mathbf{r}), \tag{5.10}$$

where $\xi_A(\mathbf{r}) \in [-1,\ 1]$ and $\xi_S(\mathbf{r}) \in [-1,\ 1]$ are random functions of $\mathbf{r}$ independent of each other and uniformly distributed in the interval $[-1,\ 1]$ such that $\langle\xi_A(\mathbf{r})\rangle_\mathbf{r} = 0$, $\langle\xi_S(\mathbf{r})\rangle_\mathbf{r} = 0$, $\langle\xi_A(\mathbf{r})\xi_A(\mathbf{r}'(\mathbf{r}))\rangle_\mathbf{r} = 0$, and $\langle\xi_S(\mathbf{r})\xi_S(\mathbf{r}'(\mathbf{r}))\rangle_\mathbf{r} = 0$ for $\mathbf{r}'(\mathbf{r}) \neq \mathbf{r}$. Moreover, these two properties remain valid for any subset of points of the transverse cross section of non-finite power (the subset with nonzero dimensions). Physically, this means that the spatial scale of noise change is much smaller than the characteristic dimensions of the subsets of interest to us, for example, arcs of the circle with lengths $2\pi r_c/m$, where r_c is the circle radius. After substitution of (5.10) into (5.6), we obtain

$$\begin{aligned} I_\mathrm{r} = 1 + I^{-1}\Big\langle A(r)^2[1 + A_{nA}\xi_A(\mathbf{r})][1 + A_{nA}\xi_A(\mathbf{r}\prime)]\cos[2\pi V_d M/m + \varphi_0 \\ + A_{nS}[\xi_S(\mathbf{r}) - \xi_S(\mathbf{r}\prime)]]\Big\rangle_\mathbf{r} \end{aligned} \tag{5.11}$$

Let us consider two special cases: $A_{nA} \neq 0$, $A_{nS} \neq 0$ and $A_{nA} = 0$, $A_{nS} \neq 0$. In the first case, from (5.11) we obtain

$$I_\mathrm{r} = 1 + I^{-1}\Big\langle A(r)^2(1 + A_{nA}\xi_A(\mathbf{r}))(1 + A_{nA}\xi_A(\mathbf{r}'))\Big\rangle_\mathbf{r}\cos[2\pi V_d M/m + \varphi]. \tag{5.12}$$

To transform (5.12), we suggest the following sequence of actions:

(a) Transition from the integral (averaging $\langle\rangle_\mathbf{r}$) in the Cartesian coordinates $(x,\ y)$ to the integral in the polar coordinates: $\langle\rangle_\mathbf{r} = \langle\langle r\rangle_\vartheta\rangle_r = (1/S_b)\int_0^{r_{\max}}\int_0^{2\pi} r\mathrm{d}r\mathrm{d}\vartheta$.

(b) Calculation (where possible) of the integral over ϑ using the properties $\langle\xi_A(\mathbf{r})\rangle_\mathbf{r} = 0$, $\langle\xi_A(\mathbf{r})\xi_A(\mathbf{r}'(\mathbf{r}))\rangle_\mathbf{r} = 0$ for $\mathbf{r}'(\mathbf{r}) \neq \mathbf{r}$ valid for any subset of points of the transverse cross section of non-finite power. When calculation is impossible, we act differently. Since the radius-vector $\mathbf{r}$, that is, r and ϑ, plays the role of time, under assumption of ergodicity, we replace averaging over ϑ by ensemble averaging. For example,

$$\left\langle\xi_A(\mathbf{r})^2\right\rangle_\vartheta = \int_{-\infty}^{\infty}\rho\{\xi_A(\mathbf{r})\}\xi_A(\mathbf{r})^2\mathrm{d}\xi_A(\mathbf{r}),$$

where $\rho\{\xi_A(\mathbf{r})\} = 1/2$ for $\xi_A(\mathbf{r}) \in [-1,\ 1]$ and $\rho\{\xi_A(\mathbf{r})\} = 0$ for $\xi_A(\mathbf{r}) \notin [-1,\ 1]$ are the probability densities (distributions of values) for the random process $\xi_A(\mathbf{r})$.

(c) Returning to the Cartesian coordinates, we obtain, for example:

$$\left\langle rA(r)^2\right\rangle_r = \left\langle rA(r)^2\langle I\rangle_\vartheta\right\rangle_r = \left\langle r\left\langle A(r)^2\right\rangle_\vartheta\right\rangle_r = \left\langle A(r)^2\right\rangle_\mathbf{r} = I.$$

Then from (5.12) we derive the compact expression:

$$I_r = 1 + \{1 + A_{nA}^2/3\}^{-1} \cdot \cos[2\pi V_d M/m + \varphi]. \quad (5.13)$$

By its meaning, $A_{nA} \leq 1$. Substituting $A_{nA} = 1$, we obtain the minimum value of the coefficient of the cosine function in (5.13). It is $\{1 + 1/3\}^{-1} = \{4/3\}^{-1} = 3/4 = 0.75$. The relative proximity of this value to unity provides a weak influence of the multiplicative white amplitude noise on I_r, but only in the absence of the phase noise (phase distortions). As a consequence, the possibility of faultless determination of the V_d value arises due to different values of I_r for different V_d. This statement is illustrated below when we describe results of computer experiments.

Let we have now $A_{nA} = 0$ and $A_{nA} \neq 0$. From (5.11) it follows that

$$I_r = 1 + I^{-1} \cdot \left\langle A(r)^2 \cos[2\pi V_d M/m + \varphi + A_{nS}[\xi_S(\mathbf{r}) - \xi_S(\mathbf{r\prime})]]\right\rangle_{\mathbf{r}}. \quad (5.14)$$

Interpreting (5.14) as a function of two independent random variables and acting by analogy with transition from (5.12) to (5.13), we obtain

$$I_r = 1 + [\sin^2(A_{nS})/A_{nS}^2] \cdot \cos[2\pi V_d M/m + \varphi]. \quad (5.15)$$

From (5.15) it follows that the amplitude of the phase noise reduces differences between I_r values for different V_d according to the law $[\sin(A_{nS})/A_{nS}]^2$, thereby complicating V_d recognition. Thus, for $A_{nS} = \pi$, the screw dislocation order cannot be identified.

Within the framework of the above simplifications, we can suggest that the more general relationship

$$I_r = 1 + \{1 + A_{nA}^2/3\}^{-1} \cdot [\sin(A_{nS})/A_{nS}]^2 \cdot \cos[2\pi V_d M/m + \varphi] \quad (5.16)$$

is valid.

Let us pay attention to one more I_r property. Unlike the case described Eq. (5.7), we consider that the phase $S(\mathbf{r})$ of the field at the detector input has no central symmetry (the term $S_r(\mathbf{r})$ is replaced by $S_{SC}(\mathbf{r})$, and in the general case $A(\mathbf{r}) \neq A(r)$:

$$E(\mathbf{r},\ t) = A(\mathbf{r}) \cos(\omega t + S(\mathbf{r})),\ S(\mathbf{r}) = S_{SC}(\mathbf{r}) + V_d \vartheta(\mathbf{r}). \quad (5.17)$$

Substitution of (5.17) into (5.6) yields

$$I_r = 1 + I^{-1} \cdot \langle [I(\mathbf{r}) I(\mathbf{r}')]^{1/2} \cos[2\pi V_d \cdot M/m + \varphi + S_{SC}(\mathbf{r}) - S_{SC}(\mathbf{r}')]\rangle_{\mathbf{r}}. \quad (5.18)$$

Let us assume that there are two light fields (5.17) that differ only by V_{d} values, and their difference is

$$V_{\mathrm{d}}'' - V_{\mathrm{d}}' \equiv \delta V_{\mathrm{d}} = m(n+0.5)/M, \text{ where } n \text{ is an integer.} \tag{5.19}$$

Substitution of (5.19) into (5.18) yields

$$\begin{gathered}(I_{\mathrm{r}}' + I_{\mathrm{r}}'')/2 = 1 \\ \text{or } [I_{\mathrm{r}}(V_{\mathrm{d}}) + I_{\mathrm{r}}(V_{\mathrm{d}} + \delta V_{\mathrm{d}})/2 = 1,\ \delta V_{\mathrm{d}} = m(n+0.5)/M.\end{gathered} \tag{5.20}$$

Here n is an integer, because it is assumed that m and M are mutually simple, and the number δV_{d} is an integer. Indeed, δV_{d} is an integer only if m/M is even, that is, if $M = 1$ and m is any arbitrary even number. We note that though it seems paradoxical, we can abandon the requirement that the increment δV_{d} and hence the dislocation order V_{d} are integers. Formula (5.20) means symmetric arrangement of I_{r}' and I_{r}'' values (corresponding to fields with charges V_{d}' and V_{d}'' that differ by δV_{d}) about the straight line $I_{\mathrm{r}} = 1$. For example, for $m = 2$ ($\Delta = \pi$ and $M = 1$), $\delta V_{\mathrm{d}} = 2n + 1$, that is, the I_{r} values, corresponding to even and odd V_{d} values, are symmetric about $I_{\mathrm{r}} = 1$.

From (5.18) it leads with evidently that consequence (1), (2) from expression (5.9) are true and the field (5.17), which is more general than the "ideal" field (5.7), i.e. the principle of structure numeration (5.3) and (5.4) is true. In particular,

$$\text{if } V_{\mathrm{d}}'' - V_{\mathrm{d}}' \equiv \delta V_{\mathrm{d}} = n \cdot m, \text{ then } I_{\mathrm{r}}' = I_{\mathrm{r}}'', \tag{5.21}$$

where n is any number. Let us note that consequence (3) from expression (5.9) is wrong for (5.17).

Finishing this Section, we note that a square of the second term in expression for $I_{\mathrm{r}} \equiv I_{\Sigma}/I$ rather reminds the Shtrell parameter Sh, if we calculate it as

$$Sh(U(\mathbf{r}),\ U_0^*(\mathbf{r})) = \left|\langle U(\mathbf{r})U_0^*(\mathbf{r})\rangle_{\mathbf{r}}\right|^2/(\langle U_0(\mathbf{r})U_0^*(\mathbf{r})\rangle_{\mathbf{r}}\langle U(\mathbf{r})U^*(\mathbf{r})\rangle_{\mathbf{r}}).$$

where $U(\mathbf{r})$, $U_0^*(\mathbf{r})$ are complex amplitudes of two light fields.

5.3.2 Simulation of Rozhdestvenskiy's Interferometer Operation as a Vortex Detector and Its Characteristics Analysis at Presence of the White (Phase and Amplitude) Noise

To check the above conclusions, mainly conclusion (5.16), to visualize them, and to complement the analytical calculations, we constructed a numerical model of the vortex charge detector with phase (additive white) and amplitude (multiplicative white) noise components having amplitudes A_{nS} and A_{nA}, respectively, imposed on field (5.8) for $\rho_V = 8r_0$. In the detector model, the field $\mathbf{E}_\Sigma(\mathbf{r}, t) = (0.5^{1/2}[\mathbf{E}(\mathbf{r}, t) + \mathbf{E}(\mathbf{r}', t - \delta t)])$ is calculated from formulas (5.8) and (5.10) using a pseudo-random number generator (built in the programming language Pascal) and then the intensities $I_\Sigma \equiv \langle I_\Sigma(\mathbf{r}, t)\rangle_\mathbf{r}$, $I \equiv \langle I(\mathbf{r}, t)\rangle_\mathbf{r}$, and $I_r \equiv I_\Sigma/I$ are calculated. Hereinafter, the parameter M of the rotation angle Δ is set equal to unity, and the length of the propagation path is equal to zero.

Let us check the efficiency of the vortex detector model in the noise presence. For visual control of recognition process, we systemize distributions: the phase $S(\mathbf{r})$ and amplitude $A(\mathbf{r})$ of distorted input signal, intensity $A_\Sigma(\mathbf{r})^2$ of the interference field in the detector, as well value of relative intensity of interference field $I_r \equiv I_\Sigma/I = \langle A_\Sigma^2\rangle_\mathbf{r}/\langle A^2\rangle_\mathbf{r}$ of one realization, which can be grouped in a view in the Fig. 5.12 [2].

At a cell size of the computing network $(r_0/8)$, which is chosen by us, the error of intensity calculation is the value of the order 6 %. In spite of this, the calculation error of I_r for field rotation by angle 180° is absent, but for $\Delta = 120°$ it has values 0.001 (0.6 %), 0.0034 (0.2 %), 0.006 (0.6 %) для $V_d = -1$, $V_d = -2$, $V_d = -3$, relatively. By whom we can explain the fundamental difference of error values for the case $\Delta = 180°$ and $\Delta = 120°$? Apparently, it can be explained by the fact that error of I_r is caused only by the error of the field interpolation (presented by the array of values of intensity and phase in the square grid) at its rotation by 120° In its turn, the interpolation error is a consequence of inevitable spatial non-coincidence of node structure of the square grid of initial and rotated fields. Spatial limited nature of the grid, excluding interference in its corners: peripheral areas, where approximately $r > 2r_0$, introduces its contribution. At rotation by the angle $\Delta = 180°$, mentioned factors are absent in principle, hence, the error as well.

Comparison of values of I_r ratio calculated according to *single* realization presented (Fig. 5.12) leads to the conclusion about efficiency of the vortex detector. Really, at noise absence, values of I_r for this or that topological charges V_d adequately differs in order to identify them with the help of beam intensity measurement. With noise appearance, difference in I_r values naturally decreases. Differences can be easily distinguished for $A_{nS} = 0.5\pi$, $A_{nA} = 0.5$. We can expect that for $A_{nS} = \pi$, $A_{nS} = 0.5$ identification of the V_d value is impracticable: for all V_d we obtain the value $I_r \approx 1$. We remind that the same situation is typical for the natural light [2].

To obtain more reliable estimates, a greater number of realizations were analyzed for different noise levels. Results of analysis of 150 realizations (for each set of three values A_{nS}, A_{nA}, and V_d) in the form of the probability density (distribution) of the relative intensity I_r are shown in Figs. 5.13 and 5.14. The parameters of numerical experiments illustrated by Figs. 5.13 and 5.14 differ by the field rotation angles Δ and shifts of the phase φ. The φ value was chosen so that to provide a maximum difference between $I_r(V_d)$ values for the given angle Δ; thus, $\varphi = 0$ for $\Delta = 180°$ and $\varphi = -90°$ for $\Delta = 120°$. The general tendencies established in [2] are confirmed.

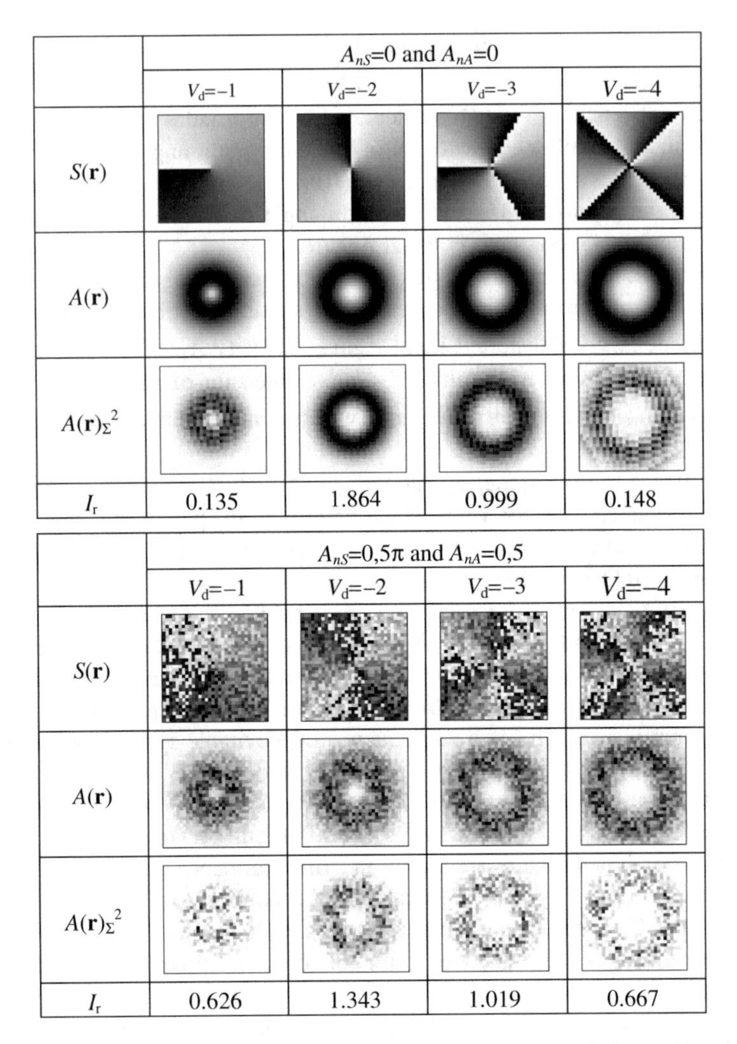

Fig. 5.12 Distributions: of phase $S(\mathrm{r})$, amplitude $A(\mathrm{r})$ of the input signal distorted by phase and amplitude noises with amplitudes A_{nS}, A_{nA}, and the intensity $A(\mathrm{r})^2_\Sigma$ of the interference field in the vortex detector as well as the ratio $\langle \xi_S(\mathbf{r})\xi_S(\mathbf{r}'(\mathbf{r}))\rangle_{\mathbf{r}} = 0$—for $\Delta = 120°$, $\varphi_0 = \pi/2$

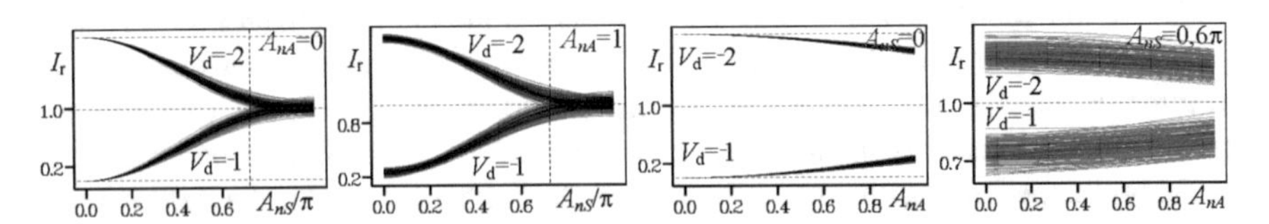

Fig. 5.13 Influence of the phase and amplitude noise components with amplitudes A_{nS} and A_{nA} on the probability density of the relative interference field intensity I_r (on the possibility of vortex recognition) for $\Delta = 180°$ and $\varphi = 0$. The *vertical dotted line* is for $A_{nS} = 0.72\pi$, and the *horizontal dotted lines* are for I_r values from the series 0, 1, and 2, respectively

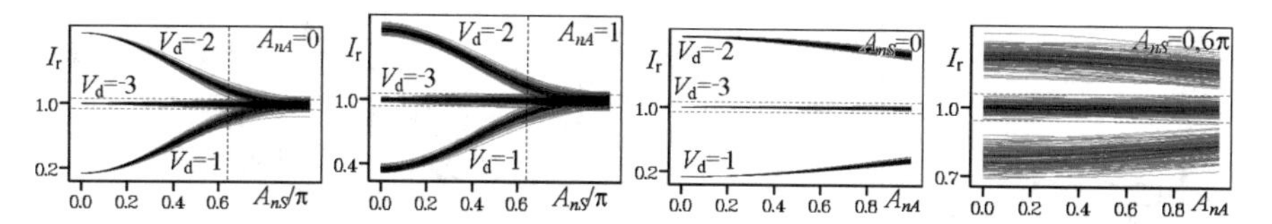

Fig. 5.14 Influence of the phase and amplitude noise components with amplitudes A_{nS} and A_{nA} on the probability density of the relative interference field intensity $I_r \equiv \langle A_\Sigma^2 \rangle_r / \langle A^2 \rangle$ (on the possibility of vortex recognition) for $\Delta = 120°$ and $\varphi = \pi/2$. The *vertical dotted line* is for $A_{nS} = 0.64\pi$, and the *horizontal dotted lines* are for $I_r = 0.94$ and 1.06

The main specific features characterizing the detector can be systematized as follows:

1. The amplitude noise component has virtually no effect on the detector performance, whereas the phase noise component has considerable effect. This thesis is in agreement with conclusions made when analyzing (5.13) and (5.15).
2. The topological charge V_d can be determined from the results of observations of only one realization by the rule

$$V_d = \begin{cases} 1+2i, & I_r \le 1 \\ 0+2i, & I_r > 1 \end{cases} \quad \text{for } \Delta = 180° \text{ and}$$
$$V_d = \begin{cases} 2+3i, & I_r < 0.94 \\ 0+3i, & I_r \in [0.94,\ 1.06] \\ 1+3i, & I_r > 1.06 \end{cases} \quad \text{for } \Delta = 120°, \tag{5.22}$$

where i is an integer. The rule is valid for $A_{nS} \le 0.64\pi$ (for $\Delta = 120°$) and $A_{nS} \le 0.72\pi$ (for $\Delta = 180°$). These boundaries are shown by vertical dotted lines in Figs. 5.13 and 5.14.
3. For a greater number of realizations, this rule allows the V_d value to be determined even for $A_{nS} \le 0.85\pi$ (both for $\Delta = 180°$ and 120°). However, the intensity $\langle I_r \rangle$ rather than I_r must be estimated here. This conclusion does not contradict the conclusion made when analyzing formula (5.15) for $A_{nS} < \pi$.

In Fig. 5.13, by analogy with Fig. 5.14, the probability density (distribution) of the relative interference field intensity I_r has a finite width, which is also manifested through non-coincident curves for the minimum and maximum in Fig. 5.15. It seems likely that the reason for this is relatively small number of points of the calculation grid (32 × 32) for which distributions of intensities in the cross-sectional plane were calculated. Probably for this reason, the conditions $\langle \xi_A(\mathbf{r}) \rangle_\mathbf{r} = 0$, $\langle \xi_S(\mathbf{r}) \rangle_\mathbf{r} = 0$, $\langle \xi_A(\mathbf{r}) \xi_A(\mathbf{r}'(\mathbf{r})) \rangle_\mathbf{r} = 0$, and $\langle \xi_S(\mathbf{r}) \xi_S(\mathbf{r}'(\mathbf{r})) \rangle_\mathbf{r} = 0$ for $\mathbf{r}'(\mathbf{r}) \ne r$ used in the derivation of formulas (5.13), (5.15), and (5.16) are fulfilled only approximately. This computational error can be eliminated completely by additional I_r averaging (for example, over 150 realizations); this follows from a comparison of the $\langle I_r \rangle_{Nr}$ value with the I_r value calculated from formula (5.16).

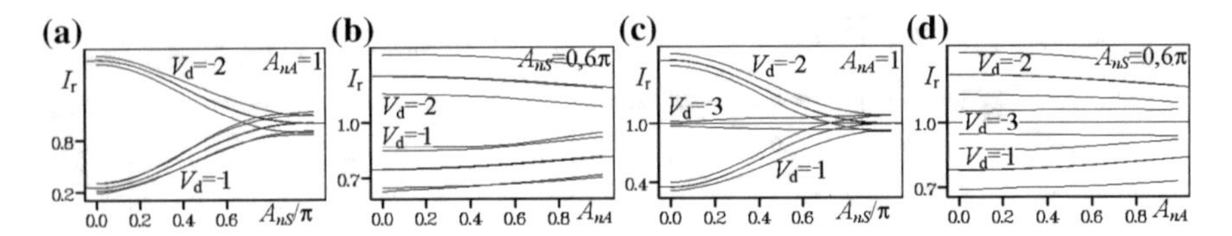

Fig. 5.15 Influence of the phase and amplitude noise components with amplitudes A_{nS} and A_{nA} on the extreme and average values of relative interference field intensities I_r calculated for 150 realizations for $\Delta = 180°$, $\varphi = 0$ (**a** and **b**) and $\Delta = 120°$, $\varphi = \pi/2$ (**c** and **d**). Here $A_{nA} = 0.5$ (**a** and **c**) and $A_{nS} = 0.6\pi$ (**b** and **d**). These extreme and average values have been calculated in the numerical experiment shown in Figs. 5.13 and 5.14. The average values were also calculated from formula (5.16); in the figure, they coincide with results of our computer experiment

The graphical comparison of these intensities in Fig. 5.15 demonstrates the validity of formulas (5.13), (5.15) and (5.16).

Let us add that in Fig. 5.13—seemingly contrary to property (5.19)—there is no complete symmetry of the dependencies of probability density for relative intensity I_r on A_{nS} for $A_{nA} = 0$. The matter is that, according to (5.8), the distribution $A(r)$ depends on the V_d value, but such possibility was excluded in the derivation of (5.19).

Hence, theoretical conclusions on the influence of white noise are in agreement with the data of field transformation modeling in the detector.

5.3.3 *Influence of the Optical Axes Displacement of the Source and Receiver Beam upon the Relative Intensity Value. Possibility of Optical Vortex Position Finding*

The results presented above were interpreted by us first of all as allowance for possible distortions of uncertain origin but reduced to the presence of the white noise in the beam. For example, this white noise can be caused by roughness of the optical source and beam receiver surfaces, that is, by their imperfection. Another important characteristic of system (im)perfection is the displacement S_{hx} of the source and beam receiver optical axes.

To elucidate the influence of S_{hx} on I_r, we modeled the processes of field transformation in the detector. Unlike modeling principles in the preceding section, it was assumed that beam distortions were absent, and the expression for the field amplitude in (5.8) was taken either in the full form or without the multiplier $(r/\rho_V)^{|V_d|}$. As before, $\varphi = 0$ for $\Delta = 180°$ and $\varphi = -90°$ for $\Delta = 120°$ (see Fig. 5.11a or b). The dependencies $I_r(S_{hx}/r_0)$ for different cases are shown in Fig. 5.16.

In these computer experiments, field (5.8) with or without multiplier $(r/\rho_V)^{|V_d|}$ can be treated both as a field in the output plane of the source and as a field in the photodetector plane at the exit from the interferometer. The first treatment is

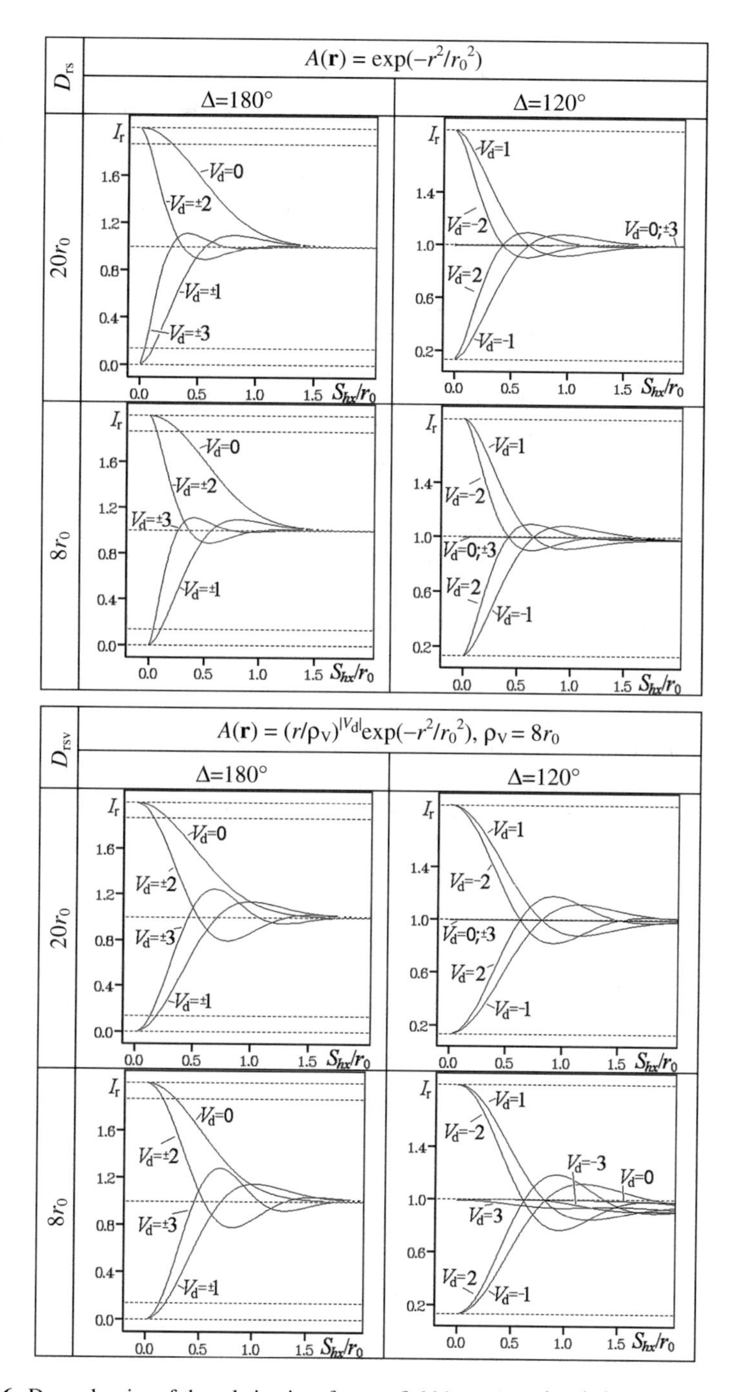

Fig. 5.16 Dependencies of the relative interference field intensity $I_r(S_{hx}/r_0)$ on the displacement S_{hx} of the optical source and receiver axes normalized by the beam radius r_0 for the indicated initial amplitude profiles, receiving apertures D_{rsv}, and Δ and V_d values

possible due to independence of I_r on the path length. The second treatment is correct if there is a method of delivering the field in the above-indicated form to the photodetector that is not beyond the framework of the formulated problem. Therefore, the r_0 value by which the displacement S_{hx} is normalized is interpreted by us mainly as an output beam size of the source.

Attention is attracted to the following regularities. Until the limitation of the receiver aperture D_{rsv} is manifested, the increasing degree of misalignment of the axes causes the I_r intensity values that differ from unity to approach to 1. In this case, the I_r values oscillate when $V_d \neq 0$.

For $\Delta = 120°$, the limitation of the aperture is manifested through the displacement of the dependencies (their oscillation center) from unity to zero with increasing S_{hx}/r_0. This is due to the fact that because of the field rotation in one of the interferometer arms, say, by $\Delta = 120°$, a portion of the beam incident at the input interferometer aperture (*squared* in the model due to the application of the squared calculation grid) does not fall within the output aperture of the same form.

The above-indicated oscillatory tendency of I_r to unity leads to the intersection of the dependencies $I_r(S_{hx}/r_0)$ for different V_d (providing different positions of $I_r(0)$ relative to unity). That is, there are such displacements at which the possibility of simple charge identification (according to the threshold rules described by (5.22)) vanishes. The coordinates of these intersection points for the dependencies $I_r(S_{hx}/r_0)$ shown in Fig. 5.16 are given in Table 5.1.

The oscillation frequency and amplitude (and hence the slope of dependencies $I_r(S_{hx}/r_0)$) the higher, the larger $|V_d|$. Therefore, the dependencies $I_r(S_{hx}/r_0)$ for different pairs of V_d values are intersected the sooner, the greater the modulus of each of V_d value. For example (see Table 5.1), in case of field rotation through $\Delta = 180°$, the pair $(V_{d0},\ V_{d1}) = (1,\ 2)$ is intersected later than the pair $(3,\ 2)$, but sooner than the pair $(3,\ 0)$.

The multiplier $(r/\rho_V)^{|V_d|}$ in the expression for the field amplitude increases the amplitude, but decreases the oscillation frequency. For $V_d \neq 0$, this causes the intersection points to displace along the S_{hx}/r_0 axis by 25–56 % (on average, by 35 %): from $S_{hx} = 0.5r_0$ to $S_{hx} = 0.7r_0 \approx 1.377 \cdot 0.5r_0$ (Table 5.1). The curve for

Table 5.1 Coordinates of the intersection points for functions $I_r(S_{hx}/r_0)$ shown in Fig. 5.16

		$\Delta = 120°$			
$A(\mathbf{r})$, see (5.8)	V_{d0} V_{d1}	−1 +1	2 1	−1 −2	+2 −2
$\exp(-r^2/r_0^2)$	S_{hx}/r_0 I_r	0.6548 1	0.5549 1.0857	0.5549 0.9137	0.42 1
$\exp(-r^2/r_0^2) \times (r/\rho_V)^{\lvert Vd \rvert}$	S_{hx}/r_0 I_r	0.818 1	0.714 1.095	0.714 0.905	0.6258 1
$\exp(-r^2/r_0^2)$	S_{hx}/r_0 I_r	0.654 1	0.5538 1.086	0.5534 0.9127	0.4195 1
$\exp(-r^2/r_0^2) \times (r/\rho_V)^{\lvert Vd \rvert}$	S_{hx}/r_0 I_r	0.8149 0.99	0.710 1.095	0.702 0.887	0.617 0.99

$V_d = 0$, naturally, remains unchanged; therefore, for $\Delta = 180°$, the absence of the multiplier $(r/\rho_V)^{|V_d|}$ causes other dependencies to move away from this curve. This tendency is most clearly expressed for the pair $(V_{d0}, V_{d1}) = (\pm 3, 0)$ (see Fig. 5.16).

The above-indicated intersections occur for relatively small S_{hx} values ($S_{hx} \approx 0.3r_0 \ldots 0.8r_0$). Therefore, trying to intercept the entire beam (the most part of its power), we do not need large receiver aperture D_{rsv}. Indeed, for $S_{hx} \in [0, 0.8r_0]$, the behavior of the dependencies $I_r(S_{hx}/r_0)$ for $D_{rsv} = 20r_0$ is almost indistinguishable from that for $D_{rsv} = 4r_0$, and intersection occurs well before a considerable portion of the beam power will fall beyond the input interferometer aperture. Moreover, for $S_{hx} = 0$, the D_{rsv} value does not influence I_r in any way. After all, field (5.8) for the *circular* receiver aperture is described by more general formula (5.7) for which (5.9) is valid: $I_r(V_d) = 1 + \cos[2\pi M V_d/m + \varphi]$. This statement remains also valid in the presence of white noise: see formulas (5.10) and (5.16).

It is obvious that the closer the dependencies $I_r(S_{hx}/r_0)$ for the expected V_d values, the more difficult to identify the V_d value. These complications of information applications for authorized users can cause the error in data transfer to increase. On the contrary, it may be impossible for an unauthorized user to read foe message. That is, similar dependencies $I_r(S_{hx}/r_0)$ create prerequisites for high degree of protection of the communication channel. In practice, these opposite tendencies will force the designer to choose compromise values of the communication system parameters. The following combinations of $(\Delta V_{d0}, V_{d1}) : (180°, -1, 0)$, $(120°, -1, 1)$, $(120°, -1, -2)$, and $(180°, -1, 2)$, arranged in the decreasing order of the displacement of axes S_{hx}/r_0 in Table 5.1, can be mentioned in this context.

In conclusions of this section, we point out the possibility of constructing *a coordinate detector* (position finder) of an optical vortex. For the simplest variant of its construction, it is suggested to split the beam registered by the receiving system into several sub-beams and to record each of them with a separate detector—interferometers shown in Fig. 5.11. It is assumed that the optical axes of these interferometers are parallel to the sub-beam axes, and their continuation (along the sub-beams in the direction of the field source) intersect the cross sectional plane of the initial beam at points such that any set of these three points do not lie on one straight line.

The output signals I_r of each interferometer bear information not only on the vortex charge but also (owing to the dependence $I_r(S_{hx}/r_0)$) on the distance from the optical interferometer axis to the beam centre (to the vortex) in the cross sectional plane.

Having pair values of such distances, we can determine the vortex position in the cross sectional plane to within the half-plane accuracy. That is, knowledge of the pair distances allows us to consider that the vortex is at a certain concrete point or at the point mirror symmetric about the straight line connecting the intersection points of the cross-sectional plane of the beam with the axes of the two chosen interferometers. The data on one more value of the distance (to the axis of the additional interferometer) lift the indicated uncertainty, that is, such three distances completely determine the vortex position in the plane, in complete analogy with the position on the plane set by three distances to the set of three points that do not lie on one straight line. Here there exists the strict analogy with problems and methods of position finding.

Of course, the interferometer axes must be spaced at distances of the order of $S_{hx} \approx 0,\ 3r_0 \ldots 0.8r_0$, for example, at the vortices of an equilateral triangle whose center coincides with that of the region of expected vortex position. If this region is larger than this triangle, it must be covered with such triangles. Naturally, the interferometers additional to the above-indicated ones will increase the reliability of vortex position finding in the presence of noise. It seems likely that in more complicated situations with a network of interferometers and many vortices with different topological charges in one beam, neural network control and recognition methods will be promising for operation of multiple position finders. For miniaturization and increase of resolution of the suggested class of devices, the element base must involve nanostructures (for example, see [63, pp. 261–276]).

5.3.4 Determination of the Screw Dislocation Order in the Presence of Beam Distortions Caused by Turbulence

White noise is convenient for modeling, but it is difficult to relate it with the statistical properties of the atmosphere. Therefore, below we analyze the influence of the turbulence on the detector performance. The capability of the detector to determine the charge V_d of dislocations in the wave front is studied for a Gaussian amplitude distribution of the beam at the beginning of the propagation path and a screw dislocation in the beam phase profile (that is, the field amplitude in (5.8) is considered without the multiplier $(r/\rho_V)^{|V_d|}$. The beam is propagated through a thin phase screen that models atmospheric distortions and then through a homogeneous path of length L_t.

The intensity of atmospheric distortions is characterized by the Fried radius L_F and outer, M_{outer}, and inner, M_{outer}, turbulence scales. Here L_F values changed from $50r_{0P}$ to $0.05r_{0P}$, $M_{inner} = 0.15625r_{0P} \approx 0.11049r_0$, and $M_{outer} = 5r_{0P} \approx 3.5355r_0$ or $M_{outer} = 20r_{0P} \approx 14.142r_0$, where $r_{0P} = 2^{-0.5}r_0$ is the Gaussian beam radius in the expression $I(\mathbf{r}) = C^2\exp(-2r^2/r_0^2) = C^2\exp(-r^2/r_{0P}^2)$. The path length L_t was 0.05 and 0.5 of the diffraction lengths, which corresponded to 5.9 and 59 km for $r_{0P} = 10$ cm and radiation wavelength of 0.5 μm. For each combination of the physical parameters, we seek for minimum, maximum, and average I_r values for 100 realizations of the field corresponding to different turbulent phase screens.

For a path length of 59 km, the computation grid size at the reception point was close to $r'_{0P}/16$, where r'_{0P} is the beam radius in the receiving plane. The error in the intensity calculation, estimated by processing with the detector of the field with the *reference* vortex (see (5.8)), did not exceed 3 %. The error in calculation of I_r for the field rotated through an angle of 180° was insignificant; for $\Delta = 120°$ it was equal to 0.0003 (0.22 %), 0.0019 (0.1 %), and 0.0012 (0.12 %) for $V_d = -1$, –2, and –3, respectively. To make the model experiments more realistic, the effect of misalignment of the source and receiver optical axes was taken into account. In the source plane, we set $S_h \equiv \left|(S_{hx}, S_{hy})\right| = |-0.07873 r_{0P}(1,\ 1)| = 0.11 r_{0P}$, which in the receiving plane led to misalignment of their optical axes by about $S'_h = 0.044 r'_{0P}$. The error caused by the last circumstance was decisive: it increased the error of I_r calculation for $\Delta = 180°$ up to 0.03, 0.06 (3 %), and 0.09; for $\Delta = 120°$, the error increased to 0.014 (10 %), 0.06 (3 %), and 0.04 (4 %) for $V_d = -1$, -2, and -3, respectively. This error is considered to be systematic. It seems likely that its influence can be further considered by empirical selection of threshold in (5.22). The calculated dependencies for sets of three I_r values on the Fried radius are shown in Fig. 5.17. It is remarkable that for weak turbulence (with $L_F / r_{0P} = 50$), for example, for $\Delta = 120°$ and $V_d = -3$, I_r differed from unity by 0.014 (1.4 %), that is, was close enough to the above estimate equal to 0.04 (4 %).

An analysis of the results of modeling partially illustrated by Fig. 5.17 demonstrates that:

1. For $L_F > 10 r_{0P}$ (weak turbulence), the relative interference field intensity I_r remains virtually unchanged. Moreover, I_r values are close to those registered in the system with white noise of low intensity: $I_r = 0$ and 2 (see Fig. 5.13 for

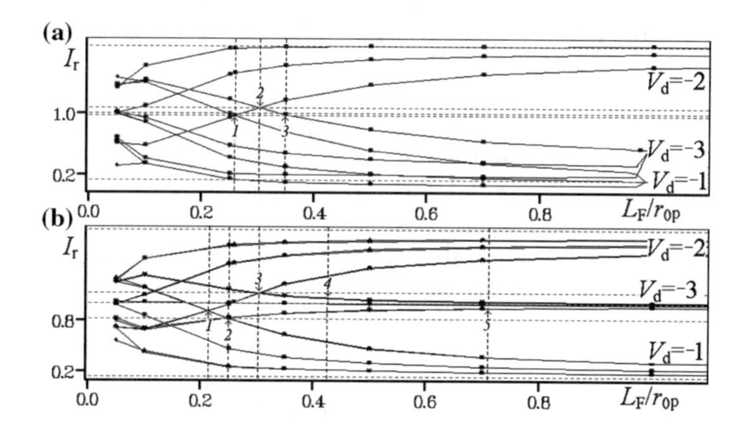

Fig. 5.17 Calculated (for a hundred of realizations and $M_{outer} = 5 r_{0P}$) extreme and average relative interference field intensities I_r versus the Fried radius L_F for $\Delta = 180°$ (**a**) and 120° (**b**) and the indicated V_d values. The *dashed straight lines* at the edges of the figures are for $I_r = 0.135$ and 1.864 registered without white noise for $\Delta = 120°$ (Fig. 5.14). The horizontal *dashed line* in the middle of Fig. 5.17a is for $I_r = 1$ and threshold values $I_r = 0.97$ and 1.063. The *horizontal dashed line* in the middle of Fig. 5.17b is for $I_r = 1$ and threshold values $I_r = 0.818$ and 1.12. The *vertical dashed lines* in Fig. 5.17a are for $L_F / r_{0P} = 0.260$ and 0.304, respectively. The *vertical dashed lines* in Fig. 5.17b are for $L_F / r_{0P} = 0.216$, 0.250, 0.304, 0.425, and 0.710, respectively

$\Delta = 120°$) and $I_r = 1.864$ and 0.135 (see Fig. 5.14 for $\Delta = 120°$). The differences are caused by the presence of the turbulence (though weak) and, to a greater extent, by the displacement $S_{hx} = 0.11r_{0P}$ (see Fig. 5.16).

2. The coincidence of curves for path lengths of 5.9 and 59 km (Fig. 5.17) confirms the theoretical conclusion that the relative intensity I_r is independent of the path length. That is, I_r depends on the screen that has formed the field rather than on the path length.
3. In the determination of charge V_d from only one realization (according to (5.22)), the following special features are observed:

   - for $\Delta = 180°$, the vortices with $V_d = -1$ and -2 are distinguishable for $L_F > 0.26r_{0P}$, and the vortices with $V_d = -2$ and -3 were distinguishable for $L_F > 0.35r_{0P}$ (see the 1st and 3rd vertical dashed lines in Fig. 5.17a);
   - for $\Delta = 120°$, the vortices with $V_d = -1$ and -3 are distinguishable for $L_F > 0.710r_{0P}$, and the vortices with $V_d = -2$ and -3 are distinguishable for $L_F > 0.425r_{0P}$ (see the 5th and 4th vertical dashed lines in Fig. 5.17b).

4. If the angle $\Delta = 180°$, rule (5.22) of dislocation recognition for vortices having only charges $V_d = -1$ and -2 or $V_d = -2$ and -3 must be reduced to the form

$$V_d = \begin{cases} 1, & I_r \le 0.97, \\ 2, & I_r > 0.97, \end{cases} \text{ or } V_d = \begin{cases} 3, & I_r \le 1.063, \\ 2, & I_r > 1.063. \end{cases} \quad (5.23)$$

   Rule (5.23) is valid for $L_F > 0.26r_{0P}$ or $L_F > 0.304r_{0P}$ (see the 1st and 2nd vertical dashed lines in Fig. 5.17a).

5. For $\Delta = 120°$, threshold values in Eq. (5.22) can be changed to derive the following rule of charge determination:

$$V_d = \begin{cases} 2 + 3i, & I_r < 0.818, \\ 0 + 3i, & I_r \in [0.818,\ 1.12], \\ 1 + 3i, & I_r > 1.12. \end{cases} \quad (5.24)$$

   The rule for distinguishing vortices with $V_d = -1$ and -3 or with $V_d = -2$ and -3 is valid for $L_F > 0.25r_{0P}$ or $L_F > 0.304r_{0P}$ (see the 2nd and 3rd vertical dashed lines in Fig. 5.17b). When it is necessary to distinguish only vortices with $V_d = -1$ and -2 for $L_F > 0.216r_{0P}$, the following rule is valid:

$$V_d = \begin{cases} 1, & I_r \le 0.904, \\ 2, & I_r > 0.904 \end{cases} \quad (5.25)$$

   (the 1st vertical dashed line in Fig. 5.17b).

6. For a great number of realizations, rule (5.22) allows the charge V_d to be determined for $L_F = 0.1r_{0P}$ and even smaller (both for $\Delta = 180°$ and 120°). But, in this case, the average intensity $\langle I_r \rangle$ rather than I_r should be measured. We note

that initial expression (5.22) for this problem yields a better result than rules (5.23) and (5.24) with refined (asymmetrical about 1) thresholds.

Recall that the empirically selected thresholds (items 3–5) are optimum for a concrete complex of situations beyond which they can lose this optimality. Thus, the thresholds may change when the displacement changes to $S_h = 0.11 r_{0P}$ for $\Delta = 180°$, but remain unchanged for $\Delta = 120°$. Since above-mentioned conclusions made on the base of analysis of hundred realizations, the statements of a type "orders of dislocations are recognized" should understand as a statement: probability of error of this recognition is not more than 0.5 %.

The Tables 5.2 and 5.3 confirm qualitatively correctness of conclusions 1–6. In these Tables $V_{\mathrm{d\,i}}$ is the order of screw dislocation, which is assigned by the detector to the field, having the order V_{d} before turbulence impact. When the identification fraction for $L_{\mathrm{t}} = 59$ km differs from the fraction in the case $L_{\mathrm{t}} = 5.9$ km, then it is indicated in brackets.

Figure 5.18 demonstrates the influence of one and the same phase screen simulating turbulence on the accuracy of recognition vortices with the indicated charges V_{d} (that is, I_{r} values). The upper, bottom, and middle curves in Fig. 5.18 are for the vortex charges $V_{\mathrm{d}} = -2, -1$, and -3, respectively. We can see that the curves in Fig. 5.18b, d differ very slightly, similar to the dependencies for $V_{\mathrm{d}} = -1$ and -3 in

Table 5.2 A fraction of true (boldface font) and false identifications of the vortex topological charge by a detector with $\Delta = 180°$ (for $M_{\mathrm{outer}} = 5 r_{0P}$, $S_h = 0.11 r_{0P}$) depending of the Fried radius and rules (5.22) and (5.23)

L_{F}/r_{0P}	V_{d}	(5.22)			(5.23) for $V_{\mathrm{d}} = 1$, $V_{\mathrm{d}} = 2$			(5.23) for $V_{\mathrm{d}} = 2$, $V_{\mathrm{d}} = 3$		
		$V_{\mathrm{d\,i}}$			$V_{\mathrm{d\,i}}$			$V_{\mathrm{d\,i}}$		
		–1	–2	–3	–1	–2	–3	–1	–2	–3
0.05	–1	**45** **(48)**	55 (52)	0	**39** **(45)**	61 (55)	–	–	–	–
	–2	55	**45**	0	45	**55**	–	–	**30**	70
	–3	0	56	**44**	–	–	–	–	37	**63**
0.1	–1	**64**	36	0	**59**	41	–	–	–	–
	–2	43	**57**	0	37	**63**	–	–	**51**	49
	–3	0	45	**55**	–	–	–	–	37	**63**
0.25	–1	**100**	0	0	**99**	1	–	–	–	–
	–2	3	**97**	0	1	**99**	–	–	**94**	6
	–3	0	6	**94**	–	–	–	–	3	**97**
0.26	–1	**100**	0	0	**–**	–	–	–	–	–
	–2	1	**99**	0	1	**99**	–	–	**95**	5
	–3	–	–	**–**	–	–	–	–	–	**–**
0.35	–1	**100**	0	0	**100**	0	–	–	–	–
	–2	0	**100**	0	0	**100**	–	–	**100**	0
	–3	0	0	**100**	–	–	–	–	0	**100**

Table 5.3 A fraction of true (boldface font) and false identifications of the vortex topological charge by a detector with $\Delta = 120°$ (for $M_{outer} = 5r_{OP}$, $S_h = 0.11r_{OP}$) depending o the Fried radius and rules (5.22), (5.24) and (5.25)

L_F/ρ_{OP}	V_d	(5.22)			(5.24)			(5.25) for V_d = 1, V_d = 2		
		$V_{d\ i}$			$V_{d\ i}$			$V_{d\ i}$		
		–1	–2	–3	–1	–2	–3	–1	–2	–3
0.05	–1	**32 (41)**	23 (31)	45 (28)	**9 (13)**	5 (19)	86 (68)	**19 (28)**	81 (72)	–
	–2	32 (34)	**22 (19)**	46 (47)	5	**10 (9)**	85 (86)	19 (24)	**81 (76)**	–
	–3	23 (25)	30 (24)	**47 (51)**	1 (4)	11 (8)	**88**	–	–	–
0.1	–1	**65 (67)**	14 (15)	21 (18)	**43 (46)**	7 (8)	50 (46)	**56 (55)**	44 (44)	–
	–2	19 (18)	**57 (58)**	24	9 (8)	**42 (44)**	49 (48)	14	**86**	–
	–3	28 (29)	31 (32)	**41 (39)**	5 (4)	18 (20)	**77 (76)**	–	–	–
0.25	–1	**100**	0	0	**100 (99)**	0	0 (1)	**100**	0	–
	–2	0	**99**	1	0	**97**	3	0	**100**	–
	–3	11 (12)	11	**78 (77)**	0	4 (2)	**96 (98)**	–	–	–
0.26	–1	**100**	0	0	–	–	–	**100**	0	–
	–2	0	**99**	1	0	**98**	2	0	**100**	–
	–3	–	–	–	0	0	**100**	–	–	–
0.35	–1	**100**	0	0	**100**	0	0	**100**	0	–
	–2	0	**100**	0	0	**100**	0	0	**100**	–
	–3	7	6 (5)	**87 (88)**	0	0	**100**	–	–	–
0.5	–1	**100**	0	0	**100**	0	0	**100**	0	–
	–2	0	**100**	0	0	**100**	0	0	**100**	–
	–3	3	0	**97**	0	0	**100**	–	–	–
0.7	–1	**100**	0	0	**100**	0	0	**100**	0	–
	–2	0	**100**	0	0	**100**	0	0	**100**	–
	–3	0	0	**100**	0	0	**100**	–	–	–

Fig. 5.18a. In the first case, this testifies to a high level of turbulence for this method of V_d identification. In the second case, this confirms the correctness of the (5.21) and the principle of numbering structures (5.3). At that, dependencies for $V_d = -1$ and $V_d = -3$ for $\Delta = 180°$ (Fig. 5.18a) would be coincided, when $S_h = 0$—see (5.21). We note that diagrams are constructed for path lengths $L_t = 5.9$ km and $L_t = 59$ km, but they, as before, are indistinguishable (see conclusion 2).

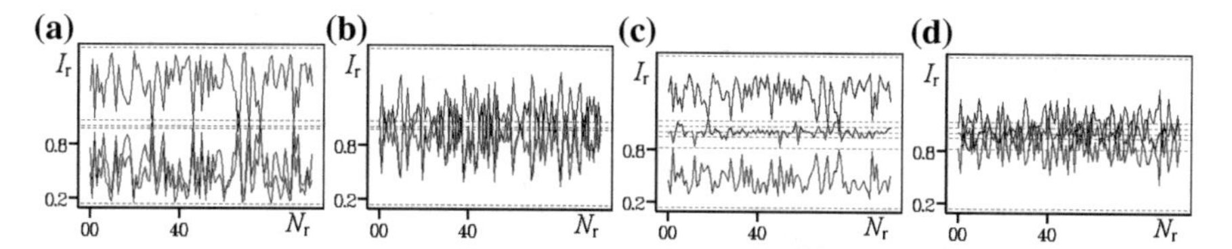

Fig. 5.18 Dependencies of the I_r value on the number N_r of phase screen realizations ($N_r \in [0, 99]$). The calculated points are connected by curves (for convenience of perception). The angles of field rotation are $\Delta = 180°$ (**a** and **b**) and 120° (**c** and **d**), and the Fried radius is $L_F = 0.25r_{0P}$ (**a** and **c**) and $0.1r_{0P}$ (**b** and **d**). The *horizontal dashed lines* at the edges of the figures are for $I_r = 0.135$ and 1.864. The *horizontal dashed lines* in the middle of Fig. 5.18a, b are for $I_r = 1$ and threshold values $I_r = 0.97$ and 1.063; the *horizontal dashed lines* in the middle of Fig. 5.18c, d are for $I_r = 1$ and threshold values $I_r = 0.818$, 0.94, 1.06, and 1.12

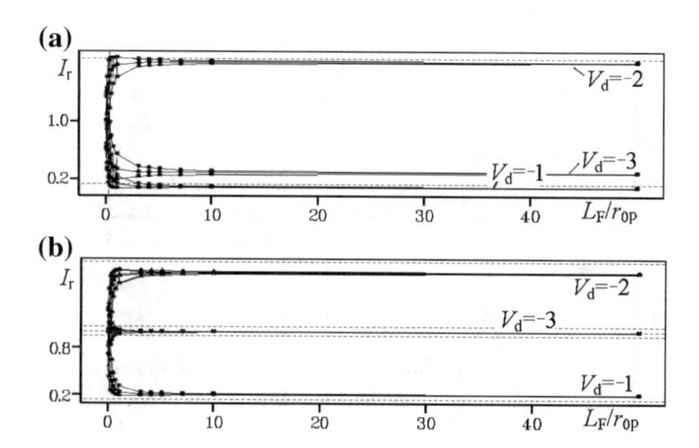

Fig. 5.19 Calculated (for a hundred of realizations and $M_{outer} = 5r_{0P}$, $S_h = 0.11r_{0P}$) extreme and average relative interference field intensities I_r *versus* the Fried radius L_F for $\Delta = 180°$ (**a**) and 120° (**b**) and the indicated V_d values. These graphs correspond to track lengths $L_t = 5.9$ km, and $L_t =$ 59 km, but in the figure they are indistinguishable. The *dashed straight lines* at the edges of the figures are for $I_r = 0.135$ and 1.864 registered without white noise for $\Delta = 120°$ (Fig. 5.14). The *dashed straight lines* in the middle of Fig. 5.19b correspond to *dashed straight lines* in Fig. 5.14 (for $I_r = 1$ and threshold values $I_r = 0.94$ and 1.06).

The mirror symmetry of the dependencies $I_r(N_r)$, though incomplete, is clearly seen for pairs $V_d = -2$ and –1; $V_d = -2$ and –3. Recall ones more that in the absence of displacement (for $S_{hx} = 0$), for the field rotated through $\Delta = 180°$, the symmetry is strict according to (5.20). On the contrary, for $\Delta = 120°$, the presence of certain easily distinguishable symmetry (see Fig. 5.18c, d) is unexpected. As far as possible, in which the "over-realization", *sit venia verba*, symmetry, i.e. diagram symmetry $I_r(N_r)$, it is present in Figs. 5.17 and 5.19. But, the inverse statement is wrong.

One more regularity is also observed: as a rule, $I_r(N_r, V_{d0}) \neq I_r(N_r, V_{d1})$ of course, for $V_{d0} \neq V_{d1}$. This circumstance and the symmetry property are vividly illustrated by Figs. 5.20, 5.21, 5.22, 5.23 and 5.24 in which the one-dimensional, $\rho(I_r)$, and two-dimensional, $\rho(I_{r0}, I_{r1})$, probability densities (distributions) are

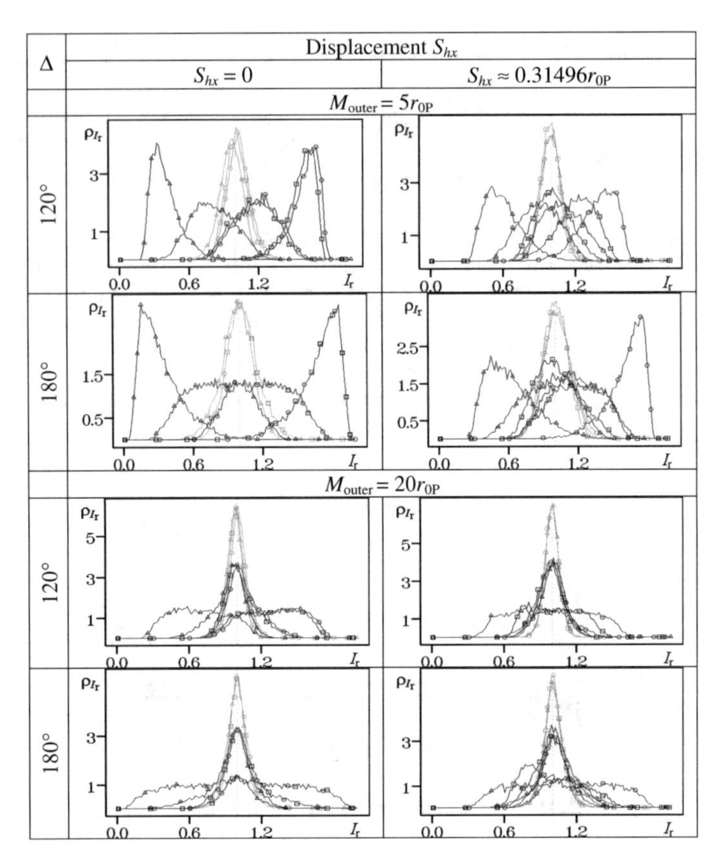

Fig. 5.20 One-dimensional probability density $\rho_{I_r} \equiv \rho(I_r)$ for the indicated values of M_{outer} and S_{hx}. The Fried radius is $L_F/r_{0P} = 0.25$, 0.1, and 0.05 for groups of curves with maxima located at the edges, closer to the middle, and in the middle (*grey color*), respectively. The charge V_d takes values –1, –2, and +1 or 0 (for $\Delta = 120°$ or 180°) for the curves with symbols Δ, □, and ○

drawn for the indicated values of L_F, M_{outer}, S_{hx}, and V_d, where $I_{r0} \equiv I_r(V_{d0})$ and $I_{r1} \equiv I_r(V_{d1})$. To describe them, we tentatively number the quadrants of the plane $(I_{r0},\ I_{r1})$ given that the origin of coordinate is at the point (1, 1).

Our calculations and curves of $(\Delta V_{d0},\ V_{d1})$ for $S_{hx} = 0$ demonstrate that the probability density $\rho(I_{r0},\ I_{r1})$ are visually undistinguishable for sets of three values of the parameters (120°, –1, 1), (120°, –1, –2); (180°, –1, 0), (180°, –1, –2). In addition, with increasing turbulence intensity, the probability density $\rho(I_{r0},\ I_{r1}) \neq 0$ only on a certain segment of the straight line $I_{r1} = 2 - I_{r0}$ (see (5.20)) displaced from the second to the fourth quadrant and tends to become symmetric about the point $(I_{r0},\ I_{r1}) = (1,\ 1)$. Considering explanation to Fig. 5.21, Fig. 5.23 illustrates only some of the above-mentioned situations. The same properties (small difference or complete coincidence for $\Delta = 180°$) can be easily established (see Fig. 5.20) for

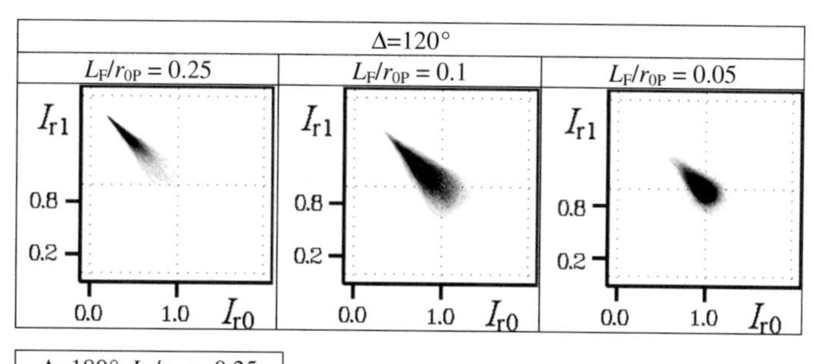

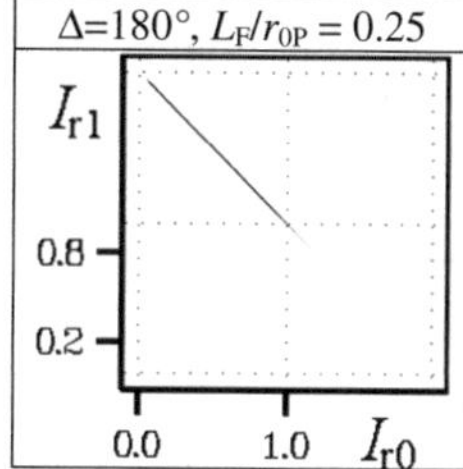

Fig. 5.21 Two-dimensional probability densities $\rho(I_{r0},\ I_{r1})$ for $(V_{d0},\ V_{d1}) = (-1,\ -2)$, $S_{hx} = 0$, $M_{outer} = 5r_{0P}$, and indicated values of L_F and Δ

one-dimensional densities $\rho(I_r)$ as well, but already for pairs of two parameters ΔV_{d1}: (120°, 1), (120°, –2); (180°, 0), (180°, –2).

According to Figs. 5.20, 5.21, 5.22, 5.23 and 5.24, with increasing turbulence intensity (decreasing L_F), the probability density initially localized far from the point (1, 1) is displaced from the second quadrant in the direction of the fourth quadrant, concentrating around the point (1, 1) and becoming more symmetric about it. As a result, the function $\rho(I_{r0},\ I_{r1})$ resembles more closely a Gaussian curve for $L_F/r_{0P} = 0.05$. Figuratively speaking, such transformation occurs through overflowing: the density $\rho(I_{r0},\ I_{r1})$, elongated and elevated from the one hand, leveling and broadening, acquires a plateau shape. Then the maximum starts to increase, but at that time in the vicinity of the point (1, 1). Similar metamorphoses with decreasing L_F are also characteristic of the one-dimensional density $\rho(\ I_r)$ (see Fig. 5.20).

Visual analysis of Figs. 5.20, 5.21, 5.22, 5.23 and 5.24 demonstrates that the increase in the outer turbulence scale M_{outer} from $5r_{0P}$ to $20r_{0P}$ changes the probability density to the form which would be caused if the Fried radius L_F decreased from $0.25r_{0P}$ to $L_F \approx 0.18r_{0P}\ldots0.1r_{0P}$, from $0.1r_{0P}$ to $L_F \approx 0.07r_{0P}\ldots0.05r_{0P}$, or from $0.05r_{0P}$ to values smaller than $0.05r_{0P}$.

The displacement of the transmitter and receiver optical axes $S_{hx} \approx 0.31496r_{0P}$ for weak turbulence ($L_F = 0.25r_{0P}$) causes the probability density to broaden as well as (except the case (120°, –1, 1)) leads to its displacement as a whole to the first and third quadrants. In this regard, it is useful to compare curves for $\Delta = 180°$. With increasing turbulence intensity, the displacement accelerates the transformation of the function $\rho(I_{r0},\ I_{r1})$ to the form similar to the Gaussian curve.

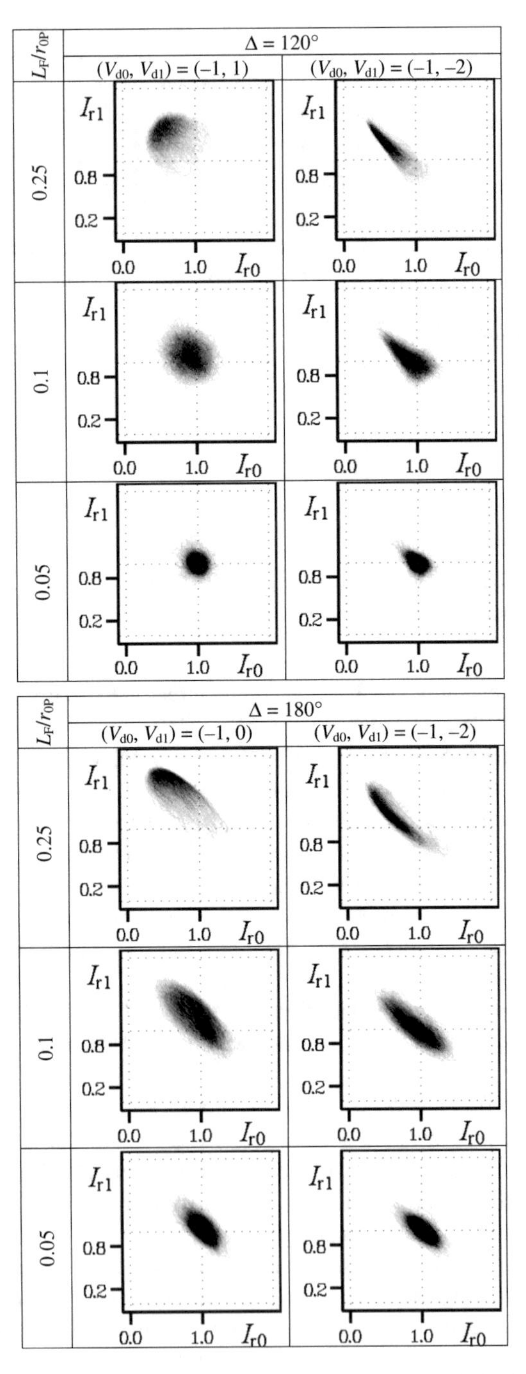

Fig. 5.22 Two-dimensional probability densities $\rho(I_{r0}, I_{r1})$ for $S_{hx} \approx 0.31496 r_{0P}$, $M_{outer} = 5 r_{0P}$, and indicated values of L_F, Δ, V_{d0}, and V_{d1}

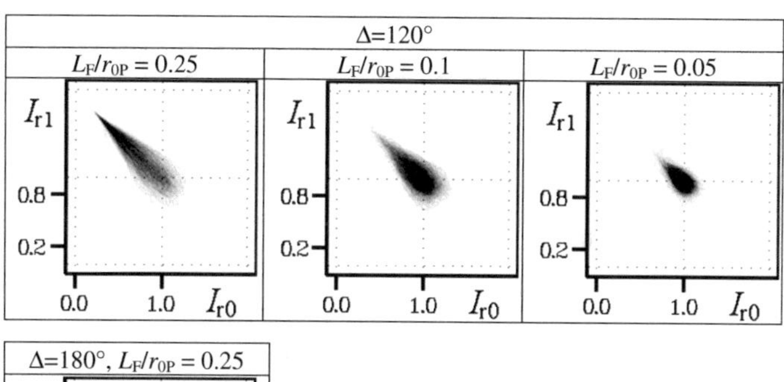

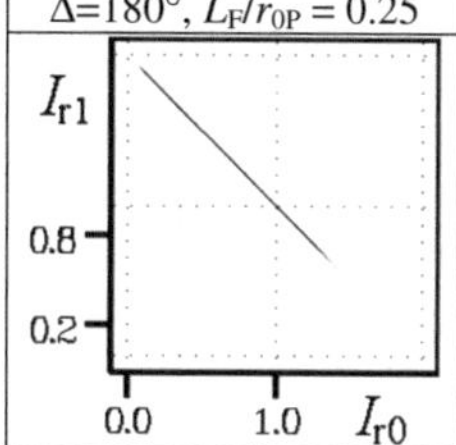

Fig. 5.23 Two-dimensional probability densities $\rho(I_{r0},\ I_{r1})$ for $V_{d0}, V_{d1} = (-1, -2)$, $S_{hx} = 0$, $M_{outer} = 20r_{0P}$, and indicated values of L_F, Δ

Below we demonstrate that in singular-optical communication systems, the property of symmetry provides correct data transfer even when the charge of the transmitted vortex has been determined with an error. Because of this peculiarity, the identification failure can be considered as a natural event, which means that it can be taken into account in the coding algorithm (which we call differential below) and, possibly, in the data transfer protocol. And the rule: $I_r(N_r,\ V_{d0}) \neq I_r(N_r,\ V_{d1})$ for $V_{d0} \neq V_{d1}$ together with the symmetry property necessitates the design of the communication systems with an adaptive (floating) threshold that allows the error in data transfer caused by the atmospheric turbulence to be minimized.

However, we first present Table 5.4 which illustrates the influence of the phase screen and detector parameters as well as of the dislocation orders on the average difference $\langle I_r(N_r,\ V_{d1}) - I_r(N_r,\ V_{d0})\rangle_{N_r} \equiv \langle I_{r1} - I_{r0}\rangle$ and on the overlap $I_{r0.max} - I_{r1.min}$ of ranges of variation of functions $I_{r0}(N_r)$ and $I_{r1}(N_r)$. This table contains quantitative arguments that confirm the correctness of our visual analysis of the behavior of the probability densities $\rho(I_r)$ and $\rho(I_{r0},\ I_{r1})$.

It is obvious that the greater the $\langle I_{r1} - I_{r0}\rangle$ value, the easier the V_d value can be identified (V_{d1} can be distinguished from V_{d0}), and a positive value of the overlap $I_{r0.max} - I_{r1.min}$ demonstrates that this identification is not always successful. According to Table 5.4, it is not always successful even for $L_F = 0.25r_{0P}$. This conclusion does not contradict with items 3–5 presented above and obtained for $S_h \approx 0.11r_{0P}$ and $M_{outer} = 5r_{0P}$. It seems that the exception is the conclusion about applicability of (5.25) for $L_F > 0.216r_{0P}$. The contradiction source can be explained

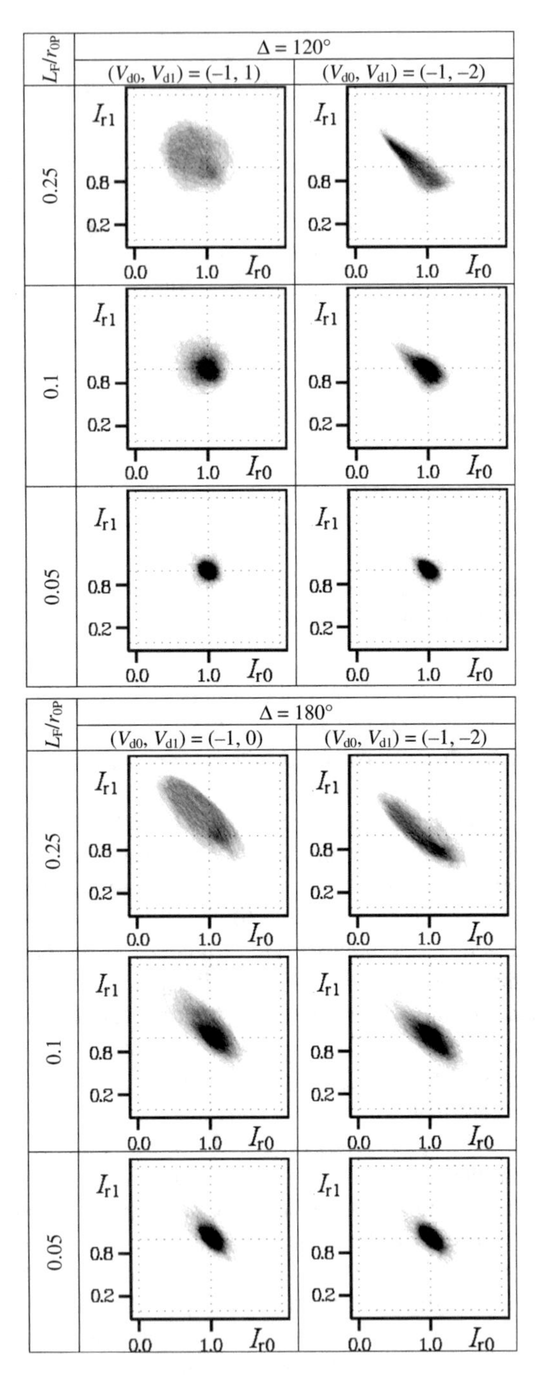

Fig. 5.24 Two-dimensional probability densities $\rho(I_{r0},\ I_{r1})$ for $S_{hx} \approx 0.31496 r_{0P}$, $M_{outer} = 20 r_{0P}$, and indicated values of L_F, Δ, V_{d0}, and V_{d1}

Table 5.4 Comparison of average (over 12,800 realizations) differences $\langle I_{r1} - I_{r0} \rangle$, overlaps $I_{r0.\max} - I_{r1.\min}$ of the ranges of variation of functions $I_{r1}(N_r)$ and $I_{r0}(N_r)$ for indicated values of the Fried radius L_F/r_{0P}, angles of field rotation Δ, displacements S_{hx} of the source and receiver optical axes, pairs of charges (V_{d0}, V_{d1}), and outer turbulence scales M_{outer} (here $S_{hx} \approx 0.31496 r_{0P} \approx 0.2227 r_0$)

L_F/r_{0P}		$\Delta = 180°$			
		$S_{hx} = 0$		$S_{hx} \approx 0.31496 r_{0P}$	
	V_{d0} V_{d1}	–1 0	–1 –2	–1 0	–1 –2
		$M_{outer} = 5 r_{0P}$			
0.25	$\langle I_{r1} - I_{r0} \rangle$	1.35		0.997	0.563
	$I_{r0.\max} - I_{r1.\min}$	0.352		0.565	0.828
0.10	$\langle I_{r1} - I_{r0} \rangle$	0.391		0.236	0.0765
	$I_{r0.\max} - I_{r1.\min}$	1.01		1.00	1.023
0.05	$\langle I_{r1} - I_{r0} \rangle$	0.0563		0.00926	–0.0159
	$I_{r0.\max} - I_{r1.\min}$	0.816		0.791	0.802
		$M_{outer} = 20 r_{0P}$			
0.25	$\langle I_{r1} - I_{r0} \rangle$	0.593		0.362	0.0874
	$I_{r0.\max} - I_{r1.\min}$	0.818		0.908	1.03
0.10	$\langle I_{r1} - I_{r0} \rangle$	0.109		0.0266	–0.0335
	$I_{r0.\max} - I_{r1.\min}$	0.987		0.979	1.02
0.05	$\langle I_{r1} - I_{r0} \rangle$	0.0323		0.00632	–0.00294
	$I_{r0.\max} - I_{r1.\min}$	0.750		0.719	0.732
L_F/r_{0P}		$\Delta = 120°$			
		$S_{hx} = 0$		$S_{hx} \approx 0.31496 r_{0P}$	
	V_{d0} V_{d1}	–1 –2	–1 +1	–1 –2	–1 +1
		$M_{outer} = 5 r_{0P}$			
0.25	$\langle I_{r1} - I_{r0} \rangle$	1.16	1.20	0.604	0.786
	$I_{r0.\max} - I_{r1.\min}$	0.206	0.178	0.554	0.450
0.10	$\langle I_{r1} - I_{r0} \rangle$	0.353	0.382	0.111	0.186
	$I_{r0.\max} - I_{r1.\min}$	0.747	0.723	0.796	0.759
0.05	$\langle I_{r1} - I_{r0} \rangle$	0.0493	0.0703	–0.00824	0.00559
	$I_{r0.\max} - I_{r1.\min}$	0.588	0.573	0.602	0.591
		$M_{outer} = 20 r_{0P}$			
0.25	$\langle I_{r1} - I_{r0} \rangle$	0.582	0.616	0.185	0.317
	$I_{r0.\max} - I_{r1.\min}$	0.608	0.585	0.816	0.777
0.10	$\langle I_{r1} - I_{r0} \rangle$	0.107	0.133	–0.00825	0.0216
	$I_{r0.\max} - I_{r1.\min}$	0.720	0.707	0.741	0.710
0.05	$\langle I_{r1} - I_{r0} \rangle$	0.0234	0.0415	–0.000517	0.0000511
	$I_{r0.\max} - I_{r1.\min}$	0.600	0.574	0.512	0.543

Table 5.5 Influence of the displacement S_{hx} and outer turbulence scale M_{outer} on average (for Table 5.4) values of the differences $\langle I_{r1} - I_{r0} \rangle$ and overlaps $I_{r0.max} - I_{r1.min}$ of ranges of variation of functions $I_{r1}(N_r)$ and $I_{r0}(N_r)$

Averaged over Table 5.4	Displacement S_{hx}/r_{0P}			
	0	0.31496	Increase: 0 → 0.31496	
			Change, %	
			Range	Average
$\langle I_{r1}-I_{r0} \rangle$	0.408	0.188 ≈ (1–0.54) · 0.408	–131…–26	–77
$I_{r0.max} - I_{r1.min}$	0.68	0.76 ≈ (1 + 0.11) · 0.68	–17…63	12
Averaged over Table 5.4	Outer turbulence scale M_{outer}/r_{0P}			
	5	20	Increase: 5 → 20	
			Change, %	
			Range	Average
$\langle I_{r1}-I_{r0} \rangle$	0.432	0.164 ≈ (1–0.62) · 0.432	–144…–31.7	–70
$I_{r0.max} - I_{r1.min}$	0.77	0.67 ≈ (1–0.14) · 0.77	–229…+15	–34

by the fact that according to our estimates, the probability of detecting the realization of the turbulent screen with $L_F = 0.25r_{0P}$ and $M_{outer} = 5r_{0P}$ for which $I_r(V_{d1} = -2) < I_r(V_{d0} = -1)$ is only 1.6 % for $S_h = 0.11r_{0P}$. To draw Fig. 5.17, only one hundred realizations were used; therefore, the probability that there is no such realization among them is high. Meanwhile, the data presented in Table 5.4 were obtained for 12,800 realizations.

The data presented in Table 5.4 allow us to conclude the following. For $\Delta = 180°$ and $S_{hx} = 0$, values of $\langle I_{r1} - I_{r0} \rangle$ and $I_{r0.max} - I_{r1.min}$ for different pairs (V_{d0}, V_{d1}) exactly coincide; this is a consequence of property (5.20), that is, of the mirror symmetry of the corresponding pairs of dependencies $I_r(N_r)$. For $\Delta = 120°$ and $S_{hx} = 0$, these values differ, but weakly. Obviously, an increase in the displacement S_{hx} breaks the (non)strict symmetry and the indicated (non)exact coincidence of $\langle I_{r1} - I_{r0} \rangle$ and $I_{r0.max} - I_{r1.min}$ for different pairs (V_{d0}, V_{d1}).

In addition, an increase of both S_{hx} and the outer turbulence scale M_{outer} also significantly decreases $\langle I_{r1} - I_{r0} \rangle$ (Table 5.5). The first tendency is in agreement with Fig. 5.16, and the second tendency demonstrates that the detector is less efficient for large-scale beam distortions, but successfully compensates for small-scale distortions. In this respect, it complements adaptive optical systems that better correct the large-scale distortions. The influence of S_{hx} and M_{outer} on the overlap $I_{r0.max} - I_{r1.min}$ is less pronounced. It is intensified with increasing S_{hx} and weakens with increasing M_{outer}. To summarize, this suggests that the increase in S_{hx} (in the above-indicated limits) complicates the identification of the dislocation order V_d stronger than the increase in M_{outer}.

It is remarkable that the arrangement of combinations (Δ, V_{d0}, V_{d1}) in the decreasing order of $\langle I_{r1} - I_{r0} \rangle$ in Table 5.4 for $S_{hx} \approx 0.31496r_{0P}$ coincides with the arrangement of these sets of three parameter values in the decreasing value of the coordinate S_{hx}/r_0 of the intersection points for the dependencies $I_r(S_{hx}/r_0,\ V_{d0})$ and $I_r(S_{hx}/r_0,\ V_{d1})$ (see Table 5.1): (180°, –1, 0), (120°, –1, 1), (120°, –1, –2), and (180°, –1, –2).

5.4 The Data Transmission System on the Basis of the Optical Vortex Detector: The Operation Principle, a Model, Simulation of Turbulence or Noise Influence

5.4.1 *Coding of the Information Bit by the Relative Intensity Value* I_r *or Its Change. Theoretical Backgrounds for Calculations of the Probability of Error in Data Transfer*

As already indicated above, one of the purposes of the present Chapter is to study the possibilities for designing a communication system based on the V_d detector limited mostly by the probability of error in data transfer P_{er}. We restrict ourselves to the examination of a binary communication system. Then it makes sense to consider two coding algorithms: *absolute* algorithm, when "0" or "1" is coded by the V_d value, and *differential* algorithm, when "0" or "1" is coded by the absence or presence of a change in the V_d value. We stress that the quality of the absolute algorithm is directly determined by recognizability of the V_d charge. The second algorithm remains efficient until different V_d values located on different sides of the threshold $I_{r.th}$ that assigns the straight line $I_r = \text{const} = I_{r.th}$ correspond to the intensity I_r in curves similar to those shown in Fig. 5.18. The advantage of the differential algorithm would mean the symmetry of the dependencies $I_r(N_r)$.

To elucidate the influence of random errors on the probability of error in data transfer using the vortex detector, we assume that the measured relative intensity has the form

$$I_{r.th}(t) = I_r(t) + \xi(t) = I_{r.id}(t) + [I_r(t) - I_{r.id}(t)] + \xi(t), \tag{5.26}$$

where $I_r(t)$ is the exact value of the relative intensity obtained by processing of the optical beam *distorted by the turbulent atmosphere*, $I_{r.id}(t)$ is the relative intensity $I_r(t)$ without noise and distorting effect of the atmosphere (ideal relative intensity), and $\xi(t)$ is the random process imitating the presence of the (high-frequency) noise component that could not be removed, for example, by filtration. The reasons for the occurrence of $\xi(t)$ are the atmospheric noise (that was not taken into account in $[I_r(t) - I_{r.id}(t)]$) and the photodetector noise. For the sake of unambiguity of the interpretation and simplicity of further consideration, we consider below that $\xi(t)$ is the photodetector noise.

Coding of the data bit $b(t) \in \{0,\ 1\}$ reduced to the noiseless receiver output by the $I_{\mathrm{r}}(t)$ *value* and *decoding* (using the absolute algorithm of (de)coding and data transfer) is described by the following formulas

$$\begin{gathered} I_{\mathrm{r.id}}(t) = b(t)I_{\mathrm{r.id}} + [1 - b(t)]I_{\mathrm{r.id0}},\ I_{\mathrm{r}}(t) = b(t)I_{\mathrm{r1}}(t) + [1 - b(t)]I_{\mathrm{r0}}(t), \\ b'(t) = I_{\mathrm{r.m}}(t) > I_{\mathrm{r.th}}(t), \end{gathered} \tag{5.27}$$

where $I_{\mathrm{r.id0}}$, $I_{\mathrm{r.id1}}$, and $I_{\mathrm{r0}}(t)$, $I_{\mathrm{r1}}(t)$ are values of $I_{\mathrm{r.id}}(t)$ and $I_{\mathrm{r}}(t)$ corresponding to the transfer of logic zero or unity, and $I_{\mathrm{r.id}}$ is a certain threshold value of the relative intensity $I_{\mathrm{r}}(t)$. Hereinafter, it is accepted $\{0,\ 1\} = \{\text{false, true}\}$. Values of $I_{\mathrm{r0}}(t)$ and $I_{\mathrm{r1}}(t)$ depend on time because of the distorting effect of the atmosphere on the received signal.

Coding of the data bit $b(t) \in \{0,\ 1\}$ by *changing* of the relative intensity and decoding (the differential (de)coding and data transfer algorithm) reduced to the output of the noiseless detector is performed by the following formulas:

$$\begin{aligned} I_{\mathrm{r.id}}(t) &= b(t) \cdot \mathrm{not}[I_{\mathrm{r.id}}(t - \tau_{\mathrm{b}})] + [1 - b(t)]I_{\mathrm{r.id}}(t - \tau_{\mathrm{b}}), \\ I_{\mathrm{r}}(t) &= b(t) \cdot \mathrm{not}[I_{\mathrm{r}}(t - \tau_{\mathrm{b}})] + [1 - b(t)]I_{\mathrm{r}}(t - \tau_{\mathrm{b}}), \\ b'(t) &= [I_{\mathrm{r.m}}(t) > I_{\mathrm{r.th}}(t)][I_{\mathrm{r.m}}(t - \tau_{\mathrm{b}}) \le I_{\mathrm{r.th}}(t - \tau_{\mathrm{b}})] \\ &\quad \vee [I_{\mathrm{r.m}}(t) \le I_{\mathrm{r.th}}(t)][I_{\mathrm{r.m}}(t - \tau_{\mathrm{b}}) > I_{\mathrm{r.th}}(t - \tau_{\mathrm{b}})], \\ \text{or} \quad b'(t) &= \mathrm{not}\{[I_{\mathrm{r.m}}(t) > I_{\mathrm{r.th}}(t)] \Leftrightarrow [I_{\mathrm{r.m}}(t - \tau_{\mathrm{b}}) > I_{\mathrm{r.th}}(t - \tau_{\mathrm{b}})]\}, \\ \text{or} \quad b'(t) &= \{[I_{\mathrm{r.m}}(t) > I_{\mathrm{r.th}}(t)] \ne [I_{\mathrm{r.m}}(t - \tau_{\mathrm{b}}) > I_{\mathrm{r.th}}(t - \tau_{\mathrm{b}})]\}, \end{aligned} \tag{5.28}$$

where

$$\begin{aligned} \mathrm{not}[I_{\mathrm{r.id}}(t - \tau_{\mathrm{b}}) &\equiv I_{\mathrm{r.id0}} \cdot [I_{\mathrm{r.id}}(t - \tau_{\mathrm{b}}) = I_{\mathrm{r.id1}}] \\ &\quad + I_{\mathrm{r.id1}} \cdot [I_{\mathrm{r.id}}(t - \tau_{\mathrm{b}}) = I_{\mathrm{r.id0}}] \equiv \begin{cases} I_{\mathrm{r.id0}}, & I_{\mathrm{r.id}}(t - \tau_{\mathrm{b}}) = I_{\mathrm{r.id1}}, \\ I_{\mathrm{r.id1}}, & I_{\mathrm{r.id}}(t - \tau_{\mathrm{b}}) = I_{\mathrm{r.id0}}, \end{cases} \\ \mathrm{not}[I_{\mathrm{r}}(t - \tau_{\mathrm{b}}) &\equiv I_{\mathrm{r0}} \cdot [I_{\mathrm{r.id}}(t - \tau_{\mathrm{b}}) = I_{\mathrm{r.id1}}] + I_{\mathrm{r1}}(t) \cdot [I_{\mathrm{r.id}}(t - \tau_{\mathrm{b}}) = I_{\mathrm{r.id0}}] \\ &\equiv \begin{cases} I_{\mathrm{r0}}, & I_{\mathrm{r.id}}(t - \tau_{\mathrm{b}}) = I_{\mathrm{r.id1}}, \\ I_{\mathrm{r1}}, & I_{\mathrm{r.id}}(t - \tau_{\mathrm{b}}) = I_{\mathrm{r.id0}}. \end{cases} \end{aligned}$$

We assume: (1) The rate of change $(1/\Delta t_{\mathrm{trb}})$ of $I_{\mathrm{r0}}(t)$ and $I_{\mathrm{r1}}(t)$ is much less than that for $\xi(t)$, and the latter is comparable with the rate of change of the (binary) information signal $b(t)$. Filtration of the measured signal $I_{\mathrm{r.m}}(t)$ does not allow us to separate $I_{\mathrm{r}}(t)$ from the mixture $I_{\mathrm{r.m}}(t) = I_{\mathrm{r}}(t) + \xi(t)$. Therefore, the statement made at the beginning of the item is correct: $\xi(t)$ is the high-frequency noise component that cannot be removed by filtration. (2) $\langle \xi(t) \rangle_{\Delta t} = 0$ and $\Delta t \ll \Delta t_{\mathrm{trb}}$, where Δt_{trb} is the characteristic time of change of $I_{\mathrm{r0}}(t)$ and $I_{\mathrm{r1}}(t)$, that is, the characteristic time of change of $I_r(t)$ due to the effect of the atmospheric turbulence. (3) The occurrence of logic zero and unity in the information signal $b(t)$ is equiprobable: $P_0 = P_1 = 0.5$.

Under these assumptions, it can be demonstrated that the probability of error in data transfer using absolute (5.27) or differential (5.28) coding algorithms at the moment of time t when values $I_{r0}(t)$ and $I_{r1}(t)$ are realized obeys the expression

$$P_{\text{er.abs}}(t) = 0.5\{1 + P_{>}[I_{r0}(t)] - P_{>}[I_{r1}(t)]\}, \tag{5.29}$$

$$\text{or } P_{\text{er.dif}}(t) = 0.5\{1 + [P_{>}[I_{r0}(t)] - P_{>}[I_{r1}(t)]]^2\} = 2[1 - P_{\text{er.abs}}(t)]P_{\text{er.abs}}(t), \tag{5.30}$$

where $P_{>}(I_{r0}) \equiv P(I_{\text{r.m}} > I_{\text{r.th}}) \equiv P(I_r + \xi > I_{\text{r.th}}) \equiv P(\xi > I_{\text{r.th}} - I_r) \equiv \int_{I_{\text{r.th}} - I_r + 0}^{+\infty} \rho(\xi)\text{d}\xi$ is the probability that the measured value $I_{\text{r.m}}$ is greater than the threshold value $I_{\text{r.th}}$ for a certain exact value of the relative intensity I_r.

In practice it is important to know the integral P_{er} value (the P_{er} value independent of time and values of the intensities I_{r0} and I_{r1}) that adequately characterizes the data transfer system as a whole. It can be estimated by averaging $P_{\text{er}}(t)$ given by Eq. (5.29) over a representative sample comprising N values of I_{r0} and I_{r1}; for smoothly changing I_{r0} and I_{r1}, it can be estimated from the sample of N realizations of $I_{r0}(t)$ and $I_{r1}(t)$:

$$P_{\text{er.abs}} \approx \frac{1}{N}\sum_{i=1}^{N}\frac{1}{\Delta t_i}\int_0^{\Delta t_i} P_{\text{er.abs}}(t)\text{d}t = 0.5\frac{1}{N}\sum_{i=1}^{N}\frac{1}{\Delta t_i}\int_0^{\Delta t_i}\{1 + P_{>}[I_{r0}(t)] - P_{>}[I_{r1}(t)]\}\text{d}t, \tag{5.31}$$

$$P_{\text{er.dif}} \approx 0.5\frac{1}{N}\sum_{i=1}^{N}\frac{1}{\Delta t_i}\int_0^{\Delta t_i}\{1 - [P_{>}[I_{r0}(t)] - P_{>}[I_{r1}(t)]]^2\}\text{d}t, \tag{5.32}$$

where Δt_i is the duration of the ith realizations $I_{r0}(t)$ and $I_{r1}(t)$.

In calculations of Eqs. (5.29)–(5.32), it is necessary to recall that $I_{\text{r.th}}$ in expressions for $P_{>}$ cannot be constant. Here two strategies of choosing of and controlling over the $I_{\text{r.th}}$ values are suitable. The first strategy is the simplest one, it obeys the expression

$$I_{\text{r.th}} = \text{const} = 0.5(I_{\text{r.id0}} + I_{\text{r.id1}}). \tag{5.33}$$

In the second strategy, $I_{\text{r.th}}(t) \neq \text{const}$, it *is adaptively adjusted* to $I_{r0}(t)$ and $I_{r1}(t)$ values, for example, as follows:

$$I_{\mathrm{r.th}} \approx 0.5[I_{\mathrm{r0}}(t) + I_{\mathrm{r1}}(\mathrm{t})]. \tag{5.34}$$

In practice, signal of type (5.34) can be generated by applying the voltage proportional to $I_{\mathrm{r.th}}(t)$ to a first-order low-frequency filter with time constant $\tau_{\Delta t} : \Delta t \ll \tau_{\Delta t} \ll \Delta t_{\mathrm{trb}}$.

Let the threshold be chosen *adaptively*: $I_{\mathrm{r.th}} \approx 0.5[I_{\mathrm{r0}}(t) + I_{\mathrm{r1}}(\mathrm{t})]$, that is, I_{r0} and I_{r1} be equidistant (and lie on opposite sides) from $I_{\mathrm{r.th}}$, and let the distribution density $\rho(\xi)$ be an even function ($\rho(\xi) = -\rho(\xi)$). Then Eqs. (5.29), (5.31) and (5.30), (5.32) for $P_{\mathrm{er.abs}}(t)$, $P_{\mathrm{er.abs}}$, and $P_{\mathrm{er.dif}}(t)$, $P_{\mathrm{er.dif}}$ can be reduced to the form

$$P_{\mathrm{er.abs}}(t) = P_{+}[I_{\mathrm{r0}}(t)], \tag{5.35}$$

$$P_{\mathrm{er.abs}} \approx \frac{1}{N}\sum_{i=1}^{N}\frac{1}{\Delta t_i}\int\limits_{0}^{\Delta t_i} P_{+}[I_{\mathrm{r0}}(t)]\mathrm{d}t, \tag{5.36}$$

$$P_{\mathrm{er.dif}}(t) = 2\{1 - P_{+}[I_{\mathrm{r0}}(t)]\}P_{+}[I_{\mathrm{r0}}(t)], \tag{5.37}$$

$$P_{\mathrm{er.dif}} \approx 2\frac{1}{N}\sum_{i=1}^{N}\frac{1}{\Delta t_i}\int\limits_{0}^{\Delta t_i} \{1 - P_{+}[I_{\mathrm{r0}}(t)]\}P_{+}[I_{\mathrm{r0}}(t)]\mathrm{d}t, \tag{5.38}$$

where $P_{+}(I_{\mathrm{r}}) \equiv P_{>}(I_{\mathrm{r}}) + 0.5P_{=}(I_{\mathrm{r}})$, and $P_{=}(I_{\mathrm{r}}) \equiv P(I_{\mathrm{r.m}} = I_{\mathrm{r.th}}) \equiv P(I_{\mathrm{r}} + \xi = I_{\mathrm{r.th}}) \equiv P(\xi = I_{\mathrm{r.th}} - I_{\mathrm{r}}) \equiv \int_{I_{\mathrm{r.th}}-I_{\mathrm{r}}-0}^{I_{\mathrm{r.th}}-I_{\mathrm{r}}+0} \rho(\xi)\mathrm{d}\xi$ is the probability that the measured $I_{\mathrm{r.m}}$ value is equal to the threshold value $I_{\mathrm{r.th}}$ for a certain exact value of the relative intensity I_{r}. It is obvious that if the distribution density $\rho(I_{\mathrm{r.th}} - I_{\mathrm{r}})$ is finite, $P_{=}(I_{\mathrm{r}}) = 0$ and $P_{+}(I_{\mathrm{r}}) = P_{>}(I_{\mathrm{r}})$.

An analysis of Eq. (5.30) allows us to compare the behavior of the instantaneous error probabilities for the absolute ($P_{\mathrm{er.abs}}(t)$) and differential ($P_{\mathrm{er.dif}}(t)$) algorithms:

(1) $P_{\mathrm{er.dif}}(t)$ as a function of $P_{\mathrm{er.abs}}(t)$ has a maximum of 0.5 for $P_{\mathrm{er.abs}}(t) = 0.5$,
(2) $P_{\mathrm{er.dif}}(t) = 0$ for $P_{\mathrm{er.abs}}(t) = 0$ and even for $P_{\mathrm{er.abs}}(t) = 1$,
(3) $P_{\mathrm{er.dif}}(t) - P_{\mathrm{er.abs}}(t) = P_{\mathrm{er.abs}}(t)[1 - 2P_{\mathrm{er.abs}}]$; therefore,

- $P_{\mathrm{er.dif}}(t) > P_{\mathrm{er.abs}}(t)$ for $P_{\mathrm{er.abs}}(t) < 0.5$ and
- $P_{\mathrm{er.dif}}(t) < P_{\mathrm{er.abs}}(t)$ for $P_{\mathrm{er.abs}}(t) > 0.5$. Moreover, the integral loss of the differential algorithm for $P_{\mathrm{er.abs}}(t) < 0.5$ ($P_{\mathrm{er.abs}} = 1/8$ as compared to $P_{\mathrm{er.dif}} = 1/6$) is less significant than its gain for $P_{\mathrm{er.abs}}(t) > 0.5$ ($P_{\mathrm{er.abs}} = 3/8$ as compared to $P_{\mathrm{er.dif}} = 1/6$). Therefore, say, if the probabilities that $P_{\mathrm{er.abs}}(t)$ falls within these two regions are equal, the application of the differential algorithm will provide a smaller error in data transfer ($P_{\mathrm{er.abs}} = 1/2$ as compared to $P_{\mathrm{er.dif}} = 1/3$).

Let $\xi(t)$ be the white noise with amplitude δ (that is, $\xi(t) \in [-\delta, \delta]$) and its distribution density $\rho(\xi)$ be uniform and even:

$$\rho(\xi) = \begin{cases} 1/(2\delta), & \xi \in [-\delta, \delta], \\ 0, & \xi \notin [-\delta, \delta]. \end{cases} \tag{5.39}$$

Then

$$P_>(I_r) = \begin{cases} 0, & I_r \le I_{r.th} - \delta, \\ 1, & I_r > I_{r.th} + \delta, \\ (I_r + \delta - I_{r.th})/(2\delta), & I_{r.th} - \delta < I_r \le I_{r.th} + \delta, \end{cases}$$
$$P_+(I_r) = \begin{cases} 0.5, & \delta = 0,\ I_r = I_{r.th}, \\ 0, & I_r < I_{r.th} - \delta, \\ 1, & I_r > I_{r.th} + \delta, \\ (I_r + \delta - I_{r.th})/(2\delta), & I_{r.th} - \delta < I_r \le I_{r.th} + \delta. \end{cases} \tag{5.40}$$

From Eq. (5.40) it follows that if $\delta = \infty$, $P_+(I_r) = P_>(I_r) = 0.5$; if $\delta = 0$, $P_+(I_r) \in \{0, 0.5, 1\}$ and $P_>(I_r) \in \{0, 1\}$; in this case, $P_+(I_r) = 0.5$ can be neglected, since it is obeyed only when $I_r = I_{r.th}(t)$. Taking this into account and using formulas (5.29) and (5.35) and expressions $P_{er.abs} = \langle\langle P_{er.abs}(t)\rangle_t\rangle_N$, $P_{er.dif}(t) = 2[1 - P_{er.abs}(t)]P_{er.abs}(t)$, $P_{er.dif} = \langle\langle P_{er.dif}(t)\rangle_t\rangle_N$ following from Eqs. (5.29)–(5.32) and (5.35)–(5.38), we can compile Table 5.6.

From Table 5.6 it follows that if the receiver noise is extremely large ($\delta = \infty$), all data transfer variants have identical values $P_{er} = 0.5$. If the receiver is noiseless ($\delta = 0$), the differential algorithm and adaptive threshold are preferable. The

Table 5.6 Comparison of probability of error in data transfer P_{er} using the absolute and differential algorithms with constant (5.33) and adaptive (5.34) threshold $I_{r.th}(t)$ for ideal ($\delta = 0$) and extremely noisy ($\delta = \infty$) photodetectors

	$I_{r.th}(t)$ = const $\Longrightarrow P_{er.abs}(t) = 0.5\{1 + P_>[I_{r0}(t)] - P_>[I_{r1}(t)]\}$		$I_{r.th}(t) \approx 0.5[I_{r0}(t) + I_{r1}(t)]$, Eq. (5.34) $\Longrightarrow P_{er.abs}(t) = P_+[I_{r0}(t)]$	
	$\delta = 0 \Longrightarrow P_>(I_r) \in \{0, 1\}$	$\delta = \infty \Longrightarrow P_>(I_r) = 0.5$	$\delta = 0 \Longrightarrow P_+(I_r) \in \{0, 1\}$	$\delta = \infty \Longrightarrow P_+(I_r) = 0.5$
$P_{er.abs}(t) \in$	$\{0, 0.5, 1\}$	0.5	$\{0, 1\}$	0.5
$P_{er.abs}$	≥ 0[a]	0.5	≥ 0[a]	0.5
$P_{er.dif}(t) \in$	$\{0, 0.5, 0\}$	0.5	0[b]	0.5
$P_{er.dif}$	$\le P_{er.abs}$	0.5	0[c]	0.5

[a]Equality to zero here is possible due to the weak distorting effect of the atmosphere on $I_{r0}(t)$ and $I_{r1}(t)$ values, such that inequality $I_{r0}(t) < I_{r.th}(t) < I_{r1}(t)$ is fulfilled

[b]This is true when $I_{r0}(t) \neq I_{r1}(t)$, for example, when $\langle I(r)\cos[2\pi V_d \cdot M/m + \varphi + S_{sc}(\mathbf{r}) - S_{sc}(\mathbf{r}')]\rangle_{\mathbf{r}}$ (Eq. (16)) depends on V_d, that is, almost always when the level of turbulence *is finite*

[c]This is true due to [b] and the definition $P_{er.dif} = \langle\langle P_{er.dif}(t)\rangle_t\rangle_N$

possibility of *faultless* transfer with $P_{\mathrm{er}} = 0$ should be specially indicated when both these conditions are satisfied.

As to the notes to Table 5.6, we can add the following. Even if the equality $I_{\mathrm{r}0}(t) = I_{\mathrm{r}1}(t)$ is sometimes satisfied, as a rule, this occurs for different realizations $S_{sc}(\mathbf{r})$ and different rotation angles Δ. Therefore, to decrease the error probability in the presence of photodetector noise, we can: design two (or more) interferometers with different angles Δ and to give priority (to trust) to the interferometer for which the ratio $|I_{\mathrm{r.m1}}(t) - I_{\mathrm{r.m0}}(t)|/\delta$ or $|I_{\mathrm{r.m}}(t) - I_{\mathrm{r.th}}(t)|/\delta$ at that time moment was greater. However, this modernization of the communication system is beyond the scope of this book.

It is also important to note that the equality $I_{\mathrm{r.th}}(t) = 0.5[I_{\mathrm{r}0}(t) + I_{\mathrm{r}1}(t)]$ can be satisfied not only through the adaptive threshold adjustment, but also due to the mirror symmetry of the dependencies $I_{\mathrm{r}0}(t)$ and $I_{\mathrm{r}1}(t)$ about a certain straight line $I_{\mathrm{r}} = \mathrm{const}$. Then the value $I_{\mathrm{r.th}}(t) = I_{\mathrm{r}} = \mathrm{const}$, obviously, will also satisfy the desirable relationship $I_{\mathrm{r.th}}(t) = 0.5[I_{\mathrm{r}0}(t) + I_{\mathrm{r}1}(t)]$. It is clear that the efficiencies of the absolute and absolute adaptive differential and differential adaptive algorithms having these properties will be the same, because the threshold $I_{\mathrm{r.th}}(t)$ is constant and equally spaced from $I_{\mathrm{r}0}$ and $I_{\mathrm{r}1}$. In addition, if the V_{d} change (from $V_{\mathrm{d}0}$ to $V_{\mathrm{d}1}$) influences only the phase distribution of the emitted field (for example, the field described by Eq. (5.8) without multiplier $(r/\rho_{\mathrm{V}})^{|V_{\mathrm{d}}|}$), it does not matter what pair $(V_{\mathrm{d}0},\ V_{\mathrm{d}1})$ we have chosen. Recall that all these properties are inherent, for example, in the detector with $\Delta = 180°$ (see Eq. (5.20)). For a communication system constructed on its basis, different pairs $(V_{\mathrm{d}0},\ V_{\mathrm{d}1})$, say, (–1, 0) and (–1, –2), are equivalent. However, we must bear in mind that the above-discussed symmetry is broken by the nonzero displacement S_{hx} of the transmitter and receiver optical axes (see comments to Fig. 5.18 and Table 5.4).

To obtain more detailed conclusions (for example, for $\delta \in (0,\ \infty)$), statistical modeling on a computer is required. To perform it correctly, we now pay attention to integrals in formulas (5.31), (5.32), (5.36), and (5.38). They have the form $\int_0^{\Delta t_i} F(I_{\mathrm{r}0}(t), I_{\mathrm{r}1}(t))\mathrm{d}t$, where F is a function (an integral of $\rho(\xi)$), and $I_{\mathrm{r}0}(t)$ and $I_{\mathrm{r}1}(t)$ cannot be set analytically. We know only some values $I_{\mathrm{r}k}(j\delta t_i)$ at discrete time moments $t_{ij} = 0\delta t_i,\ 1\delta t_i, \ldots,\ J_i\delta t_i$ (where $J_i\delta t_i \equiv \Delta t_i/\delta t_i$ is an integer), and these values $I_{\mathrm{r}k}(j\delta t_i)$ differ relatively strongly. Therefore, the dependencies $I_{\mathrm{r}k}(t)$ must be interpolated based on $I_{\mathrm{r}k}(j\delta t_i)$. This situation is observed, for example, when $I_{\mathrm{r}k}(t)$ are obtained by imitating the influence of the smoothly varying turbulence on $I_{\mathrm{r}}(t)$.

It is well known that in the step h_t of numerical integration, the integrand function must change weakly. This imposes a restriction on the step size h_t and causes the resource-intensive operation. If interpolation of $I_{\mathrm{r}k}(t)$ is linear, $\rho(\xi)$ obeys Eq. (5.39), and hence Eq. (5.40) is valid, these integrals in Eqs. (5.31), (5.32), (5.36), and (5.38) can be taken analytically, thereby significantly decreasing the computation time. Therefore, for convenience of calculations by formulas (5.31), (5.32), (5.36), and (5.38), we further consider that ξ is white noise and Eq. (5.39) for $\rho(\xi)$ is valid.

5.4.2 *Analysis of the Influence of the Turbulent Screen and Communication System Parameters on the Error in Data Transfer*

Leaning upon the theoretical results stated above and the data of modeling, we now elucidate the special features in the behavior of the probability of error in data transfer P_{er} (for four coding algorithms) for the suggested communication system accompanying changes of the amplitude δ of photodetector noise, parameters L_{F} and M_{outer} of smoothly varying turbulence, displacement S_{hx} of the beam source and receiver optical axes, and employed combinations $(\Delta,\ V_{\text{d0}},\ V_{\text{d1}})$.

Let us emphasize that the performance of the absolute algorithm is directly connected with recognizability of the V_{d} value. The differential algorithm remains efficient until the intensities I_{r} located on opposite sides of the threshold $I_{\text{r.th}}$ correspond to different V_{d} values and the intensities I_{r} located on the same side of the threshold $I_{\text{r.th}}$ correspond to identical V_{d} values. For curves shown in Fig. 5.18, the threshold is shown by the straight line $I_{\text{r}} = \text{const} = I_{\text{r.th}}$. The advantage of the differential algorithm is the symmetry of dependencies $I_{\text{r}}(N_{\text{r}})$. The foregoing provides the basis for prediction of the properties of algorithms using the one-dimensional, $\rho(I_{\text{r}})$, and two-dimensional, $\rho(I_{\text{r0}},\ I_{\text{r1}})$, probability densities drawn in Figs. 5.20, 5.21, 5.22, 5.23 and 5.24. Indeed, we now consider the evident geometrical treatment of P_{er} values with the help of $\rho(I_{\text{r0}},\ I_{\text{r1}})$ and $\rho(I_{\text{r}})$.

Let us consider four quadrants Qu_q on the plane $(I_{\text{r0}},\ I_{\text{r1}})$, where $q = 1\ldots4$ is the serial quadrant number (Fig. 5.25). We first assume that the photodetector is noiseless ($\xi(t) = 0$, in particular, $\delta = 0$) and the threshold is fixed: $I_{\text{r.th}} = 1$, that is, only absolute differential algorithm is used. Then the relationship between the measured value of the pair of relative intensities $(I_{\text{r0}}(t),\ I_{\text{r1}}(t))$ and the arising error of this or that symbol transfer is formalized in the form of Table 5.7. Lines drawn in the table indicate faultless data transfer. This means that the error character depends on the serial number of the quadrant in which the pair $(I_{\text{r0}}(t),\ I_{\text{r1}}(t))$ is located and on the employed algorithm. It is obvious that the relative number N_q/N_Σ of events corresponding to quadrant No. q and event $(I_{\text{r0}}(t),\ I_{\text{r1}}(t)) \in Qu_q$ depends on the form and localization of the probability density $\rho(I_{\text{r0}},\ I_{\text{r1}})$. Namely, $N_q/N_\Sigma \to \iint_{Qu_q} \rho(I_{\text{r0}},\ I_{\text{r1}})\mathrm{d}I_{\text{r0}}\mathrm{d}I_{\text{r1}}$ when $N_\Sigma \to \infty$, where N_q is the number of events $(I_{\text{r0}},\ I_{\text{r1}}) \in Qu_q$ and $N_\Sigma = N_1 + N_2 + N_3 + N_4$. The weight with which these events are considered in calculations of the error probability P_{er} is given in Table 5.7 under assumption of equal probabilities of transfer of symbols “0” and “1.”

The one-dimensional probability density $\rho(I_{\text{r}})$ shown in Fig. 5.20 is also capable of forming geometrical image of the probability P_{er}, but only for the absolute algorithm. The matter is that the event $I_{\text{r0}} > I_{\text{r.th}}$ or $I_{\text{r1}} \le I_{\text{r.th}}$ corresponds to the error of transfer of symbol “0” or “1,” and the probability of error in transfer of symbol “0” is $P_{\text{er0.abs}} = \int_{I_{\text{th}}+0}^{\infty} \rho(I_{\text{r0}})\mathrm{d}I_{\text{r0}}$; that for symbol “1” is

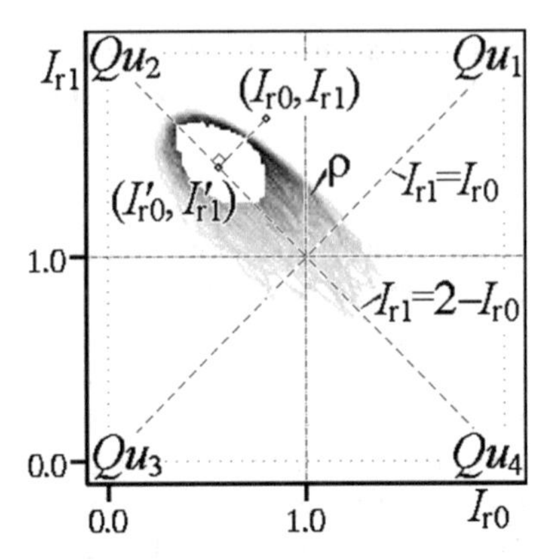

Fig. 5.25 Subdivision of the plane $(I_{r0},\ I_{r1})$ or $(I'_{r0},\ I'_{r1})$ into the quadrants Qu_q. The main $I_{r0} = I_{r1}$, and the descending, $I_{r1} = 2 - I_{r0}$, diagonals are shown. The *points* $(I_{r0},\ I_{r1})$ and $(I'_{r0},\ I'_{r1})$ and the *dashed line* connecting them symbolize transformation of values $(I_{r0},\ I_{r1})$ into $(I'_{r0},\ I'_{r1})$, that is, the event of projection of the probability density $\rho(I_{r0},\ I_{r1})$ onto the straight line $I'_{r1} = 2 - I'_{r0}$ with the formation of $\rho(I'_{r0},\ I'_{r1})$

$P_{er1.abs} = \int_{-\infty}^{I_{th}+0} \rho(I_{r1})dI_{r1}$. Moreover, $P_{er.abs} = 0.5(P_{er0.abs} + P_{er1.abs})$, $(N_1 + N_4)/N_\Sigma \to P_{er0.abs}$, $(N_3 + N_4)/N_\Sigma \to P_{er1.abs}$ when $N_\Sigma \to \infty$.

The special feature of application of these geometrical considerations to the algorithms with *adaptive* threshold $I_{r.th}(t) = 0.5[I_{r0}(t) + I_{r1}(t)]$ is that the plane subdivision into the quadrants must depend on the measured value of the dyad $(I_{r0}(t),\ I_{r1}(t))$. In fact, this makes interpretation of P_{er} with the help of $\rho(I_{r0},\ I_{r1})$ impossible. We can bypass the above-indicated effect of quadrant displacement proceeding to the new system of coordinates $(I'_{r0},\ I'_{r1})$: $I'_{r0}(t) = 1 + I_{r0}(t) - I_{r.th}(t)$, $I'_{r1}(t) = 1 + I_{r1}(t) - I_{r.th}(t)$ (see Fig. 5.25). Now the quadrants (1, 1) are joint by the point $(I_{r.th}(t),\ I_{r.th}(t))$, floating relative to $(I_{r0},\ I_{r1})$ rather than by the fixed point (1, 1) of the old system of coordinates $(I_{r0},\ I_{r1})$. The intensities $I'_{r0}(t)$ and $I'_{r1}(t)$ and the probability density $\rho(I'_{r0},\ I'_{r1})$ for the algorithms with adaptive threshold play the same role as $I_{r0}(t)$, $I_{r1}(t)$, and $\rho(I_{r0},\ I_{r1})$ for algorithms with $I_{r.th} = \text{const}$. Therefore, the interpretation suggested above can be used for them.

By definition, the intensities $I'_{r0}(t)$ and $I'_{r1}(t)$ are symmetric about $I_{r.th}(t)$ irrespective of the parameters of turbulence, detector, and misalignment S_{hx}. It can be demonstrated that transition from $\rho(I_{r0},\ I_{r1})$ to $\rho(I'_{r0},\ I'_{r1})$, that is, from constant to adaptive threshold is reduced to *the projection* of $\rho(I_{r0},\ I_{r1})$ onto the descending diagonal $I'_{r1} = 2 - I'_{r0}$ (see Fig. 5.25). Therefore, the probability density $\rho(I'_{r0},\ I'_{r1}) \neq 0$ only on a certain segment of the straight line $I'_{r1} = 2 - I'_{r0}$ (Fig. 5.25), in complete agreement with the case of $\rho(I_{r0},\ I_{r1})$ for $\Delta = 180°$ and $S_{hx} = 0$ (Figs. 5.21 and 5.23). Moreover, this segment is displaced from Qu_2 to Qu_4 with decreasing L_F, becoming more and more symmetric about $(I'_{r0},\ I'_{r1}) = (1,\ 1)$.

Owing to such projection, a number of events $(I_{r0}(t),\ I_{r1}(t)) \in Qu_1$ and $(I_{r0}(t),\ I_{r1}(t)) \in Qu_3$ from the half-plane below the diagonal $I_{r0} = I_{r1}$ will be

Table 5.7 Relationship of values of the pair $(I_{r0}(t),\ I_{r1}(t))$ with instantaneous errors of transfer of symbols "0" and "1" and integral probability of error in data transfer P_{er} for ideal photodetector $(\delta = 0)$

Event: $(I_{r0}(t), I_{r1}(t))$ belong to quadrant No. q, that is, $(I_{r0}(t), I_{r1}(t)) \in Qu_q$				
q	Absolute algorithm		Differential algorithm	
	Wrongly transfers	Weight of the event	Wrongly transfers	Weight of the event
1	"0"	0.5	"1"	0.5
2	–	0	–	0
3	"1"	0.5	"1"	0.5
4	Both "0" and "1"	1	–	0
P_{er}	$[0.5(N_1 + N_3) + N_4]/N_\Sigma$		$0.5(N_1 + N_3)/N_\Sigma$	

projected onto the fourth quadrant of the plane $(I'_{r0},\ I'_{r1})$, and a number of events from the half-plane above the diagonal $I_{r0} = I_{r1}$ will be projected onto the second quadrant. This allows us to compare the characteristics of the algorithms (both absolute and differential) with constant and adaptive thresholds. Indeed, projection onto Qu_4 (Table 5.7) causes the weight of the events corresponding to values $(I_{r0},\ I_{r1})$ (equal to 0.5) to change: (1) it increases (by 0.5) for the absolute algorithm and (2) it decreases (by 0.5) for the differential algorithm. The projection onto Qu_2 causes the weight to decrease (by 0.5). As a result, $P_{er.dif}$ for the noiseless photodetector *does not increase*, whereas the probability $P_{er.abs}$ can both increase and decrease. It is obvious that if the relative numbers of the above-mentioned events are equal, values $P_{er.abs}$ for the algorithms with constant and adaptive thresholds will be identical. For example, this is the case when the function $\rho(I_{r0},\ I_{r1})$ within Qu_1 and Qu_3 is symmetric about $I_{r0} = I_{r1}$. As a rule, this symmetry is approximately observed (Figs. 5.21, 5.22, 5.23 and 5.24).

If we do not restrict ourselves by the condition $\delta = 0$ (which leads to $\xi(t) = 0$), in the preceding reasoning we must replace (except expression $I_{r.th}(t) = 0.5[I_{r0}(t) + I_{r1}(t)]$ the intensity $I_r(t)$ by the measured quantity $I_{r.m}(t) = I_r(t) + \xi(t)]$, see Eq. (5.26). Semantic nuances are distinguishable here as well. For decoding, *absolute* algorithm (5.27) envisages the application of the $I_{r.m}(t)$ value at only one moment of time. Therefore, in the preceding reasoning we must proceed from $I_{r0}(t)$ and $I_{r1}(t)$ to the intensities $I_{r.m0}(t) = I_{r0}(t) + \xi(t)$ and $I_{r.m1}(t) = I_{r1}(t) + \xi(t)$, where the photodetector noise components $\xi(t)$ are identical.

For decoding by means of *differential* algorithm (5.28), the intensity values $I_{r.m0}$ and $I_{r.m1}$ at different time moments (delayed by τ_b), for example, $I_{r.m0}(t) = I_{r0}(t) + \xi(t)$ and $I_{r.m1}(t - \tau_b) = I_{r1}(t - \tau_b) + \xi(t - \tau_b)$ are used. Owing to the slowly varying turbulence (small changes of $I_{r0}(t)$ and $I_{r1}(t)$ during $\tau b \ll \Delta t_{trb}$, see comments to Eq. (5.28)), we can set $I_{r.m1}(t) \approx I_{r1}(t) + \xi(t - \tau_b)$, and owing to the delta-correlation of the chosen random process $\xi(t)$—white noise—we arrive at the expressions $I_{r.m0}(t) \approx I_r(t) + \xi_0(t)$ and $I_{r.m1}(t) \approx I_r(t) + \xi_1(t)$, where $\xi_0(t) \equiv \xi(t)$ and $\xi_1(t) \equiv \xi(t - \tau_b)$ are two uncorrelated random processes with identical distribution functions. Ergo, in the case of the differential algorithm, the intensities

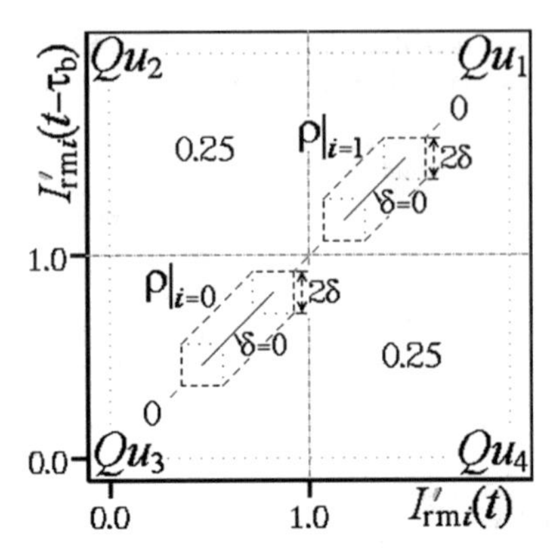

Fig. 5.26 Subdivision of the plane $(I_{\mathrm{r.m}i}(t),\ I_{\mathrm{r.m}i}(t-\tau_\mathrm{b}))$ or $(I'_{\mathrm{r.m}i}(t),\ I'_{\mathrm{r.m}i}(t-\tau_\mathrm{b}))$ into the quadrants Qu_q, $I \in \{0,\ 1\}$, $q \in \{1,\ 2,\ 3,\ 4\}$. For $\delta = 0$ and $\delta \neq 0$, the boundaries of the regions in which the probability density $\rho(I_{\mathrm{r.m}i}(t),\ I_{\mathrm{r.m}i}(t-\tau_\mathrm{b})) \neq 0$ or $\rho(I'_{\mathrm{r.m}i}(t),\ I'_{\mathrm{r.m}i}(t-\tau_\mathrm{b})) \neq 0$ are schematically shown. Values 0 and 0.25 indicate weights of the events $(I_{\mathrm{r.m}i}(t),\ I_{\mathrm{r.m}i}(t-\tau_\mathrm{b})) \in Qu_q$ or $(I'_{\mathrm{r.m}i}(t),\ I'_{\mathrm{r.m}i}(t-\tau_\mathrm{b})) \in Qu_q$ for the error of transfer of "0" by the differential algorithm

$I_{\mathrm{r}0}(t)$ and $I_{\mathrm{r}1}(t)$ in our reasoning must be replaced by these $I_{\mathrm{r.m}0}(t)$ and $I_{\mathrm{r.m}1}(t)$ values.

In addition, the error of transfer of logic zero arises in the differential algorithm when $\delta \neq 0$ (with the probability $P_{\mathrm{er.dif0}}(t)$), whereas there was no reason even to discuss it for $\delta = 0$, because $P_{\mathrm{er.dif0}}(t) = 0$ for $\delta = 0$. This error arises, for example, when symbol "0" is transferred by $I_{\mathrm{r.m}0}(t)$ and $I_{\mathrm{r.m}0}(t-\tau_\mathrm{b})$ values arranged on different sides of the threshold $I_{\mathrm{r.th}}$ Analogous statement is true for $I_{\mathrm{r.m}1}(t)$ and $I_{\mathrm{r.m}1}(t-\tau_\mathrm{b})$ values. Since $I_{\mathrm{r}0}(t-\tau) \approx I_{\mathrm{r}0}(t)$ and $I_{\mathrm{r}1}(t-\tau) \approx I_{\mathrm{r}1}(t)$, such errors arise only if $\xi(t-\tau_\mathrm{b}) \neq 0$ (that is, $\delta \neq 0$).

To form the geometric image of errors of "0" transfer, the probability densities $\rho(I_{\mathrm{r}0}(t),\ I_{\mathrm{r}0}(t))$, $\rho(I_{\mathrm{r}1}(t),\ I_{\mathrm{r}1}(t))$ and $\rho(I_{\mathrm{r.m}0}(t),\ I_{\mathrm{r.m}0}(t-\tau_\mathrm{b}))$, $\rho(I_{\mathrm{r.m}1}(t),\ I_{\mathrm{r.m}1}(t-\tau_\mathrm{b}))$ corresponding to them (see Fig. 5.26) must be considered. For the letter, the following equalities are valid:

$$\begin{aligned}
&\rho(I_{\mathrm{r.m}0}(t),\ I_{\mathrm{r.m}0}(t-\tau_\mathrm{b})) \equiv \rho(I_{\mathrm{r}0}(t)+\xi_0(t),\ I_{\mathrm{r}0}(t-\tau_\mathrm{b})+\xi_0(t-\tau_\mathrm{b}))\\
&\quad\approx \rho(I_{\mathrm{r}0}(t)+\xi_0(t),\ I_{\mathrm{r}0}(t)+\xi(t-\tau_\mathrm{b})) = \rho(I_{\mathrm{r}0}(t)+\xi_0(t),\ I_{\mathrm{r}0}(t)+\xi_1(t)),\\
&\rho(I_{\mathrm{r.m}1}(t),\ I_{\mathrm{r.m}1}(t-\tau_\mathrm{b})) \equiv \rho(I_{\mathrm{r}1}(t)+\xi_1(t),\ I_{\mathrm{r}1}(t-\tau_\mathrm{b})+\xi_1(t-\tau_\mathrm{b}))\\
&\quad\approx \rho(I_{\mathrm{r}1}(t)+\xi_1(t),\ I_{\mathrm{r}1}(t)+\xi(t-2\tau_\mathrm{b})) = \rho(I_{\mathrm{r}1}(t)+\xi_1(t),\ I_{\mathrm{r}1}(t)+\xi_2(t)),
\end{aligned}$$

where $\xi_i(t) \equiv \xi(t-i\tau_\mathrm{b})$ are uncorrelated processes (due to delta-correlation of $\xi(t)$).

On the planes $(I_{\mathrm{r}0}(t),\ I_{\mathrm{r}0}(t))$, $(I_{\mathrm{r}1}(t),\ I_{\mathrm{r}1}(t))$ for $\delta = 0$ (Fig. 5.25) and on the planes $(I_{\mathrm{r.m}0}(t),\ I_{\mathrm{r.m}0}(t-\tau_\mathrm{b}))$, $(I_{\mathrm{r.m}1}(t),\ I_{\mathrm{r.m}1}(t-\tau_\mathrm{b}))$ for $\delta \neq 0$ (Fig. 5.26), the intensity falling within the first or third quadrant demonstrates faultless transfer of "0" (the weight of the event is zero), and their falling within the second or fourth quadrant demonstrate erroneous transfer of “0” (the weight of the event is 0.25). Obviously, the intensity values lying on the planes $(I_{r0}(t),\ I_{r0}(t))$ and $(I_{r1}(t),\ I_{r1}(t))$, by definition, lie on the diagonals, and errors are absent.

For algorithms with adaptive threshold $I_{r.th}(t)$, the role of pairs $(I_{r0},\ I_{r1})$ is played by pairs $(I'_{r0},\ I'_{r1})$. Therefore, the replacement of $(I_{r0},\ I_{r1})$ by $(I'_{r0},\ I'_{r1})$ in the above reasoning still provides the transfer of the properties of algorithms with $I_{r.th}(t) =$ const to the adaptive algorithms, but already for $\delta \neq 0$.

With allowance for the established differences in treatment of intensities $I_{r.m0}(t)$ and $I_{r.m1}(t)$ for $\delta \neq 0$, we must consider the probability density $\rho(I_{r.m0},\ I_{r.m1})$ or $\rho(I'_{r.m0},\ I'_{r.m1})$ as well as $\rho(I_{r.m0}(t),\ I_{r.m0}(t-\tau_b))$, $\rho(I_{r.m1}(t),\ I_{r.m1}(t-\tau_b))$ or $\rho(I'_{r.m0}(t),\ I'_{r.m0}(t-\tau_b))$, $\rho(I'_{r.m1}(t),\ I'_{r.m1}(t-\tau_b))$. Naturally, they are broadened in comparison with $\rho(I_{r0},\ I_{r1})$ or $\rho(I'_{r0},\ I'_{r1})$ as well as $\rho(I_{r0}(t),\ I_{r0}(t))$, $\rho(I_{r1}(t),\ I_{r1}(t))$ or $\rho(I'_{r0}(t),\ I'_{r0}(t))$, $\rho(I'_{r1}(t),\ I'_{r1}(t))$ as follows:

- For the absolute algorithm, $\rho(I_{r.m0},\ I_{r.m1})$ or $\rho(I'_{r.m0},\ I'_{r.m1})$ are broadened in proportion to $2 \cdot 2^{0.5}\delta$ along the direction of the main diagonal ($I_{r.m0} = I_{r.m1}$ or $I'_{r.m0} = I'_{r.m1}$), and the degree of broadening is the same along the direction $I_{r.m0} = -I_{r.m1}$ or $I'_{r.m0} = -I'_{r.m1}$.
- For the differential algorithm, all probability densities are broadened in proportion to 2δ along the direction of each of their coordinate axes.

More strictly, for the absolute algorithm, for example, the $\rho(I_{r0},\ I_{r1})$ value at the *point* $(I_{r0},\ I_{r1})$ contributes to $\rho(I_{r.m0},\ I_{r.m1})$ values on the *segment* of length $2 \cdot 2^{0.5}\delta$ parallel to the line $I_{r.m0} = I_{r.m1}$ and centered at $(I_{r.m0},\ I_{r.m1}) = (I_{r0},\ I_{r1})$. For the differential algorithm, for example, the density $\rho(I_{r0},\ I_{r1})$ influences $\rho(I_{r.m0},\ I_{r.m1})$ values within the *square* centered at $(I_{r.m0},\ I_{r.m1}) = (I_{r0},\ I_{r1})$ with sides 2δ parallel to the coordinate axes. The same statements are true for the algorithms with adaptive threshold $I_{r.th}(t) = 0.5[I_{r0}(t) + I_{r1}(t)]$. In particular, the shape of $\rho(I'_{r.m0},\ I'_{r.m1})$ is not the segment of the line of length l. It is a stripe $2 \cdot 2^{0.5}\delta$ wide: with length l for the absolute algorithm and with length l along the edges and length $l + 2 \cdot 2^{0.5}\delta$ along the stripe axis for the differential algorithm. Obviously, broadening (with increasing δ) leads to a redistribution of the relative number N_q/N_Σ of events $(I_{r.m0}(t),\ I_{r.m1}(t)) \in Qu_q$ together with the relative number of events $(I_{r.m0}(t),\ I_{r.m0}(t-\tau_b)) \in Qu_q$ and $(I_{r.m1}(t),\ I_{r.m1}(t-\tau_b)) \in Qu_q$ and to a change of the error probability P_{er}.

Alternative (to the transition from $I_r(t)$ to $I_{r.m}(t)$) explanation of the reason for changing the error probability P_{er} is also possible. For $I_{r0}(t)$ and $I_{r1}(t)$ values, the noise $\delta \neq 0$ causes the necessity to distinguish not only the four events (with three weights, see Table 5.7 and Fig. 5.25) for $(I_{r0}(t)$ and $I_{r1}(t))$, but also the continuum of events and the corresponding continuum of weights for $\rho(I_{r0},\ I_{r1})$ (Fig. 5.27). They are considered to be the instantaneous probability of error in data transfer $P_{er}(t)$ as a function of I_{r0} and I_{r1}, since the integral probability of error $P_{er} = \int_0^{+\infty}\int_0^{+\infty} \rho(I_{r0},\ I_{r1}) P_{er}(I_{r0},\ I_{r1}) dI_{r0} dI_{r1}$. Here $\int_0^{+\infty}\int_0^{+\infty} \rho(I_{r0},\ I_{r1}) dI_{r0} dI_{r1}$

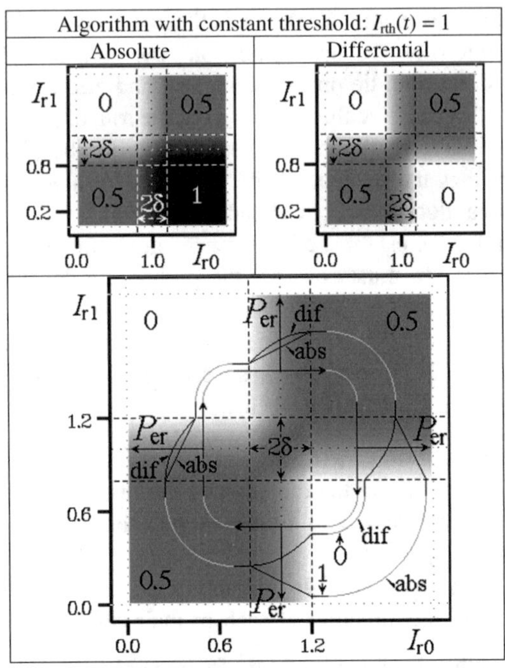
Algorithm with constant threshold: $I_{rth}(t) = 1$
Absolute
Differential
I_{r1}
I_{r0}
0
0.5
1
2δ
0.8
0.2
0.0
1.0
0.6
1.2
P_{er}
dif
abs

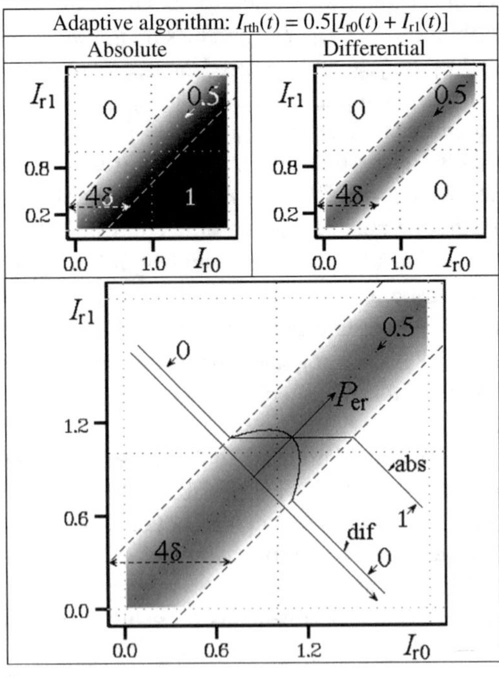
Adaptive algorithm: $I_{rth}(t) = 0.5[I_{r0}(t) + I_{r1}(t)]$
Absolute
Differential
I_{r1}
I_{r0}
0
0.5
1
4δ
0.8
0.2
0.0
1.0
0.6
1.2
P_{er}
abs
dif

◀ **Fig. 5.27** Forms of the instantaneous probabilities of errors in data transfer $P_{\text{er.abs}}(t)$ and $P_{\text{er.dif}}(t)$ as functions of I_{r0} and I_{r1} (see Eqs. (5.29), (5.30), (5.35), and (5.37)) for the photodetector noise with amplitude δ ($\delta = 0.2$ in contrast with Table 5.7 for which $\delta = 0$) for two (de)coding algorithms. The *darkest region* is for $P_{\text{er}}(I_{\text{r0}}, I_{\text{r1}}) = 1$, and the white region is for $P_{\text{er}}(I_{\text{r0}}, I_{\text{r1}}) = 0$. The *lower figures* show diagrams against the background of the distribution $P_{\text{er.dif}}(I_{\text{r0}}, I_{\text{r1}})$. The diagrams explain the rules of change $P_{\text{er.abs}}(I_{\text{r0}}, I_{\text{r1}})$ and $P_{\text{er.dif}}(I_{\text{r0}}, I_{\text{r1}})$ when moving in the plane $(I_{\text{r0}}, I_{\text{r1}})$ around the point (1, 1) at sufficient spacing from it ($I_{\text{r.th}}(t) = 1$) and from the half-plane above the diagonal $I_{\text{r0}} = I_{\text{r1}}$ in the half-plane below it ($I_{\text{r.th}}(t) \neq \text{const}$)

plays the role of the operator $\frac{1}{N}\sum_{i=1}^{N}\frac{1}{\Delta t_i}\int_0^{\Delta t_i} \mathrm{d}t$ in Eqs. (5.31), (5.32), (5.36), and (5.38). As can be seen, this alternative explanation retains a certain two-dimensional density (in our case, $\rho(I_{\text{r0}}, I_{\text{r1}})$) independent of δ, supplementing it with the continuum of weights for the events depending on δ. Previously, on the contrary, the number and values of weights were fixed, and the density ρ was broadened in proportion to δ.

The general rule of δ influence is that: the greater (on average) the derivative of $\rho(I_{\text{r0}}, I_{\text{r1}})$ or $\rho(I'_{\text{r0}}, I'_{\text{r1}})$ in the direction of ρ broadening in the δ-vicinity of the quadrant boundaries, the stronger the influence. To be more exact, the greater the scalar product of the derivative into the normal to the boundary. The normal is directed toward the quadrant in which the event has a larger weight (see Table 5.7), and a negative value of the product demonstrates the negative influence of δ, that is, an increase in P_{er}. For the differential algorithm, the special feature is that broadening occurs in two directions parallel to the axes. Therefore, important is the degree of proximity of the $(I_{\text{r0}}, I_{\text{r1}})$ or $(I'_{\text{r0}}, I'_{\text{r1}})$ value to the point (1, 1). In addition, the form of functions $\rho(I_{\text{r0}}(t), I_{\text{r0}}(t))$, $\rho(I_{\text{r1}}(t), I_{\text{r1}}(t))$ or $\rho(I'_{\text{r0}}(t), I'_{\text{r0}}(t))$, $\rho(I'_{\text{r1}}(t), I'_{\text{r1}}(t))$ used to predict the probability of error in transfer of symbol "0" (Fig. 5.26). Therefore, the derivative of ρ with respect to the normal to the boundary in its δ-vicinity (on average) becomes actual in this case. In particular, when the probability density ρ is localized far from the quadrant boundaries, noise does not influence the P_{er} value. As a rule, this situation corresponds to the weak turbulence and small misalignment S_{hx}.

It can be easily seen that geometrical interpretation for the differential algorithm based on the assumption about broadening of the probability density is more complicated than for the absolute algorithm. Therefore, it seems more efficiently to use the alternative approach considering the transformation of weights of events for $\rho(I_{\text{r0}}, I_{\text{r1}})$ or $\rho(I'_{\text{r0}}, I'_{\text{r1}})$ (see Fig. 5.27). It leans on the analytical expressions and is more strict assuming the express estimate of the influence of δ and other factors on the probability $P_{\text{er}} = \int_0^{+\infty}\int_0^{+\infty} \rho(I_{\text{r0}}, I_{\text{r1}}) P_{\text{er}}(I_{\text{r0}}, I_{\text{r1}}) \mathrm{d}I_{\text{r0}} \mathrm{d}I_{\text{r1}}$. The assumptions on the influence of the photodetector noise as a whole are also valid here.

When the threshold $I_{\text{r.th}}(t) = \text{const}$ (5.33) and the probability density $\rho(I_{\text{r0}}, I_{\text{r1}})$ is axially symmetric about the point (1, 1), all relative numbers are equal: $N_q = 0.25$, and according to Table 5.7, $P_{\text{er.abs}} = 0.5$, but $P_{\text{er.dif}} = 0.25$. Thus, the photodetector noise ($\delta \neq 0$) does not influence the $P_{\text{er.abs}}$ value; moreover, it can be demonstrated that it does not influence even when $\rho(I_{r0}, I_{r1})$ is symmetric about the main diagonal $I_{r0} = I_{r1}$ (Fig. 5.27). Nevertheless, irrespective of the shape of $\rho(I_{r0}, I_{r1})$, the noise increases $P_{er.dif}$ to 0.5. Indeed, when the noise stretches $\rho(I_{r.m0}, I_{r.m1})$ strongly along the coordinate axes, it does not change the relative numbers $N_q = 0.25$, retaining the probability of erroneous transfer of "1" at a level of 0.25. However, the noise increases the relative numbers N_2/N_Σ and N_4/N_Σ of events $(I_{r.mi}(t), I_{r.mi}(t - \tau_b)) \in Qu_2$ and $(I_{r.mi}(t), I_{r.mi}(t - \tau_b)) \in Qu_4$ from zero to 0.25 on the planes $(I_{r.m0}(t), I_{r.m0}(t - \tau_b))$ and $(I_{r.m1}(t), I_{r.m1}(t - \tau_b))$ (Fig. 5.26), thereby increasing the probability of erroneous transfer of "0" from zero to 0.25. As a whole, the probability $P_{er.dif}$ approaches 0.5.

This geometrically deduced thesis is in agreement with analytically derived results given in Table 5.6. It is important to note that the axial symmetry of $\rho(I_{r0}, I_{r1})$ is caused primarily by the strong turbulence, displacement of the axes ($S_{hx} \neq 0$), and to a lesser degree by the increase in the outer scale of turbulence (Figs. 5.21, 5.22, 5.23 and 5.24). Therefore, we can state that the probability of error in data transfer $P_{er.dif} \leq 0.25$ for $\delta = 0$ even for very strong turbulence ($L_F \to 0$).

For the adaptive threshold $I_{r.th}(t) \neq \text{const}$, even very strong turbulence does not change the fact that values $\rho(I'_{r0}, I'_{r1}) \neq 0$ belong to the line $I'_{r1} = 2 - I'_{r0}$, though it causes the concentration of the distribution and the increase in the density $\rho(I'_{r0}, I'_{r1})$ in the vicinity of point (1, 1), making it more symmetric with respect to the main diagonal. In turn, the derivative of $\rho(I'_{r0}, I'_{r1})$ also increases. Therefore, by analogy with $I_{r.th}(t) = \text{const}$, the photodetector noise ($\delta \neq 0$) has virtually no effect on the $P_{er.abs}$ value for small L_F values. For the differential algorithm (by virtue of $\rho(I'_{r0}, I'_{r1}) \neq 0, I'_{r1} = 2 - I'_{r0}$), the probability $P_{er.dif} = 0$ for $\forall L_F$. By virtue of concentration of $\rho(I'_{r0}, I'_{r1})$ at the point (1, 1), the degree of influence of δ is the greater, the stronger the turbulence. This conclusion is also true for the differential algorithm with constant threshold $I_{r.th}(t) = \text{const}$ owing to the concentration of density $\rho(I_{r0}, I_{r1})$, this time Gaussian one. For the same reason as in case of $I_{r.th}(t) = \text{const}$, that is, because of stretching of $\rho(I_{r.m0}, I_{r.m1})$, $\rho(I_{r.m0}(t), I_{r.m0}(t - \tau_b))$ and $\rho(I_{r.m1}(t), I_{r.m1}(t - \tau_b))$ along the coordinate axes, the probability $P_{er.dif} \to 0.5$ when $\delta \to \infty$.

By virtue of a certain similarity of the effect of increasing S_{hx} and M_{outer} with the effect of decreasing L_F on ρ (overflow of ρ to the point (1, 1) and concentration in its vicinity), their effects on the probability of error in data transfer P_{er} are also similar. The special feature of the effect of axis displacement S_{hx} for the large Fried radii ($L_F > 0.25 r_{0P}$) and $|V_{d0}| \neq |V_{d1}|$ is that the moderate S_{hx} values ($S_{hx} \approx 0.31496 r_{0P}$) reduce the degree of symmetry of $\rho(I_{r0}, I_{r1})$ about the line $I_{r1} = 2 - I_{r0}$. Recall that the assumption about one of the reasons for the advantage of the differential algorithm (with $I_{r.th} = \text{const}$) over the absolute one is based exactly on this symmetry. In some cases, exactly this decrease weakens its advantage.

In the foregoing reasoning, we considered the point (1, 1) because we implicitly assumed that the constant threshold is $I_{r.th}(t) = 1$ (and for $I_{r.th}(t) \neq \text{const}$, the point

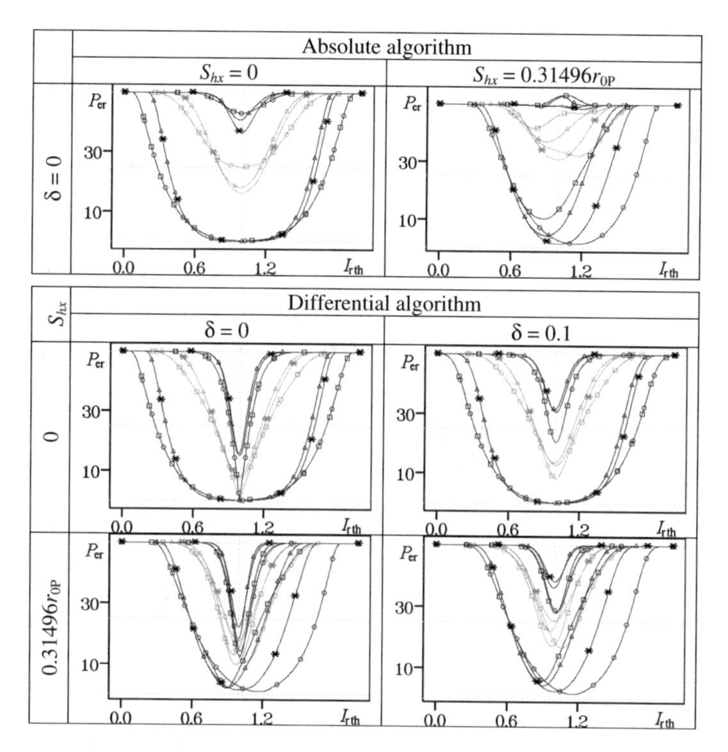

Fig. 5.28 Dependence of the probability of error in data transfer P_{er} (in %) on the established threshold $I_{r.th} = \text{const}$ for $M_{outer} = 5r_{0P}$ and the indicated S_{hx} values and algorithms. The Fried radius L_F/r_{0P} is 0.25, 0.1, and 0.05 for the lower, middle (*grey*), and upper groups of curves. Sets of three values (Δ, V_{d0}, V_{d1}) take values (180°, –1, 0), (180°, –1, –2), (120°, –1, 1), (120°, –1, –2) and are shown by curves with symbols O, □, *, and Δ, respectively

(1, 1) formed the origin of coordinates for the subdivision into the quadrants based on the transformation $I'_{r0}(t) = 1 + I_{r0}(t) - I_{r.th}(t)$ and $I'_{r1}(t) = 1 + I_{r1}(t) - I_{r.th}(t)$). The choice of $I_{r.th}(t) = 1$ seems logical, because these values without distortions are equally spaced from unity. When the axes are displaced ($S_{hx} \neq 0$), the property of equal spacing, generally speaking, is lost (Fig. 5.16). This means that the choice of the threshold is not indisputable: from Figs. 5.28 and 5.29 it can be seen that the error probability P_{er} reaches a minimum when the threshold values differ from unity no more than by 20 % (see also Tables A.1–A.4 in Appendix A in [62]). The sign of this difference and its value correlate with the value of I_r deviation from unity at the intersection points of dependencies $I_r(S_{hx}/r_0)$ (see Fig. 5.16 and Table 5.1). This suggests that in search for an optimum, it is expedient to consider these dependencies. In this case, the increased level of turbulence in some situations can decrease or increase the difference $\left|I_{r.th.optim} - 1\right|$, and the probabilities $P_{er.abs}$ and $P_{er.dif}$ can decrease by a factor of 2–4 at the expense of the chosen value $I_{r.th} = I_{r.th.optim}$.

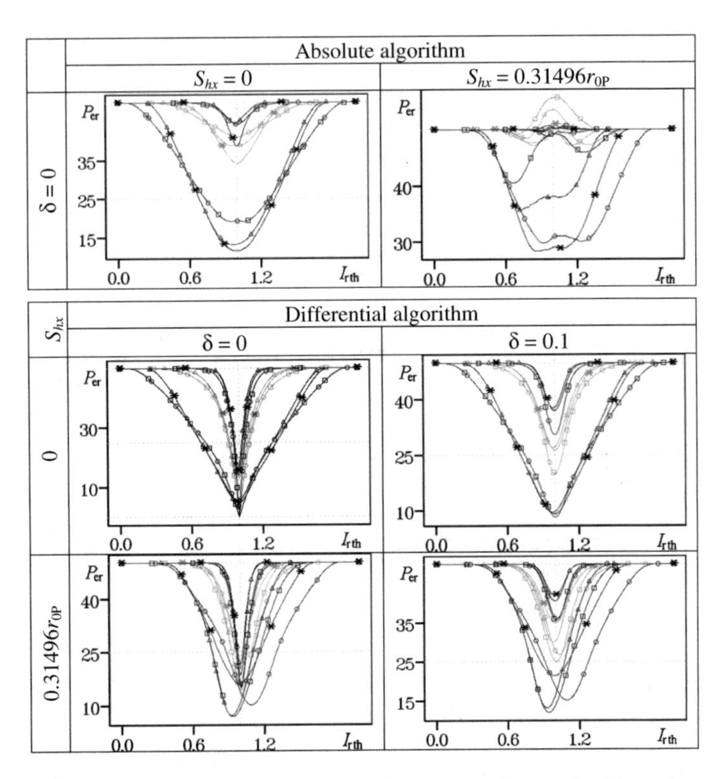

Fig. 5.29 Dependence of the probability of error in data transfer P_{er} (in %) on the established threshold $I_{r.th} = \text{const}$ for $M_{outer} = 20r_{0P}$ and the indicated S_{hx} values and algorithms. The Fried radius L_F/r_{0P} is 0.25, 0.1, and 0.05 for the lower, middle (*grey*), and upper groups of curves. Sets of three values (Δ, V_{d0}, V_{d1}) take values (180°, –1, 0), (180°, –1, –2), (120°, –1, 1), (120°, –1, –2) and are shown by curves with symbols O, □, *, and Δ, respectively

From the above-mentioned tables and figures obtained by direct statistical modeling on a computer of Eqs. (5.31), (5.32), (5.36), and (5.38), it can be seen that generally the optimum threshold level $I_{r.th.optim}$ is sensitive to misalignment, turbulence intensity, and photodetector noise. It makes no sense to solve the multi-parameter problem of choice of such threshold (under conditions of variable factors influencing it), since its control algorithm will be not simpler than that considered above ($I_{r.th}(t) = 0.5[I_{r0}(t) + I_{r1}(t)]$), and its efficiency will be deliberately lower (Tables A.1–A.4 in Appendix A in [62]; Figs. 5.28 and 5.29). We add that the dependencies of the probability P_{er} of error in data transfer on the established threshold $I_{r.th} = \text{const}$ for the absolute algorithm on the segment $\delta \in [0,\ 0.1]$ are visually identical; therefore, Figs. 5.28 and 5.29 illustrate only the situation with $\delta = 0$. It is important to note that with increasing turbulence level, the deeps in the plots $P_{er}(I_{r.th})$ are significantly narrowed, and with increasing noise δ, their depths decrease. This decrease is the stronger, the smaller the L_F value. These statements

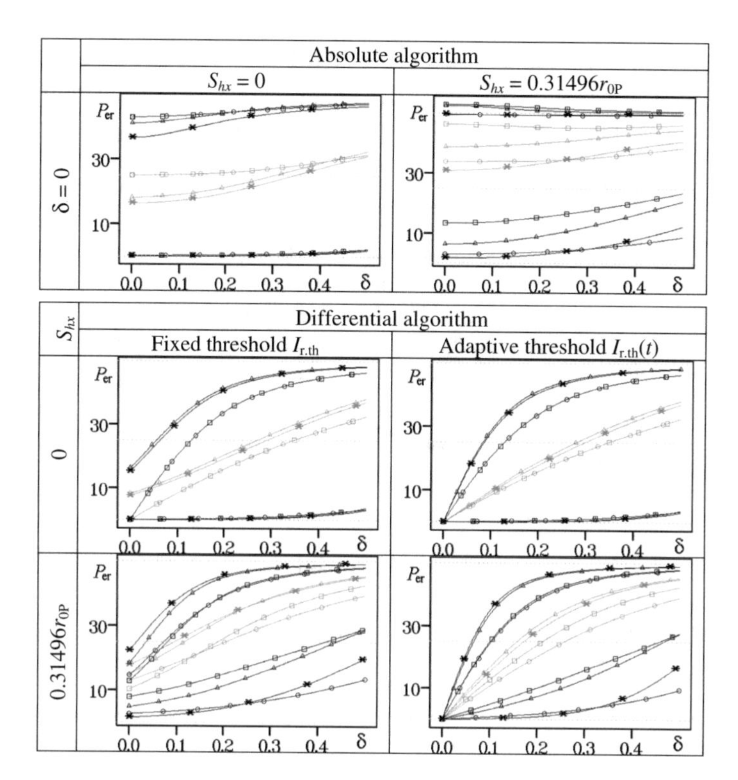

Fig. 5.30 Dependence of the probability of error in data transfer P_{er} (in %) on the photodetector noise amplitude δ for $M_{outer} = 5r_{0P}$. The Fried radius L_F/r_{0P} is 0.25, 0.1, and 0.05 for the lower, middle (*grey*), and upper groups of curves. Sets of three values (Δ, V_{d0}, V_{d1}) take values (180°, –1, 0), (180°, –1, –2), (120°, –1, 1), and (120°, –1, –2) and are shown by curves with symbols ○, □, *, and Δ, respectively

obtained as a result of modeling are in agreement with the above predictions based on the analysis of the two-dimensional probability density ρ.

Predictions are also illustrated and are confirmed by Figs. 5.30 and 5.31 that show the results of simulation based on Eqs. (5.31), (5.32), (5.36), and (5.38). In addition, they vividly and quantitatively illustrate values and tendencies of changing of the probability of error P_{er}. In these figures, we omitted the results for the absolute adaptive algorithm, because they (according to our predictions and actually), as a rule, are very similar to those for the absolute algorithm. The requirements to the modern communication systems involve very small values of P_{er}. On the plots, the range of P_{er} variations is extremely wide: from 0 to ≈50 %. Therefore, sometimes it is difficult to estimate the degree of proximity of P_{er} to zero; this is provided by Tables A.1–A.4 from Appendix A in [62].

Naturally, without distorting factors (turbulence, photodetector noise, and misalignment), when the detector is capable of identifying the charge V_d, all methods of coding must be equally efficient (provide faultless data transfer). These

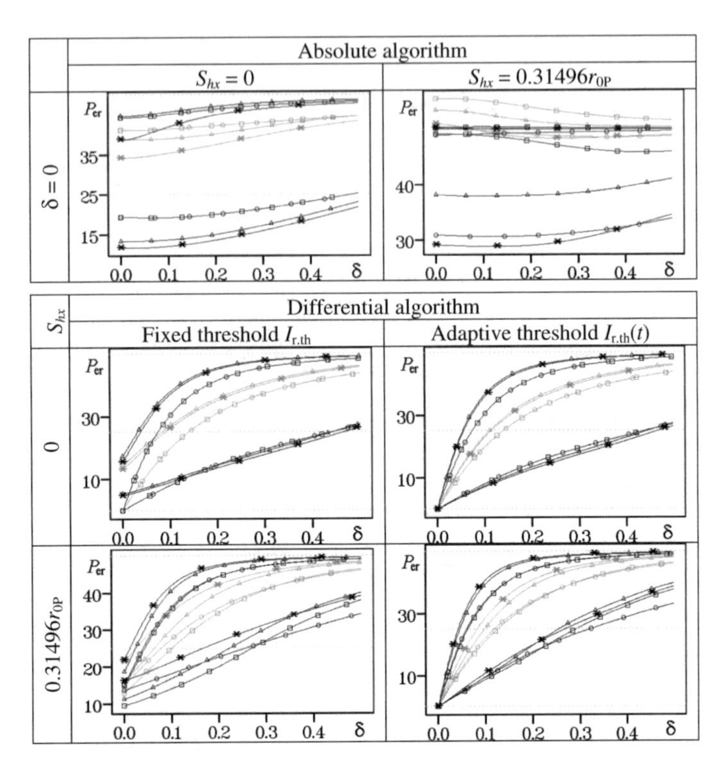

Fig. 5.31 Dependence of the probability of error in data transfer P_{er} (in %) on the photodetector noise amplitude δ for $M_{outer} = 20r_{0P}$. The Fried radius L_F/r_{0P} is 0.25, 0.1, and 0.05 for lower, middle (*grey*), and upper groups of curves. Sets of three values (Δ, V_{d0}, V_{d1}) take values (180°, –1, 0), (180°, –1, –2), (120°, –1, 1), and (120°, –1, –2) and are shown by curves with symbols O, □, *, and Δ, respectively

tables and figures demonstrate that the presence of only one weak turbulence ($L_F = 0.25r_{0P}$) causes the detector to wrongly identify V_d in 0.0285...0.328 % of cases (see values of $P_{er.abs}$ and $P_{er.abs.adp}$). Then the efficiency of the algorithms becomes unequal. The degree of this difference (on average for sets of three values (Δ, V_{d0}, V_{d1})) is also characterized by Table A.5 in Appendix A in [62] that demonstrates the capability of algorithms *to decrease* the probability of error in data transfer in comparison with the probability of error in V_d recognition, that is, in comparison with $P_{er.abs.Ith1}$.

As to the probability of error in data transfer, the adaptive differential algorithm gains considerably in comparison with the differential one (with increasing noise δ, the gain of $P_{er.abs.adp}$ decreases from 100 % of $P_{er.dif}$ to 0). According to Table A.5 in Appendix A in [62], the average capability to decrease P_{er} (in comparison with $P_{er.abs.Ith1}$), for example, for $\delta = 0.1$, $S_{hx} \approx 0.31496$, $L_F = 0.10r_{0P}$, and $M_{outer} = 5r_{0P}$ is 0.379 as compared to 0.586, and for $M_{outer} = 20r_{0P}$; it is equal to 0.472 as compared to 0.570 for the adaptive differential and differential algorithms. For weak turbulence ($L_F = 0.25r_{0P}$), this capability is illustrated by the following pairs of values: 0.316 as compared to 1.069 ($M_{outer} = 5r_{0P}$) and 0.279 as compared to 0.517 ($M_{outer} = 20r_{0P}$).

In turn, the absolute algorithm, as a rule, is less efficient than the differential one, because the letter is less sensitive to the error in V_d identification. In particular, this follows from $\langle P_{er.dif.r}\rangle$ and $\left\langle P_{er.dif.adp.r}\right\rangle$ values presented in Table A.5 in Appendix A in [62] and from the corresponding set of four values $\left\langle P_{er.abs.adp.r}\right\rangle$: 0.945, 1.00, 0.820, and 0.946. Values $\left\langle P_{er.abs.adp.r}\right\rangle$ are typically close to unity, thereby emphasizing the low efficiency of application of the adaptive threshold for the absolute algorithm.

We emphasize that the differential algorithm remains sometimes efficient in the situation when the detector does not recognize V_d, that is, when $P_{er.abs}$ and/or $P_{er.abs.adp}$ are close to 50 %. However, the reverse situations can be realized when $M_{outer} = 5r_{0P}$, $L_F = 0.25r_{0P}$, $S_{hx} = 0$, and $\Delta = 120°$ (Table A.1 in Appendix A in [62]); $S_{hx} \approx 0.31496r_{0P}$, $\Delta = 180°$, $V_{d0} = -1$, $V_{d1} = 0$; or $\Delta = 120°$, $V_{d0} = -1$, $V_{d1} = -2$ (Table A.2 in Appendix A in [62]). Such "inversion" is possible when $\delta \neq 0$, but the relative number of events $(I_{r.m0}(t),\ I_{r.m1}(t)) \in Qu_4$ is relatively small, and the total number of events $(I_{r.m0}(t),\ I_{r.m0}(t-\tau_b)) \in Qu_2$, $(I_{r.m0}(t),\ I_{r.m0}(t-\tau_b)) \in Qu_4$, $(I_{r.m1}(t),\ I_{r.m1}(t-\tau_b)) \in Qu_2$, $(I_{r.m1}(t),\ I_{r.m1}(t-\tau_b)) \in Qu_4$ is so large that exceeds more than four times that of the events $(I_{r.m0}(t),\ I_{r.m1}(t)) \in Qu_4$. Indeed, the first number of events is responsible for the loss of the absolute algorithm (Table 5.7 and Fig. 5.25), and the second one is responsible for the loss of the differential algorithm (Fig. 5.26).

We predicted this inversion when analyzed expression $P_{er.dif}(t) = 2[1 - P_{er.abs}(t)]P_{er.abs}(t)$ in Eq. (5.30). Its reason can be easily understood if we look at Fig. 5.27 which, in particular, visualizes this expression. It can be seen that on the plane $(I_{r0},\ I_{r1})$, there are regions where $P_{er.abs}(I_{r0},\ I_{r1}) > P_{er.dif}(I_{r0},\ I_{r1})$ and where $P_{er.abs}(I_{r0},\ I_{r1}) < P_{er.dif}(I_{r0},\ I_{r1})$. Exactly the dominance of the second situation over the first one (with allowance for the shape of $\rho(I_{r0}(t),\ I_{r1}(t))$) causes this inversion. It seems likely that this is usually the case for the localization of the density ρ in the second quadrant and in the initial stage of its penetration into other quadrants, that is, for small noise, weak turbulence with small outer scale, and close arrangement of the transmitter and receiver optical axes. Because of convex character of contour lines for density ρ, inversion is more probable for algorithms with constant threshold. This is in agreement with the data presented in Tables A.1 and A.2 of Appendix A in [62]. These advantages of the differential algorithm prove the existence of symmetry (though incomplete) of the dependencies $I_r(N_r)$ shown in Fig. 5.18 and hence, the necessity of its account to increase the reliability of the communication system.

Judging by Tables A.1–A.4 of Appendix A in [62], the assumption of advantageous combinations (Δ, V_{d0}, V_{d1}): (120°, –1, 1) over (120°, –1, –2) and (180°, –1, 0) over (180°, –1, –2) is confirmed everywhere for the absolute algorithm of data transfer (both for $I_{r.th}(t) = \text{const}$ and $I_{r.th}(t) \neq \text{const}$). This means that it is desirable to use vortices with minimal (by the modulus) topological charges.

For the differential algorithm, this assumption is justified only without displacement of the axes ($S_{hx} = 0$). For $S_{hx} \approx 0.31496 r_{0P}$ and $M_{\text{outer}} = 5 r_{0P}$, on the contrary, the combinations (120°, –1,–2) and (180°, –1, –2) are optimal. If we compare detectors with different rotation angles Δ and $M_{\text{outer}} = 20 r_{0P}$, for the absolute algorithm it is preferable to use the detector with $\Delta = 120°$ (as well as for $M_{\text{outer}} = 5 r_{0P}$); for the differential algorithm, the detector with $\Delta = 180°$ is preferable. For $M_{\text{outer}} = 5 r_{0P}$ and the differential algorithm, it is difficult to chose between $\Delta = 120°$ and 180°. In addition, Tables A.1–A.4 of Appendix A in [62] and Figs. 5.30 and 5.31 confirm the above prediction that in the case of detector with $\Delta = 180°$ and $S_{hx} = 0$, the efficiencies of the absolute, absolute adaptive, differential, and differential adaptive algorithms are the same irrespective of the choice of sets of pair values $(V_{\text{d}0}, V_{\text{d}1})$.

As addition, illustration, and partial checking of above-offered statements, we construct the phase $S(\mathbf{r})$ and amplitude $A(\mathbf{r})$ of the input signal.

5.5 The Visual Analysis of Phase and Amplitude Distributions of the Input Signal of Vortex Topologic Charge Detector at Presence of the Turbulence

To obtain the general conclusions concerning the Rozhdestvenskiy's interferometer as an identifier of the V_{d} value, it is enough to examine only several significant cases. We are guided only by the rule of V_{d} determination according to the only realization (5.22) for $S_h = 0.11 r_{0P}$, $\delta = 0$, $\Delta = 180°$ и $\Delta = 120°$, choosing the path with a length $L_{\text{t}} = 59$ km, screens with $M_{\text{outer}} = 5 r_{0P}$, $L_{\text{F}} = 0.25 r_{0P}$ and several scenes when $L_{\text{F}} = 0.1 r_{0P}$.

For the Fried radius $L_{\text{F}} = 0.25 r_{0P}$ and for the vortex charge $V_{\text{d}} = -1$ the detector correctly $(V_{\text{d}} = V_{\text{d i}})$ reconstructs the distorted field history (a vortex presence with $V_{\text{d}} = -1$ in **non**-distorted field) with a 100 % probability (Fig. 5.32). For visual analysis convenience on spatial distributions of intensity $I(\mathbf{r})$ and phase $S(\mathbf{r})$ the dotted line are drawn with an angle 120°: their intersection marks the optical axis center Oz (Fig. 5.11) of the interferometer. Examining structures for $V_{\text{d}} = -1$ in Fig. 5.32, we can establish that sure in correct identification is relevant to connect with remoteness of I_{r} values from the threshold values $I_{\text{r th}}$ in the rule (5.22): $I_{\text{r th}} = 1$ for $\Delta = 180°$ and $I_{\text{r th}} = 0.94$ for $\Delta = 120°$. Or, on the contrary, to connect with the nearness of I_{r} values to values in Fig. 5.19 in the case of the weak turbulence (or to values estimated in the computing experiments with the white noise of the low level—see Figs. 5.13 and 5.14). Hence, realization № 0 according this criterion can be recognized by the detector more reliable than the realization № 1.

The visual analysis confirms propriety of this circumstance: intensity distribution for realization № 0 (on the contrast to realization № 1) rather remoteness reminds the usual view $I(\mathbf{r})$ of the field with vortex. In distribution of the phase $S(\mathbf{r})$ in both cases the dominating vortex is clearly seen (a circle with an arrow) localized near the optical axis Oz of the interferometer but slightly from it. Exactly this vortex is

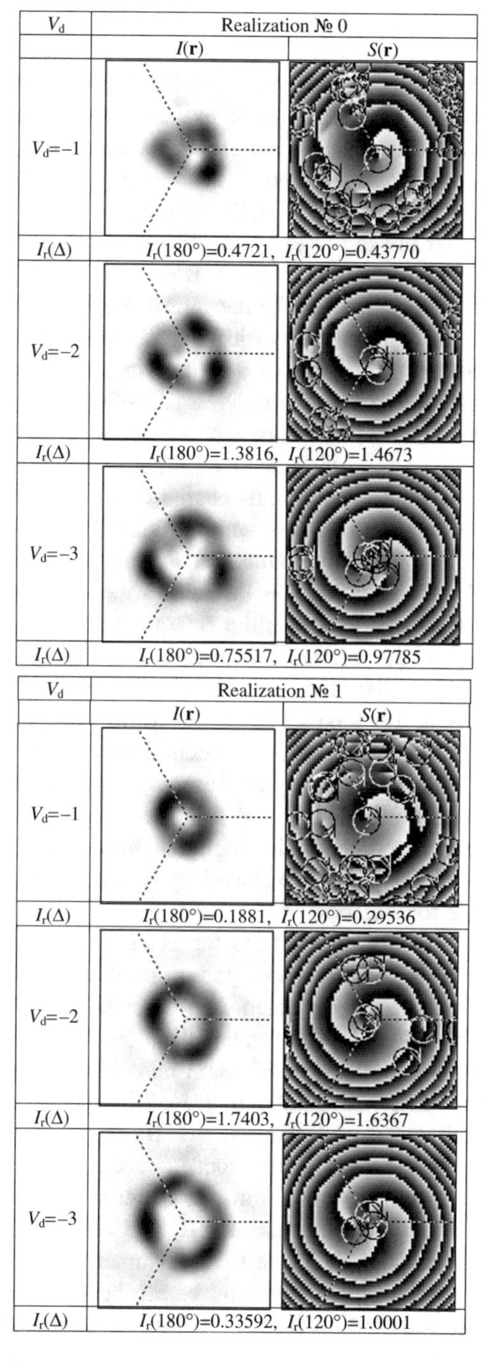

Fig. 5.32 Examples of correctly recognized realizations for different Δ and V_d (the Fried radius $L_F = 0.25r_{0P}$, $M_{outer} = 5r_{0P}$, $L_t = 59$ km, $S_h = 0.11r_{0P}$)

recognized by the detector in the same manner as initially transmitted. Plenty of vortices-"satellites" throws in eyes, which are formed during the diffraction of distorted (by the phase screen) of he beam. But, being rather remote from the axis and mainly paired (with opposite charge signs), they slightly influence on the value of I_r and correctness of sensor operation. At vortex charges $V_d = -2$, $V_d = -3$, in the case of successful recognition, above-mentioned observations and conclusions keep the force (taking into consideration the variation of $I_{r\,th}$)—see Fig. 5.32. Here, vortices of higher order fall to pieces to appropriate number of vortices with $V_d = -1$ (demonstrating the well-know fact [64]). But, in discussing realizations this fall to the pieces is not the interference for detector operation. Probably, due to beam expansion at V_d growth, a number of vortices-"satellites" in the area of investigation decreases. Remarkably, that for $V_d = -3$ in the vicinity of vortices number triple arising by the initial vortex of higher order, a pair of closed vortices appears with the opposite charges. Owing to two last circumstances, these vortices do not contradict to identification in spite to their vicinity with Oz axis.

We would like to remind that at vortex charge $V_d = -2$ and the field rotation by angle $\Delta = 180°$ the detector recovers without error the history of distorted field with probability 97 %. And with probability 3 % it mistakenly maintains that it was a vortex with $V_d = -1$ (or with $V_d = -3$). For $V_d = -3$ a part of truly identification is 94 %, at that, with a probability 6 % the value $V_d = -2$ (Table 5.2) is mistakenly arrogated to the vortex. When $V_d = -2$, but $\Delta = 120°$, the detector correctly reconstructs the distorted field evolution in 99 % of cases. With probability 1 % out falsely affirm a presence of the vortex with $V_d = -3$ in the initial field. If $V_d = -3$, then a fraction of truly identification is 77 %, but with probabilities 12 % and 11 % the detector unwarrantedly indicates values $V_d = -1$ и $V_d = -2$, relatively (Table 5.3).

Figures 5.33 and 5.34 serve as an illustration to above-mentioned. In them cases of wrong identification $V_d \neq \mathbf{V_{d\,i}}$ are shown by boldface. Figure 5.34 arrange the structures responsible for situation of birth in near-axis area of the vortices pair for $V_d = +1$, $V_d = -1$. Observation for the forms of $S(\mathbf{r})$ easily discovers a set of vortices of the first order. This is disintegration products of vortices for $|V_d| > 1$, which are present in non-distorted beam. Nevertheless, the detector does not adequately determine the value V_d. The same realization, depending on the detector's parameter Δ, can lead to both successful and false identification of the value V_d. However, there are such realizations (e.g. № 76), which does not permit V_d determination for both values of Δ and $|V_d| > 1$. By the way, for given realization it is typical that light energy localization is concentrated in the one spot of the regular shape, which remotes from the interferometer center on a distance closed to r_{0P}.

Assumed reasons of the detector errors: (A) vortices displacement of the mentioned complex with respect to Oz, at that, asymmetric; (B) essentially different (compared to axial-symmetric) distribution of the beam intensity, appearance of areas of energy concentration in it. Reasonably to suggest that all this implies deflections of phase distribution from the "etalon" ones, at that, in the same area intensity is still no small. Visual analysis of Figs. 5.32, 5.33 and 5.34 prompts to the following questions (emergent from the limit of our work). How much do

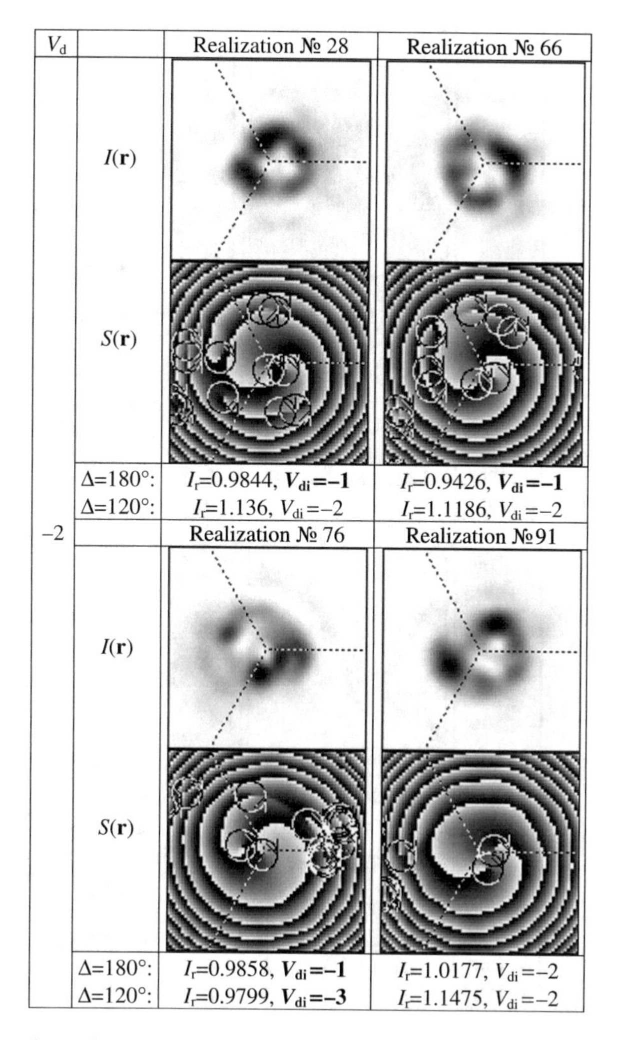

Fig. 5.33 Examples of (non)correctly recognized realizations for different Δ and V_d ($L_F = 0.25r_{0P}, M_{outer} = 5r_{0P}, L_t = 59$ km, $S_h = 0.11r_{0P}$)

differences of value I_r with respect to $I_{r\,th}$ correlate with the visual perception of structures $I(\mathbf{r})$ and $S(\mathbf{r})$ of the field with a vortex? Is it possible to introduce simple mathematical criteria, in particular, adequate to this perception, but following also from geometric and physical features of $I(\mathbf{r})$, $S(\mathbf{r})$ distributions for transmitted field?

Formulated assumption provokes the relatively simple method of it checking. Assuming that mentioned areas are near to the Oz axis, it is logic to imitate a presence of the non-transparent circular screen (a mask) with radius R_m. Simulation shows that mask optimality with more sharp boundaries: super-Gaussian profile of

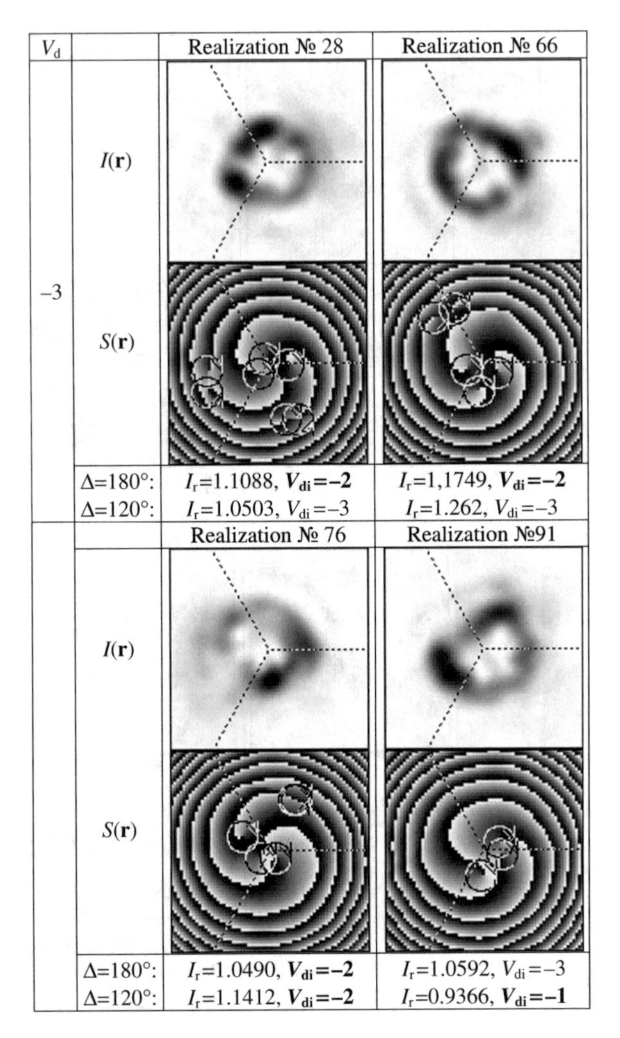

Fig. 5.33 (continued)

its absorption coefficient is more preferable than the Gaussian one. And more better—∏-type profile. The optimal radius is obtained empirically: $R_m = 1.55r_{0P}$ for $\Delta = 180°$ and $R_m = 1.3r_{0P}$ for $\Delta = 120°$. Its boundary can be easily distinguished on the distribution $I(\mathbf{r})$ on the background of $S(\mathbf{r})$ it is the contrast circle with a center in the point O (Fig. 5.35).

Figure 5.35 contains examples both successful and non-successful application of the mask. So, in the case $V_d = -2$ earlier false predictions becomes true, but for realization the prediction № 91, on the contrary, becomes wrong (Fig. 5.33). In the case $V_d = -3$ the mask insignificantly improves the value I_r, but does not lead to

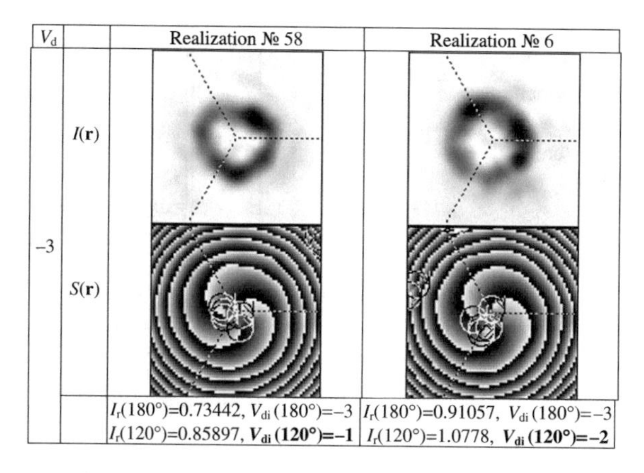

Fig. 5.34 Birth of additional near-axis pairs of vortices: examples of false (for $\Delta = 120°$) and truly (for $\Delta = 180°$) recognition ($V_d = -3$, $L_F = 0.25r_{0P}$, $M_{outer} = 5r_{0P}$, $L_t = 59$ km, $S_h = 0.11r_{0P}$)

recognition correctness of realizations № 28 and 66. At the same time, the mask "inverted" the recognition correction of realizations № 76 and 91 (Fig. 5.33).

Total estimations of recognition variation V_d at the expense of mask using characterize the Table 5.8 and Fig. 5.36. Calculations of mean values (over realizations) $\langle I_r \rangle_{Nr}$ convince in the dual influence of the mask: in the case of blameless recognition ($V_d = -1$) of the mask decreases (not more than 20 %) the value $\langle I_r \rangle_{Nr}$ and it means that it slightly decreases a fraction of true predictions. But at the same time, in the case of detector operation with errors ($|V_d| > 1$) of the mask improves (by several percent) values $\langle I_r \rangle_{Nr}$, promoting to more sure recognition.

It should note that often values I_r of non-recognized realizations are rather close to threshold $I_{r\,th}$. Hence, inevitable approximation of the criterion (5.22), i.e. the value $I_{r\,th}$, and the calculation error of intensity I_r, generally speaking, require causion in recognition interpretation as (non)correct, when $I_r \approx I_{r\,th}$.

Figures 5.19 and 5.17 convince in detector operation possibility for $L_F < 0.25r_{0P}$ in the case of realization array analysis (i.e. the analysis of $\langle I_r \rangle_{Nr}$), and Figs. 5.28, 5.29, 5.30 and 5.31 and Tables A.1–A.4 in [62]—with acceptable probability of data transmission errors for $L_F < 0.25r_{0P}$. Hence, it is logic to find out the typical distribution $I(\mathbf{r})$ and $S(\mathbf{r})$ in the presence of strong, which is imitated by the screen with $L_F = 0.1r_{0P}$ (Fig. 5.37).

When creating Fig. 5.37 we used predominantly the same realizations that are above-mentioned for more weak turbulence ($L_F = 0.25r_{0P}$). Including in it the realization № 4 is caused by a wish to express the case, when $V_d = -2$, $V_{d\,i} = -3$ for $\Delta = 120°$. Evidently, for this strong turbulence the detector cannot mistake more often. However, its "misfortunes" in a series of situations are explainable. So, for realization № 66 in Fig. 5.37, when $V_d = -1$, the detector diagnoses $V_{d\,i} = -2$ for $\Delta = 180°$ and $V_{d\,i} = -3$ for $\Delta = 120°$. These values $V_{d\,i}$ due to formula (5.3)

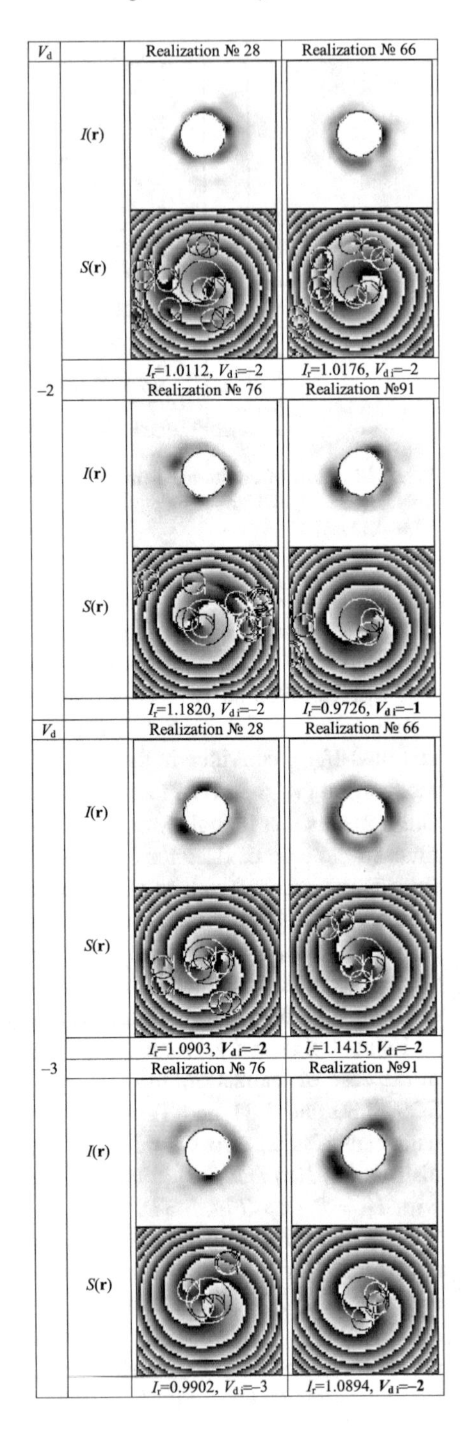

Fig. 5.35 Examples of the mask effect on the vortex charge recognizability when $\Delta = 180°$ $M_{\text{outer}} = 5r_{0P}$, $L_t = 59$ km, $S_h = 0.11r_{0P}$, and V_d has various values

Table 5.8 Influence of the mask to ratio of truly (bold) and false type identifications of the vortices topological charge with detector with $\Delta = 180°$ and $\Delta = 120°$ depending on the Fried radius when using the rule (5.22), $L_t = 59$ km, $M_{outer} = 5r_{0P}$, $L_t = 59$ km, $S_h = 0.11r_{0P}$

$L_F/_{OP}$	V_d	$\Delta = 180°$						$\Delta = 120°$					
		Without mask			Mask with $R_m = 1.55r_{0P}$			Without mask			Mask with $R_m = 1.3r_{0P}$		
		$V_{d\,i}$			$V_{d\,i}$			$V_{d\,i}$			$V_{d\,i}$		
		−1	−2	−3	−1	−2	−3	−1	−2	−3	−1	−2	−3
0.1	−1	**64**	36	0	**67**	33	0	**67**	15	18	**58**	18	24
	−2	43	**57**	0	38	**62**	0	18	**58**	24	12	**63**	25
	−3	0	45	**55**	0	44	**56**	29	32	**39**	26	29	**45**
0.25	−1	**100**	0	0	**97**	3	0	**100**	0	0	**99**	0	1
	−2	3	**97**	0	2	**98**	0	0	**99**	1	0	**100**	0
	−3	0	6	**94**	0	5	**95**	12	11	**77**	15	7	**78**

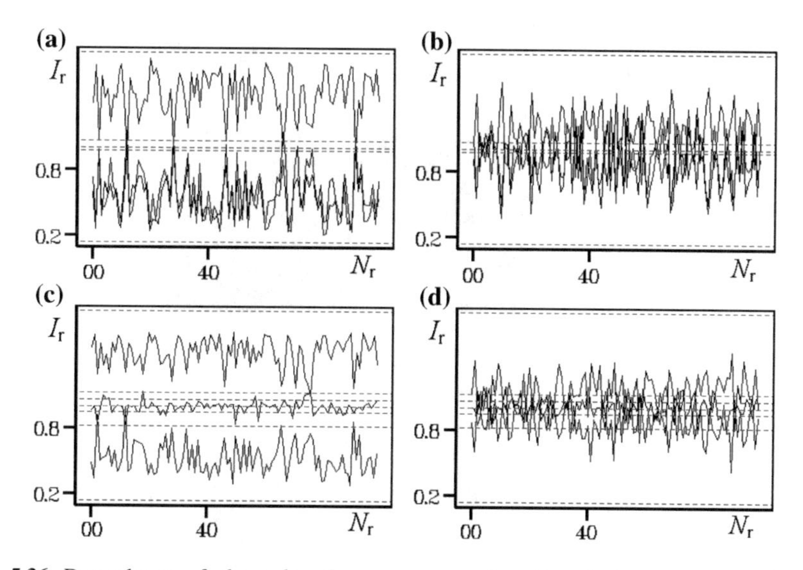

Fig. 5.36 Dependence of the value I_r versus a number N_r of phase screen realization ($N_r \in [0,\ 99]$) with mask presence, when $M_{outer} = 5r_{0P}$, $L_t = 59$, km, $S_h = 0.11r_{0P}$. For convenience perception, the calculated points are connected with lines. The field rotation angle $\Delta = 180°$ (**a**, **b**), $\Delta = 120°$ (в, г). The Fried radius $L_F = 0.25r_{0P}$ (**a**, **c**), $L_F = 0.1r_{0P}$ (**b**, **d**). At the ends of the figure the *horizontal dotted line* has the same sense that in Fig. 5.17 ($I_r = 0.135, I_r = 1.864$). In the *center* of figures **a**, **b** the *horizontal dotted line* corresponds to $I_r = 1$ and the threshold values of $I_r = 0.97$, $I_r = 1.063$; in the *center* of figures **c**, **d** corresponds to $I_r = 1$ and the threshold values of $I_r = 0.818$, $I_r = 0.94$, $I_r = 1.06$, $I_r = 1.12$

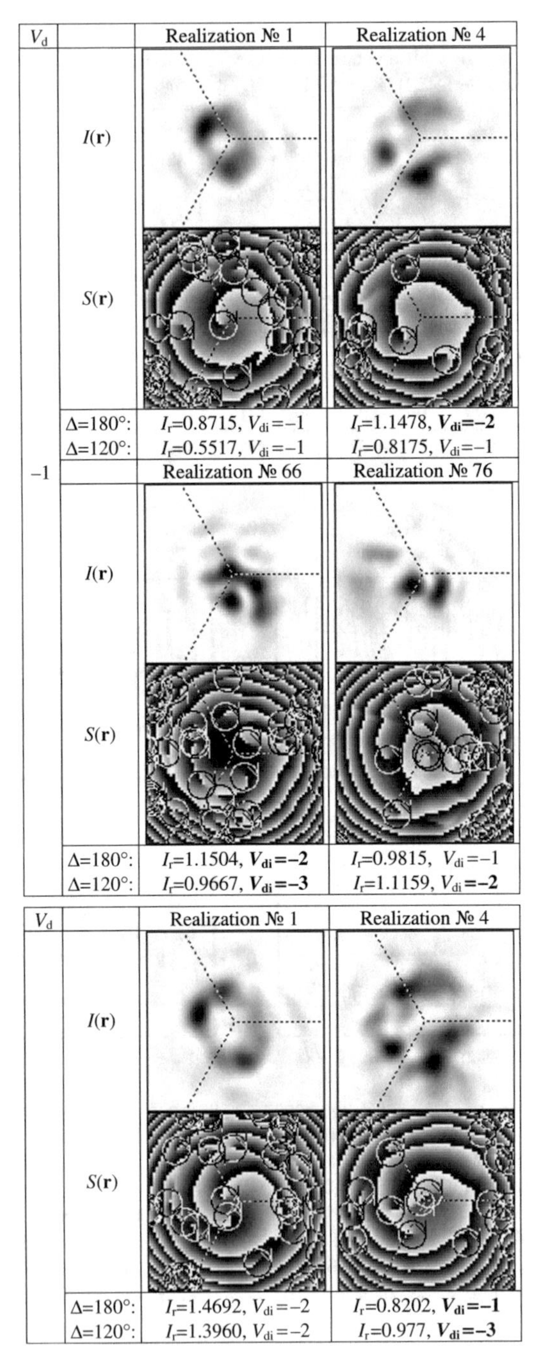

Fig. 5.37 Examples of (non)correctly recognized realizations for different Δ and V_d ($L_F = 0.1r_{0P}$, $M_{outer} = 5r_{0P}$, $L_t = 59$ km, $S_h = 0.11r_{0P}$)

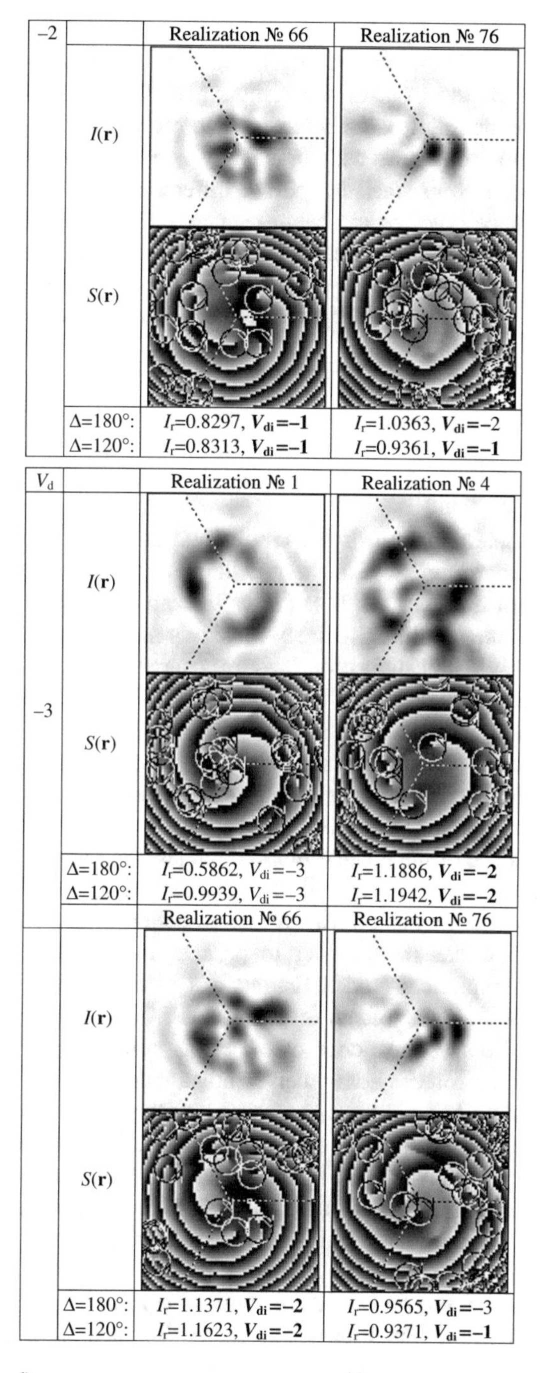

Fig. 5.37 (continued)

are equivalent to zero. Really, the bright light spot is situated on the interferometer axis Oz (intersection of three dotted lines), and distribution $S(\mathbf{r})$ in this area of the beam highly reminds the front of the plane wave. Two highly-intensive parts of $I(\mathbf{r})$, remoted from the interferometer axis at field rotation in the plain xOy by angle Δ, which “colliding” on areas with practically zero intensity. Visibility of such interference pattern is very small in spite of difference in the phase shift. In other words, the turbulence gives of the beam near axis Oz the properties of the plane wave. Hence, the diagnosis “formulated” is true. Probably, the similar conclusion is true with respect to realization № 4, when $V_d = -1$. Fulfillment of such visual studying of the reasons of detector errors for all 600 subjects is not expedient; at that, its success is not guaranteed.

5.6 Conclusions

The principle of determination of the topological charge of the optical vortex is offered based on measurement of the light field intensity, and the construction of the appropriate detector is discussed. It can be built both on the base of nonlinear ring interferometer and the modified Rozhdestvenskiy’s interferometer. In the first case, change of the screw dislocation order (vortex charge) is capable to influence not only to beam mean intensity, but to dynamic mode in the interferometer, i.e. to initiate bifurcation.

A parallel action for artificial neuron networks is carried out. The mathematical model of the detector operation for vortex topological charge is given. Results of numerical experiments are presented, which imitate the vortex recognition in the presence of turbulence or the noise (amplitude, phase) in the radiation investigated, as well as at displacement of optical axes of the beam source and the detector. Principles of creation of the optical vortex direction-finder are formulated (i.e. detector of its coordinates), allowing to pass to its realization in the form of mathematical and numerical model. Estimations of conditions of effectiveness of the vortex detector and constructed on its base the “singular-optical” communication line is carried out. Dependence of the error probability of data transmission *versus* the turbulence force, photo-receiver noise amplitude and displacements of the optical axes when using different coding algorithms (“absolute”, “differential”, with fixed or adaptive threshold) is described. Distributions of phase and amplitude of the detector input signal are built and presented at the turbulence presence for visual check of detector operation success. Mask influence is investigated, which is mounted in the beam center, on the vortex recognition error. Simulation data confirms a series of analytical calculations.

The next chapter returns us to discussion of cryptological aspects of chaotic systems. We shall pay an attention to *divercification* of nonlinear elements in the chaotic oscillators and the structure of communication systems as a strategy of increase a number of keys (increase a resistance of the communication system).

References

1. Paterson C. Atmospheric turbulence and orbital angular momentum of single photons for optical communication. Phys Rev Lett 2005;94:153901.
2. Aksenov VP, Izmailov IV, Kanev FYu, Poizner BN. Adaptive system for data transmission with the help of optical vortices. In: Atmospheric and oceanic optics. abstracts of the 16-th joint international symposium “Atmospheric and Ocean Optics. Atmospheric Physics”. Tomsk, Russia. 13–15 October 2009. Tomsk, 2009. pp. 166–168.
3. Hartmann J. Objektivuntersuchungen. Z. Instrum. 1904;1:1–21, 33–47, 97–117.
4. Shack RB, Platt BC. Production and use of a lenticular Hartmann screen. J. Optics Soc. Am. 1971;61:656–662.
5. Voitsekhovich V, Sanchez L, Orlov V, Cuevas S. Efficiency of the Hartmann test with different subpupil forms for the measurement of turbulence-induced phase distortions. Appl Opt. 2001;40(9):1299–1304.
6. Voliar AB, Zhilaitis VZ, Fadeeva TA. Optical vortexes in the small-mode fibers. III. Dislocation reactions, phase transitions and topological two-beam-refraction. Opt Spectrosc. 2000;88(3):446–455.
7. Aksenov VP, Izmailov IV, Poizner BN, Tikhomirova OV. Wave and beam spatial dynamics of the light field at birth, evolution and annihilation of phase dislocation. Opt Spectrosc. 2002;92 (3):452–461.
8. Aksenov VP, Izmailov IV, Kanev FYu, Starikov FA. Localization of optical vortices and reconstruction of wavefront with screw dislocations. Proc SPIE. 2005;5894:68–78.
9. Aksenov VP, Izmailov IV, Kanev FYu. Algorithms of a singular wavefront reconstruction. In: Jiang W, editor. The 5-th international workshop on adaptive optics for industry and medicine (29 August–1 September 2005, Beijing, China), Proceedings of SPIE, vol. 6018.—SPIE, Bellingham, WA, 2005. p. 60181B-1-60181B-11.
10. Aksenov VP, Izmailov IV, Kanev FYu, Starikov FA. Algorithms for the reconstruction of the singular wave front of laser radiation: analysis and improvement of accuracy. Quant Electron. 2008;38(7):673–677.
11. Ragazzoni R. Pupil plane wavefront sensing with an oscillating prism. J Mod Opt. 1996;43:289–293.
12. Ragazzoni R, Ghedina A, Baruffolo A, Marchetti E, et al. Testing the pyramid wavefront sensor on the sky. Proc SPIE. 2000;4007:423–429.
13. Ghigo M, Crimi G, Perennes F. Construction of a pyramidal wavefront sensor for adaptive optics compensation. In: Proceedings of International conference “Beyond Conventional Adaptive Optics”. 2001. p. 465–472.
14. Kolosov VV. Current lines of energy in the vicinity of dislocation of three-dimension wave field. Atmos Oceanic Opt. 1996;9(12):1631–1638.
15. Aksenov VP, Kolosov VV, Tartakovskii VA, Fortes BV. Optical vortex in non-uniform media. Atmos Oceanic Opt. 1999;12(10):952–958.
16. Bobrov BD. Visualization of the phase surface in the vicinity of screw dislocation. Spiral interferometric structures. Opt Spectrosc. 1991;70(2):436–438.
17. Bobrov BD. Screw dislocations of laser speckle-fields on the interferometer patterns with circular line structure. Quant Electron. 1991;18(7):886–890.
18. Korolenko VP. Optical vortices. Soros Educ J. 1998;6:94–99.
19. Mansuripur M, Wrignt E. Linear optical vortices. Opt Photonics News. 1999;10(2):40–44.
20. Kanev FYu, Lukin VP, Makenova NA. Detection of dislocations as branching points of interference pattern. Proc SPIE. 2001;4357:231–235.
21. Rockstuhl C, Ivanovskyy AA, Soskin MS, Sal MG, Herzig HP, Dandliker R. High-resolution measurement of phase singularities produced by computer-generated holograms. Opt. Commun. 2004;242:163–182.

22. Angelsky OV, Maksimyak AP, Maksimyak PP, Hanson SG. Interference diagnostics of white-light vortices. Opt Express. 2005;13:8179–8183.
23. Soskin MS, Polyanskii PV, Arkhelyuk OO. Computer-synthesized hologram-based rainbow optical vortices. New J Phys. 2004;6:196–204.
24. Berry MV. Colored phase singularities. New J Phys. 2002;4:66–73.
25. Berry MV. Exploring the colors of dark light. New J Phys. 2002;4:74–80.
26. Bogatyryova GV, Felde ChV, Polyanskii PV, Ponomarenko SA, Soskin MS, Wolf E. Partially coherent vortex beams with a separable phase. Opt Lett. 2003;28:878–880.
27. Angelsky OV, Hanson SG, Maksimyak AP, Maksimyak PP. On the feasibility for determining the amplitude zeroes in polychromatic fields. Opt. Express. 2005;13:4396–4405.
28. Fried DL. Branch point problem in adaptive optics. J Optics Soc Am. 1998;15 (10):2759–2768.
29. Fried DL. Adaptive optics wave function reconstruction and phase unwrapping when branch points are present. Opt Commun. 2001;200:43–72.
30. Ghiglia DC, Pritt MD. Two-dimensional phase unwrapping: theory, algorithms, and software. A Wiley-interscience publication, Wiley; 1998. 512 p.
31. Aksenov V, Izmailov I, Kanev F, Starikov F. Performance of a wavefront sensor in the presence of singular points. In: Slangen P, Cerruti C, editors. Speckle06: Speckles, from grains to flowers (13–15 September 2006, Nimes, France), Proceedings of SPIE, vol. 6341. SPIE, Bellingham, WA, 2006. P. 634133-1–634133-6.
32. Starikov FA, Kochemasov GG, Kulikov SM, Manachinsky AN, Maslov NV, Ogorodnikov AV, Sukharev SA, Aksenov VP, Izmailov IV, Kanev FYu, Atuchin VV. Wave front reconstruction of an optical vortex by Hartmann–Shack sensor. Opt Lett. 2007;32 (16):2291–2293.
33. Izmailov IV. Development of measurent system for the light field phase: algorithmic aspect. In: Proceedings of international scientific conference "problems of development of natural, engineering and social systems" (Apr. 12–14 2007, Taganrog, Russia). Part 3. Taganrog: Anton Publications, 2007. pp. 16–22 (in Russian).
34. Kanev FYu, Aksenov VP, Izmailov IV, Starikov FA. Features of the phase recognition of the vortex beam at growth of a number and order of singular points. In: Proceeding of Tomsk Polytechnical University. 2009. vol. 315, no 2. Mathematics and mechanics. Physics. pp. 44–48 (in Russian).
35. Baker KL, Stappaerts EA, Wilks SC, Gavel D, Young PE, Tucker J, Olivier SS, Silva DA, Olsen J. Open- and closed-loop aberration correction by use of a quadrature interferometric wave-front sensor. Opt Lett. 2004;29(1):47–47.
36. Baker KL, Stappaerts EA, Wilks SC, Gavel D, Young PE, Tucker J, Olivier SS, Silva DA, Olsen J. Performance of a phase-conjugate engine implementing finite-bit phase correction. Opt. Lett. 2004;29(9):980–982.
37. Dandliker R, Marki I, Salt M, Nesci A. Measuring optical phase singularities at subwavelength resolution. J Opt A: Pure Appl Opt. 2004;6:189–196.
38. Smith CP, McDuff R. Charge and position detection of phase singularities using holograms. Opt Commun. 1995;114:37–44.
39. Zhu B, Ueda K. Real-time wavefront measurement based on diffraction grating holography. Opt Commun. 2003;225:1–6.
40. Aksenov VP, Banakh VA, Valuev VV, Zuev VE, Morozov VV, Smalikho IN, Zvyk PI. Powerful laser beams in random-non-uniform atmosphere. Under edition of V.A. Banakh. Novosibirsk: Siberian branch of RAS Publication, 1998. 341 p (in Russian).
41. Lukin VP, Fortes BV. Adaptive beam and images formation in the atmosphere. Novosibirsk: SB RAS Publications, 1999. 214 p. (in Russian).
42. Kanev FYu, Lukin VP. Adaptive optics. Numerical and experimental investgations. Tomsk: Institute of atmospheric optics Publications, 2005. 250 p. (in Russian).
43. Fried DL. Using the noise-variance-weighted complex exponential reconstructor with measurements from a standard Hartmann sensor. *Report No. TN-103*. December 1999.

44. Kanev F, Aksenov V, Ustinov A, Izmailov I, Poizner B. Reconstruction of a singular wavefront using shack-hartmann sensor. In: Abstracts of the 11-th joint international symposium "Atmospheric and Ocean Optics. Atmospheric Physics". Tomsk, Russia. 23–26 June 2004. Tomsk, 2004. p. 70.
45. Kanev F, Aksenov V, Ustinov A, Izmailov I, Poizner B. Comparative analysis of the algorithms for reconstruction of wave front of optical field under the conditions of strong fluctuations. In: Abstracts of the 11-th joint international symposium on"Atmospheric and Ocean Optics. Atmospheric Physics". Tomsk, Russia. 23–26 June 2004. Tomsk, 2004. p. 79.
46. Starikov FA, Kochemasov GG, Koltygin MO, Kulikov SM, Manachinsky AN, Maslov NV, Sukharev SA, Aksenov VP, Izmailov IV, Kanev FYu, Atuchin VV, Soldatenkov IS. Correction of vortex laser beam in a closed-loop adaptive system with bimorph mirror. Opt Lett. 2009;34(15):2264–2266.
47. Aksenov VP, Kanev F, Izmailov I, Starikov F. Singular wavefront reconstruction with the tilts measured by Shack–Hartmann sensor. In: Book of summaries. 16-th international symposium on gas flow and chemical lasers & High power laser conference. Gmunden, Austria, September 4–8, 2006. Gmunden: Stanzell druck, 2006. pp. 158–159.
48. Izmailov IV, Aksenov VP, Kanev FYu, Starikov FA. Modifications of Fried's algorithm: increase of precision and phase unwrapping. In: Abstracts of the 14-th international symposium "Atmospheric and Ocean Optics. Atmospheric Physics" (24–30 June 2007, Buryatiya, Russia). Tomsk: IAO SB RAS, 2007. pp. 81–82.
49. Izmailov IA, Magazannikov AL, Poizner BN. Modeling of processes in the ring interferometer with nonlinearity, delay and diffusion at non-monochromatic radiation. Russ Phys J. 2000;2:29–35.
50. Izmailov IV, Lyachin AV, Poizner BN. The deterministic chaos in models of the nonlinear ring interferometer. Tomsk: Tomsk State University Publications; 2007. 258 p (in Russian).
51. Izmailov IA, Magazannikov AL, Poizner BN. Identification of the screw dislocation of the wave front and compensation of its influence on the structure-formation in the ring interferometer. Atmos Oceanic Opt. 2000;13(9):805–812.
52. Izmailov IA, Poizner BN. Imitation of by nonlinear-optical neuro-network optical vortices recognition. In: Scientific session of MIFI—2003. Scientific collection of 5 All-Russia scientific-technical conference "Neuro-informatics-2002" (Moscow, Jan. 29–31, 2003). In 2 parts. Moscow: MIFI, 2003. Part 1. pp. 77–84 (in Russian).
53. Izmailov IV, Poizner BN, Ravodin VO. The elements of non-linear optics and synergetics in opto-informatics context: the manual. Tomsk: TML-Press; 2007. 92 p (in Russian).
54. Izmailov IA. Structure-formation in the models of nonlinear ring interferometer in the case of monochromatic linear-polarized with screw dislocation of the wave front. Russ Phys. J. 1999;11:96 (in Russian).
55. Izmailov IV, Poizner BN. Ring cavity containing liquid crystal as a means of identification of a singular light beam. In: Abstracts of international scientific conference "optics of crystals". Republic of Belarus, Mozyr. 26–30 September 2000. Mozyr: MSPI; 2000. p. 49 (in Russian).
56. Izmailov IV, Poizner BN. Ring interferometer with liquid crystal as a means of identification of a singular light beam. In: Shepelevich VV, Egorov NN, editors. International scientific conference "optics of crystals" (26–30 September 2000, Mozyr, Republic of Belarus), Proceedings of SPIE. vol 4358, pp. 227–235 (2001). 9 p.
57. Izmailov IV, Makukha NE, Poizner BN. Structures self-organization in the model of the ring interferometer under action of the optical vortex. In: Problems of evolution of open system. Collection of papers of international conference. "Organization and evolution of open systems" (Almaty, Kazakhstan. Sept. 24–27, 2001). Issue 4. Almaty: Evero Publications, 2002. pp. 73–81 (in Russian).
58. Zubinskiy A. Return of analog computers. Comput Rev. 2009;(12). URL:http://ko.com.ua/node/42013. (in Russian).
59. Aksenov VP, Izmailov IV, Kanev FY, Poizner BN. Determination of topological charge of optical vortex according to measurement of the signal intensity at interferometer output: principles and modeling. Atmos Oceanic Opt. 2010;23(11):1036–1041.

60. Aksenov VP, Izmailov IV, Kanev FY, Poizner BN. Influence of the random phase screen imitating the atmosphere turbulence on operation of the interferometric detector of the topological charge of the optical vortex. Atmos Oceanic Opt. 2010;23(12):1132–1136.
61. Aksenov V, Izmailov I, Kanev F, Poizner B. Detector of optical vortices as the main element of the system of data transfer: Principles of operation, numerical model, and influence of noise and atmospheric turbulence. Int J Opt. 2012;2012(Article ID 568485):14 p. (doi:10.1155/2012/568485; URL:http://www.hindawi.com/journals/ijo/2012/568485/).
62. Aksenov VP, Izmailov IV, Kanev FYu, Poizner BN. Optical vortex detector as a basis for a data transfer system: Operational principle, model, and simulation of the influence of turbulence and noise. Opt Commun. 2012;285:905–928. (doi:10.1016/j.optcom.2011.10.060).
63. Nanostructures in electronics and photonics. Ed. Rahman F, editor. World Publication Co. Pte. Ltd, 2005. 312 p.
64. Dennis MR. Rows of optical vortices from elliptically perturbing a high-order beam. Opt Lett. 2006:31(9):1325–1327.

Chapter 6
Variety of Nonlinear Type in the Chaotic Oscillator and Structure Organization of the Chaotic Communication System as a Way to Increase the Confidence Degree

Let us pass to the last chapter. We shall devote it to *diversification* as some strategy of number of keys increase (the system resistance increase). In the one case, the diversification object will be the structural organization of the confidential communication system. In this aspect, we shall begin from classification of nonlinear-dynamic cryptology system predicting on its base of the possible new devices. In other case, such an object of diversification is reasonably to make the functional view of the nonlinear transfer characteristic.

For this, we shall analyze the concept of nonlinearity. We shall add it by our concept of controllable ("formed") and self-control nonlinearity and discover the unity of nonlinear elements structure. We give a series of examples from the history of physics, which demonstrate the latent existence of "formed" nonlinearity in practice of researchers and engineers of XIX–XX centuries.

6.1 A Variety of Structural Organization of Nonlinear-Dynamic Systems of Confidential Communication and Its Classification

From the content of previous chapters, it is clear that structural organization of confidential communication system is the one from the key component, since without its knowledge the enemy cannot decipher the message. Since we can consider the *diversification* of structural organization and possibility of urgent change we can consider (and make) the one of the strategies to improve resistance of communication system. So we saw the single- and double-channel systems, as well as much more complicate devices in configuration (see, for instance, p. 1.4.5 and Fig. 1.37, p. 4.1.2). Let us say, these configurations in the case of DNRI are quickly and easily changeable. But, in the case of radio electronic devices, such a reconstruction today can be seen as possible but pregnant to system inconvenience.

I. Izmailov et al., *Cryptology Transmitted Message Protection*,
Signals and Communication Technology, DOI 10.1007/978-3-319-30125-9_6

It is clear, that a choice of system organization inevitably influences on a quality, noise immunity, cost and other characteristics of the communication system. Therefore, this problem of choice is many-aspect and, hence, is non-trivial. Mentioned theme is open for engineering-technology and technology innovations. In combination with the wideness of the application areas of communication systems differ by its specific, it inevitably leads he diversification of offered methods and devices of nonlinear-dynamic cryptography.

In its turn, this, naturally, causes a necessity in its classification. And selection of classification signs allows systematically and completely to diversify the object of development discovering and occupying the “lacunes”. In this book, we offer a series of systems, which features create the base for extending the classification, presented in Fig. 1.54 in the Sect. 1.4.5, and its additions by the Table 6.1. The table is illustrated by Figs. 6.1, 6.2, 6.3, 6.4 and 6.5 [1].

For unification of structural features description of discussed cryptosystem, we take the following abbreviations in Table 6.1 and in Figs. 6.1, 6.2, 6.3, 6.4 and 6.5:

IC—the information channel,

CS—the channel of synchronization,

SIC—the synchro-information channel,

Table 6.1 Chaos generator Addition to cryptosystem classification on the position of the chaotic oscillator, operation mode of the decipherer, a number of communication channels and their functions

Operation mode of the decipherer	Destination of the chaotic oscillator (its belonging to the system)		
	Parameters of the chaotic oscillator (noise) do not serve as key (SSI outside the system)	Parameters of the chaotic oscillator serve as a key (chaotic oscillator is inside the system)	Parameters of internal chaotic oscillator serve as a key, and external chaotic oscillator (oscillator of noise) is SSI, SCI or SSCI
Double-channel with *separate* channel of synchronization (Figs. 6.1 and 6.2)			
Chaotic response	Figure 6.1a	Figure 6.1b	Figure 6.1c
Active synchronization	–	Figure 6.2	Figure 6.2
Double-channel *without* separate channel of synchronization (Fig. 6.3)			
Chaotic response	–	Figure 6.3a	–
Active synchronization	–	Figure 6.3b	–
m-channel (type of m, M_1, M_2) with a fraction n/m of purely synchronization channels (Fig. 6.4)			
Chaotic response	+, $n \neq 0$	Figure 6.4a	+, $n \neq 0$
Active synchronization	–	Figure 6.4b	+, $n \neq 0$

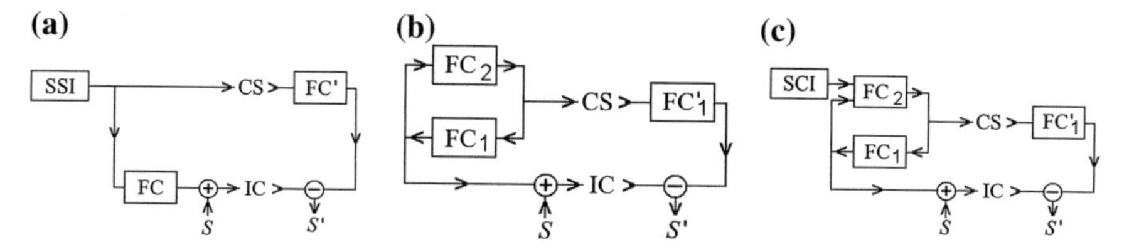

Fig. 6.1 Circuits of double-channel systems of the nonlinear-dynamical cryptographic with the separate synchronization channel and with deciphering in the chaotic response mode: with external SSI (**a**), with internal chaotic oscillator (**b**), internal chaotic oscillator and external SCI (**c**)

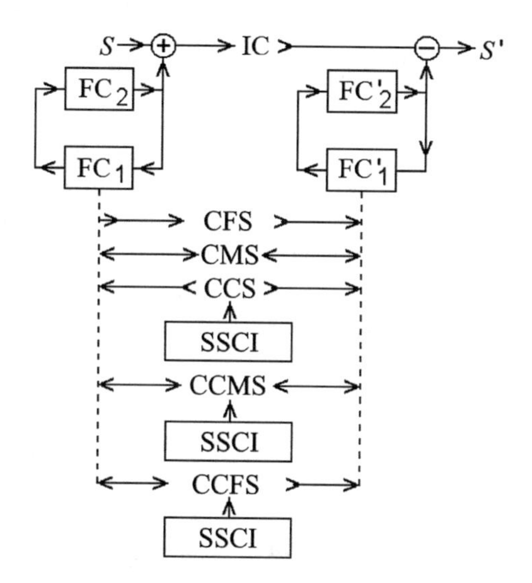

Fig. 6.2 Circuits of double-channel systems of nonlinear-dynamic cryptography with separate synchronization channel with internal chaotic oscillator. The main types of active synchronization are shown: mutual (*CMS*), forced (*CFS*), coordinating (*CCS*)—and combined: "coordinating-mutual" (*CCMS*) и "coordinating-forced" (*CCFS*). Here *SSCI* is *SSI* + *SCI*

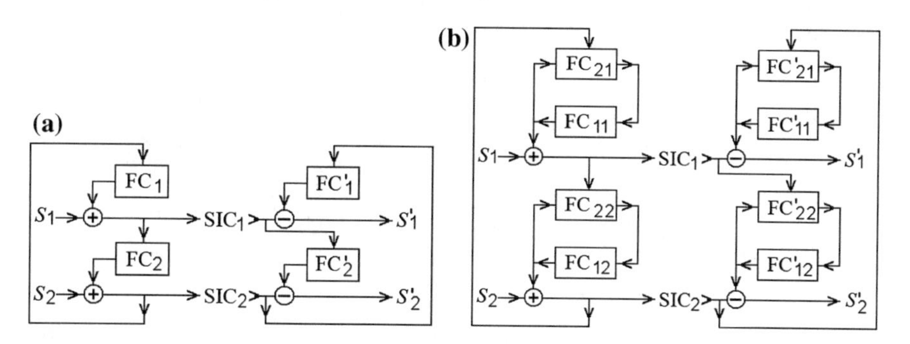

Fig. 6.3 Circuits of double-channel systems of nonlinear-dynamic cryptography without the separate synchronization with internal chaotic oscillator. The chaotic response (**a**) is used in the decipherer or coordination active synchronization (**b**)

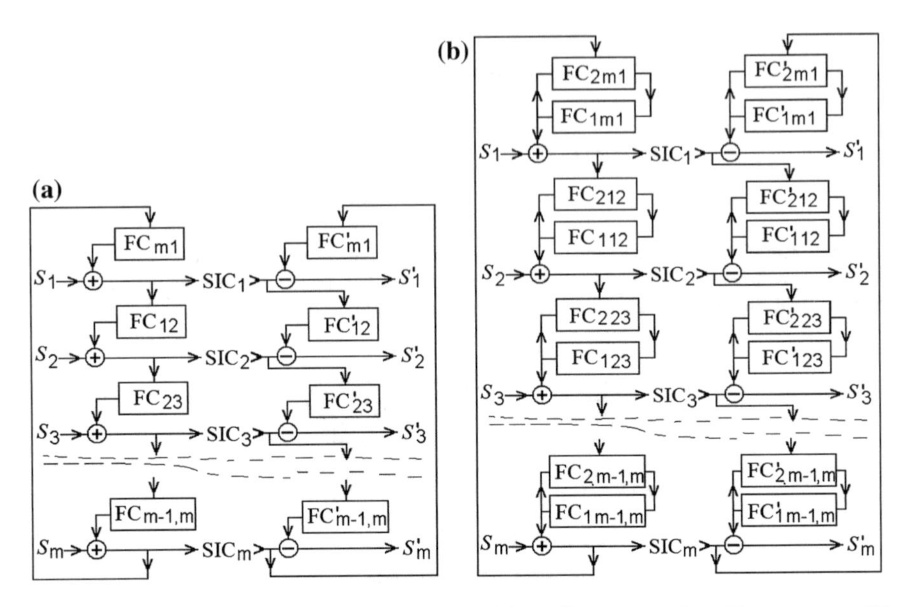

Fig. 6.4 Circuits of *m*-channel systems of nonlinear-dynamic cryptography without separate CS with internal chaotic oscillator. The chaotic response (**a**) is used in decipherer or coordinating active synchronization (**b**)

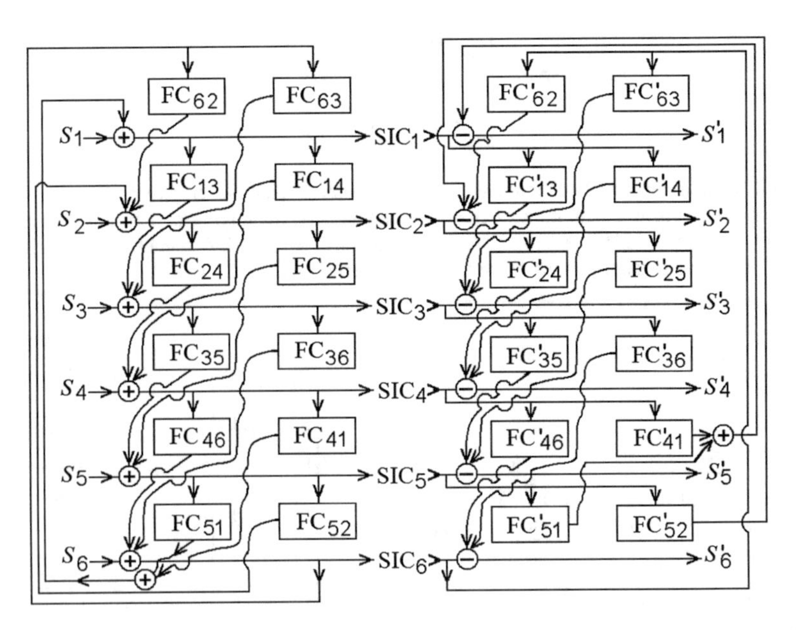

Fig. 6.5 The circuit of the cryptosystem corresponding to DNRI, when field is rotated in FBL_1 and FBL_2 by 180° and 120° angles pro tanto

SCI—the source of chaotic influence (noise oscillator or chaotic oscillator),

SSI—the source of synchronization influence, i.e. the synchronizer (chaotic oscillator, noise oscillator, or oscillator of the regular signals),

SSCI—the source of synchronized and chaotic influence (chaotic oscillator or noise oscillator),

FC—the functional converter of signals without arranged internal feedback (elementary FC),

CMS—the channel of mutual synchronization,

CFS—the channel of forced synchronization,

CCS—the channel of coordinating synchronization,

CCMS—the channel of "coordinating-mutual" synchronization,

CCFS—the channel of "coordinating-forced" synchronization,

S—the information signal,

The symbol of touch (′) designates belonging to decipherer; the converter FC_i' is considered as nominally identical to the appropriate converter FC_i of the cipherer.

Figure 6.1 shows circuits of double-channel systems with the separate synchronization channel. Operation principles of two of them (*a* and *b*) were discussed in Chaps. 1 and 4 (see Sects. 1.4.5, 4.1.2 and Fig. 1.37). These circuits in Figures *a* and *b* are similar to circuits in Figs. 3.11a in [1], 5.30a in [2], Fig. 1.37 and, accordingly, the circuits in Figs. 3.12a in [1], 5.31a in [2].

The circuit in Fig. 6.1c differs by a presence both external SCI, and the internal chaotic oscillator. It is, obviously, capable for working and its advantages and shortcomings merit the future discussion. For example, the case of "under-excited" operation mode of the internal dynamic system on FC_1 and FC_2. For is, the set of values of FC or FC_1 parameters is the key.

The structure of double-channel systems with separate channel of active synchronization is shown in Fig. 6.2. It is characterized by the fact that a transmitter and a receiver equitably influence each other ensuring mutual synchronization. In the case of forced synchronization unilateral influence performs by a transmitter. The idea of "coordinating" synchronization consists in the fact that a transmitter and receiver are synchronized between themselves owing to external and independent signal. Since coordinating synchronization does not exclude a presence of mutual or forced, then there are possible two types combined synchronization, shown in Fig. 6.2. We underline that when using CFS and CCFS, the converter in the decipherer FC_1' inevitably differs from FC_1 in the cipherer due to asymmetry of interconnection of the chaotic oscillator of the cryptosystems. Factually, the circuit in Fig. 6.2 generalizes the circuits in Fig. 3.7 in [1] and in Fig. 5.26 in [2].

Probably, it is impossible to construct a cryptosystem with the separate channel of active synchronization, in which there is only external chaotic oscillator (noise oscillator)—see Table 6.1. You see, according to definition of active synchronization, the decipherer should contain the chaotic oscillator, which parameters serves as a key and, hence, the cipherer should have the similar component.

The idea of construction double-channel without separate synchronization channel consists in that signal transmitted through IC on cryptosystem is used for the synchronization of another system. At that, IC and CS become SIC, and functional converters and/or the chaotic oscillator of the cipherer are covered by the mutual feedback, i.e. stop to be independent, forming now the unite dynamic system. Construction of cryptosystems, in which chaotic oscillator parameters serve as a chaotic response chaotic oscillator inside the system, and in decipherer it is used both as the chaotic response and coordinating active synchronization are shown in Fig. 6.3. The cryptosystem on the base of NRI (or DNRI) a rotation of optical field in its FBL by the angle 180°, when chains of transposition points (CTP) consist in two points (see Chap. 2 and Fig. 2.32, Table 2.3), operates on the principle illustrated by Fig. 6.3a.

Probably, in the cryptosystem we may use not only coordinating active synchronization, but the types of combined synchronization. Due to incompatibility of two conditions: a presence external chaotic oscillator and an absence the separate synchronization channel it is impossible to realize the cryptosystems *without* separate CS with external chaotic oscillator (noise oscillator)—see Table 6.1.

Relying on the above-described approach to construct of cryptosystems, we may synthesize m-channel system without the separate CS with internal chaotic oscillator—see Fig. 6.4. It is evident, that replacement of any input information signal S_i by the signal of the external chaotic oscillator i (noise oscillator i) leads to the fact that appropriate SIC_i becomes CS_i. Hence, we can obtain the cryptosystem with *non*-zero fraction of CS. At that, the external chaotic oscillator (noise oscillator) plays the double role.

Now, if to break off the feedback in the cipherer in the system with the chaotic response (Fig. 6.4a) by removing $FC_{i-1,\,i}$, then we obtain the cryptosystem without the internal chaotic oscillator (in which SSI is outside the system). The cryptosystem on the base NRI (or DNRI) at optical field in its FBL by angle $\Delta = \pi M/m$ operates on the principle illustrating by Fig. 6.4a. Really, in this case CTP consist of m points and has the closed structure of type m 1 1 (example of CTP see Sect. 2.4.2, Tables 2.2, 2.3 and 2.4).

This is easy to see that separately taken FC_{ij} (in Fig. 6.4a) or internal chaos oscillator (CO_{ij}), consisting from $FC_{1\,ij}$ and $FC_{2\,ij}$ (in Fig. 6.4b), connects the adder output i with the input of adder j. Evidently, that a number of such CO_{ij} and FC_{ij} may form from $m - 1$ to m^2. And they themselves are capable to form the closed or non-closed or combined complicate configuration of interconnections: linear, of type "convergent" or "divergent" star, circular etc. The configuration, in its turn, may serves as a key. For instance, for DNRI in the case of field rotations in FBL_1 and FBL_2 by angle 180° и 120° (CTP has a structure of type 6 3 2—see Fig. 2.35) the cryptosystem circuit is presented in Fig. 6.5. It is clear, that the resource of crypto-resistance increase is the most relevant for systems, where high values of m are achievable, assume for spatial distributed devices.

The thoughtful reader, probably, already notices the Fig. 6.4a in its contain is equivalent to Fig. 6.4b, if the elementary FC in it to replace onto chaotic oscillator, consisting in two FC. In other words, we can state that the system with active synchronization is the system *with chaotic response of the chaotic oscillators*. Checking of this property universality is a subject of independent investigations.

Let us underline that n the case of m-channel cryptosystems, the cipherer can be interpreted as the *unite* dynamical system with l dynamic variables (where $l \gtreqless m$). And therefore, the cryptosystem can be interpreted as *single-channel*, but with l-dimension communication channel (in the sense of simultaneous transmission of l signals).

We assume above that the external chaotic oscillator or the noise oscillator (i.e. its parameters or the signal itself) is not a key. But, we should imagine such a situation, when transmitting and receiving sides receive (independently of each other) the signal from chosen in advance mutual chaotic oscillator (noise oscillator) and according to stated principle. Then unknown to the enemy chaotic oscillators (noise oscillators) and/or a principle (rather, "removable" part of the last) serves as a key. There is a reason to consider that offered approach is promising in the practical sense because there are many natural or artificial chaotic oscillators (noise oscillators), say, stars and its constellations [1]. The precedent of using such a principle may detect in publication [3]—see Fig. 1.45.

And now, it is a time to discover a possibility of form diversification of the transfer characteristic of the nonlinear element in the chaotic oscillator.

6.2 Elements with Nonlinear Transfer Characteristic: Universality of Its "Constructions" and a Concept of Self-controlled Nonlinearity

Here we require essentially widen a context and remember that our world is occupied by *nonlinear systems*. On the language "influence v—response $N(v)$" nonlinearity is a *disproportion* of the second to the first. Or more strict, it is an infringement of the superposition principle. At first, the role of nonlinearity in radio electronics become aware of, probably, Russian physicist Papaleksi (1911) and Dutch scientist B. van der Pol (the van der Pol equation, 1920), in optics Russian physicist (1928), Vavilov (1950) and Khokhlov (1959–1960) [4–6]. As it is known, the nature of the nonlinear systems is extremely multifarious. They unexpectedly discovered everywhere: from spiral Galaxies to nano-dimension objects, from especially radio electronic and optical devices to the neuro-network bio-information substations, as well as in economics, management, sociology, psychology of creative work, ecology. In such systems, in spite of its simplicity, *nontrivial phenomena* are possible: instability, generation of oscillations and waves, self-organization, bifurcations, deterministic chaos (see, for instance, [1, 4–10]). All these phenomena are widely applicable.

As absolutely clear from previous chapters, in chaotic communication systems the form of the transfer characteristic of nonlinear element in the chaotic oscillator can be considered as a "complicated parameter". It is capable to be the one of the key of such a system [1, 11, 12]. Therefore, in competition with the enemy the owner of larger variety of transfer characteristics of devices (i.e. their nonlinearities) will win. This thesis agrees with the necessary variety principle of Ashby: "The variety of controlled should be not less than variety of controlling" [13]. At that, we need the effective and relatively simple approach of *nonlinearity change* (a key), which ensure the maximal variety of keys [11].

Hence, in order to satisfy these requirements, it is necessary to solve a problem of *nonlinearity formation*, i.e. control of its functional view. Developing such an approach, reasonably to state the more scaled problem of *self-variable* (self-controllable) nonlinearity. Then it becomes one more dynamic variable of the system. And, therefore, it embarrasses opening the ciphered message for the enemy, i.e. improve resistance of the confidential communication systems. Besides, self-varying nonlinearity is capable to attach to the technical devices the patterns of living organisms. The reason of this consists in the fact that nonlinearity in significant measure causes properties of real systems, in particular, the chaotic oscillator, and its change is capable to convert one dynamic system into another, i.e. differing by properties of its behavior [11, 14].

But a question "how do form a nonlinearity?" obviously, assumes the knowledge an answer to another question: "how the "ready" nonlinearity is arranged?" You see, the procedure of formation means that nonlinearity "is made from something". Moreover, the generalized approach to solution of these problems pushes to a guess about universality of "construction" of different elements, which transfer characteristics are nonlinear. A guess, in its turn, causes a necessity to formalize description of this hypothetic "construction".

General principle of nonlinearity formation. Let us begin with the fact that in accordance with axiomatics [11] there are only four types of transfer characteristics (TC). Really, let some modifier (converter) $\mathbf{M}'$ converts the input flow (signal) $\mathbf{f}_{\text{in}}$ of some form $\mathbf{F}_{\mathbf{f}\,\text{in}}$ with parameters $\mathbf{\Pi}_{\mathbf{f}\,\text{in}}$ into the output flow (signal) $\mathbf{f}_{\text{out}}$, having the form $\mathbf{F}_{\mathbf{f}_{\text{out}}}$, with parameters $\mathbf{\Pi}_{\mathbf{f}\,\text{out}}$. It is assumed here that (1) the flow is defined into four-dimension space-time: $\mathbf{x} \equiv (\mathbf{r},\ t) \equiv (x,\ y,\ z,\ t)$; (2) the form $\mathbf{F}_{\mathbf{f}}$ of the flow, modifier etc. imagines as the mathematical *name form* describing the flow, a modifier etc.

Then, three types of transfer characteristics correspond to this converter:

(1) in the space of forms $\mathbf{F}{:}\mathbf{F}_{\mathbf{f}\,\text{in}} \to \mathbf{F}_{\mathbf{f}\,\text{out}}$;
(2) in the space of parameters $\mathbf{\Pi}{:}\mathbf{\Pi}_{\mathbf{f}\,\text{in}} \to \mathbf{\Pi}_{\mathbf{f}\,\text{out}}$ (we can speak about it only at given transfer characteristic (TC) in the space of forms);
(3) in the space of flow values $\mathbf{f}$:

$$\mathbf{f}_{\text{out}}(\mathbf{x}_{\text{out}}) = \mathbf{f}_{\text{out}}(\mathbf{f}_{\text{in}}(\mathbf{x}_{\text{in}}),\ \mathbf{r}_{\text{in}}),\ \mathbf{x}_{\text{out}} = \mathbf{x}_{\text{out}}(\mathbf{x}_{\text{in}}), \tag{6.1}$$

where $\mathbf{x} \equiv (\mathbf{r},\ t) \equiv (x,\ y,\ z,\ t)$, time t may be a part of $\mathbf{f}_{\text{in}}$. This TC is static and, therefore, it exists not always: say, it is absent in the case of oscillator, which converts an energy of DC into the AC energy [12].

(4) Any signal (a flow) $\mathbf{f}(\mathbf{x}, \mathbf{\Pi_f})$ has the transfer characteristic of the flow: $\mathbf{f} = \mathbf{f}(\mathbf{x}, \mathbf{\Pi_f})$, i.e. $(\mathbf{x}, \mathbf{\Pi_f}) \rightarrow \mathbf{f}$.

So, square-ware oscillator for harmonic signal at output has a signal of the *similar* form (in the sense [11]). In other words, $\mathbf{F}{:}A_{in} \cdot \sin(\omega_{in}t + \varphi_{in}) \rightarrow A_{out}[1 - \cos(\omega_{out}t + \varphi_{out})]$. Its two parameters linear depend on parameters of input signal, but the one parameter depends nonlinear: $(A_{in}, \omega_{in}, \varphi_{in}) \rightarrow (A_{out}, \omega_{out}, \varphi_{out}) = (0.5A_{in}^2, 2\omega_{in}, 2\varphi_{in})$. TC ($\mathbf{f}{:}f_{in} \rightarrow f_{out} = f_{in}^2$) in the space of signal values is nonlinear.

Another example is an adder of two ($i = 1, 2$) single-frequency signals of sine form. Transfer characteristics of this linear device, obviously, are:

—in the space of forms $\mathbf{F}$:

$(A_{in\,1} \cdot \cos(\omega_{in}t + \varphi_{in\,1}), A_{in\,2} \cdot \cos(\omega_{in}t + \varphi_{in\,2})) \rightarrow A_{out} \cdot \cos(\omega_{out}t + \varphi_{out})$;

—in the space of parameters $\mathbf{\Pi}$:

$$(\omega_{in}, A_{in\,1}, \varphi_{in\,1}, A_{in\,2}, \varphi_{in\,2}) \rightarrow (A_{out}, \omega_{out}, \varphi_{out}) = ((Ac^2 + As^2)^{0.5}, \omega_{in}, \arg(Ac, As)),$$

where $Ac = \Sigma_i A_{in\,i} \cos(\varphi_{in\,i})$, $As = \Sigma_i A_{in\,i} \sin(\varphi_{in\,i})$,

$$\arg(Ac, As) = \begin{cases} (\pi/2) - \mathrm{arctg}(Ac/As), & As > 0 \\ (3\pi/2) - \mathrm{arctg}(Ac/As), & As < 0 \\ 0, & As = 0,\ Ac \geq 0 \\ \pi, & As = 0,\ Ac < 0 \end{cases} \in [0;\ 2\pi];$$

—in the space of signal values $\mathbf{f}{:}(f_{in\,1}, f_{in\,2}) \rightarrow f_{out} = f_{in\,1} + f_{in\,2}$.

In physics, oscillation and wave theory, we assume usually under nonlinearity a view of (static) TC in the space of flow values $\mathbf{f}$ (6.1). It is rightful to state that nonlinearity may be "prepared" or "forming". We can treat with "prepared" nonlinearity as a black box (in the sense of Ashby [13]), having known nonlinear static transfer characteristics. But "forming" nonlinearity is some product of signals and several converters interaction. We underline that nonlinearity opposition on the sigh "prepared" and "forming" is conditional enough. It relates with the researcher aims, with his cognitive position with respect to the specific nonlinear element: is it interesting for researcher by whom this or that TC conditioned, how do the appropriate converter (modifier) construct etc.

Let us return to the problem of nonlinear formation. Evidently, we can often create the required nonlinearity N in the form of superposition of "prepared" $N_i{:}f_{out} = N(f_{in}) = N_{super}(f_{in}) = N_1(N_2(f_{in}) + N_3(f_{in}))$ etc. But this specific approach is deprived of transformation simplicity of the TC view and it is not flexible. Applied physicist and engineer must anticipate in advance all TC types, which will be required, and "put" them into the material construction of the element. In addition, the fundamental question remains make clear: how are these nonlinearity

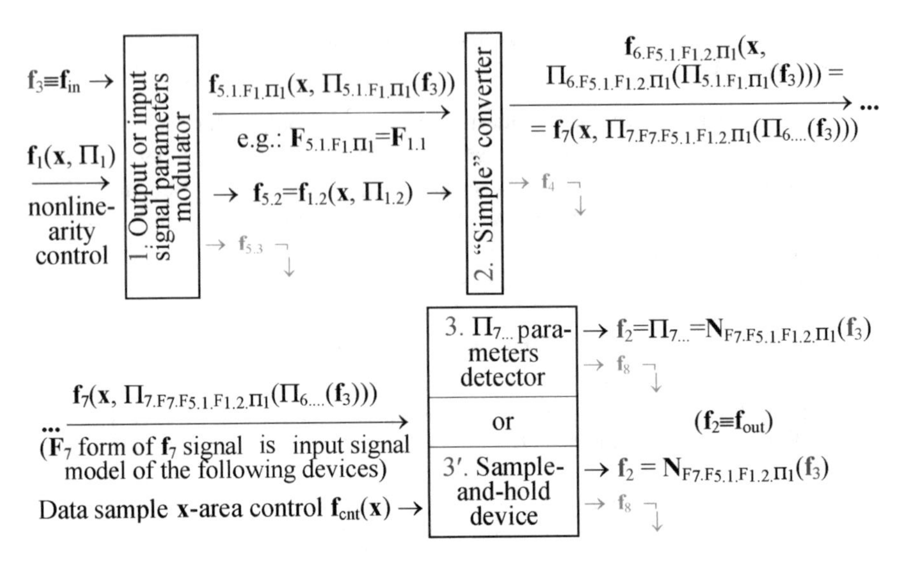

Fig. 6.6 The diagram opening the one from two principles ($f_3 \to f_2$) of nonlinearity formation [12]

N_i constructed, from which N_{super} has got? Of course, they may be constructed according to the same superposition principle. And so to the infinity?

We offer the second approach—to get nonlinearity from *two* transfer characteristics: TC of the *signal* $\mathbf{f}_7$ or TC $\mathbf{\Pi}_7(\mathbf{\Pi}_6(\mathbf{\Pi}_{5.1}(\mathbf{f}_3)))$ in the space of *parameters* (Fig. 6.6). At that, both TC are controllable by the form $\mathbf{F}_1$ and parameters $\mathbf{\Pi}_1$ of some signal $\mathbf{f}_1$.

Figure 6.6 contains the device circuit with the forming conversion nonlinearity $\mathbf{f}_3$ into $\mathbf{f}_2$ *under control* of the signal $\mathbf{f}_1$, i.e. under control of its form $\mathbf{F}_1$ and parameters $\mathbf{\Pi}_1$. In Fig. 6.6 the unit 1 is a modulator of parameters $\mathbf{\Pi}_{5.1}$ of the output (if the form $\mathbf{F}_{5.1.\text{F}_1\Pi_1} \neq \mathbf{F}_{1.1}$) or input (if $\mathbf{F}_{5.1.\text{F}_1\Pi_1} = \mathbf{F}_{1.1}$) signal $\mathbf{f}_{5.1}$ allows to the signal $\mathbf{f}_3$, i.e. the "nonlinearity argument", to influence on the signal parameters $\mathbf{f}_{5.1}$. The unit 2 (converter "simple") is the not obligatory component intended for additional variation (complication) of functional dependences in the space of parameters: $\mathbf{\Pi}_{5.1.\text{F}_1\Pi_1}(\mathbf{f}_3)$ $\mathbf{\Pi}_{6.\text{F}_{5.1}\text{F}_{1.2}\Pi_1}(\mathbf{\Pi}_{5.1.\text{F}_1\Pi_1}(\mathbf{f}_3))$, $\mathbf{\Pi}_{7.\text{F}_7\text{F}_{5.1}\text{F}_{1.2}\Pi_1}(\mathbf{\Pi}_{6\ldots}(\mathbf{f}_3))$.

The unit 3 (detector or "extractor" of parameters $\mathbf{\Pi}_{7\ldots}$) together with the parameter modulator (unit 1) allows "to transform" TC in the space of parameters $\mathbf{\Pi}_{7\ldots}(\mathbf{\Pi}_{6\ldots}(\mathbf{\Pi}_{5.1\ldots}))$ taking into account the modulation characteristics $\mathbf{\Pi}_{5.1\ldots}(\mathbf{f}_3)$ in TC in the space of signal values: $\mathbf{f}_2 = \mathbf{\Pi}_{7\ldots} = \mathbf{N}_{\text{F}_7.\text{F}_{5.1}.\text{F}_{1.2}.\Pi_1}(\mathbf{f}_3)$. So, we can get the nonlinearity from the transfer characteristic in the space of *parameters*.

The alternative unit 3′ is a device of sample-storage (data about signal characteristics). It "transfers" TC of the signal $\mathbf{f}_7$ into TC in the space of signal values: $\mathbf{f}_2 = \mathbf{N}_{\text{F}_7.\text{F}_{5.1}.\text{F}_{1.2}.\Pi_1}(\mathbf{f}_3)$. At that, the signal $\mathbf{f}_3$ becomes the "nonlinearity argument" or owing to a presence of the modulator (unit 1) or owing to passage a part of $\mathbf{f}_3$ directly on input $\mathbf{f}_{\text{cnt}}(\mathbf{x})$ of control to $\mathbf{x}$-area of data sample. In the last case, we can refuse from the unit 1. So, we obtain the nonlinearity from the transfer characteristics of the *signal* $\mathbf{f}_7$.

Let us make a logical stress on the fact that in two alternatives described in the last paragraphs, the sense of offered method of nonlinearity formation is.

At first glance, the element construction in Fig. 6.6 seems to be intricate and strained. However, we may show that a variety of known "prepared" nonlinear elements (if to look narrowly to each of them) is arranged in accordance to Fig. 6.6 [11]. So, all simple nonlinear elements (metallic cathode emitting electrons) satisfy this principle, so as more complicate (nonlinear ring interferometer in the static mode, which can be interpreted as a nonlinear element). About processes in the last devices we spoke in detail in Chaps. 2 and 4.

As an illustration, we examine the classical example: the emission of electrons (with a mass m and a charge e) at temperature T through the potential barrier $A - eV$ from metal with the work function $A = \mathbf{\Pi}_7$ at bias voltage $V = \mathbf{f}_{\text{cnt}} = \mathbf{f}_3$. In metal the dependence of a number n electrons on values of its impulse $\mathbf{p} \equiv (p_x, p_y, p_z)$ obeys to Fermi–Dirac distribution function

$$n(\mathbf{p}) = (2/h^3)\{1 + \exp[(E(p_x, p_y, p_z) - E_f)/kT]\}^{-1}. \tag{6.2}$$

Here the potential barrier literally realizes the *sample* of moving outside electrons near the metal surface (unit 3′ in Fig. 6.6), which energy E exceeds the barrier $A - eV$ (sample of *flow* $\mathbf{f}_7$). The form of Fermi–Dirac distribution function $n(\mathbf{p})$ turn out, at bottom, the *form* $\mathbf{F}_{\mathbf{f}_7}$ of the flow $\mathbf{f}_7$. Indeed, the function: (1) describes the flow on the statistics language; (2) permits to judge which fraction of electrons at given voltage V is capable to leave the cathode and participate in the charge transfer forming the current I.

Thereby, distribution (6.2), i.e. the form $\mathbf{F}_{\mathbf{f}_7}$, influencing on the kind $I(V)$ by determining manner, transforms into the "forming" nonlinearity $\mathbf{f}_2 \equiv \mathbf{N}\ldots(\mathbf{f}_3) \equiv I(V)$ in TC in the space of voltages and current. Given nonlinear current dependence upon the voltage $I(V)$ has a view

$$I = (4\pi mek^2/h^3)T^2 \exp[-(A - eV)/kT] \tag{6.3}$$

and known as Richardson–Deshman law [15, p. 757]. In the exponential law (6.3) we can easily distinguish a track of the form $\mathbf{F}_{\mathbf{f}_7}$ (6.2), containing an exponent. It is easy to see that such classical nonlinearity as volt-ampere characteristic (i.e. TC in the space of voltage-current) of the semiconductor diode, forms in the similar way.

Nonlinearities $\mathbf{f}_2 \equiv \mathbf{N}\ldots(\mathbf{f}_3)$ from these examples are "formed" (at low, conditionally speaking, level of fundamental phenomena in the substance) by physics and technology of electronic devices manufacture. At higher level, which is associated with radio electronics, radio-physics etc., nonlinearities (as the distinctive TC attribute) are used only as "prepared" and therefore, they are perceived exactly so. Then the deduction suggests itself: if this "prepared" and for a long time usual for physicist and engineer nonlinearity (6.3) factually "formed", then almost full variety of known nonlinear elements is the same in its nature. How true this is?

The methods of nonlinearity formation, which illustrated by Fig. 6.6, is *complete* in that regard that for TC obtaining in the space of flow values, it mobilizes *all* other types of transfer characteristics: TC in the space of parameters (together with TC in the space of forms) and TC of the flow. The above-mentioned approach basing on superposition construction of "prepared" nonlinearities is additional to this method. To obtain TC $N(f_{\text{in}}) = N_{\text{super}}(f_{\text{in}})$ in the space of flow values we can use the TC variety N_i in the same space of flow values, from which the superposition is constructed as from elements.

Thus, we put forward the *general principle of nonlinearity formation*. It consists in mutual use of the nonlinearity formation method (disclosed in Fig. 6.6) sand superposition of "prepared" nonlinearities. This principle, evidently, mobilizes all possible TC to create TC in the space of flow values [11, 12].

Methods of nonlinearity formation, according to Fig. 6.6, i.e. formation on the *lower level* (in the sense of fundamental physical phenomena like processes of thermal emission), are developed, to our mind, not enough. Evidently, this is caused by the fact that for a long time researchers could create the enough manifold nonlinearities with the help of superposition construction. In other words, requirements of practice satisfied activity on the *higher level*, in the sense of device combination or its units as the "black boxes" (in the sense of Ashby [13]). Therefore, till this time, the superposition method dominated. However, during perception of universality of nonlinear elements "construction", both these methods must transfer from the state of implicit competition to the state of subsidiarity. That is, we assume that time arrives to address to the low level since it has a series of advantages, which discussed at the beginning of publication [11]. Moreover, now the stormy growth of NBICS-technologies, which extend possibility of substance design with given in advance (abbreviation NBICS is formed according to first letters of prefixes: nano-, bio-, info-, cogno-, socio-) [16].

The fact that our opinion about relations between levels, says about the benefit and it agreed with the law of hierarchic compensations in the hierarchic (Sedov's law). It is true for competitive information systems operating in situations of resources limitation. In our case, resources are cognitive and manufacturing facilities in the scientific-technological sphere. The law reads: growth of diversity at higher levels leads to its reduction at lower levels [17].

Distantly-controlled, self-controlled nonlinearity and possibility of evolution. Let us examine advantages, which a possibility of nonlinearity construction $\mathbf{f}_2 \equiv \mathbf{N}\ldots(\mathbf{f}_3)$ gives at low level. We can regard the following: constructive simplicity of remote nonlinearity control and its self-control, practically unlimited *variety* of its types as well as evolution of the evolution operator in the subsystem. It should be noted that between XX–XXI centuries the role of *variofication* (from Latin vario + facio – fulfillment of difference) is deeply recognized as the one from consequence of scientific-technical revolution. Variofication is the global phenomenon of appearance of innovation ideas, which can be materialized in new types of products. Among the reasons of variofication is output of principally new product, occupying the free "ecological niche"; increase of the product variety inside manufactures family and extraction in it new kinds; complication of product itself and its components; qualitative variation of technology level, necessary for product manufacture; rigid resource limitations forcing to diversify the output and to put it in correspondence to limitation [18, pp. 27–28].

According to above-mentioned (Fig. 6.6), a view of nonlinear TC $\mathbf{f}_2 \equiv \mathbf{N}\ldots(\mathbf{f}_3)$ we *replace* by the view of external controlling flow $\mathbf{f}_1(\mathbf{x}, \prod_1)$ and/or values of its parameters $\mathbf{\Pi}_1$. Hence, if the nonlinear element is a part of some dynamic system, the view of nonlinearity $\mathbf{f}_2 \equiv \mathbf{N}\ldots(\mathbf{f}_3)$ in it may be *distantly* replaced (controlled) by the flow $\mathbf{f}_1$—see Fig. 6.7a,—and even *self-controlled* (рис. 6.7b). For self-control it is necessary to anticipate *additional* feedback: the detector-steward of parameters $\mathbf{\Pi}_1$ and controlled generator of the flow $\mathbf{f}_1$ in Fig. 6.7b. It influences on the parameters $\mathbf{\Pi}_1$ of the flow $\mathbf{f}_1$, which controls by nonlinearity in the subsystem, which is mutual for Fig. 6.7a, b. We note: if he generator $\mathbf{f}_1$ and the detector-steward $\mathbf{\Pi}_1$ are mutually-inversed, they together is the single (identical) converter. Hence, they may be absent. This case corresponds to additional feedback by means of high frequency, i.e. without extraction of parameters $\mathbf{\Pi}_1$ as the intermediate signal.

In the case of self-control, the nonlinearity $\mathbf{N}_{\mathrm{F}_1\Pi_1}$, which is parametrized by the variable value $\mathbf{\Pi}_1$, becomes the *dynamic variable*. It is equivalent with other dynamic variables at commensurable speed of variation $\mathbf{f}_2$ and $\mathbf{\Pi}_1$, i.e. for $\tau_1 \approx \tau_2$, where τ_1, τ_2 are typical times of variation $\mathbf{f}_2$ and $\mathbf{\Pi}_1$ $(\tau_1 \leq \tau_2)$. Nonlinearity $\mathbf{N}_{\mathrm{F}_1\Pi_1}$ is unusual if parameters $\mathbf{\Pi}_1$ rarely change or slowly $(\tau_1 \ll \tau_2)$ compared to the signal. It is unusual since in this case:

(1) nonlinearity has time to fully realize its influence on dynamics in the subsystem, which is mutual for Fig. 6.7a, b;
(2) nonlinearity is identification characteristic of the evolution operator.

Therefore, as a result, for $\tau_1 \ll \tau_2$ we have a right to consider that in this "fast" subsystem the *evolution of its evolution operator* happens as well. We may speak also about evolution **of** this subsystem and the evolution **in** it. In the context of capability to self-variation, it is reasonable to draw parallels with biology objects and speak about the ontogenesis of the technical system [11, 12, 14]. Hence, to our opinion, both self-controllable nonlinearity and products of NBICS-technologies evenly promote the *technique biologization*. At that, the self-controlled nonlinearity is capable to be a mechanism biologization its driver. And NBICS-technologies are

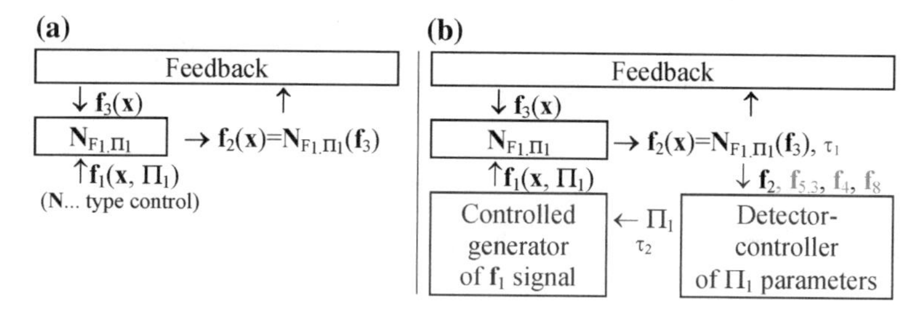

Fig. 6.7 The circuit discovering the general principle ($f_3 \rightarrow f_2$) of nonlinearity control (**a**) and self-control (**b**) [12]

not only promising and productive means for *creation* such nonlinearities, as we underlined above, but they turns out the sphere of nonlinearity *application*. The fact of such ring feedback is typical for internal dynamics of NBICS-technology as the technoscience.

In the thematic field of Sect. 6.2 it is logical to raise the question about approaches of synthesis of the nonlinear element with transfer characteristic of given form. Let the reader allows us to give without proof the variant of the device with controllable and rather manifold TC, naturally, constructed on the principle of Fig. 6.6.

The transfer characteristics as a sum of Fourier series and nonlinearity synthesis algorithm of any desired shape. Let address to Fig. 6.8, where the element is shown, which nonlinear TC $\mathbf{f}_2(\mathbf{x}) = \mathbf{N}_{\mathbf{f}_1}(\mathbf{f}_3(\mathbf{x}))$ is formed as a *sum of sums* of Fourier series. This element TC can be purposely construct, attaching the characteristic the given shape but such that it is the sum of sums of Fourier series.

Explaining the content of offered approach, we assume:

(1) the i-th component $f_{1,\,i}$ of input two-dimension signal ($m_{\mathbf{f}_1}$ is dimension vector $\mathbf{f}_1$) can be presented by the Fourier series, i.e. in it $A_{i,\,n}(\mathbf{r})$, ω, $\varphi_{i,\,n}(\mathbf{r})$ are potential controllable:

$$f_{1,\,i}(\mathbf{x}) = \sum\nolimits_{n=0}^{\infty} A_{i,\,n}(\mathbf{r}) \cos(n\omega t + \varphi_{i,\,n}(\mathbf{r})), \ i = 1\ldots m_{\mathbf{f}_1};$$

(2) the signal $\mathbf{f}_3$ is scalar: $\mathbf{f}_3(\mathbf{x}) \equiv f_3(\mathbf{x})$;
(3) the modulator of control signal parameters $\mathbf{f}_1$ contains the linear phase modulators: $\delta t_j(\mathbf{r}, \mathbf{f}_3(\mathbf{x})) = K_j \cdot f_3(\mathbf{x})$, where δt_j is delay, $K_1 = 0$;
(4) these phase modulators realize the spatial shifts $\delta \mathbf{r}_j$, which do not depend from $\mathbf{f}_3 : \delta\mathbf{r}_j(\mathbf{r},\ \mathbf{f}_3(\mathbf{x})) = \delta\,\mathbf{r}_j(\mathbf{r}),\ \delta\mathbf{r}_1(\mathbf{r}) = 0$.

Then, relying on the calculations earlier performed [14], we may show that the device, really has a nonlinearity $\mathbf{f}_2(\mathbf{x}) = \mathbf{N}_{\mathbf{f}_1}(\mathbf{f}_3(\mathbf{x}))$ in form of a sum of sums of

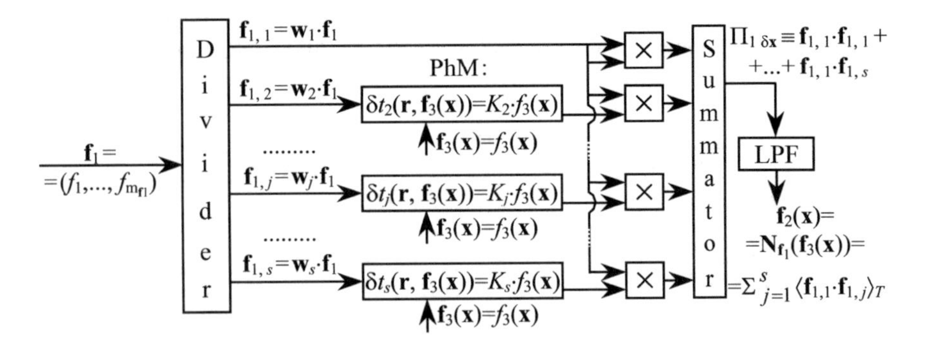

Fig. 6.8 The element with $\mathbf{f}_2(\mathbf{x}) = \mathbf{N}_{\mathbf{f}_1}(\mathbf{f}_3(\mathbf{x}))$ transfer characteristic controlled: by $\mathbf{f}_1$ type, slope values of the phase modulators K_j modulation characteristic and initial shifts in the modulators. Symbol "×" denotes the multiplier, PhM is phase modulator, LPF is low-pass filter

Fourier series: $\sum_{j=1}^{S} \langle f_{1,1} \cdot f_{1,j} \rangle_T$, $T \equiv 2\pi/\omega$. Series are added with weights $w_{1,i,i} \cdot w_{j,i,i}$, and sums of Fourier series have a view

$$\begin{aligned}
&\langle f_{1,1} \cdot f_{1,j} \rangle_T = 0.5 \cdot \sum_{i=1}^{m_{f_1}} w_{1,i,i} \cdot w_{j,i,i} \cdot \sum_{n=1}^{\infty} A_{j,i,n}(\mathbf{r}) \cdot \cos\{n \cdot \omega_{0,j} \cdot f_3(\mathbf{x}) + \delta\varphi_{j,i,n}(\mathbf{r})\} \\
&A_{j,i,n}(\mathbf{r}) \equiv A_{i,n}(\mathbf{r}) A_{i,n}(\mathbf{r} + \delta\mathbf{r}_j(\mathbf{r})), \ \omega_{0,j} \equiv \omega K_j, \\
&\delta\varphi_{j,i,n}(\mathbf{r}) \equiv \varphi_{i,n}(\mathbf{r} + \delta\mathbf{r}_j(\mathbf{r})) - \varphi_{i,n}(\mathbf{r} + 0).
\end{aligned} \tag{6.4}$$

It is clear from here that for synthesis of desired TC it is necessary to expand it in the Fourier series or in its superposition (there may be much variants of this expansion). Then amplitudes, phases, frequencies of components of these series become known, i.e. left parts of three last expressions in (6.4). From them, it is easy to determine the parameters of the control signal $\mathbf{f}_1$ and modulators, i.e. the right parts of these expressions.

We note that in the general case the phase modulators should treat as the controlled delay line and a generator as a source of not only periodic signals. This aperiodicity and generalization of the phase modulators will lead to changeableness (evolution) of the nonlinear TC. Besides, we underline that in the element in Fig. 6.8 the view of nonlinearity is determined by not only temporal but the spatial form $\mathbf{f}_1$. Obviously that a variety of variants and spatial and temporal forms $\mathbf{f}_1$, and hence, TC $\mathbf{f}_2(\mathbf{x}) = \mathbf{N}_{\mathbf{f}_1}(\mathbf{f}_3(\mathbf{x}))$ is practically infinitely.

We note that actualization of the spatial aspect of the signal f_1 for nonlinearity synthesis happens in the interferometers with principally two-dimension signals, say, containing the optical vortex. Interested reader can independently look to the material of Chap. 5 through such an unexpected prism. Interpretation (5.16) or even (5.3) can serve as a first step.

Controllable nonlinearity: historical-scientific retrospective review. Judging to physic history, two families of optical devices are *precedents* of systems with formed nonlinear TC. First, these are *interferometers*: the screen with two holes, which was offered by Young (1802 [15, p. 224]), bi-mirrors of Fresnel (1816 [15, p. 832]), double-beam interference refractometer of Jamin (1856 [15, pp. 225–226])—together with an observer's eye an its hand, varying the propagation difference $\delta\, l$ in interfering beams in the device. Then, distribution of intensity on the interference pattern is proportional to $\cos^2(\omega n \delta\, l/c + \text{const})$, where ω is a circular frequency, n is the refraction index, c is the light speed [15, p. 224]. In other words, $\delta\, l$ is the argument of the nonlinear function (transfer characteristic) $I(\delta\, l) = \cos^2(\omega n \delta\, l/c + \text{const})$. The shape of it, evidently, is determined by the sine dependence in time of the optical wave at the input of interferometer—by signal, external with respect to the interferometer.

Second, these are *polarization devices* containing analyzers together with the observer's eye and its hand, which gives the angle α between polarization planes of the incident linear-polarized light (with intensity I_0) and the analyzer. According to Malus (1808) law, the light, emanating from the analyzer, has intensity

$I = I_0 \cos^2 \alpha$ [15, pp. 201, 391]. In other words, the analyzer with changeable orientation of passing direction α fulfills the role of parameter modulator of the output signal. Nonlinearity of modulation characteristics (owing to properties of observer's eye) transfers from the space of parameters into the space of signal values: $I = I(\alpha)$. At that, it is clear that the shape of its nonlinear TC depends on the shape of the optical signal, which passes to the analyzer. Speaking more exactly, from the fact how it is polarized: linearity, elliptically, has a natural polarization etc.

If to use by concepts of circuit in Fig. 6.6, the optical interferometers and polarizers (as the analyzer) carry out the function of the unit 1. In other words, they serve as a parameters modulator $\mathbf{\Pi}_{5.1.\mathrm{F}_1\Pi_1}$ of the input (as $\mathbf{F}_{5.1.\mathrm{F}_1\Pi_1} = \mathbf{F}_{1.1}$) light flow $\mathbf{f}_1$, and output flow $\mathbf{f}_{5.1}$ of the unit 1 has modulated parameters $\mathbf{\Pi}_{5.1.\mathrm{F}_1\Pi_1}(\mathbf{f}_3)$. The function $\mathbf{\Pi}_{5.1.\mathrm{F}_1\Pi_1}(\mathbf{f}_3)$ is nonlinear (see example of TC analysis of adder or the Malus law). Observer's eye "measures" the flow intensity $\mathbf{f}_7 = \mathbf{f}_{5.1}$ (i.e. serves as the unit 3 in Fig. 6.6—the detector of parameters $\mathbf{\Pi}_7$). And his hand changes the beam propagation difference $\delta l = \mathbf{f}_3$ in the interferometer or the angle $\alpha = \mathbf{f}_3$ between planes of polarization. Evidently, that those part of reason-consequence chain, which is between perceived eye and "modulating" hand of the observer, plays a role of the unit "Feedback" in the diagram in Fig. 6.7.

The first *prototype* of interferometers in radio range, probably, we should consider accessories of Hertz [19]. He described experiments with waves radiated by the Hertz vibrator and having the wavelength $\lambda \approx 9.6$ m. In the lecture room, the wave radiated by a vibrator was reflected from the wall, on which "the zinc plate was attached with 4 m of height and 2 m of width, forming the head see. "Owing to interference of both waves, in the air, the standing waves are formed",—explained Hertz [19, pp. 77, 80, 85]. He measured their intensity and "Hertz's eyes were the receiver detector",—notice the historian of radio electronics Shtykov [20, p. 47] (as in considered cases of optical devices). The think is that the receiving antenna ("secondary circuit", in Hertz terminology) were the "conductor bended on the circle of 35 cm in radius". Displacement of the circle from the vibrator to the zinc plate on the wall corresponded to variation of their propagation difference of δl waves. "At that, sparks (in the antenna gap—*Authors*) were strong enough to notice them in dark premise on the distance of several meters" [19, p. 78]. Later, the combinations of the phase modulator with the phase detector happen to be the device with formed nonlinearity in radio electronics [15, p. 429].

The simple "prepared" nonlinearity (types of $N(v) \approx v^2$) of the medium/control circuit allows to be *without* a man in the feedback (i.e. the medium/circuit urgently "operates for" slowly operator). According to this principle, the devices with complicate behavior were arranged, say, bistable interferometer of Fabry–Perot (1969, 1975) [21, pp. 7–13] and many system, where the chaos was observed. Among the last, the Ikeda system (four-mirror interferometer, 1979, Fig. 1.22) [4]; variety of circuits of phase-locked loop (probably, from with the middle of 1980s [22–26]) and oscillator media [27, 28], modeling on its base, as well as the ring interferometer with the field rotation on the transverse plane of the light beam (1989, Fig. 1.24) [29, 30].

Earlier, in the example with the Richardson–Deshman (6.3) law the role of signal form $\mathbf{f}_7$ was played by the shape of the Fermi–Dirac distribution function (6.2). It is possible to change it, i.e. the signal $\mathbf{f}_7$, but in practice, it is difficult. Therefore, possibilities of control of formed nonlinearity (6.3) are minimal. We underline that in presented historical subjects, nonlinearity is formed *implicitly*, due to which it seems "prepared" or its nature is not yet prepared. It is clear, in physics, including technical, and in engineering we can discover another similar examples. *Perception* of these facts in the sense of the approach expressed by Figs. 6.6 and 6.7, the revision of standard models of nonlinear optics and radio physics from positions of the concept of controlled nonlinearity developed will open prospects of successful self-control by nonlinearity. Obviously, that systems having «prepared» and formed nonlinearity must be similar. The likenesses and distinctions of radio electronic chaos generators having «prepared» and formed nonlinearity, and also their replacement are discussed in [31].

6.3 Conclusions

In the context of *diversification* of structural diagrams as some strategy of increase of a number of keys (increase of system cryptoresistance), terminological specifications and taking into consideration the devices offered by this book authors, the addition is developed to classification of cryptosystems according to the chaotic oscillator position, the operating mode of the decipherer, a number of communication channels and its functions. The addition is formed as a table and series of accompanied circuits of potentially realizable cryptosystems.

Besides, universalism of elements "constructions" with the nonlinear transfer characteristic $\mathbf{N}_{...}(\mathbf{f}_3)$ is revealed, the method of control and self-control by TC nonlinearity is suggested. Since TC in this method depends on the shape of external signal, it is distant. In the systems of nonlinear-dynamic cryptology (when TC in the chaotic oscillator serves as important component of the key) such a method will ensure the simplicity of key change and will increase its number with growth the system cryptoresistance. In other words, such a *diversification* of transfer characteristics of the nonlinear elements of chaotic oscillators is additional strategy of resistance increase of the communication systems. The example of appropriate nonlinear element is suggested for illustration. The arsenal of potential TCs of this method is inexhaustible at bottom.

Besides, the offered method of nonlinearity formation attaches a pattern of simplest bio-systems to engineering devices. It has also the methodical valuables since it brings together *in one class* of systems the classical devices of optics and radio physics unifying its understanding and description. The above-discovered historical-scientific retrospective of the concept of formed nonlinearity is evidence about it. Therefore, the further search of system precedents realizing this (it might seem, especially physical-mathematical principle of nonlinearity formation) method among natural and artificial objects of arbitrary nature is extremely important.

Both these strategies in the aggregate show ways of further development of confidential communication systems using the dynamic chaos.

Now we proceed to Conclusion, supply the brief totals and debate about terminological boundaries of the concept “cryptology”.

References

1. Vladimirov SN, Izmailov IV, Poizner BN. Nonlinear-dynamic cryptology: radio physical and optical systems/under edition by S.N. Vladimirov. FizMatLit Publ.; 2009. 208 p. (in Russian).
2. Vladimirov SN, Smolskiy SM. Non-traditional dynamics in electronics: theory and practice. Paceo Segovia Irvine (USA, CA): Scientific Research Publ. Inc.; 2011. 260 p.
3. He R, Vaidya PG. Implementation of chaotic cryptography with chaotic synchronization. Phys Rev E. 1998;57(2):1532–1535.
4. Landa PS. Nonlinear oscillations and waves in dynamical systems. Kluwer Academic Publ.: Dordrecht, Boston, London; 1996. 538 p.
5. Trubetskov DI. Oscillations and waves for humanitarian students: Textbook for universities. Saratov: “College” Publ.; 1997. 392 p. (in Russian).
6. Trubetskov DI. Science on complexities in persons, dates and destinies: how the foundations of Synergetics were laid. Moscow: LIBROKOM Publ.; 2013. 312 p. (in Russian).
7. Hoppensteadt FC. Analysis and simulation of chaotic systems. N.Y: Springer; 2000. 408 p.
8. Kuznetsov AP, Kuznetsov SP, Ryskin NM. Nonlinear oscillations: textbook for universities. Moscow: FisMatLit Publ.; 2002. 292 p. (in Russian).
9. Novikov SS, Mokrinskiy DV, Usiukevich AA. Chaotic dynamic of the system of two coupled UHF oscillators. Russ Phys J. 2013;56(8/2):321–325. (in Russian).
10. Koronovskii AA, Moskalenko OI, Hramov AE. On the use of chaotic synchronization for secure communication. Uspekhi Fizicheskikh Nauk. 2009;179(12):1281–310. doi:10.3367/UFNr.0179.200912c.1281.
11. Izmailov IV, Poizner BN. Axiomatic scheme for dynamic system investigation: from criteria of their out-of-identity to self-variation. Tomsk: STT Publ.; 2011. 574 p. (in Russian).
12. Izmailov IV, Poizner BN, Romanov IV. Forming and self-changing non-linearity as a security measure of communication system used chaos. In: Proceedings of 24th International Crimean Conference “Microwave & Telecommunication Technology” (7–13 September, Sevastopol, Crimea, Russia). Sevastopol, 2014. pp. 219–220 (IEEE Catalog Number: CFP14788).
13. Ashby WR. An introduction to cybernetics. lnd: chapman& hall ltd; 1956. 416 p.
14. Izmailov IV, Poizner BN. Nonlinearity generation for increase of a variety of systems with dynamic and static instability. Russ Phys J. 2010;53(2):18–21.
15. Physical encyclopedia/Under edition of A.M. Prokhorov. Moscow: Sovetskaya Entsiklopediya Publ.; 1983. 928 p. (in Russian).
16. Nanostructures in Electronics and Photonics. In: Rahman F, editor. World Publ. Co. Pte. Ltd; 2005. 312 p.
17. Sedov EA. Information criteria of ordering and complexity of system structure organization. In: System conception of information processes. Moscow: AISR of system investigation, 1988. Issue 3. pp. 37–46 (in Russian).
18. Philosophy of technique: classical, post-classical post-neo-classical. Dictionary/Under edition of B.I. Kudrin. Issue 37 “Cenological investigations”. Moscow: Tekhnetika Publ.; 2008. 180 p. (in Russian).
19. Hertz H. Über elektrodynamishe Wellen im Luftraume und deren Reflexion. Wied Annalen. 1888;34:610–619.
20. Shtykov VV. ABC book of young radio operator, or Introduction to radio electronics. Kiev: Osvita Ukrainy Publ.; 2012. 286 p. (in Russian).

21. Rosanov NN. Optical bi-stability and the hysteresis in distributed nonlinear systems. Moscow: Nauka Publ.; 1997. 336 p. (in Russian).
22. Osipov GV, Shalfeev VD. Transients in their chain of single-directly coupled phase-locked loop systems. Radioengineering. 1988;6:19–23 (in Russian).
23. Afraimovich VS, Nekorkin VI, Osipov GV, Shalfeev VD. Stability, structures and the chaos in nonlinear networks of synchronization. Gorky: IAP AS of USSR Publ.; 1989. 224 p. (in Russian).
24. Kapranov MV, Chernobaev VG. Controllable oscillators of chaotic oscillations on the base of PLL systems. Radiotekhnicheskie tetradi (Moscow, MPEI). 1998;15:86–90 (in Russian).
25. Matrosov VV, Shmelyov AV. Nonlinear dynamics of the ensemble from two phase-controlled oscillators with the ring type of coupling. Izvestiya VUZ. Appl Nonlinear Dyn. 2010;18 (4):67–80 (in Russian).
26. Ponomarenko VP. Nonlinear effects in the selg-oscillation system with frequency-phase control Izvestiya VUZ. Appl Nonlinear Dyn. 2012;20(4):66–84 (in Russian).
27. Mischenko MA, Shalfeev VD, Matrosov VV. Neuron-similar dynamics in the system of phase synchronization Izvestiya VUZ. Appl Nonlinear Dyn. 2012;20(4):122–130 (in Russian).
28. Kriukov AK, Osipov GV. Influence of oscillator medium properties on propagation of excitation Izvestiya VUZ. Appl Nonlinear Dyn. 2013;21(2):188–200 (in Russian).
29. Akhmanov SA, Vorontsov MA. Instabilities and structures in coherent nonlinear-optical systems, enveloped by two-dimensional feedback Nonlinear waves: dynamics and evolution. Moscow: Nauka Publ.; 1989. pp. 228–237 (in Russian).
30. Izmailov IV, Lyachin AV, Poizner BN. Deterministic chaos in models of the nonlinear ring interferometer. Tomsk: Tomsk State University Publ.; 2007. 258 p (in Russian).
31. Izmailov IV, Poizner BN, Romanov IV, Smolskiy SM. About parallels in dynamics of radio electronic chaos generators having «prepared» and formed nonlinearity. Russ Phys J. 2015; 58(8/3):109–113 (in Russian).

Chapter 7
Nonlinear-Dynamic Cryptology Versus Steganography and Cryptografics

Completing the discussion about using in cryptology of radio physical and optical systems with complicated dynamics, naturally to ask: why exactly in cryptology?

It might seem, we may achieve the desired freedom and psychological comfort during their communication by means of using the steganographics or the physical methods of information protection. Nevertheless, it is problematically that dynamic systems with complicated behavior become effective as the physical protection means. On the contrary, using the devices with chaotic dynamics (oscillators of deterministic "noise") as a base of *steganographical* systems quite *naturally* since, in it, the message is immersed into the noise, thereby, hiding the fact of its transmission [1]. In this regard, it is significantly that in the review [2] authors limit themselves exactly by the steganographic interpretation of system application with chaotic dynamics. But this method of information protection automatically falls under intention, theory and practice of application of noise-like signals, which heave a rich history. In this context, we offered and theoretically investigated of the communication system by means of optical vortices. According to our data, it has elements of physical and stenographic resistance.

Against this background, the cryptological application of radio physical and optical systems with complicated dynamics seems to be less traditional, and possibilities of dynamic systems in this aspect are less obvious. You see, in this case, we assume that the cryptosystem construction is known, and the channel of message transmission is available for everybody. Closed to the sense judgment, but expressed in terms of "masking task" and "confidentially task" we find in the review [3].

Here it is not reasonable to wait that we can strictly proof that some device is the cryptosystem. For instance, even for ciphers applied in the traditional cryptology, a question about existence of functions with the secret remains open yet. And this, as a matter of fact, does not give the right to call their as cryptosystems. In this book

I. Izmailov et al., *Cryptology Transmitted Message Protection*,
Signals and Communication Technology, DOI 10.1007/978-3-319-30125-9_7

we considered nothing but some oscillators of deterministic chaos and systems of message transmission (nonlinear dynamic cryptosystems) on their base. Also, we outline their following ways:

- of revealing parallels between the categorical cryptography instrument and elements of languages for processes description in these devices;
- of feature clarification of cipher cryptoresistance estimation, which we can interpret pairs of nonlinear dynamic systems (cipherer/decipherer) realizing by hardware or in the view of differential equations or discrete maps.

Evidently, nonlinear dynamic cryptosystems could be put in the one series with classical ones, if we succeed to prove: at unknown parameters of the cipherer there is not the polynomial algorithm for inversing of the ciphering algorithm; According to message and cryptogram, the key is calculated with exponential complexity. Then, the problem raises to obtain for nonlinear dynamic cryptosystems the correct prototypes of known definitions of algorithms polynomial and exponential complexity. So nontrivial task will be necessary to solve in the future.

After its solution, we could have a possibility the following:

- to unify principles and methods of cipher cryptoresistance estimation, which are given by the system of differential equation or by discrete maps, as well as, probably, by the traditional ciphers;
- to develop methods of cipher generation with necessary resistance.

Openness of this theme in combination with the width of application widening, which differs by its specificity, inevitably leads to diversification of suggested methods and devices of nonlinear-dynamic cryptography. It plays into somebody hands to increase the communication system quality in particular, their resistance. Chapter 6 is devoted to problems of diversification.

At last, we note that nevertheless, many from existing devices of confidential communication have advantages of steganographical systems: for example, in the mode of addition mixing the message masking by the chaotic signal. On the contrary, according to the conception itself, the steganographical systems not necessary must to keep confidentiality at operation under conditions of determined by the main "postulate" of cryptography (Sect. 1.1.). Naturally, if the communication confidentiality decreases, they are not cryptosystems [1].

If the reader permits, we limit this laconic conclusions on results, not to wish to repeat conclusions of each chapters. Here we only indicate the potential ways of further fundamental investigations. In particular, they appeals the *eliminate the break* between nonlinear-dynamic and traditional cryptographic paradigms.

References

1. Vladimirov SN, Izmailov IV, Poizner BN. Nonlinear-dynamic cryptology: radio physical and optical systems. Under edition of Vladimirov CN. Moscow: FizMatLit Publication; 2009. 208 pp. (in Russian).
2. Koronovskii AA, Moskalenko OI, Hramov AE. On the use of chaotic synchronization for secure communication. Uspekhi Fizicheskikh Nauk. 2009;179(12):1281–310. doi:10.3367/UFNr.0179.200912c.1281.
3. Dmitriev AS, Starkov SO. Message transmission using the chaos and classical theory of information. Foreign Radio Electron. 1998; 11:4–31. (in Russian).

Index

I. Izmailov et al., *Cryptology Transmitted Message Protection*,
Signals and Communication Technology, DOI 10.1007/978-3-319-30125-9

D

E

F

G

H

N

O

Printed in the United States
By Bookmasters